Element	Symbol	Atomic No.	Atomic Weight	Element	Symbol	Atomic No.	Atomic Weight
Gold	Au	79	196.967	Praseodymium	Pr	59	140.907
Hafnium	Hf	72	178.49	Promethium	Pm	61	[147]*
Helium	He	2	4.0026	Protactinium	Pa	91	[231]*
Holmium	Ho	67	164.930	Radium	Ra	88	[226]*
Hydrogen	H	1	1.00797[a]	Radon	Rn	86	[222]*
Indium	In	49	114.82	Rhenium	Re	75	186.2
Iodine	I	53	126.9044	Rhodium	Rh	45	102.905
Iridium	Ir	77	192.2	Rubidium	Rb	37	85.47
Iron	Fe	26	55.847[b]	Ruthenium	Ru	44	101.07
Krypton	Kr	36	83.80	Samarium	Sm	62	150.35
Lanthanum	La	57	138.91	Scandium	Sc	21	44.956
Lead	Pb	82	207.19	Selenium	Se	34	78.96
Lithium	Li	3	6.939	Silicon	Si	14	28.086[a]
Lutetium	Lu	71	174.97	Silver	Ag	47	107.870[b]
Magnesium	Mg	12	24.312	Sodium	Na	11	22.9898
Manganese	Mn	25	54.9380	Strontium	Sr	38	87.62
Mendelevium	Md	101	[256]*	Sulfur	S	16	32.064[a]
Mercury	Hg	80	200.59	Tantalum	Ta	73	180.948
Molybdenum	Mo	42	95.94	Technetium	Tc	43	[99]*
Neodymium	Nd	60	144.24	Tellurium	Te	52	127.60
Neon	Ne	10	20.183	Terbium	Tb	65	158.924
Neptunium	Np	93	[237]*	Thallium	Tl	81	204.37
Nickel	Ni	28	58.71	Thorium	Th	90	232.038
Niobium	Nb	41	92.906	Thulium	Tm	69	168.934
Nitrogen	N	7	14.0067	Tin	Sn	50	118.69
Nobelium	No	102	…	Titanium	Ti	22	47.90
Osmium	Os	76	190.2	Tungsten	W	74	183.85
Oxygen	O	8	15.9994[a]	Uranium	U	92	238.03
Palladium	Pd	46	106.4	Vanadium	V	23	50.942
Phosphorus	P	15	30.9738	Xenon	Xe	54	131.30
Platinum	Pt	78	195.09	Ytterbium	Yb	70	173.04
Plutonium	Pu	94	[242]*	Yttrium	Y	39	88.905
Polonium	Po	84	[210]*	Zinc	Zn	30	65.37
Potassium	K	19	39.102	Zirconium	Zr	40	91.22

Introduction to Analytical Chemistry

Introduction to Analytical Chemistry

2nd edition

GEORGE H. SCHENK
Wayne State University

RICHARD B. HAHN

ARLEIGH V. HARTKOPF
Mobil Chemical Corporation

ALLYN AND BACON, INC.
Boston, London, Sydney, Toronto

Cover illustration: Hippuric acid, photograph by James Bell.

Library of Congress Cataloging in Publication Data

Schenk, George H
  Introduction to analytical chemistry.

  First ed. published in 1977 under title: Quantita-
tive analytical chemistry.
  Includes bibliographical references and index.
  1. Chemistry, Analytic—Quantitative. I. Hahn,
Richard Balser, 1913–   joint author. II. Hartkopf,
Arleigh V., joint author. III. Title.
QD101.2.S33 1981   545   80-39986
ISBN 0-205-07236-4

Printed in the United States of America.
10 9 8 7 6 5 4 3 2 1   85 84 83 82 81

To my wife Isabelle, whose love, comfort, strength, and cheerfulness are more precious to me than gold and without whose daily support this book could never have been written.

G.H.S.

Contents

PART IV SELECTED EXPERIMENTS

APPENDIXES

Preface

In this Second Edition we have tried to improve the text for the benefit of the student. We have reorganized the chapter sequence, expanded topic coverage and depth, and significantly increased the number of end-of-chapter problems. We have also updated the Second Edition by including a new chapter on Radiochemistry and Radioimmunoassay.

Chapter 2 is now a survey of measurement devices, showing that the analytical balance is but one of many such devices; following the survey, the balance is discussed in detail as an aid to instructors who introduce the laboratory with a brief overview of the balance. Because of this change, the discussion of data handling, now revised and strengthened, has been moved back to Chapter 3. Sampling is now discussed in Chapter 4.

Gravimetric analysis remains in Chapter 5, which has been improved by the addition of more material on particle size, digestion, and washing, as well as more problems. Material on gravimetric analysis of biological and pharmaceutical substances has been moved to the back of the chapter to allow for omission of this material if the instructor desires.

Chapter 6 still emphasizes the fundamentals of volumetric analysis, although we have added the deciliter (dL) concept for those clinical students who are using it instead of 100 ml; this is also the case for the concentration unit of mg/dL, which replaces mg%.

We feel that Chapter 7, "Introduction to Titrations," will be a strong chapter for chemistry students as well as for life science students. It has been expanded from a brief survey of all types of reactions to include a discussion of the precipitation of chloride ion to illustrate all aspects of titrations. The concept of quantitative titration reactions is retained in this chapter, but we have also introduced the calculation of titration curves. We hope that all students will appreciate a background in titration methods for the chloride ion, the most important anion in body fluids.

Chapter 8, "Precipitation and Complexation Titrations," has been moved forward from its position in the First Edition to follow the theoretical discussion of precipitation titrations in Chapter 7 and to be closer to the gravimetric methods in Chapter 5. This chapter contains all practical details of precipitation titrations. It also has been strengthened with more discussion of equilibrium and with the inclusion of numerous end-of-chapter problems.

Chapter 9, "Elementary Acid-Base Equilibria," has retained its form from the First Edition, but Chapter 10, "Acid-Base Titrations," has been improved with an early introduction of the calculation of a titration curve of a weak acid, a topic missing from the First Edition. Chapter 11, "Oxidation-Reduction Methods," has been improved with more emphasis on inorganic reactions and expansion of the discussion on the determination of vitamin C.

The introduction to instrumental methods has been combined with the introduction to spectrophotometry in the new Chapter 12. The result is a better chapter on spectrophotometry, with new material and more problems. The material in Chapter 13, "Applications of Spectrophotometry," has been strengthened with additional topics, such as multicomponent analysis, and with more problems. The other chapters remain essentially as they were.

Since we anticipate that many nonchemists will use this text, we have included a number of features helpful to them as well as to chemistry students. Some of these are as follows:

1. Self-tests on topics that might require extra work are numerous.
2. Important principles or rules are boxed off in the text for easy access.
3. Worked-out examples for better understanding are numerous.
4. The equations and reactions are for the most part numbered for easy referencing between chapters.
5. Problems are divided into logical groups with a heading for each group. Answers to most even-numbered problems are given in Appendix 5.

The self-tests are important because they help students to delve a little more deeply into a topic, and help them review the principles from the preceding material.

We have been selective in including experiments because we feel that most instructors have accumulated experiments of their own. We have tried to include mostly experiments related to health science, along with a few standard experiments.

Finally, the authors wish to acknowledge several people who were helpful in preparing the Second Edition: Jim Smith, the present editor, who was so encouraging and helpful at every step; Rachel Meisler, who patiently handled many, many phone calls to help make the deadlines; Barbara Tookoian, who brightened the long commuting hours with good conversation; Bill Roberts, the editor of the First Edition, who provided the idea for the book; and Jane Hoover and Sally Lifland, who brought style and sense to the manuscript at all stages. We also wish to acknowledge the contributions of the reviewers of this new edition: Larry R. Field, University of Washington; Robert L. Grob, Villanova University; Robert E. Kirby, Queens College; John C. MacDonald, Fairfield University; James D. Winefordner, University of Florida; and Alfred M. Wynne, University of Massachusetts. Thank you!

G.H.S.

Introduction to Analytical Chemistry

PART
ONE

Fundamentals

This part of the book is a treatment of the fundamentals of analytical chemistry. In Chapter 1 we introduce the student to chemical analysis from both qualitative and quantitative viewpoints. Chapter 2 surveys measurement devices and covers the use of the analytical balance and the measurement of density and specific gravity. In Chapter 3 we present certain fundamentals of data handling that are necessary in all of the succeeding chapters. Sampling and preparation of the sample are discussed in Chapter 4. Gravimetric analysis is described in Chapter 5.

Chapters 6–11 are concerned with various aspects of volumetric, or titrimetric, analysis, concluding the discussion of fundamentals.

1 Introduction to Chemical Analysis

"There is our result—and a very workmanlike little bit of analysis it was," said Holmes.

ARTHUR CONAN DOYLE
The Valley of Fear, Chapter 1

Chemical Analysis and Detective Work

This and following chapters begin with a quotation from the cases of Sherlock Holmes for a good reason. Chemical analysis *is* essentially detective work, and the analytical chemist is surely a scientific detective. If you have ever read any of Sherlock Holmes's cases, you know that detective work consists of observations, testing, interpretation, and the final glorious deduction. The same is true for chemical analysis. Analytical chemists are usually faced with the question of Why? or What is wrong? To answer such a question, they first make certain observations to see what tests or analyses must be conducted. Then they judiciously choose the proper sample and test it. Finally, they interpret results to someone, and may even get to make the final deduction, although it may not be as glamorous as those made by Mr. Holmes.

In this book we are concerned with all of the above four steps. However, observation is difficult to teach; it must be learned by experience in the laboratory or in the real world. We will therefore be more concerned with understanding how to use data in analyses (Ch. 3), choosing a sample (Ch. 4), testing the sample (Chs. 5–22), and interpreting the results (Ch. 3) for that final deduction.

1-1 | INTRODUCTION TO CHEMICAL ANALYSIS

What Kinds of Chemical Analysis Are There?

Chemical analysis encompasses numerous types of measurements. We can identify four unique areas of chemical analysis as follows:

1. Measurements of physical and chemical properties
2. Qualitative analysis

3. Quantitative analysis
4. Diagnostic analysis based on a combination of qualitative and quantitative analyses

We will discuss each of these areas to give a feeling for chemical analysis in its broadest sense.

Measurement of Physical and Chemical Properties

New compounds, crystals, polymers, and natural substances are being discovered and synthesized all the time. One of the most important things chemical analysts are concerned with is the characterization of these substances by measuring their physical and chemical properties. If possible, the analyst will try to measure the melting point, boiling point, density, index of refraction, and so on. If the substance is an acid or a base, the ionization constant should be measured. If the substance is insoluble in water, its solubility should be measured and/or a solubility product constant arrived at. In a similar manner, the analyst may try to evaluate other chemical properties.

Qualitative Analysis

Qualitative analysis involves the *identification* of a compound whose identity is unknown or only partially known, as well as *screening* on a routine basis for the presence or absence of significant amounts of a compound whose presence is only suspected. One of the most powerful tools for qualitative analysis is spectroscopy; Chapters 13 and 14 both give examples of the application of spectroscopy to qualitative analysis. Chapter 13 discusses the application of ultraviolet, visible, and infrared spectrophotometry to qualitative analysis, and Chapter 14 discusses the use of fluorescence in qualitative analysis. It is important to employ several different measurements for rigorous proof in qualitative analysis. These measurements may consist of a combination of spectroscopic measurements with measurements of physical properties such as melting point, or of a number of spectroscopic measurements alone.

Quantitative Analysis

Quantitative analysis involves the determination of the *amount* of a substance present, either its percentage or its concentration in a solution. Quantitative analysis of a sample may be performed just once, or it may be performed continuously to monitor changes in percentage purity or concentration with time. Quantitative analysis may be performed by the classical gravimetric (Ch. 5) or titration methods (Chs. 6–11), as well as by spectroscopic (Chs. 12–15) and electrical means (Chs. 16 and 17). Different types of quantitative measurements give different kinds of information. When speed is important, it is imperative that the analyst measure only what is necessary and transmit the results quickly. Some of the different types of quantitative analysis are described below.

A *complete analysis* is an analysis in which each compound in a sample is identified and the percentage or concentration measured. For example, the complete analysis of an aspirin tablet would include the percentage of acetylsalicylic acid (ASA), the percentage of impurities such as salicylic acid (and possibly any acetic acid which has not volatilized), and the amount of inert material used as binder to hold the ASA particles together in tablet form.

An *elemental analysis* is an analysis in which a mixture or pure compound is analyzed for the percentage or concentration of each element present. This type of analysis may be used together with a molecular weight determination to establish the identity of a compound, or it may be used to characterize a complex sample. Some samples, such as petroleum products, are such complex mixtures that it is not worth running a complete analysis. It is just as useful to determine the percentage of certain key constituents, such as sulfur, oxygen, and phosphorus, to characterize the petroleum.

A *partial analysis* is an analysis in which the concentration or percentage of one or more impurities is all that must be determined. For example, the main impurity in aspirin tablets is salicylic acid, which arises from the hydrolysis of acetylsalicylic acid (ASA):

$$[C_6H_4(OH)]C-O-C-CH_3 + H_2O \longrightarrow [C_6H_4(OH)]C-OH + CH_3C-OH$$

$$\overset{\|}{O} \qquad\qquad\qquad\qquad\qquad\qquad\qquad\qquad \overset{\|}{O}$$

(ASA)   (Salicylic acid)   (Acetic acid)

Acetic acid is also produced in the reaction, but it gradually evaporates from the aspirin tablets. Therefore, to evaluate the purity of the aspirin, it is simpler to determine only the salicylic acid rather than measure the ASA present. In fact, the U.S. Food and Drug Administration (FDA) does just that. It has established a *tolerance* of 0.15% salicylic acid for aspirin tablets. If a company's aspirin exceeds this tolerance, the FDA issues a warning to them to improve the product.

Diagnostic Analysis

Diagnostic analysis may involve a combination of qualitative and quantitative tests. It may also involve quantitative measurements that establish primarily whether a certain concentration is within or outside a certain normal *range* of values. The best examples of this are of course the standard clinical tests that are run on blood and/or urine samples.

Since most clinical measurements are performed on automated instrumentation, the results are frequently plotted by a recorder across a standard graph paper displaying the range of values possible for each test. A section of such a graph paper for blood analysis is shown in Figure 1-1. The shaded area in each vertical column of values represents the normal range of concentration for each chemical species. For example, for calcium(II) ion, a normal blood sample should contain from 7.9 to 10.5 mg% of Ca^{+2}.

An example of the usefulness of a combination of diagnostic tests follows. Suppose that a certain blood sample is found to be low in calcium(II) ion from a protein deficiency. Since the calcium(II) ion concentration controls the phosphate ion, then we would expect the phosphate ion concentration to be raised above normal. If instead the phosphate concentration is lower than normal (Fig. 1-1), this may indicate a kidney dysfunction resulting in retention of phosphate in the kidney. The increase in blood urea nitrogen (BUN) also indicates a kidney dysfunction (Fig. 1-1).

Sample Size

Analytical methods can be classified according to a number of criteria, one of which is sample size. Table 1-1 contains a summary of the classification according to the weight and the concentration of the sample.

As we will discuss below, most analytical techniques are not suitable for measuring more than one or two sample sizes, unless the sample can be diluted or concentrated. It is important for the analytical chemist to become experienced with analytical techniques so that the proper technique can be chosen to obtain an accurate answer.

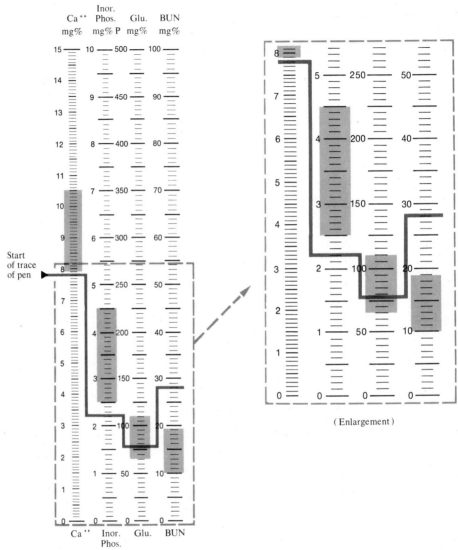

**FIGURE 1-1. A section of a graph of an automated clinical analysis (the term mg%
is the same as mg/dL).**

**Analytical
Measure-
ment
Techniques**

Quantitative analysis (and to some extent qualitative analysis) is performed by using
one of the following measurement techniques:

1. Gravimetric (precipitation)
2. Volumetric (titrimetric)
3. Optical (spectroscopy)
4. Electrical
5. Gasometric
6. Separation

TABLE 1-1. Sample Size Classification of Analytical Methods

Name of method	Approximate sample weight	Approximate molarity: sample of 100 ml
Macro method	100–1000 mg	0.01–0.1M
Semimicro method	10–100 mg	10^{-3}–$10^{-2}M$
Micro method	1–10 mg	10^{-4}–$10^{-3}M$
Ultramicro method	0.001–1 mg	10^{-7}–$10^{-4}M$
Submicrogram method	10^{-5}–10^{-3} mg	10^{-9}–$10^{-7}M$

Gravimetric Analysis

Gravimetric analysis involves isolating and weighing a pure compound, which is or which contains the desired constituent. It is mainly a macro or semimicro method, since the solubility of most compounds determined by gravimetry is not small enough for micro determination. Micro samples have to be concentrated to be analyzed gravimetrically.

Volumetric Analysis

Volumetric analysis is primarily performed by titrating the desired constituent with a solution of known concentration. It is mainly used as a macro method, but can be adapted to the semimicro or micro level for reactions with large equilibrium constants. Such reactions include most acid-base reactions, most complex-forming reactions, a few precipitation reactions, and most oxidation-reduction reactions.

Optical Analysis (Spectroscopy)

Optical analysis involves the measurement of light or other radiation after it has interacted, or failed to interact, with ions or molecules in solution, or with atoms in the vapor phase. It is usually very sensitive to small concentrations and is used as a micro method (absorption spectroscopy), as an ultramicro method (emission or absorption by gaseous atoms), or as a submicrogram method (fluorescence in solutions). Macro samples have to be diluted to be analyzed by optical methods.

Electrical Analysis

Electrical analysis involves the measurement of electrical potential, or the electrical output from an electrochemical reaction. It is usually used as a semimicro or a micro method. However, electrical potential measurements can be used over a wide range of concentration from ultramicro to the macro sample level. A few electrical methods of analysis can be used as ultramicro methods also.

Gasometric Analysis

We will arbitrarily define a gasometric method as one in which a *molecular* sample is measured in the gaseous form. This could include measurement of the volume of gas in a gas buret, but mainly refers to the gas chromatographic method in which gases are separated and measured at the same time on the same instrument. This latter method covers a wide range of sample sizes, and no dilution or concentration is needed for most samples.

Separations

Separation primarily involves isolating the various constituents in the sample so that the amount of each constituent can be measured. Separation methods include solvent extraction, liquid-liquid chromatography, thin-layer chromatography, gas chromatography, dialysis of small water-soluble molecules, and so on. All such separations make qualitative and quantitative analysis much easier to perform by removing possible interferences from the constituent to be identified or determined.

1-2 | REVIEW OF SOME USEFUL CONCEPTS

In this section we will review some very useful concepts that you will need through the rest of the book. We will begin with a survey of the nomenclature of inorganic and organic compounds. It is taken for granted in the book that you understand the naming of these compounds.

Inorganic Nomenclature

Following will be a short review of the naming of metal ions, anions, molecules, compounds, acids, and complex ions.

Metal Ions

The Stock nomenclature for cations involves the use of the oxidation state of the ion in parentheses immediately following the name of the element. For example, Fe^{+2} and Fe^{+3} are named

$$Fe^{+2}: \text{iron(II) ion} \qquad Fe^{+3}: \text{iron(III) ion}$$

We prefer this system, although occasionally we will use the common or trivial names, such as ferrous ion or ferric ion, where it is convenient. Even in the case of the Hg_2^{+2} ion the name used is simply mercury(I) ion. It is understood that this name implies a dimeric cation.

In the case of the alkali and alkaline earth metal ions it is usually not necessary to write the oxidation state after the element's name. However, we will occasionally use it to be sure that nonchemists are aware of the oxidation state. Thus the Na^+ ion may be named either way,

$$Na^+: \text{sodium ion or sodium(I) ion}$$

Anions and Molecules

When anions and molecules exist alone or in simple compounds they are named differently than when they are bonded to metal ions in complex ions. A list of a few representative anions and molecules and their names follows:

Formula	Alone (or in simple compound)	Complex ion
H_2O	Water	Aquo
NH_3	Ammonia	Ammine
$NH_2(CH_2)_2NH_2$	Ethylenediamine	(Ethylenediamine)
Cl^-	Chloride	Chloro
OH^-	Hydroxide	Hydroxo
CN^-	Cyanide	Cyano
SO_4^{-2}	Sulfate	Sulfato
PO_4^{-3}	Phosphate	Phosphato

When the molecule attached to a metal is complex, such as ethylenediamine, its name is used unchanged but parentheses are placed around its name in the name of the complex ion.

Compounds When compounds are named, the cation is named first using the Stock system, and the anion is named next. No prefix is necessary to indicate the number of anions, since this is assumed to be understood from the oxidation state in the name. Some examples:

Formula	Name
$FeSO_4$	Iron(II) sulfate
$Fe_2(SO_4)_3$	Iron(III) sulfate
Hg_2Cl_2	Mercury(I) chloride
Na_2CO_3	Sodium carbonate

Acids Acids are referred to by their common names. Some examples:

Formula	Name
H_2CO_3	Carbonic acid
HCl	Hydrochloric acid
HNO_3	Nitric acid
$HClO_4$	Perchloric acid
H_3PO_4	Phosphoric acid

Compounds that are acid salts, such as $NaHCO_3$, are formally named using the name of the metal ion followed by *hydrogen* with the proper prefix for the number in the formula. Thus $NaHCO_3$ is sodium hydrogen carbonate, although it is also known as sodium carbonate. The compound NaH_2PO_4 would be named sodium dihydrogen phosphate.

Complex Ions In *positively* charged complex ions, the ligand is named first, preceded by a prefix (di-, tri-, tetra-, penta-, or hexa- for simple ligands; bis-, tris-, etc., for complex ligands) to indicate the number of ligands. The metal ion is named next. Any anions not complexing the metal ion are named last. Some examples:

Formula	Name
$Ag(NH_3)_2{}^+$	Diamminesilver(I) ion
$FeCl_2{}^+$	Dichloroiron(III) ion
$Cu(NH_2C_2H_4NH_2)_2^{+2}$	Bis(ethylenediamine)copper(II) ion
$[Fe(SO_4)^+]Cl^-$	Sulfatoiron(III) chloride

In *negatively* charged complex ions, the name of the metal ion is followed by the suffix -*ate*; elements whose names end in *y* drop the *y* when the name of the metal ion is formed. In addition, certain metal ions are named according to their Latin names. The important metal ions for which this is done are listed below:

Formula	Name
$AgCl_2{}^-$	Dichloroargentate(I) ion
$Fe(CN)_6^{-3}$	Hexacyanoferrate(III) ion; also, ferricyanide
$CuBr_4^{-2}$	Tetrabromocuprate(II) ion
$PbCl_4^{-2}$	Tetrachloroplumbate(II) ion
$SbCl_6^{-3}$	Hexachlorostibnate(III) ion

Organic Nomenclature Rather than discuss organic nomenclature in a complete formal manner, we will simply list each important class of compounds and name one or two examples of that class that may be encountered in later chapters. You should consult an organic text for a complete review if needed.

Alkanes   Alkanes are straight chain or branched chain aliphatic hydrocarbons, such as butane, $CH_3CH_2CH_2CH_3$.

Alkenes   Alkenes are olefins with one or more double bonds, such as ethylene, $H_2C{=}CH_2$, or 1,3-butadiene, $H_2C{=}CH{-}CH{=}CH_2$.

Alkynes   Alkynes are hydrocarbons with one or more triple bonds. An example is acetylene, $HC{\equiv}CH$.

Aromatic Hydrocarbons   Aromatic hydrocarbons are cyclic conjugated unsaturated hydrocarbons, such as benzene (C_6H_6) or naphthalene ($C_{10}H_8$). Their structures are

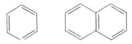

Halogenated Hydrocarbons   Halogenated hydrocarbons include halo-substituted alkanes, alkenes, aromatic hydrocarbons, etc. Examples are chloroform, $CHCl_3$, and chlorobenzene, C_6H_5Cl.

Hydroxyl Compounds   Hydroxyl-substituted compounds include alcohols, such as ethyl alcohol (ethanol), CH_3CH_2OH, and phenols, such as phenol itself, C_6H_5OH.

Carbonyl Compounds   Carbonyl compounds are compounds containing a carbon-oxygen double bond, or carbonyl group, bonded to other carbon or hydrogen atoms. Examples include aldehydes, such as formaldehyde, $H_2C{=}O$, and ketones, such as acetone, $CH_3{-}\underset{\underset{O}{\|}}{C}{-}CH_3$.

Carboxylic Acids and Esters   Carboxylic acids are aliphatic acids, such as acetic acid, $CH_3\underset{\underset{O}{\|}}{C}{-}OH$, or aromatic acids, such as phthalic acid, $C_6H_4(COOH)_2$. Phthalic acid consists of two carboxylic groups bonded to a benzene ring. It is half neutralized with potassium hydroxide to make potassium acid phthalate, $C_6H_4(COOH)COO^-K^+$, an important primary standard for basic titrants.

Esters are compounds made from carboxylic acids and alcohols. A common example is ethyl acetate, $CH_3\underset{\underset{O}{\|}}{C}{-}O{-}CH_2CH_3$.

Sugars   Sugars as a class of compounds contain hydroxyl groups and an aldehyde and/or a ketone group. Examples are glucose,

$$HC{-}CHOH{-}CHOH{-}CHOH{-}CHOH{-}CH_2OH$$
$$\underset{O}{\|}$$

and fructose,

$$CH_2OH{-}C{-}CHOH{-}CHOH{-}CHOH{-}CH_2OH$$
$$\underset{O}{\|}$$

These compounds usually exist in cyclic forms as well as the straight chains shown above, but for our purposes the straight chain is sufficient.

Atomic Weights, Moles, and Concentration

The following discussion may be useful as a review preceding Chapters 5 and 6, where the calculation of analysis results will be presented.

Atomic Weights and Moles

Although chemists use atomic weights as absolute quantities, atomic weights are really the weights of atoms relative to the weight of the carbon isotope of mass number 12. Thus the atomic weight of 24.312 for magnesium denotes that it is a little over twice the weight of the carbon 12 isotope. A table of atomic weights is included on the inside front cover of this book.

The terms *molecular weight* and *formula weight* are sometimes used interchangeably, although a molecular weight refers strictly to the sum of the atomic weights of the atoms in a molecule and a formula weight refers to the sum of the atomic weight(s) of the atom(s) in an element, an ionic compound, or a molecule.

To simplify working with these terms, we can use the *mole* as a fundamental quantity to specify amounts of any species. A mole is defined as 6.023×10^{23} atoms, ions, or molecules. The working definition of a mole is that it is a formula weight of any species expressed in grams. Thus a mole of chloride ions is 35.45 g, and a mole of sodium chloride is 58.44 g. In addition to the gram and the mole, the milligram (mg) or microgram (μg) and the millimole (mmole) or micromole (μmole) are used. ($1 \mu g = 10^{-6}$ g; 1μmole $= 10^{-6}$ mole.) Thus a millimole of Cl^- is 35.45 mg, and a micromole of sodium chloride is 58.44 μg.

The number of moles or millimoles is calculated as follows:

$$\text{mole } Cl^- = \frac{g\ Cl^-}{35.45\ g/\text{mole } Cl^-\ (\text{formula weight})}$$

$$\mu\text{mole NaCl} = \frac{\mu g\ \text{NaCl}}{58.44\ \mu g/\mu\text{mole NaCl (formula weight)}}$$

$$\text{mmole NaCl} = \frac{mg\ \text{NaCl}}{58.44\ mg/\text{mmole NaCl (formula weight)}}$$

Note in the above calculations that the formula weight can be expressed in grams per mole, milligrams per millimole, or micrograms per micromole.

Concentration

Concentration will be discussed fully in Section 6-2; we will provide only a brief introduction here. The most general unit of concentration used in this text is molarity, M. The molarity of a solution can be expressed using moles and liters or millimoles and milliliters (ml).

$$M = \frac{\text{mole}}{\text{liter}} = \frac{\text{mmole}}{\text{ml}}$$

Thus a $0.10M$ solution of sodium chloride means that the solution contains 0.10 mole of sodium chloride per liter, or 0.10 mmole of sodium chloride per ml.

An alternate method of expressing concentration is to use the normality, N. Normality depends on the type of reaction involved and will be fully defined in Section 6-2.

Clinical chemists frequently use the concentration term of milligram per deciliter (mg/dL). This is defined as milligrams of a substance per 100 g of solvent, or in the case of water, per 100 ml of water. The term can also be used to describe the milligrams of a substance present in 100 g of a solid.

Another concentration term that is frequently used is parts per million (ppm), which can be defined as

$$\text{ppm} = \frac{\text{mg of solute}}{\text{kg of solution or solid}} = \frac{\mu g \text{ of solute}}{\text{g solution}}$$

Since a kilogram is almost the same as a liter for a dilute aqueous solution, this term can also be defined for aqueous solutions as

$$\text{ppm} = \frac{\text{mg of solute}}{\text{liters of aq soln}} = \frac{\mu g \text{ of solute}}{\text{ml of aq soln}}$$

An Optional Look at Activity

It is necessary for an understanding of Chapter 16 and helpful for reading Chapter 8 to master the concept of *activity* as a measure of concentration. Activity is a measure of concentration *not calculated directly* from the number of moles of a solute weighed into a certain number of liters of solution. It is really a measure of the effective molarity of an ion, corrected for interionic attraction.

Activity may be calculated from molarity as follows:

$$a_i \cong f_i[I] \tag{1-1}$$

where a_i is the activity of an individual ion, f_i is the activity coefficient of the ion, and [I] is the molarity of the ion I. The activity coefficient is generally a fraction except at infinite dilution, where it approaches unity. Thus in very dilute solution, the activity of an ion is essentially equal to the molarity of an ion.

Ionic Strength

Activity coefficients can be estimated by a calculation involving the ionic strengh, μ. The magnitude of the ionic strength depends on the concentrations and charges of the ions in the solution. For compounds of $+1$ cations and -1 anions, the ionic strength is the same as the molarity of the compound. Thus the ionic strength of a solution containing 0.01 mole per liter of potassium nitrate and 0.030 mole per liter of sodium perchlorate (NaClO$_4$) is 0.040. For compounds with ions of higher charges than one, the ionic strength must be calculated using this equation:

$$\mu = \frac{1}{2}\sum C_i(Z_i)^2 \tag{1-2}$$

where C_i is the concentration of an ion with charge Z_i.

EXAMPLE: Calculate the ionic strength of a solution of $0.030M$ magnesium(II) nitrate and $1 \times 10^{-5}M$ hydrochloric acid.

Solution: First, note that the molarity of the hydrochloric acid is not *significant* compared to the molarity of the magnesium(II) nitrate (see Sec. 3-1). Since the magnesium nitrate concentration is known only to the third digit to the right of the decimal, the total concentration of the ions must be limited to that digit also, making the amount of HCl negligible. Second, calculate the molarity of each ion:

$$[Mg^{+2}] = M_{Mg(NO_3)_2} = 0.030M$$

$$[NO_3^-] = 2M_{Mg(NO_3)_2} = 0.060M$$

Third, use Equation 1-2 to calculate the ionic strength:

$$\mu = \frac{1}{2}\sum 0.030M \ Mg^{+2}(+2)^2 + 0.060M \ NO_3^-(-1)^2$$

$$\mu = 0.090$$

At ionic strengths below 0.10, activity coefficients for univalent ions may be estimated using the abbreviated form of the DeBye-Huckel equation:

$$-\log f = 0.509(Z_i)^2\sqrt{\mu} \tag{1-3}$$

Equation 1-3 can also be used for divalent ions at ionic strengths below about 0.04.

Equation 1-3 yields a value for a *mean* activity coefficient for both cations and anions of the same charge magnitude. More accurate values for individual ions are given in Table 1-2 for selected ionic strengths.

TABLE 1-2. Individual Ion Activity Coefficients as a Function of Ionic Strength

Ion	Activity coefficient at				
	$\mu = 0.001$	$\mu = 0.005$	$\mu = 0.01$	$\mu = 0.05$	$\mu = 0.10$
H^+	0.967	0.933	0.914	0.86	0.83
Li^+	0.965	0.929	0.907	0.835	0.80
Na^+, IO_3^-, HSO_4^-	0.964	0.927	0.901	0.815	0.77
OH^-, F^-, ClO_4^-	0.964	0.926	0.900	0.81	0.76
K^+, Cl^-, Br^-, I^-	0.964	0.925	0.899	0.805	0.755
NH_4^+, Ag^+	0.964	0.924	0.898	0.80	0.75
Mg^{+2}, Be^{+2}	0.872	0.755	0.69	0.52	0.45
$Ca^{+2}, Cu^{+2}, Zn^{+2},$ $Mn^{+2}, Ni^{+2}, Co^{+2}$	0.870	0.749	0.675	0.485	0.405
Ba^{+2}, Cd^{+2}	0.868	0.744	0.67	0.465	0.38
Pb^{+2}	0.867	0.742	0.665	0.455	0.37
SO_4^{-2}, HPO_4^{-2}	0.867	0.740	0.660	0.445	0.355
$Al^{+3}, Fe^{+3}, Cr^{+3}$	0.738	0.54	0.445	0.245	0.18
PO_4^{-3}	0.725	0.505	0.395	0.16	0.095
$Th^{+4}, Zr^{+4}, Ce^{+4}$	0.588	0.35	0.255	0.10	0.065

From J. Kielland, *J. Am. Chem. Soc.* **59**, 1675 (1937).

Once the activity coefficient is known, it can be substituted into 1-1 to calculate the activity of an ion or ions.

EXAMPLE: Calculate the activity of both the hydrogen ion and the chloride ion of $1.0 \times 10^{-5}M$ HCl in (a) $0.030M$ magnesium(II) nitrate and (b) $9.0 \times 10^{-5}M$ sodium(I) chloride.

Solution to a: As we noted in the preceding example, the contribution of the hydrochloric acid to the ionic strength is negligible compared to that of the magnesium(II) nitrate. The ionic strength (as calculated in the preceding example) is 0.090. Using **1-3**, the activity coefficient for the hydrogen ion and for the chloride ion is estimated.

$$-\log f = 0.509\sqrt{0.090} = 0.1527$$

$$\log f = -0.15_{27} = +0.84_{73} - 1$$

$$f = 0.70 \text{ (estimate)}$$

Note that the more accurate values in Table 1-2 are significantly different; for hydrogen ion f is about 0.83, and for chloride ion f is about 0.76. We therefore use these values to calculate activities:

$$a_{H^+} = 0.83[1.0 \times 10^{-5}M] = 8.3 \times 10^{-6}$$

$$a_{Cl^-} = 0.76[1.0 \times 10^{-5}M] = 7.6 \times 10^{-6}$$

Solution to b: The ionic strength here is the sum of that of the hydrochloric acid and the sodium chloride, or 1.0×10^{-4}. Using **1-3**, the mean activity coefficient for both ions is

$$-\log f = 0.509\sqrt{1.0 \times 10^{-4}} = 0.00509$$

$$\log f = -0.00509 = +0.9949 - 1$$

$$f = 0.99 \text{ (estimate)}$$

This is so close to unity that the activity and the molarity of each ion can be said to be equal (1.0×10^{-5}).

| QUESTIONS AND PROBLEMS

(Answers to most even-numbered problems are in Appendix 5.)

Definitions and Concepts
1. Define the following terms:
   a. Qualitative analysis
   b. Quantitative analysis
   c. Diagnostic analysis
   d. Macro method
   e. Micro method
2. Contrast the following pairs of methods as to the size of sample involved:
   a. Volumetric analysis and gravimetric analysis
   b. Optical analysis and electrical analysis
   c. Optical analysis and gas chromatographic analysis

3. Name the following compounds using the Stock nomenclature:
   a. Hg_2Cl_2
   b. $HgCl_2$
   c. $[Cu(NH_3)_4^{+2}]SO_4^{-2}$
   d. $K_4^+[Fe(CN)_6^{-4}]$

Concentration Calculations
4. Calculate the number of millimoles and micromoles in 50 mg of sodium chloride. Also calculate the molarity of a solution containing 500 μg of sodium chloride in 10 ml of solution.

5. A solution contains 100 mg of potassium(I) ion and 100 mg of chloride ion in 50 ml of water. Are the molarities of each ion the same? Show by calculation.

6. A common level of water hardness is 100 ppm calcium(II) carbonate. Calculate the mg and the μg of $CaCO_3$ in 0.01 liter. Also calculate the molarity.

Activity Calculations

7. Calculate the activity of both H^+ and NO_3^- in a $1.0 \times 10^{-6} M$ HNO_3 solution in $0.030 M$ potassium(I) sulfate and in $9.0 \times 10^{-6} M$ potassium(I) chloride.

8. Calculate the activity of both Mg^{+2} and SO_4^- in a $0.01 M$ $MgSO_4$ solution.

2 Survey of Measurement Devices; The Analytical Balance

(Holmes) explained, ". . . the skilful workman . . . will have nothing but the tools which may help him in doing his work . . . and all in the most perfect order."

ARTHUR CONAN DOYLE
A Study in Scarlet, Chapter 2

Chapter 1 described the entire process of analysis—deciding what analysis is necessary, choosing the sample, making the measurements, and interpreting the answer. The measurement step will require the most effort, chiefly because there are so many *measurement devices*. So this chapter will first survey the types of measurement devices and then will describe fully the most frequently used measurement device, the analytical balance.

2-1 | SURVEY OF TYPES OF MEASUREMENT DEVICES

The important types of measurement devices determine weight, volume, photon flow (light intensity), electrical potential, electrical current, electrical conductance, and thermal conductivity. Of course it is generally known that the balance is used to measure weight, and that pipets and burets are useful for volume manipulations. Some of the other types of measurement devices that are not as generally known are spectrophotometers and spectrometers for photon flow (light intensity), potentiometers for electrical potential, coulometers for electrical current, conductance bridges for electrical conductance, and thermal conductivity detectors for heat measurements.

Weight Measurements

The analytical balance is the principal device used for measuring weight in connection with quantitative analysis. There are three types of balances, one for each of the three main sample sizes listed in Table 1-1: the standard analytical balance (readable to

16

0.1 mg), the semimicro analytical balance (readable to 0.01 mg), and the micro analytical balance (readable to 0.001 mg, or 1 μg). Analytical balances are usually used to measure solid and liquid samples prior to analysis; gaseous samples are more often determined by volume than by weight. The standard analytical balance is used for weighing samples for all macro methods, as well as for the actual measurement step in gravimetric analyses (Ch. 5). The semimicro balance is used for semimicro methods needing a smaller amount of sample, and the micro balance is used for micro analysis, where the sample weights are the smallest commonly weighed. These balances will be discussed further in following sections.

Volume Measurements

The volumetric flask, the buret, and the pipet are the three devices used for measuring liquid volume. A brief discussion is presented here; pictures and a more complete discussion will follow in Chapter 6.

Volumetric flasks are employed to prepare an accurately known volume of solution from 1 ml to perhaps 2 liters. The maximum allowable error of these flasks is very small; for example, for a 100-ml flask, it is 0.08 ml.

Burets readable to $\pm$ 0.01 ml are used for volumetric measurement of the amount of a known concentration of a reagent added to a titration flask. Burets are available which measure maximum volumes of 5, 10, 25, 50, and 100 ml; smaller semimicro and micro burets are also made. So macro, semimicro, and micro titrations can be performed (Ch. 6) on solid or liquid samples dissolved in a suitable solvent.

Volumetric pipets are available which hold volumes from 0.5 to 100 ml. Semimicro and micro pipets can of course be obtained in much smaller volumes, down to 0.001 ml (1 μL). The latter are supplemented by syringe pipets, which allow much more rapid delivery.

Measurement of Photon Flow (Light Intensity)

The spectrophotometer (Ch. 12) and the spectrometer (Chs. 14 and 15) are the usual devices used for measuring the flow of photons of light or other types of radiation. The flow of photons is generally expressed as the *intensity* of the radiation involved. The spectrophotometer source emits photons, and its detector measures the fraction of photons absorbed by a sample, which may be gaseous, liquid, solid, or a solution of any of these. The spectrophotometer measures *concentration* (rather than weight or volume), usually from $10^{-3} M$ down to perhaps $10^{-7} M$. It can thus perform analyses in the micro to ultramicro regions (Table 1-1).

The term "spectrometer" refers to a type of instrument that can measure different kinds of interactions of radiation with chemical species, such as atomic absorption, atomic emission, molecular fluorescence, etc. Such an instrument, like the spectrophotometer, measures a flow of photons from the sample and can measure concentration from $10^{-3} M$ down to perhaps $10^{-9} M$. Thus it can perform analyses in the micro, ultramicro, and submicrogram regions.

Electrical Measurements

The voltage of an electrochemical cell is subject to too many factors to be used as an indication of concentration. However, by use of an electrical measuring device called a potentiometer (Ch. 16), the potential, or voltage, available for a chemical reaction can be measured without allowing a reaction to occur significantly. The potentiometer, like the spectrophotometer, indicates concentration, but in a wider range—from as low as $10^{-10} M$ to as high as $10^{-1} M$.

In contrast, the total current passing through an electrochemical cell can be measured accurately with much less interference from other factors. This is done by measuring the coulombs of electrical current passing through an electrochemical cell with a coulometer. The coulometer determines the moles of a sample species directly, but can indirectly indicate the concentration. The coulometer can measure from 0.001 to 10 mmoles of an electroactive species.

The potentiometric and coulometric types of measurements are the most useful measurements for the life sciences, but electrical conductance can also be measured. All three measurements are restricted to solutions of a sample, rather than gaseous or solid samples.

Thermal Conductance Measurements

Although gaseous samples can be analyzed by certain types of spectrophotometers, the more general approach is to measure the thermal conductance of a gas using a thermal conductivity detector. These detectors generally consist of four heat-sensing elements connected in a Wheatstone-bridge arrangement (Ch. 19). One pair of elements senses the conductance of a reference gas stream, and the other senses that of a sample gas stream. The detector responds to the flow of molecules and can detect as little as 5×10^{-6} g/ml of gas.

2-2 | INTRODUCTION TO THE USE OF THE BALANCE

Weight Readout and Sample Weighing Methods

This section contains general information on how to use the balance. The following sections provide a more detailed discussion of principles and construction and further directions for using the balance. The analytical balance is such a widely used tool in laboratories that it is necessary to learn to use it properly and also to appreciate its limitations. One rule of proper use is that powders and liquids must be weighed in special ways to protect the pan of the balance. One limitation to the use of the balance is that if the weight of an object to be weighed on it is below a given weight, the resulting weight readout will be highly uncertain.

Readout

A typical standard analytical balance has a weighing range of 0–160 g, and is readable to ± 0.0001 g (± 0.1 mg). Since the balance readout is in g rather than mg, it must be read to four significant figures to the right of the decimal point. Depending on the sample, its weight may typically be read to five to seven significant figures. Many balances give a complete digital readout, as shown in Figure 2-1, while others have a digital readout to just the first decimal place and a scale readout for the rest.

FIGURE 2-1. Typical four-decimal readout of a standard analytical balance. The readout is 118.7325 g; seven is the maximum number of significant figures.

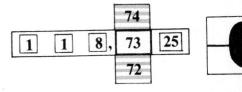

Sample Weighing Methods Solid objects can be weighed by a method different from that used for powdered solids or liquids. The two methods and their applications are given in the box:

Single-weighing method: most solid (non-powdered) objects
Weighing-by-difference method: powders and liquids

Samples that are not powders or liquids can be weighed by placing them directly on the pan of a balance and obtaining the weight by a *single weighing*. This is the case, for example, for objects such as a volumetric flask or a solid like a penny (see Exp. 1). This method is suitable as long as the sample does not contaminate or corrode the balance pan, and as long as the sample is not hygroscopic (does not absorb water from the air).

The *weighing-by-difference method* is used for most powders and liquids, and for solids that may contaminate the pan or are hygroscopic. When a sample is weighed by difference, the weight of the container with the sample in it is determined first. Then some of the sample is withdrawn from the container, and the container plus the remaining sample is weighed again, and so on. In this manner, only four weighings are necessary for three samples, rather than contaminating the balance pan or using six weighings for three samples.

Uncertainty of Weighing Since most analytical samples are weighed only once, there is no such thing as the precision of a single balance weighing. However, in the absence of qualifying information, we can say that a single weighing is uncertain by ± 1 in the last digit. Thus the *absolute uncertainty* (A.U.) of a single weighing on a standard analytical balance is ± 0.0001 g (0.1 mg), and, in general, the absolute uncertainty for any type of balance is the same as its readability. For weighing by difference, the total *possible* absolute uncertainty for a given sample is ± 0.0002 g (0.2 mg). This is due to the fact that for each sample one weighing may be high by $+0.0001$ g and the other low by -0.0001 g. The total uncertainty may in fact be less than 0.0002 g, but we must proceed on the assumption that any given weighing-by-difference operation could have the largest possible uncertainty.

To compare the uncertainties of weighing two samples of different weights, it is common to calculate the *relative uncertainty*—that is, the uncertainty relative to the weight of the sample. This is usually expressed in pph, parts per hundred (or in ppt, parts per thousand), because the quotient is multiplied by one hundred:

$$\text{R.U., pph} = \frac{\text{A.U.}}{\text{sample wt}} (100) \tag{2-1}$$

It is desirable that the relative uncertainty in most analytical work be no more than 0.1 to 0.2 pph. Suppose we are weighing a 0.5000-g (500.0-mg) sample, using a single weighing, and want to know if the relative uncertainty of the weighing is within the desired uncertainty of 0.1 pph. The relative uncertainty is calculated:

$$\text{R.U.} = \frac{0.0001 \text{ g}}{0.5000 \text{ g}} (100) = \frac{0.1 \text{ mg}}{500.0 \text{ mg}} (100) = 0.02 \text{ pph}$$

Minimum Sample Size It is important to be able to calculate the minimum size, in g or mg, of sample that can be accurately weighed, given the desirable relative uncertainty and the absolute uncertainty of a single balance weighing. This can be done by using a rearranged form of **2-1**:

$$\text{min sample wt} = \frac{\text{total A.U.}}{\text{R.U.}} (100) \qquad (2\text{-}2)$$

For a desirable relative uncertainty of 0.1 pph on the standard analytical balance using a single weighing, we substitute into **2-2**:

$$\text{min sample wt} = \frac{0.0001 \text{ g}}{0.1 \text{ pph}} (100) = 0.1 \text{ g (100 mg)}$$

For the same situation, except that the sample is weighed by difference, the total absolute uncertainty is 0.0002 g, and thus the minimum sample weight is 0.2 g (200 mg). These results are summarized in the box.

On the standard analytical balance (0.1 mg uncertainty for one weighing):
Minimum sample weight for single-weighing method = 0.1 g (100 mg)
Minimum sample weight for weighing-by-difference method = 0.2 g (200 mg)

The minimum sample weights are, of course, different for semimicro and micro balances and for special cases where a standard analytical balance does not have an absolute uncertainty of 0.1 mg for one weighing.

2-3 | THE BALANCE AS A LEVER

Although most modern balances are single-pan balances, the double-pan balance provides a better illustration of how the balance operates as a lever. In Figure 2-2 the pans and knife-edge of a double-pan balance are pictured. Since this is an unloaded balance, it is assumed that the pointer (between the pans) is at a balanced position, or rest point, equidistant from both pans. This is the *original rest point*. The discussion that follows is concerned with forces exerted by putting weights on both pans of the analytical balance.

In principle the analytical balance is a lever of the first class; its fulcrum lies between the points of application of the forces. As shown in Figure 2-2, the fulcrum, K, is the point at which the knife-edge rests on the agate plate. The fulcrum lies halfway between the points of application of the forces, F_L and F_R, exerted by weights on the pans of the balance.

Suppose that the force F_L just balances the force F_R (this is true if the balance comes to rest at its original rest point). Then

$$(F_L)(l_L) = (F_R)(l_R)$$

where l_L and l_R are the lengths of the arms of the lever from the knife-edges to the fulcrum.

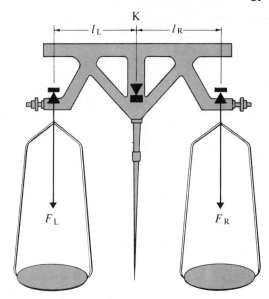

FIGURE 2-2. The pans and knife-edge as part of the lever system of the analytical balance. The fulcrum is at K, the points of application of the forces are directly above F_L and F_R and the hypothetical lever arms are l_L and l_R.

Now, the forces F_L and F_R are proportional to the mass on each pan.

> The origin of force is the attraction of the earth's gravity on an object of a given mass. Thus, *mass* is the quantity of matter in a body, and *weight* is the force exerted by gravity on that body. The relationship between force and mass is $F = (m)(g)$, where m is the mass and g is the acceleration due to gravity. Although the force of gravity varies somewhat throughout the world, it is constant for any given locality. Therefore the force or weight of an object is proportional to its mass.

As it is customary to speak of the mass of an object as its weight, the following equation holds true for the analytical balance:

$$(W_L)(l_L) = (W_R)(l_R)$$

Since the balance is so constructed that l_L and l_R are equal within an uncertainty of one part in 10^5, then $W_L = W_R$. The weight of the object on the left pan is then known directly from the sum of the weights on the right pan.

2-4 | SENSITIVITY AND CAPACITY

Sensitivity   The sensitivity of a balance is defined as the amount of deflection of the beam produced by a given unit of weight. For the standard analytical balance this unit is the milligram (mg). For a micro analytical balance it may be a microgram (0.001 mg).

Sensitivity is affected most significantly by the mass of the beam, the pans, and the weights on the pan(s). A common relationship between sensitivity and the weight on the pan(s) is that as the weight increases, the sensitivity decreases. This reduces the response of the balance needed to reach a rest point for large weights. In effect, this also slightly increases the uncertainty of weighings when samples are weighed by difference on a double-pan balance.

With the modern single-pan, direct-reading balance, weighing operations are always conducted at a constant load (Sec. 2-5) so that the sensitivity of the balance is constant no matter what the weight up to the capacity.

Capacity The capacity of many modern single-pan balances is of the order of 150 to 200 g. It is neither desirable nor necessary to weigh anything over 100 g on an analytical balance, since the relative uncertainty of weighing 100 g on a trip balance accurate to ± 0.01 g is usually satisfactory. Objects weighing less than 10 mg should be weighed on a balance having a smaller capacity—a semimicro or micro balance.

2-5 | CONSTRUCTION OF SINGLE-PAN BALANCES

We will now discuss essentially the construction of the single-pan, direct-reading balance. However, there are certain features common to both the single-pan and double-pan balances that will be described first. Refer to Figure 2-2 while reading about these.

The first feature is the *beam* which supports the pan(s) and weights. It must be rugged in construction, yet lightweight to allow for high sensitivity.

The *knife-edges* support the beam and the pan(s) and permit the balance to reach a rest point rapidly. These must be made of extremely hard material, such as agate or sapphire. Although this ensures that the knife-edges will remain sharp for a long time, it also means that the knife-edges can be easily damaged by the shock of releasing an unbalanced pan(s). Damage can be avoided by adopting the habit of never releasing the beam until the object and weights are almost equally balanced.

The *pan* (or pans) is made to hold an object or sample to be weighed. Pans are made of lightweight alloys resistant to air oxidation, but not resistant to chemical attack by reagents used in quantitative analysis. It is recommended that any reagent be weighed on a waxed paper or in a weighing bottle.

General Features of Single-Pan Balances In general most single-pan balances have the same characteristic construction. As shown in Figure 2-3, such a balance is enclosed in a casing so that only the single pan is visible; the beam and weights are completely enclosed. The weights are controlled by means of three or four knobs on the front of the balance case. Other knobs on the front or side of the case control the beam and zero adjustment.

Instead of by *direct comparison* as on the double-pan balance, an object's weight is determined by the *substitution method* on the single-pan balance. When the pan is empty, the weight of the pan plus the weight of a set of weights suspended from the front of the beam is balanced by the weight of a large *counterweight* suspended from the rear of the beam. When an object is placed on the pan, the increase in weight is compensated for by *removing* an equal weight from the set of weights suspended from the front of the beam. Since no change occurs in the weight of the counterweight suspended from the rear of the beam, the weight of any object is found at a constant weight, or load.

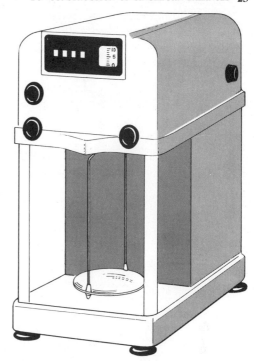

FIGURE 2-3. Front view of the Mettler H-15 balance. (Courtesy of the Mettler Instrument Corporation.)

It is important enough to reemphasize that the operation of the single-pan balance at constant load assures a constant sensitivity. A constant sensitivity ensures a uniform precision for all weighings, particularly for weighing by difference (Sec. 2-2).

A side view of a typical single-pan balance is shown in Figure 2-4. The beam arrangement is not symmetrical as it is in the double-pan balance (Fig. 2-2). The removable weights are located *above* the pan, just behind the weight control knobs on the front of the balance. The counterweight is located at the back, much farther away from the central knife-edge than the pan. Since the weights and the knife-edges are completely enclosed, they are much better protected against atmospheric corrosion than in an open double-pan balance.

Weight Readout

On many older single-pan balances, the weight readout was divided into three parts. The number of grams plus tenths of a gram were read from a digital scale. Hundredths and thousandths (mg) of a gram were read directly from a lighted optical scale. The fourth decimal (0.0001 g) was estimated from a vernier on the optical scale.

The modern weight readout is a complete digital readout (Fig. 2-1). All of the digits are read from a single scale, without a vernier to estimate the last decimal place. This prevents errors from incorrect vernier estimations as well as incorrect readings of hundredths and thousandths of a gram from an optical scale.

Special Features

Before 1975 balances were not protected from the mechanical shock caused by releasing the balance before the weights had been adjusted to balance the weight of an object. In 1975 a number of balances, such as that shown in Figure 2-5, were introduced with a special "air-release" device. This is essentially a piston which compresses air when the

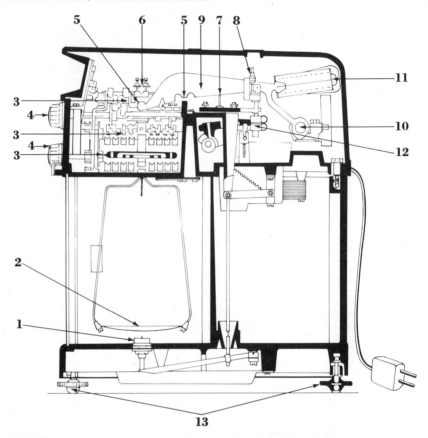

FIGURE 2-4. Cutaway (side) view of the Mettler H-15 balance. The following parts are numbered: (1) pan brake; (2) pan; (3) set of weights; (4) weight-control knobs; (5) sapphire knife-edge; (6) stirrup for hangers and weights; (7) lifting device; (8) movable weights for adjustment of sensitivity; (9) beam; (10) engraved optical scale; (11) air damper; (12) counterweight; (13) foot screws. (Not indicated are the arrest knob and the movable weight for zero adjustment.) (Courtesy of the Mettler Instrument Corporation.)

balance is released, so that complete release occurs gradually over a five-second period rather than instantly.

Other special features found on many balances include automatic preweighing, a scale-type weight-filling guide, weight locking, automatic weight printout, and automatic weight recording. The automatic preweighing feature gives an instantaneous indication of the approximate weight of the sample on the pan. The weight-filling guide shows the approximate sample weight between 0 and 100 g on a scale reading in 10 g units. The weight-locking feature prevents operation of the knobs controlling the heavier weights when the beam is completely released. Some balances provide automatic printout of weights on paper tape or provide electronic signals which can be used to record weights.

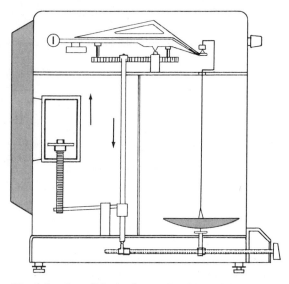

FIGURE 2-5. The "air-release" device for gentle release of the beam on balances such as the Mettler H78AR. As the beam moves down (right arrow), the piston moves up (left arrow), compressing air and cushioning the release.

2-6 | USE OF THE SINGLE-PAN BALANCE

When you finish this section, you will have read all the information needed to operate a single-pan balance. You should then take the self-test at the end of this section to review this information.

Handling of Objects

It is important to remember that your fingertips are coated with perspiration and perhaps some dirt; because of this, handling objects with your fingers will always increase the weight of an object to some degree. When handling heavy objects to be weighed, you should use a pair of tongs or finger gloves. When handling light objects, such as a weighing bottle, you may use finger gloves or a paper loop (see Sec. 2-7).

Chemicals can corrode and change the weight of the balance pan. You may weigh nonhygroscopic chemicals by difference on a waxed paper or in a weighing bottle. Always weigh hygroscopic chemicals by difference in a weighing bottle that is kept covered as much as possible.

A more detailed discussion of the weighing operations needed for various types of samples is given in Section 2-7.

Rules for Operation of a Single-Pan Balance

Read the rules below carefully before using a single-pan balance for the first time. You should also take the self-test at the end of this section as preparation.

1. Treat the balance as carefully as you would any other piece of expensive equipment. The instructor will prefer that you ask an apparently ignorant question rather than take the chance of damaging the balance.

2. Before you operate the balance or adjust its zero point, you should *always* check the following:

   a. The balance pan should be empty and free of dust and chemicals.

   b. All the weight knobs should be adjusted to the zero setting.

   c. The glass doors of the balance should be closed to prevent air currents from disturbing the equilibrium position of the pan when the balance is in use.

3. Check the zero point of the balance before you perform any weighings. This needs to be done only once for a series of weighings, but if someone uses the balance after you have set the zero point, you should clean the pan if necessary and check to see that the setting is still at zero. Adjust the zero point with the balance beam in the full-release position, not in the semirelease position.

4. To weigh an object, follow the steps below:

   a. Put the object on the pan only when the balance beam is in the *arrest*, or *secured*, position. If this is not done, the shock of the sudden change in weight can cause misalignment of or damage to the balance.

   b. Close the doors and adjust the beam to the semirelease position.

   c. To find the weight of an object quickly, develop a systematic approach of varying the weights. First, try a small weight, such as 1 to 5 g, and see if the beam moves. If not, try a larger weight, such as 90 g, and see if the beam moves. Repeat, increasing the smaller weights and decreasing the larger weights.

   d. When the weights have been adjusted to the nearest weight just less than the object's weight, adjust the beam to the *full-release* position and record the weight from the lighted scale of the balance.

5. Always record the weight of the object to *four* decimal places (0.0001 g), even if the last digit is zero. You should record the weight in a notebook, not on a piece of loose paper which may be easily lost.

6. Immediately adjust the beam to the arrest, or secured, position after finishing a weighing. Then remove the object from the pan and clean the pan if necessary. Return all weight knobs to the zero setting and close the glass doors.

SELF-TEST | **1. Preparation for Using the Single-Pan Balance**

Answers:

Directions: Cover the answers in the left column before beginning the test. After reading over the material on using the direct-reading single-pan balance, fill in each of the blanks below. Check the answers to each part before going on to the next.

A. Before adjusting the zero point of the balance, you should automatically check at least three things:

zero
   a. The weights on the optical scale should all read _____.

closed completely
   b. Both glass doors should be _____.

empty; clean
   c. The pan of the balance should be _____ and _____.

B. Before any weighings are performed, the zero point of the balance should be checked.

   a. To avoid air currents that can cause a mistake in the zero point, the check should be performed only after the

closed
      glass side doors are _____.

full-release

clean; recheck

secured

semirelease

1; 90

full-release

12.1100 (both zeros are significant)

is secured
zero
cleaned; closed

b. When the zero point is checked, the beam of the balance should be in the _____ position.
c. If someone else uses the balance after you have checked the zero point, you should _____ the pan and _____ the zero point.

C. To adjust or change weights while trying to find the weight of an object, you should follow the steps below:
a. Before the object is put on the pan, the beam of the balance should be in the _____ position.
b. Weights may be adjusted after the beam of the balance has been turned to the _____ position.
c. To find the weight of an object quickly, first try a small weight, such as __.0 g, and then a heavy weight, such as __.0 g. Repeat until the weight is between two consecutive readings, such as 11 and 12 g.

D. To obtain the exact weight from the balance, you should:
a. Adjust the beam to the _____ position and read the optical scale.
b. Since the balance is accurate to four decimal places, record a weight such as 12.1100 g as __ g.

E. After the weighing is finished and the weight has been read:
a. The object cannot be removed from the pan until the beam _____.
b. All weights are then returned to the _____ positions.
c. The pan is _____ if necessary and the glass side doors are _____.

2-7 | WEIGHING OPERATIONS

Drying the Sample

The time and temperature for drying depends on the nature of the sample. Samples of organic and biological materials are not usually dried, because drying would partially decompose them. Many inorganic samples should properly be heated at high temperatures (300°C), as, for example, with sodium carbonate to change the sodium bicarbonate impurity back to sodium carbonate. Still other samples, such as primary-standard arsenic(III) oxide that has not been appreciably exposed to air, are not hygroscopic and need not be dried.

If drying is needed, place the samples in a drying oven for one to two hours at 110°C. Put the sample in an open weighing bottle in a beaker. Cover the beaker with a watch glass on glass hooks, as shown in Figure 2-6. If the sample decomposes, oxidizes, or sublimes at 110°C, dry it by placing it in a desiccator containing a drying agent efficient enough to absorb water vaporized from the sample.

Desiccators

After the sample has been oven-dried, allow it to cool somewhat and then place it in a desiccator (Fig. 2-7). Leave the lid ajar for about ten minutes so that a partial vacuum will not form in the desiccator as the air inside cools. Allow the sample to cool in the desiccator for at least thirty minutes before weighing.

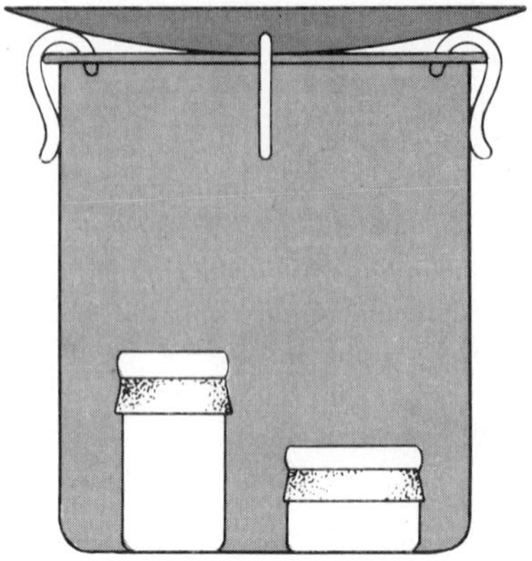

FIGURE 2-6. Beaker covered with a watch glass and weighing hooks.

FIGURE 2-7. A dessicator.

To ensure a tight seal and efficient drying, grease the contacting surfaces of the desiccator and the lid lightly with petroleum jelly. To open or close the desiccator, slide the lid sideways with a steady pressure. Never jerk or lift the lid directly upwards. Keep one hand on the lid when carrying the desiccator to prevent a loose lid from sliding off.

Before using the desiccator, charge it with fresh desiccant. Although calcium chloride is often used, Drierite ($CaSO_4 \cdot \frac{1}{2} H_2O$) and anhydrous magnesium perchlorate absorb water more efficiently and either is recommended. Also, a colored indicating form of Drierite is available.

A vacuum desiccator is a special type of desiccator that can be used for samples that cannot be dried by heating. Simply place the sample in the vacuum desiccator and connect a vacuum pump. Evacuate to the desired vacuum and close the port of the desiccator to retain the vacuum.

The freeze-drying technique may be used to dry samples that decompose even at room temperature. The sample is cooled to or below the freezing point, and water is removed by continuously evacuating with a vacuum pump.

Weighing the Sample   Three methods for weighing solid samples and one method for weighing liquids are given below.

FIGURE 2-8. Handling a weighing bottle with a paper loop.

Method 1   Solid samples in weighing bottles: After the sample is cooled, weigh it by difference from the weighing bottle. This is especially recommended for deliquescent samples. Throughout the weighing process handle the weighing bottle only with a paper loop, as shown in Figure 2-8; moisture from the fingers will change the weight of the bottle and cause errors. Another technique for handling glassware is to use finger gloves on the thumb and first two fingers.

 To weigh by difference, first weigh the bottle plus sample and record the weight in your notebook. Then carefully remove the estimated amount of sample from the bottle with a clean spatula and place in a marked beaker or flask. Weigh the bottle plus reduced sample to the nearest 50 mg to see whether the amount of sample withdrawn is within the desired weight range. If necessary, withdraw more sample and again weigh the bottle approximately. When enough sample has been transferred, weigh the bottle plus remaining sample to the nearest 0.0001 g and record the weight. Subtract the final weight from the initial weight to obtain the exact weight of the sample.

 As an alternative to using a spatula, insert a small aluminum scoop into the weighing bottle and weigh the scoop along with the bottle and sample for both the initial and final weighings. Handle the scoop with paper or finger gloves when transferring the sample.

Method 2   Solid sample in a scoop: Weigh accurately a large, lightweight scoop, such as shown in Figure 2-9. With a clean spatula add sample to the scoop until a weighing to the nearest 50 mg indicates that the sample weight is within the desired range. Then weigh the scoop plus sample to the nearest 0.0001 g and record the weight. Carefully transfer the sample from the scoop to a marked beaker, handling the scoop with a piece of paper. Brush the last traces of sample into the beaker with a small camel's-hair brush.

Method 3   Solid sample with waxed paper or polyethylene-coated weighing paper: Weigh the weighing paper alone on the balance. Then add the sample to the weighing paper, using a clean spatula. Weigh the sample plus weighing paper. Subtract the weights to obtain the weight of the sample. This method is only valid for samples that do not absorb water from the air while standing.

FIGURE 2-9. Lightweight scoop for handling solid samples.

Method 4   Liquid samples: Most liquid samples can be weighed out with a small bottle fitted with a medicine dropper and a ground-glass stopper. Weigh the bottle, sample, and medicine dropper accurately and record the weight. Use the dropper to transfer part of the liquid from the bottle to a marked flask or beaker. Replace the stopper, being careful not to get any liquid on the ground-glass surface of the bottle. When the proper amount of sample has been removed, as indicated by a weighing to the nearest 50 mg, weigh the bottle, dropper, and remaining sample to the nearest 0.0001 g and record the weight. The difference is the weight of the liquid sample.

2-8 | MEASUREMENT OF DENSITY AND SPECIFIC GRAVITY

Definitions   The density, and therefore the specific gravity, of a liquid is accurately and easily measured by weighing a known volume on the analytical balance. The density of an object is defined as its weight per unit volume at a specified temperature. In scientific work density is expressed in units of g/ml or g/cm^3, but in other areas it may be expressed in units such as lb/gal. The density of water at 20°C is 0.99823 g/ml or 8.330 lb/gal. (The density of water is at its maximum of 1.0000 g/ml at 4°C.)

The specific gravity of a liquid is the ratio of its weight to the weight of an equal volume of water. Specific gravity thus has no units, but it may be expressed either as the density at 20°C relative to the density of water at 4°C $\left(\dfrac{20°}{4°}\right)$, or as the density at 20°C relative to the density of water at 20°C. Thus carbon tetrachloride has a specific gravity of 1.594 $\left(\dfrac{20°}{4°}\right)$, as well as a specific gravity of 1.606 $\left(\dfrac{20°}{20°}\right)$.

The density of a solvent is increased by dissolving a solute in it; that is, the solution resulting from adding a solute to a solvent has a greater density than the pure solvent. For example, the density of sea water is 1.025 g/ml at 20°C; the dissolved sodium chloride increases the density above that of pure water. As the concentration of the salt increases, the density also increases. The measurement of the density of a solution will therefore give an estimate of the concentration of the salt in the solution.

Measure-   The density of a liquid is readily measured by carefully weighing a known volume of the
ment of   liquid at a constant temperature. Depending on the accuracy required in measuring
Density   the volume, the volumetric apparatus used may be either a volumetric flask or a pycnometer (Fig. 2-10).

Volumetric   The measurement of density with a volumetric flask is the less accurate method of the
Flask   two, but it is more convenient and slightly faster. There are two factors that limit the
Method   accuracy of the method. One is how closely the meniscus is adjusted to correspond to
the mark on the neck of the flask (Fig. 2-10). This error cannot be evaluated because it
will vary according to who is making the measurement. Because the neck of the flask
is relatively wide, even a slight error may limit the accuracy so that it is only moderately
good. The other factor that limits the accuracy of this method is the tolerance, or
maximum allowable error of the volume of the volumetric flask. This can be evaluated.

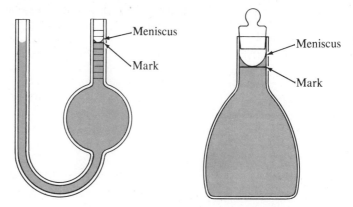

FIGURE 2-10. Two volumetric containers for the determination of density. Left: the pycnometer with a narrow capillary for a highly accurate volume measurement. Right: a volumetric flask for moderately accurate volume measurement. Note that a small discrepancy between the meniscus and the mark will cause a larger error in the volumetric flask because of its larger diameter.

For example, a 25-ml volumetric flask has tolerance of ± 0.030 ml. This is the absolute value; the relative maximum allowable error is

$$\frac{0.030 \text{ ml}}{25.000 \text{ ml}} (100) = 0.12 \text{ pph}$$

A relative error of the order of 0.1 pph is typical for most volumetric flasks. This means that the density will also have a relative error of ± 0.1 pph. For a liquid such as water, with a density near 1 g/ml, the density may only be reported with an absolute uncertainty of ± 0.001 g/ml. This means that water may be reported as 1.000 g/ml or 0.999 g/ml, but not 1.0000 g/ml. The density of liquids in the range from 1.5 to 2.0 g/ml must be reported with an uncertainty of ± 0.002 g/ml. This means that it may be reported as 1.500 ± 0.002 g/ml or 1.50 g/ml, but not as 1.500 g/ml. To report the latter would imply that the last significant figure is only uncertain by ± 0.001 g/ml, which is not true.

Pycnometer Method

The measurement of density with a pycnometer is more accurate than with a volumetric flask, but it is slightly slower because the adjustment of the volume is more time-consuming. The pycnometer has a capillary device (Fig. 2-10) for accurate measurement of the volume; it requires careful adjustment to match the meniscus to the mark. The narrow diameter of the capillary compared to the wide diameter of the neck of the volumetric flask reduces the error arising from an inexact matching of the meniscus and the mark.

Because the pycnometer is more accurate, densities measured with it can be reported to four (or five) decimal places; for example, water can be reported as 1.0000 g/ml or 0.99823 g/ml, instead of 1.000 g/ml or 0.998 g/ml. The same is true for specific gravities. Most data reported in handbooks contain four significant figures to the right of the

TABLE 2-1. The Density of Aqueous Sodium Chloride Solutions

Percentage NaCl	Density, g/ml at 20°C
1.0%	1.0053
2.0%	1.0125
4.0%	1.0268
6.0%	1.0413
10.0%	1.0707
12.0%	1.0857
14.0%	1.1009
16.0%	1.1162
18.0%	1.1319
20.0%	1.1478

decimal point; the data for water are expressed to five significant figures to the right of the decimal point. Apparently a more accurate type of pycnometer was used for water.

Estimation of Concentration Measurement of the density of a solution can be used to estimate the concentration of a solute. For example, consider the data in Table 2-1. As the percentage of sodium chloride increases, the density of the sodium chloride solution increases from 1.0053 to 1.1478 g/ml. Thus a sample may be analyzed for its percentage of sodium chloride by measuring its density (Exp. 3A). One approach would be to estimate the density to the nearest 0.1% by interpolation, using the data in Table 2-1. Since smaller differences cannot be estimated so easily, a better approach would be to plot the density of the sodium chloride solutions against the concentration, as shown in Experiment 3A. The concentration of an unknown solution can then be easily read by entering its density on the plot.

| QUESTIONS AND PROBLEMS

(Answers to most even-numbered problems appear in Appendix 5.)

Definitions and Concepts

1. What is the absolute uncertainty on the usual analytical balance for:
   a. A single weighing?
   b. Weighing by difference?
2. Explain weighing by difference.
3. Explain the difference between the operation of a double-pan balance and a single-pan balance.
4. Explain the advantage of the substitution method of weighing used on single-pan balances.
5. What is the purpose of the air-release device on single-pan balances?

6. What are some ways to handle samples without touching them during the weighing process?
7. Define:
   a. Density
   b. Specific gravity
   c. Pycnometer
8. Summarize the important rules for operation of a single-pan balance.

Balance Problems

9. Calculate the relative uncertainty of weighing each of the following objects on a standard analytical balance, using first a single weighing and then weighing by difference. Decide

whether each relative uncertainty is within the usual desirable value (you should know the value).

   a. A 3.3-g penny

   b. A 0.150-g salt sample

   c. A 100-mg salt sample

10. Suppose that a semimicro balance has an absolute uncertainty of 0.01 mg. Calculate the minimum weight for a sample using a relative uncertainty of 0.1 pph for:

   a. A single weighing.

   b. Weighing by difference.

11. Calculate the minimum weight of a sample that is to be weighed under the conditions specified below, assuming the usual desirable relative uncertainty (which you should know):

   a. Standard analytical balance, single weighing.

   b. Standard analytical balance, weighing by difference.

   c. Analytical balance with absolute uncertainty of 0.0002 g for each weighing, single weighing.

   d. Same balance as in c, weighing by difference.

12. During World War I a clerk in a French intelligence bureau became curious about a suspected German agent who carried many pencils. He suspected that something was being carried inside the pencils. How could he have checked each pencil without breaking it open? (See A. A. Hoehling, *Women Who Spied*, Dodd, Mead & Co., New York, 1967, p. 58.)

13. Discover the maximum weight that can be weighed on the analytical balance in your laboratory and calculate the relative uncertainty (pph) of weighing that weight once.

14. Calculate the minimum weight in mg of a sample that is to be weighed by difference on a standard analytical balance, if the desirable relative uncertainty is 0.2 pph.

15. Calculate the minimum weight in mg of a sample that is to be weighed on an analytical balance with an absolute uncertainty of 0.0002 g, if the desirable relative uncertainty is 0.2 pph.

Density/Specific Gravity Problems

16. A 5.000-ml volume of carbon tetrachloride weighs 7.970 g at 20°C. Calculate its density in g/ml.

17. Calculate the specific gravity $\left(\dfrac{20°}{4°}\right)$ of the carbon tetrachloride in the previous problem. Also calculate the specific gravity $\left(\dfrac{20°}{20°}\right)$.

18. A solution of unknown percentage sodium chloride has a density of 1.056 g/ml at 20°C. Determine its percentage sodium chloride to the nearest 0.5% by plotting the data in Table 2-1.

Challenging Problems

19. One of a set of five 1-g weights is either too heavy or too light. Devise a method that will identify that weight, using a sixth correct weight and only three weighings on a double-pan analytical balance.

20. A combination of seven weights excluding the 50-g weight in a set of weights will give any weight up to 60 g. Devise a scheme to give any weight up to 63 g with the use of only six weights of any denomination and a double-pan balance.

3 | Data Handling

"What is the meaning of it all, Mr. Holmes?"
"Ah. I have *no data*. I cannot tell," he said.

ARTHUR CONAN DOYLE
The Adventure of the Copper Beeches, 1892

The most important reason for performing an analytical measurement is to obtain useful data. If you are unable to obtain such data, you will not be able to make any quantitative conclusions (even though you may be able to draw a few qualitative conclusions from visual observation).

To report useful, meaningful data, it is necessary to use numbers, both exact and measured, in correct scientific fashion. A careful scientist reports in the final data only as many digits as are justified by the *measured*, not the exact, numbers in his or her data. Since this is an important difference, you should know the difference between these two types of numbers.

Exact numbers have no uncertainty; they include arbitrarily defined reference values, factors, exponents, logs to the left of the decimal point, and small counted numbers. An arbitrarily defined reference value is, for example, the $E°$ of 0.0 V assigned to the reduction of hydrogen ion to hydrogen gas; this value could be expressed as 0.0 V, 0.00 V, etc., since it is exact. A typical factor is 100; it is used to calculate percentage. Exponents, as in 10^{-7}, are exact since they arise from manipulation of the decimal point, not from measurement of numbers. Similarly, in a log value, such as a pH of 7.0, digits to the left of the decimal are exact since they arise from taking the log of an exponent, rather than of a measured number.

Measured numbers are uncertain because any measurement involves some uncertainty. For example, a balance capable of weighing only to the nearest gram is used to measure a person's daily intake of protein. One weighing gave a weight of 141 g of protein. The last digit in the weight is uncertain, however, since the balance can only be read to ± 1 g. Hence we should say that the weight of the protein is 141 ± 1 g.

The above example is fairly typical; measured numbers are usually assumed to be uncertain by ± 1 in the last digit. This is true whether the last digit is to the left or to the right of the decimal point. Thus the number 1.83 is assumed to imply 1.83 ± 0.01, unless otherwise specified. We will describe the treatment of measured numbers in greater depth throughout this chapter.

A single number or set of numbers may be characterized by its *accuracy*, or closeness to the true value. A set of numbers may also be characterized by its *precision*, or agreement among the numbers in the set. For a single number, there is no such thing as its precision, but we may refer to its *uncertainty*. All three of the above concepts—accuracy, precision, and uncertainty—can be expressed in mathematical terms only by calculating the error (for accuracy) or the deviation. Both error and deviation can be expressed in *absolute* and in *relative* terms.

Absolute and Relative Error

The absolute error (A.E.) of any measurement, X, is the difference between that measurement and the true value, μ. Stated mathematically,

$$\text{A.E.} = X - \mu$$

Absolute errors cannot be compared with one another; instead the *relative error* (R.E.) of each measurement must be calculated and used for comparison. The mathematical definition of relative error in parts per hundred (pph) is

$$\text{R.E.} = \frac{(X - \mu)}{\mu}(100) = \frac{\text{A.E.}}{\mu}(100)$$

Note that relative error is simply the absolute error divided by the true value; this is then converted to pph by multiplying by one hundred.

For example, suppose an erroneous analysis of vitamin C in a urine sample gives a value of 402 mg for 24 hours when the true value is 400 mg for 24 hours. The absolute error is calculated as follows:

$$\text{A.E.} = 402 - 400 = 2$$

The relative error of the analysis is then:

$$\text{R.E.} = \frac{402 - 400}{400}(100) = 0.5 \text{ pph}$$

Absolute and Relative Deviation

The deviation of a number or a set of results is used to describe the precision of the measurement(s) mathematically. In general, deviation is defined as the difference between a measurement and the mean, $\bar{X}$. Later on, you will be introduced to the standard deviation of a set of results.

Both deviation and standard deviation can be expressed in absolute and relative terms, just as error can. Absolute deviation (A.D.), like absolute error, is just an *absolute difference* between two numbers; for absolute deviation this is the absolute difference between the measurement and the mean. Relative deviation (R.D.) is the absolute deviation divided by the mean and multiplied by one hundred:

$$\text{R.D.} = \frac{\text{A.D.}}{\bar{X}}(100)$$

This is a mathematical definition of relative deviation in parts per hundred (pph).

For example, suppose the first weighing of an aspirin tablet gives a value of 0.5011 g. Further weighings are made and the mean of all the weighings is 0.5010 g. The absolute deviation of the first weighing is

$$\text{A.D.} = 0.5011 \text{ g} - 0.5010 \text{ g} = 0.0001 \text{ g}$$

The relative deviation of the first weighing is

$$\text{R.D.} = \frac{0.5011 \text{ g} - 0.5010 \text{ g}}{0.5010 \text{ g}} (100) = 0.02 \text{ pph}$$

Absolute and Relative Uncertainty
When something must be said about the deviation of a single result, all that can be done is to calculate its uncertainty. In the absence of qualifying information, it may be assumed that the last digit is uncertain by ± 1. Another way of saying this is to state that the absolute uncertainty is ± 1. The relative uncertainty, or relative deviation, of a single measurement is calculated as above. As an example, consider counting the number of aspirin tablets in a bottle supposed to contain 200 tablets. The relative uncertainty (R.U.) of counting this number, assuming an absolute uncertainty of ± 1, is

$$\text{R.U.} = \frac{1}{200} (100) = 0.5 \text{ pph}$$

This raises the question of whether all such counting operations are uncertain by ± 1. The answer is certainly no. Where the sample size is small enough to preclude an error, such as in a sample of 10, the absolute uncertainty is usually zero. In an important situation, such as counting the number of capsules of a potentially dangerous prescription, the pharmacist will check the count so that the uncertainty is also zero.

3-1 | SIGNIFICANT FIGURES

Not all data recorded during a measurement are significant. Although it is not necessary to decide which digits from a measurement are significant at the time, it is necessary to decide this when *reporting* the final result. To make this decision it is necessary to understand the concept of significant figures and the rules of significant figures that govern the handling of measured numbers. (Of course, rules of significant figures do not apply to exact numbers, since they contain no errors from measurements.) After reading this section, take the self-test at the end.

Concept of Significance
Significant figures are defined as those digits in a number that are known with certainty, *plus* the first uncertain digit. All other digits are termed *nonsignificant figures*. In scientific work it is desirable either to report data showing which digits are significant and which are not or to report data using only significant figures. For scientific work, use the following simple rules.

1. Usually only the final result, or answer, to a calculation need be expressed with the correct number of significant figures. It is best to write such data in semiexponential form—that is, write a result such as 0.0123 as 1.23×10^{-2} to show that three significant figures are justified.

2. Use appropriate symbolism to designate which digits are nonsignificant figures if reporting digits that are not significant. Throughout this book nonsignificant figures will be written as subscripts; for example, if there are only two significant figures in a number such as 1.544, it will be written as 1.5_{44}, or 1.5_4.

3. If the last digit of a result is uncertain by more than ± 1, write the uncertainty after the number, for example, 3.1 ± 0.2. All results written without such qualifying information should be assumed to have a last digit that is uncertain by ± 1.

Zeroes Special attention should be given to whether zeroes are significant or not. The following rules are helpful.

1. Zeroes that appear between other digits, as in 10.04, are significant.

2. Initial zeroes, such as in 0.104 or 0.0014, are usually not significant, since they serve only to locate the decimal point and are not "certain" digits. Logs are an exception —initial zeroes to the right of the decimal in a log are significant. For example, the log of $1.010 = 0.0043$; both zeroes to the right of the decimal point in 0.0043 are just as significant as those in 1.0043, which is the log of 10.10.

3. Terminal zeroes are generally considered to be significant. If a terminal zero is not significant but is only used to fix the decimal point, it should be eliminated and the number written using powers of ten. For example, 10,100 as so written is considered to have five significant figures. If only three digits are meant to be significant, then it should be written as 1.01×10^4.

Rounding Off When using an electronic calculator, you should be especially careful in rounding off. It's a good idea to carry along at least two nonsignificant figures until obtaining the final result, to avoid error in the last significant figure of that result. This is particularly important when there is more than one calculation sequence, or when data are re-entered on a calculator.

Because there are at least two different approaches to rounding off, you may wish to avoid it as much as possible and instead report one or two nonsignificant figures. We have listed two alternative rules below. The first is to be used if you wish to round off whenever possible, and the second is to be used if you wish to avoid rounding off as much as possible.

1. *Rounding off:* Eliminate nonsignificant figures by rounding off. Increase the last retained figure by 1 if the adjacent discarded digit is 5 or greater. If it is less than 5, do not change the last retained figure.

2. *Reporting nonsignificant digits:* Avoid rounding off by reporting one or two non-significant figures as subscripts. If there are more than two nonsignificant figures, discard all but two and report two nonsignificant figures. Usually rounding off is not necessary if two nonsignificant figures are given; for example, if 1.4989 is to be reported with one significant figure, it is written as $1._{49}$. This method of reporting also indicates that the trend of the measurements is toward a number higher than 1.0 rather than toward a number between 0.5 and 1.0.

Addition and Subtraction The absolute uncertainty of each number controls the number of significant figures in the sum or the difference of the numbers. To apply this concept you should first write all the numbers in semiexponential form and *adjust all numbers so that they have the* *same exponent.* (For numbers close to one, an all-digital form is recommended, for example, $9.0 - 0.8 = 8.2$.)

To express a sum or difference to the correct number of significant figures use the following rule.

Rule 1   Use the same number of significant figures to the right of the decimal in a sum or difference as the *number with the fewest number of significant figures to the right of the decimal.*

Do not round off any of the numbers before adding or subtracting; round off afterwards.

> **EXAMPLE:**   Saliva contains about 4×10^{-7} mole/liter of H^+.   Find its $[H^+]$ after ingestion of a food acid adding 5.55×10^{-6} mole/liter of H^+ to the mouth.
>
> *Solution:*   Adjust the exponents of each number to match, then add.
>
> $$5.55 \times 10^{-6} \text{ mole } H^+/\text{liter} = 5.55 \times 10^{-6} M \text{ } H^+ \text{ from food acid}$$
> $$4 \times 10^{-7} \text{ mole } H^+/\text{liter} = 0.4 \times 10^{-6} M \text{ } H^+ \text{ from saliva}$$
> $$[H^+] = \overline{5.9_5 \times 10^{-6} M}, \text{ or } 6.0 \times 10^{-6} M \text{ (rounded off)}$$

Multiplication and Division   The relative uncertainty of each number controls the number of significant figures in the product or the quotient of two or more numbers.   Comparing relative uncertainties is time-consuming, and a shorter, but sometimes less accurate, method is recommended instead.   To apply this method you should first write all the numbers in semiexponential form, using only one digit to the left of the decimal point.   (This is not necessary for numbers between one and ten.)

To express a product or quotient to the correct number of significant figures or to take a root such as a square root, use the following rule.

Rule 2   Use the *same number* of significant figures in a product, quotient, or root as in the number with the fewest number of significant figures.   (Note that this is different from Rule 1 above.)

> **EXAMPLE:**   Find the $[H^+]$ of 0.020 liter of gastric juice containing 4.00×10^{-4} mole H^+.
>
> *Solution:*   Divide as follows:
>
> $$[H^+] = \frac{4.00 \times 10^{-4} \text{ mole}}{2.0 \times 10^{-2} \text{ liter}} = 2.0 \times 10^{-2} M \text{ (two sig. figs.)}$$

Log Terms   Log terms, such as pH, should be expressed with the same number of significant figures to the right of the decimal as the total number of significant figures in the semiexponential number from which they are calculated [1].   This is because the digits to the left of the decimal come from the exponent, which is an *exact* number, not a *measured* number.   The following examples will illustrate this rule.

EXAMPLE: The equilibrium constant, K, for an oxidation-reduction reaction is 2.7×10^4. Calculate log K.

Solution: Take the log of the number and the exponent separately and add:

$$\log(2.7 \times 10^4) = \log 2.7 + \log 10^4 = 0.4314 + 4 = 4.43_1$$

Note that the log of 2.7 was expressed to four digits initially because it was found in the four-place log table on the inside back cover. After adding, it was rounded off to the proper number of significant figures, two to the right of the decimal.

EXAMPLE: Calculate the pH of gastric juice having $[H^+] = 2.0 \times 10^{-2} M$.

Solution: Take the negative log of the number and the exponent separately and combine algebraically (see the inside back cover for four-place logs):

$$pH = -\log 2.0 + (-\log 10^{-2}) = -0.3010 + 2 = 1.69_9 \text{ (two sig. figs. to right)}$$

Note that the one is really an *exact* number rather than a *measured* number because it is the result of a subtraction involving a digit that is the log of the 10^{-2} term.

Since the absorbance, A, is defined as the negative log of transmittance, T, on a spectrophotometer (see Ch. 12), it should be calculated from transmittance in the same manner as pH is. See the example below.

EXAMPLE: The transmittance of a solution is 0.10. Calculate the absorbance, A.

Solution: Write transmittance, T, in semiexponential form. Then take the negative log:

$$A = -\log T = -\log(0.10) = -\log(1.0 \times 10^{-1}) = -\log(1.0) + -\log(10^{-1}) = 1.00$$

Note that absorbance has three significant figures, compared to two for transmittance, only because the first digit comes from the exponent.

SELF-TEST | **2. Significant Figures**

Answers:

Directions: Cover the answers in the left column before beginning the test. After reading the material on significant figures you should work these problems. Since it is important to understand each problem before proceeding further, after completing each one, consult the correct answer given to the left before proceeding to the next. If an answer is wrong, you should reread the appropriate section before going on.

A. Express the following numbers in scientific notation, using the correct number of significant figures.

2.02×10^{-3}
2.000×10^1
1.10 (or 1.10×10^0)

    a. 0.00202
    b. 20.00
    c. 1.10

B. Using a five-digit calculator the calculation of percent chloride from three analyses of a single sample gives these results: 2.0510%, 2.0481%, and 2.0440%. Only two significant figures are justified for each calculated result.

2.1, 2.0, 2.0 (Rounding off makes the differences appear larger than they are.)

    a. Round off each result without using a subscript.

$2.0_5, 2.0_4, 2.0_{48}$ (Rounding to 2.0_5 is wrong; it implies further rounding to 2.1.)

b. Round off each result using a subscript to avoid the ambiguities in a.

C. Add each of the following pairs of numbers, using the correct number of significant figures.

   a. $1 \times 10^{-9} M \ OH^- + 1.0 \times 10^{-7} M \ OH^-$ from water

$1.0 \times 10^{-7} M$ (1×10^{-9} isn't significant.)

$1.1 \times 10^{-7} M$

$5.0 \times 10^{-5} M$ (H^+ from water is not significant.)

   b. $1 \times 10^{-8} M \ H^+ + 1.0 \times 10^{-7} M \ H^+$ from water

   c. $5.0 \times 10^{-5} M \ H^+Cl^- + 1.0 \times 10^{-7} M \ H^+$ from water

D. Using $K_w = 1.0 \times 10^{-14} = [H^+][OH^-]$, calculate each concentration using that given.

   a. If $[OH^-] = 2.0 \times 10^{-9} M$, $[H^+] = ?$

   b. If $[H^+] = 3.00 \times 10^{-7} M$, $[OH^-] = ?$

   c. If $[H^+] = 4 \times 10^{-4} M$, $[OH^-] = ?$

$[H^+] = 5.0 \times 10^{-6} M$

$[OH^-] = 3.3_3 \times 10^{-8} M$

$[OH^-] = 2._5 \times 10^{-11} M$

E. Calculate the pH of each solution below using the correct number of significant figures. Subtract a four-digit mantissa (0.3010, etc.) and round off after subtraction.

   a. $[H^+] = 3.00 \times 10^{-11} M$

$pH = 11 - 0.4771 = 10.523$ or 10.522_9

   b. $[H^+] = 3 \times 10^{-11} M$

$pH = 11 - 0.4771 = 10.5$ or 10.5_2

   c. $[H^+] = 3.0 \times 10^{-10} M$

$pH = 10 - 0.4771 = 9.52$ or 9.52_3

   d. $[H^+] = 2.000 \times 10^{-10} M$

$pH = 10 - 0.3010 = 9.6990$

3-2 | THE ARITHMETIC MEAN, THE MEDIAN, AND THE MODE

We have already discussed the true value, μ. In an analytical context, this is the arithmetic mean of an infinite number of results obtained by an experienced analyst using a reliable method. In most laboratory situations, time and expense limit the number of analytical results that can be obtained so that only an *estimate* of μ can be obtained. Commonly this is done by calculating the mean ($\overline{X}$) of the results. In many situations, the median (M), or middle value, is more effective and is preferred. A third estimate that is not as commonly used is the mode. Since it will not be used extensively in this book, it will be discussed only briefly. It is stressed that *all* of these are only estimates of the true value; there is no statistical axiom that demands that the mean of a small number of results be used all of the time.

The Mean   The arithmetic mean, simply called the mean, is an estimate of μ that is calculated from *all* the results in a sample. The formula for calculating the mean from a set of n results arranged in increasing order ($X_1, X_2, \ldots, X_n$) is

$$\overline{X} = \frac{X_1 + X_2 + \cdots + X_n}{n} \tag{3-1}$$

It can be seen that each analytical result contributes *equally* to the value of the mean. If, as shown in the upper half of Figure 3-1, the results are symmetrically (or normally)

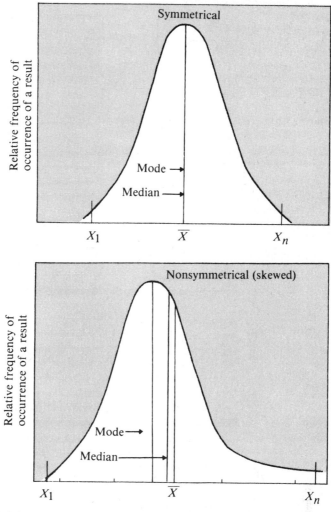

FIGURE 3-1.   Distribution curves for two different large samples.   Above: a symmetrical distribution where the mean is exactly halfway between the smallest value, X_1, and the largest value, X_n.   Below: a nonsymmetrical (skewed) distribution, showing the mean closer to the smallest value.   In such a distribution, the median and the mode are to the left of the mean, the mode always farther away from the mean.

distributed about the mean, the mean will be exactly halfway between the largest and smallest values.   In this situation, the mean is the most effective, or efficient, estimate of the true value.   Indeed, as the value of n increases above 10, the mean becomes an increasingly better estimate of μ.   There is little point in using the median for values of n above 10.

If, as shown in the lower half of Figure 3-1, the distribution of the results is not symmetrical, then the mean will not be halfway between the largest and smallest results.   Instead, the mean will be displaced away from the median in the direction of the most extreme result (X_n in Fig. 3-1). For samples of 10 or fewer measurements, the effect of a gross (large) error, such as that in X_n, becomes larger as n becomes smaller. If the precision of a set of between 3 and 10 measurements indicates the presence of a gross error in either X_n or X_1, then it is desirable to use the median. In fact, it has been shown [2] that for three results from a symmetrical population, it is more effective to use the median rather than the mean of the two closest results.

The Median The median, M, is the middle result of a set of results arranged in increasing order. The median is only defined when n is odd; if n is even, the median is undefined, strictly speaking. In such a case, the average of the two middle results is *used* as the median.

   As shown in Figure 3-1, the median is the same as the mean and the mode in a symmetrical sample. If the distribution of the results is not symmetrical, as is typical in a sample of ten or fewer results, then the median is not the same as the mean or the mode. Since the line drawn for the median, as in the lower half of Figure 3-1, must divide the area under the curve into two equal parts, the median will usually fall between the mean and the mode.

   Except for $n = 2$, the median is less effective than the mean for symmetrical samples where no gross error is present [3]. Nevertheless, it is interesting to compare the effectiveness of the median to that of the mean for a symmetrical sample. This comparison is displayed in Table 3-1. Note that the median is much more effective when n is even than when it is odd (this is explored in detail in the self-test at the end of this section). This fact is important in deciding whether to use the median when reporting results, and will be discussed more fully.

The Mode The mode is the result that occurs most frequently in a given sample of results. It is undefined if no result occurs more than once in a sample. Like the median, it is not affected by gross errors in extreme results.

TABLE 3-1. Effectiveness of the Median Compared to the Mean for a Symmetrical Sample (Normal Distribution)

Number of results, n	Effectiveness of median for even n	odd n
2	1 (= effectiveness of mean)	
3		0.74[a]
4	0.84	
5		0.69
6	0.78	
7		0.67
8	0.74	

[a] A fraction of 0.74 indicates that a median calculated from three results gives as much information about the true value as a mean calculated from 0.74 × 3, or 2.2 results.

In a large sample (more than 10), where each result is known only to one or two significant figures, at least one mode usually exists. In a symmetrical large sample (Fig. 3-1), the mode is identical with the mean and the median. If the distribution of the results in a large sample is not symmetrical but skewed to the right (Fig. 3-1), the mode lies to the left of the median and the mean, or vice versa [4]. In a small sample (less than 10), where each result has three or more significant figures, the mode is usually undefined and is therefore less important than the median and the mean for our purposes.

Choosing Between the Mean and the Median

We will see that it is the *precision* of a set of results that dictates whether the mean or the median should be reported. However, since a beginning student in the laboratory does not have the experience to judge whether the precision is acceptable, we recommend for this course the simplified approach given in the box.

> In the laboratory, you should report the *mean* of your data unless advised otherwise by the instructor.

For this discussion let us put ourselves in the position of experienced analysts and consider what to do if information is available on the precision of a method, or if there is no information available on the precision.

If the precision is known and the results have as good or better precision than previously obtained, then the mean should be reported. If the precision is significantly poorer than past precision, then it is always desirable to obtain more results before reporting. If this is impossible, then we recommend the approach given in the box.

> An experience analyst should report the median rather than the mean only if the precision is significantly poorer than past precision and no additional results can be obtained.

When reporting the median, it is of course an advantage to obtain an *even* number of results (see Table 3-1). To gain experience in deciding whether to report the mean or the median, you should work both the self-test at the end of this section and Problems 11–16 at the end of this chapter as though you were an experienced analyst.

If there is no information available on the precision of a method and a result appears to have a gross error, then it is always desirable to obtain additional results. If this is impossible, then the precision of the results themselves can be used to *test* for the possibility of a gross error. One statistical test that can be used for this purpose is called the Q test; it will be discussed in Section 3-4. Until you become familiar with the Q test, you are limited in this situation to reporting the mean. Occasionally, however, a result will deviate so greatly from the other results that it is probable that the result contains a gross error. The magnitude of such a deviation will vary, but for analytical measurements yielding two or more significant figures, a difference in the first digit indicates a probable gross error. For such a situation, the generalization in the box is useful.

A result probably contains a gross error if it differs *significantly* from all the other results in the first digit (and all the other results have the same first digit). For example, there is a significant difference for a sample of 3.3, 1.9, and 1.8, but not for a sample of 3.3, 2.9, and 2.8.

When such a gross error occurs, it of course indicates poor precision, and even if no information on the precision of the method is available, the median may be reported.

The following example illustrates the more common situation of the experienced analyst who knows what the precision of a method should be.

EXAMPLE: An FDA analyst obtains results of 0.13, 0.47, 0.57, and 0.59 ppm of mercury after analyzing a large catch of fish. The precision obtainable from past experience (a range of 0.30 ppm) indicates poor precision and the possible presence of a gross error in the 0.13 ppm result. The FDA tolerance for mercury in fish is 0.5 ppm. Should the fish be certified as meeting FDA standards or not?

Solution: Normally the mean is calculated first ($\overline{X} = 0.44$ ppm), but this is useless here since the poor precision indicates the presence of a gross error. Calculation of the median gives an M of 0.52 ppm, indicating that the fish barely fail to pass FDA standards. Since it is important to avoid wasting a large amount of fish as well as to avoid poisoning people, more analyses should definitely be obtained. Suppose, however, that this is not possible and that a decision must be made. What should be done? A comparison of the precision of the four results (range = 0.46 ppm) with that from past experience (range = 0.30 ppm) indicates that the precision definitely exceeds that obtained in the past. Assuming that the analysis is good to two significant figures, it is valid to use the median and discard the fish. Since the median is calculated from the two middle results, it has an attractively large effectiveness in this case (see Table 3-1).

(Note: The typical FDA lab generally has *two* analysts analyze the same batch of fish for mercury. The mean and/or median of *both* analysts must exceed 0.5 ppm before the fish are rejected for certification. Thus in this example the results of a second analyst must also exceed 0.5 ppm before the fish are rejected.)

SELF-TEST	**3. The Arithmetic Mean and the Median**

Answers:

Directions: Cover the answers in the left column before beginning the test. Work problems one at a time and check with the answers given at the left. The problems are meant to help you compare the usefulness of the mean and the median for increasingly larger samples from $n = 2$ through $n = 5$. Assume you are an experienced analyst without any knowledge of past precision.

A. Two measurements of the pH of gastric juice give values of 1.1 and 1.50.

$\overline{X} = 1.3$

a. Calculate the mean using the correct number of significant figures.

$M = 1.3$. Median is the average of both results.

b. Calculate the median using the correct number of significant figures.

No. Both are calculated using the entire sample.

c. Is there any difference between the effectiveness of the mean and median for $n = 2$? Why?

B. Three measurements of the pH of saliva give values of 6.4, 6.6, and 4.4.
  a. Calculate the mean using the correct number of significant figures.

$\bar{X} = 5.8$

  b. Calculate the median using the correct number of significant figures.

$M = 6.4$

  c. Is there any difference between the effectiveness of the mean and median for $n = 3$? Why?
     What is the effectiveness *in general* of the median relative to the mean for $n = 3$? Why?

Yes. The mean is generally more effective because all of the results are used.
0.74. It is calculated from only one-third of the results.

  d. *In the case above*, is the mean or the median more effective? Why?

The median. It is not affected by the gross error in 4.4.

C. Four measurements yield values of 1.1, 1.2, 1.2, and 3.30.
  a. Calculate the mean using the correct number of significant figures.

$\bar{X} = 1.7$

  b. Calculate the median using the correct number of significant figures.

$M = 1.2$

  c. What is the effectiveness *in general* of the median relative to the mean for $n = 4$? Why is it higher than for $n = 3$?

0.84. It is calculated from two-fourths of the results instead of one-third for $n = 3$.

  d. *In the case above*, is the median or the mean more effective? Why?

The median. It is not affected by the gross error in 3.30.

D. Five measurements yield values of 1.10, 1.10, 1.20, 1.20, and 1.30.
  a. Calculate the mean using the correct number of significant figures; show additional uncertain digits using a subscript, if necessary.

$\bar{X} = 1.18$

  b. Calculate the median using the correct number of significant figures.

$M = 1.20$

  c. What is the effectiveness *in general* of the median relative to the mean for $n = 5$?

0.69

  d. *In the above case*, is the median or the mean more effective? Why?

The mean. There are no obvious gross errors.

E. A symmetrical distribution of 101 results yields a mean of 1.50. What should the median of these results be? Why?

Median should be 1.50. Mean and median should be equal for a symmetrical sample.

3-3 | PRECISION

The precision or agreement of a set of results is governed by two types of errors that occur in the measurements. These are determinate or systematic errors and indeterminate or random errors. Determinate or systematic errors are those that (in principle at least) can be determined or measured, while indeterminate or random errors cannot.

The former type cause one or more results to be either high or low, but usually not both. Determinate or systematic errors originate in the form of personal errors, equipment (instrumental) errors, or method errors. Personal error may be misreading a buret or instrument display; equipment error may arise from a dirty buret or unstable instrument; and method error may result from nonstoichiometric reactions or incomplete precipitation. The net effect of any of these errors is that one or more of the results will be high or low. In particular, with instrumental readings, the *first reading or result* has the largest chance for equipment error because of improper start-up, warm-up, etc.

Random errors are those that cause the results to be symmetrically scattered on either side of the true value. Random errors cannot be eliminated; hence they are always present and limit the precision of any analysis. An example of a random error is the dropwise approach to a titration end point. One time a drop too little of titrant will be added, the next time a drop too much, and so on. All of the results of the titration will have errors scattered symmetrically on either side of the true value.

Now, how do these errors affect precision? Precision is the agreement among a set of measurements, or the ability to reproduce a measurement. If there are large systematic errors in an analysis, the results will tend to be scattered to one side or the other of the true value. If there is little or no systematic error, then only random error will limit precision—that is, the results will be symmetrically scattered on either side of the true value.

Although precision is not *defined* as the deviation of a set of measurements from the mean, it may be measured by calculating one of the three measures of deviation: standard deviation, range, or average deviation. First, let us discuss the standard deviation of a *population*.

Standard Deviation of a Population

If an infinite number of results could be obtained by an analysis, then the standard deviation (σ) of this infinite number, or population, would yield a 100% reliable measure of the precision of the population. The standard deviation is defined as the square root of the average of the squares of the individual deviations from the *mean of the population*. Each individual deviation is expressed as $(X_i - \mu)$, which leads to the following equation for the standard deviation of a population:

$$\sigma = \left[\frac{\sum (X_i - \mu)^2}{n} \right]^{1/2} \tag{3-2}$$

where X_i represents all values from X_1 to X_n and where $\sum (X_i - \mu)^2$ is the sum of the squares of all the deviations from the mean. Now consider Figure 3-2. This is the so-called normal distribution curve. It is formed by plotting an infinite number of measurements on the graph. If there are no systematic errors present, the ever-present random error would cause the results to be distributed in a random fashion about the true value. Also note that 99.74% of the results would occur within $\pm 3\sigma$ units of the true value, or population mean.

Now, the population standard deviation is a true measure of the precision of an analysis. Unfortunately, time and economics force analysts to perform only a finite number of analyses, and they can only estimate precision by calculating a sample standard deviation, a sample range, or a sample average deviation. The important concept of sample standard deviation will be discussed first.

Standard Deviation of a Sample

The standard deviation of a sample is calculated using an equation similar to **3-2**, except that the average is found by dividing by $(n - 1)$, not n. The formula for the calculation is

$$s = \left[\frac{\sum (X_i - \bar{X})^2}{(n - 1)} \right]^{1/2} \tag{3-3}$$

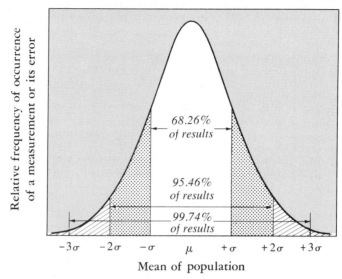

FIGURE 3-2. The normal distribution curve. (See Appendix 4 for other values.)

The reason $(n - 1)$ is used instead of n is essentially that a degree of bias is introduced by using an estimate of the true value that is derived from the very data to be used to calculate s.

A number of electronic calculators have a key that permits automatic readout of s after values of X_i have been entered. If your calculator does not have such a key, refer to Appendix 3 for instructions on how to calculate s quickly using an electronic calculator.

Equation **3-3** gives the absolute value of the sample standard deviation. From the general definition given at the beginning of the chapter, the relative value of s (s_R) is calculated:

$$ s_R = \frac{s}{\bar{X}} (100) $$

The relative sample standard deviation is expressed in pph, whereas the absolute value is expressed in the same units as the data. The following example illustrates the calculation of both values.

EXAMPLE: Babies are tested for cystic fibrosis by measuring the chloride in their sweat. One such analysis gives molarities of 2.15×10^{-2}, 2.35×10^{-2}, and 1.50×10^{-2} chloride. Calculate the mean and standard deviation and evaluate the results.
Solution: First, calculate the mean of the three results.

$$ \bar{X} = 2.00 \times 10^{-2} M $$

Second, calculate the deviations (omitting the 10^{-2} factor) and square them.

$$ \left. \begin{array}{l} (2.15 - 2.00)^2 = 0.0225 \\ (2.35 - 2.00)^2 = 0.1225 \\ (1.50 - 2.00)^2 = 0.2500 \end{array} \right\} \quad \text{(Retain at least two nonsig. figs.)} $$

Third, sum the squares.

$$0.0225 + 0.1225 + 0.2500 = 0.39_{50}$$

Fourth, divide by $(n - 1)$.

$$\frac{0.39_{50}}{(3 - 1)} = 0.19_{75} \quad \text{(Retain two nonsig. figs.)}$$

Fifth, take the square root.

$$\sqrt{0.19_{75}} = 0.44$$

The *absolute* standard deviation is thus $0.44 \times 10^{-2}M$ chloride. The *relative* standard deviation in pph is calculated as follows:

$$s_R = \frac{0.44 \times 10^{-2}M}{2.00 \times 10^{-2}M}(100) = 22 \text{ pph}$$

Since the mean chloride level for normal children is $2.0 \times 10^{-2}M$, the analysis indicates that the child does not have cystic fibrosis (Sec. 16-3).

Importance of the Relative s — The relative standard deviation of any sample is independent of the mean because of the way it is calculated. Thus, the relative standard deviation of any two sets of analytical measurements may be readily compared even though the two means may be very different. In contrast, it would be useless to compare the absolute standard deviations of sets of measurements having means of say 1.0% and 90.0% chloride.

For gravimetric and certain other reliable and precise methods of analysis, the relative standard deviation should be 0.1 pph or less. For many methods, the relative standard deviation is acceptable if it is 1 pph or less. This implies that the data in the last example (not obtained gravimetrically) are not very precise. Part of the reason for this is that sweat chloride is not always easy to measure in a baby (Sec. 16-3), but another part of the reason is that *only three analyses* were performed. As can be seen in Figure 3-3, the standard deviation is a very poor estimate of precision when $n = 3$. This is

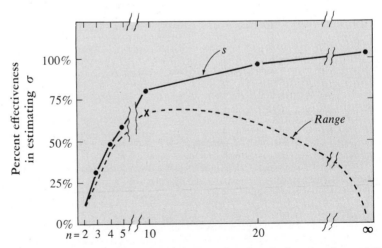

FIGURE 3-3. The percent effectiveness in the estimation of σ by using the s and the range of a sample.

because only three values are used to sample an infinite number of values in the population. For instance, the only way of evaluating the large deviation between the results of 1.50 and 2.15 in the last example is the other large deviation between the results of 2.15 and 2.35. This deviation itself is fairly large; without more measurements it is difficult to conclude that the precision is poor or that these measurements happen to have a large deviation, possibly because of two gross errors.

The Range of a Sample   The range (also called the *spread*) of a sample of results is the difference between the largest value and the smallest value, $X_n - X_1$. It is also an estimate of σ, the standard deviation of a population, and is the simplest and most rapidly calculated meausre of precision. Just as for the other measures of precision, both absolute and relative values of the range may be calculated. The relative value of the range is calculated by dividing the absolute value by the mean and multiplying by 100 to express it in pph. The example below will illustrate the use and calculation of the range.

> **EXAMPLE:** It is difficult to determine mercury in biological samples because there are so few stable materials containing known amounts of mercury. A photographic gelatin containing 2.0 ppm of mercury was developed to meet this need [5]. The following six results were obtained by flameless atomic absorption (Sec. 15-3): 1.95, 2.10, 1.90, 2.10, 2.05, and 2.00 ppm of mercury. Calculate the mean and the range and comment on the number of significant figures that are justified for the mean.
> *Solution:* First, calculate the absolute value of the range. This is $2.10 - 1.90$, or 0.20 ppm. Second, calculate the mean.
>
> $$\bar{X} = 2.00 \text{ ppm}$$
>
> Third, calculate the relative value of the range.
>
> $$\frac{0.20 \text{ ppm}}{2.02 \text{ ppm}}(100) = 10 \text{ pph}$$
>
> Fourth, consider the number of significant figures that are justified for the mean. If a mean of 2.02 ppm is reported, this implies an absolute uncertainty of ± 0.01 ppm in the last significant figure. The relative uncertainty would be
>
> $$\frac{0.01 \text{ ppm Hg}}{2.02 \text{ ppm Hg}}(100) = 0.5 \text{ pph}$$
>
> Since the range has a relative value of 10 pph, the use of three significant figures for the mean is clearly not justified. The precision of the meausrements is too poor to allow the analyst to make a definite statement about the third digit. The next question is: Can two significant figures be used to report the mean? The relative uncertainty in 2.0 ppm is easily seen to be 5 pph. Although this is still smaller than the relative value of the range, it is better to report the mean as 2.0 ppm rather than as 2 ppm. In the latter case, the relative uncertainty would be 50 pph, which would be much larger than the relative value of the range.

In general, the range can be used as an estimate of precision in preference to the ćandard deviation of a sample when $n = 3$ to 10. This is not to say that the range is a better estimate of precision than s. As can be seen from Figure 3-3, the range is in general less effective than s. However, there is very little difference between the two estimates when $n = 3$ to 10, so the *convenience* of using the range is the reason for using it. Only as n approaches 10 does s become appreciably more effective (see Fig. 3-3).

Average Deviation of a Sample

The average deviation (Av.D.) of a sample is the average of the *absolute* deviations of the individual results from the mean. It is calculated as follows:

$$\text{Av.D.} = \frac{\sum |X_i - \bar{X}|}{n} \tag{3-4}$$

The vertical lines on either side of the numerator indicate that the absolute value of each deviation is used—that is, all negative signs are dropped. Note that the average deviation is found by dividing by n, not $(n - 1)$ as is done for the standard deviation. It does not reflect values that deviate widely from the others as much as the standard deviation does and is therefore not as useful. Like s and the range, it may be given as an absolute or a relative value.

SELF-TEST

4. Average Deviation, Standard Deviation, and the Range

Answers:

Directions: Cover the answers in the left column before beginning the test. Compare your answers with those given at the left.

A. Analysis of a blood sample for inorganic phosphate gives 4.00, 4.20, 3.60, and 4.20 mg% phosphate.

0.20 mg%   a. Calculate the average deviation.
0.28₃ mg%   b. Calculate the standard deviation.
0.60 mg%   c. Calculate the range.
Yes   d. Are the results in the normal range? (See Fig. 1-1.)

B. Using the *absolute* values in A, calculate

5.0 pph   a. The *relative* average deviation in pph.
7.0₈ pph   b. The *relative* standard deviation in pph.
15 pph   c. The *relative* range in pph.

C. An analyst obtains the following values: 3.00, 4.20, 4.50, and 4.30 mg% phosphate.

0.50 mg%, 12.₅ pph   a. Calculate the average deviation and its relative value.
0.67₈ mg%, 17 pph   b. Calculate the standard deviation and its relative value.
1.50 mg%, 37.₅ pph   c. Calculate the range and its relative value.

D. Make the following comparisons between the data in A and those in C.

The means are both 4.00 mg% phosphate.   a. Compare the mean of each set of data.

3.00 deviates more than 3.60 mg% does.   b. In which set of data is there a larger deviation of one value from the other three?

The data in C   c. Which set of data has the larger average deviation? The larger standard deviation? The larger range?

The larger deviation of 3.00 mg%   d. What does the larger standard deviation reflect or indicate?

E. What measure of precision was the quickest indication of the larger deviation of the data in C? Is it as effective, more effective, or almost as effective as the standard deviation?

The range; almost as effective

3-4 | USE OF PRECISION IN TESTING FOR GROSS ERRORS

Often one of the results in a small sample where $n = 3$ to 10 appears to reflect a gross error. One approach for resolving this doubt is to employ a statistical test using the precision of the results themselves to reject the suspected result. Such a test should only be used as part of the following three-step procedure:

1. Evaluate the likelihood of a determinate error. (Always consult your notebook to do this.) If there is such an error, discard the suspected result. A determinate error might occur because the analyst makes a mistake, because of faulty equipment (buret, instrument, etc.), or because an instrument has not warmed up enough to stabilize.
2. Obtain additional results if more sample and more time are available. If two or three additional good results can be obtained, the mean of *all* of the results will usually be accurate enough to make it unnecessary to test the suspected result.
3. If necessary, employ a statistical test to determine if the result with a possible gross error should be discarded.

Some of the statistical tests available are based on s or the average deviation, but the most valid test when $n = 3$ to 10 is based on the range. This test is the Q test.

The Q Test The Q test is a convenient method for testing for a gross error in one result when the value of n is from 3 to 10. It involves identifying the result with a possible gross error, followed by dividing the *absolute value* of the difference, d, between the result and its nearest neighbor by the range. The resulting quotient is symbolized as Q.

$$Q = \frac{d}{(X_n - X_1)} \tag{3-5}$$

The calculated Q is compared with a rejection quotient, $Q_{0.90}$ (Table 3-2). If the calculated Q is larger than or the same as the rejection quotient, then the result is rejected [3]. This can be expressed as follows:

If $Q \geq Q_{0.90}$, reject the result being tested.

**TABLE 3-2. Rejection
Quotients for the Q Test**

n	$Q_{0.90}$
3	0.941
4	0.765
5	0.642
6	0.560
7	0.507
8	0.468
9	0.437
10	0.412

A result with a gross error is also termed an "outlier"; Healy [6] suggests that outliers arise most often from two typical determinate errors—from incorrect transcription of results and from sample identification.

If the Q test fails to reject a result, it may still be possible to report the median. This situation will be considered in Section 3-5.

The example below illustrates the application of the three-step rejection procedure, including the use of the Q test.

EXAMPLE: Three high-precision measurements of the pH of a sample of gastric fluid give values of 1.701, 1.601, and 1.607. Decide whether any result should be rejected: (a) assuming there is no time or sample to obtain additional results, and (b) assuming that three additional results are obtained. Calculate the mean of the results in each case.

Solution to a: In general, the largest and smallest values should be inspected for possible gross errors. In this case, the largest result of 1.701 is such a result. Then follow the three-step procedure:

1. Possible determinate error: There is a good possibility that the pH of 1.701 may have resulted from an instrumental error (improper start-up, warm-up, etc.), particularly because it was the first pH reading. An experienced analyst would have the background to make this deicision quickly, since pH readings usually agree quite closely. Rather than rejecting 1.701, assume that for some reason it cannot be rejected.
2. Additional results: This is the next best choice, but it has been stipulated above that there is no time or sample available to do this.
3. Apply the Q test, using **3-5**:

$$Q = \frac{1.701 - 1.607}{1.701 - 1.601} = 0.94_0$$

Since the calculated Q of 0.94_0 is not greater than the $Q_{0.90}$ of 0.941, the pH of 1.701 cannot be rejected, even though it deviates quite widely for a typical pH measurement. Unfortunately, then the report of the mean must use all three readings, so that pH = 1.636. Of course an experienced analyst could avoid this by finding a determinate error, by obtaining more results, or by reporting the median on the basis of knowledge of past precision (see Sec. 3-6).

Solution to b: Again, the pH of 1.701 is the result with the possible gross error. The three-step procedure is as follows:

1. Possible determinate error: As in *a*, the pH of 1.701 may be the result of an instrumental error. If this is found to be so, then this value may be rejected. Again, assume that it cannot be rejected and go on.
2. Additional results: Now assume that time and sample are available and that three additional results of 1.598, 1.597, and 1.598 are obtained. Calculation of the mean of all six results gives a mean pH of 1.617, as compared to the mean pH of 1.636 of the initial three results. The former differs by only a factor of one in the second decimal from the next largest result of 1.607.
3. Apply the Q test: If additional results have been obtained, strictly speaking the Q test should not be applied. The Q test is intended for situations where no additional results can be obtained, although a more liberal viewpoint might hold that the additional results in many cases might be considered as part of the same sample as the initial results.

SELF-TEST | **5. The Q Test**

Answers:

Directions: Cover the answers in the left column before beginning the test. When you have finished, compare your answers with those given at the left.

A. The following results are obtained: 9.00% N, 11.00% N, and 9.10% N.

Calculated Q of 0.95 is greater than 0.941, and 11.00% N is rejected.
Mean = 9.05% N

    a. Assume there is no determinate error and no time to obtain additional results; test the result with a possible gross error.
    b. Calculate the mean after applying the Q test.

B. Three vitamin-C analyses give 100.0, 110.0, and 100.8 mg/tablet.

Check for a determinate error

    a. What should be done first to check for a gross error?
    b. Assuming no time is available for obtaining additional results, apply the Q test.

Calculated Q of 0.92 is <0.941. Report the mean of 103.6 mg.

    c. Decide what value to report.

C. In a determination of mercury in fish, the following results are obtained: 1.00, 3.00, 1.10, and 1.56 ppm of Hg.

Check for a determinate error

    a. What should be done first to check for a gross error?
    b. Assuming no time is available for additional results, apply the Q test.

Calculated Q of 0.72 is <0.765. Since 3.00 ppm differs in the first digit, report M of 1.33 ppm.

    c. Decide what value to report.

3-5 | REPORTING LABORATORY RESULTS; LAB NOTEBOOK

This section will summarize the important concepts of the preceding sections as they apply to reporting laboratory results in a lab notebook. First, the suggested format of a lab notebook and a report page in the notebook will be discussed.

The Laboratory Notebook

The lab notebook has two purposes—to record raw data and to report complete results for an analysis. Left-hand pages should be reserved for raw data. These data need not be neat; the idea is to avoid writing raw data on odd pieces of paper which can easily be lost. Always make all entries, no matter how messy, in the notebook.

All reports, in contrast, should be written neatly on right-hand pages. A vertical tabular form is recommended. Each sample number should be entered at the top, and all data, in the order needed, should be entered below. See the example report form in Table 3-3. In general the weight of a dried sample (or the volume of a liquid sample) is recorded first, since this is usually the first piece of data obtained. Then the actual analytical data are recorded: the volume of titrant or, in the case of gravimetric analysis, the weights of the crucible plus precipitate and the weights of the empty crucibles (Table 3-3). Other data may be added below the actual analytical data—for example, the actual weight of the precipitate in a gravimetric analysis and, if desired, the weight of the constituent sought (see *Weight Cl, g* in Table 3-3).

The last entry under each sample heading should always be the desired result; this might be a percentage or a concentration, such as normality or g/liter. The application

TABLE 3-3. A Report Form for a Gravimetric Analysis

Title: Determination of % Chloride
Reaction: $Cl^- + AgNO_3 = AgCl(s) + NO_3^-$

	I	II	III
Weight Cl sample, g	0.5000	0.5001	0.5001
Weight crucible + AgCl	31.8002	32.8014	33.8250
Weight crucible, g	31.0001	32.0004	33.0100
Weight AgCl, g	0.8001	0.8010	0.8150
Gravimetric factor = Cl/AgCl	0.24737		
Weight Cl, g	0.1980	0.1984	0.2022
% Cl	39.60%	39.68%	40.44%
Application of Q test	—	—	$Q = 0.91$ (Retain)
Mean		39.91%	
Median		39.68%	

(The range of 0.84% above is greater than an expected range of 0.01
(1 pph) × 39.91% or 0.4%. Reporting the median is justified.)
The better estimate of μ: Median of 39.68%

of the Q test to the result with the largest deviation should be shown, if necessary. If the instructor indicates the expected range, then this can be used to decide whether to report the mean or the median. (In Table 3-3, the expected range is calculated from an expected relative range of 1 pph and the mean.) Let us discuss this in more detail.

Deciding What Value to Report in the Lab

You should review Section 3-2 and 3-4 when deciding what value to report after obtaining three or more results. If the results all appear close together and the range of the results is not large, then in general you should report the mean. If you have obtained a result in the lab that appears to have a gross error, you should follow these steps:

1. Evaluate the possibility of a determinate error. If there is one, discard the suspected result. Always consult your notebook for observations on each sample analyzed.
2. Obtain two to three additional results if sample and time are available. Then report the mean of all of the results.
3. If there is no determinate error and no time to obtain additional results, apply the Q test. If this does not reject the suspected result, you must report the *mean* of the results unless the past precision is known (see next step).
4. If the result is not rejected by the Q test and there is no time to obtain additional results, then use the precision from past experience to decide whether or not the median can be reported (see the example on p. 44). The relative range for many methods should be no larger than 1 pph; so multiplying 0.01 by the mean will give a good estimate of the absolute value of the range (see Table 3-3). If the actual range is significantly larger than this estimated value, then there is good reason to use the median, even if the Q test does not reject a suspicious result.

Occasionally the Q test will reject a reasonable result because of the agreement of the other results. This important exception to the usually absolute reliability of the Q test will be discussed next.

Two
"Lucky"
Results out
of Three
It occasionally happens that two results out of the usual three are closer together than would be expected from the precision of the measurement used. For example, suppose results of 60.00%, 60.01%, and 61.00% are obtained. The Q test would reject 61.00% because the first two results are so close together. (Note that 61.00% would not be rejected if the middle result were 60.10%.) The *relative* deviation of the first two results is in fact 0.02 pph. This is less than the relative uncertainty of 0.1 pph for buret reading error and less than the relative uncertainty of 0.04 pph for a gravimetric analysis (where the weighing error for a 0.500-g sample weighed by difference is assumed to be the least precise step). The first two results are "lucky" results because they are closer together than the minimum measuring error in any method would normally allow.

In a case such as this, the Q test should not be used to reject a value because it is not clear that the value has a gross error. It would be better to obtain additional results before applying the Q test. If there is no time available to do this, then the only alternative is to report the mean. Reporting the median would be incorrect because, in effect, a possibly valid result would be rejected.

SELF-TEST	**6. Reporting Laboratory Results**

Answers:

Directions: Cover the answers in the left column before beginning the test. Work the problems one at a time and compare your answers with those given at the left.

A. You have just completed three gravimetric analyses for chloride in the laboratory and you are ready to enter your calculations in your notebook.

a. What is the last entry you should make under each sample number heading?

The % Cl in each sample
Identify the result with possible gross error and test it using the Q test.

b. Before deciding whether to report the mean, what should you do?

If the Q test rejects a value, report the mean. If not, compare the range of results with the range from past experience and report the median if justified.

c. After applying the Q test, should you report the mean or median?

d. Assuming that a range is too large if it is greater than 1 pph, what would the absolute value of an undesirable range be if the mean were 50.00%?

$0.01 \times 50.00\% = 0.50\%$

e. Suppose you had a mean of 50.00% and had a value that could not be rejected by the Q test. If the range of your results were 0.60%, what should be reported instead of the mean? Why?

The median; the range of 0.60% is greater than 0.50% (see d).

B. A student obtains gravimetric chloride results of 30.00%, 30.10%, and 31.00% Cl.

a. In a case such as this where the results are not satisfactory, what should be done first to check the possibility of a gross error?

Check for a determinate error.

b. If there is no determinate error, the next step is to obtain two to three additional results. Is this a practical approach for a gravimetric analysis? Why or why not?

No; gravimetric analysis is too slow.

The calculated Q of 0.90 is less than 0.941, and 31.00% is retained. The actual range of 1.00% is more than $0.01 \times 30\%$; report the median of 30.10%.

c. The next step is to test the result with a possible gross error. What is the result of a Q test?

d. Assuming that a relative range of less than 1 pph is valid for this gravimetric analysis, decide whether the median or the mean of the results should be reported.

C. A student obtains titrimetric results of 50.00%, 50.10%, and 51.00% chloride, using a method where the relative range is about 2 pph.

a. What should be done first to check the possiblity of a gross error?

Check for a determinate error.

b. Since it takes much less time to obtain additional results using a titration, assume two additional results of 50.22% and 50.80% are obtained. What do you do now?

Calculate the mean of all five results (50.28%).

No; the mean of five results is accurate enough.

c. Is it necessary to perform the Q test? Why or why not?

3-6 | CONFIDENCE LIMITS FOR THE MEAN

It is possible to use statistical theory to predict within what limits around the sample mean the true value will be found. These limits are called *confidence limits*, and they are calculated using values of t, or *Student's t*, and s, the sample standard deviation.

$$\text{confidence limits} = \bar{X} \pm \frac{ts}{\sqrt{n}} \tag{3-6}$$

Unfortunately, statistical theory will not permit the limits for the true value to be calculated with 100% probability. There is always some fraction of risk, α, or percentage probability $(100 - 100\alpha)$ involved in such a prediction. The value of t varies with the fraction of risk, or percentage probability, involved and the number of results. Such values are listed in Table 3-4. (Some scientists prefer the degrees of freedom, $n - 1$, rather than n, so this is also listed in Table 3-4.)

When confidence limits for the true value are calculated using **3-6**, it may be said that the fraction of risk that the true value lies *outside these limits* is α. Since the true value may lie above or below these limits, $\alpha/2$ is the fraction of risk that the true value is *either above or below* these limits. It may also be said that the probability that the true value lies *inside these limits* is $100 - 100\alpha$.

EXAMPLE: Suppose that ten results for the mercury content in fish have been obtained with a mean of 0.44 ppm Hg and an absolute standard deviation of 0.10 ppm Hg. Calculate the confidence limits for a fraction of risk of 0.10 and interpret them.

Solution: For $\alpha = 0.10$, $t = 1.833$ at $n = 10$. The limits are calculated as follows:

$$0.44 \text{ ppm} \pm \frac{(1.833)(0.10)}{\sqrt{10}} = 0.44 \text{ ppm} \pm 0.057_9 = 0.38 \text{ to } 0.50 \text{ ppm}$$

The interpretation is that there is a .10 fraction of risk that the true value lies outside the range 0.38 to 0.50 ppm and a 90% probability that it lies inside this range. There is a 0.05 fraction of risk that the true value lies below 0.38 ppm or above 0.50 ppm (Fig. 3-4).

TABLE 3-4. Values of _t_ for Calculating Confidence Limits

Number of measurements, n	Degrees of freedom, $n-1$	Risk and probability level 0.10 90%	0.05 95%	0.01 99%
2	1	6.314	12.706	63.657
3	2	2.920	4.303	9.925
4	3	2.353	3.182	5.841
5	4	2.132	2.776	4.604
6	5	2.015	2.571	4.032
7	6	1.943	2.447	3.707
8	7	1.895	2.365	3.499
9	8	1.860	2.306	3.355
10	9	1.833	2.262	3.250
11	10	1.812	2.228	3.169
12	11	1.796	2.201	3.106
13	12	1.782	2.179	3.055
14	13	1.771	2.160	3.012
15	14	1.761	2.145	2.977
16	15	1.753	2.131	2.947
21	20	1.725	2.086	2.845
26	25	1.708	2.060	2.787
31	30	1.697	2.042	2.750
41	40	1.684	2.021	2.704
61	60	1.671	2.000	2.660
$\infty + 1$	∞	1.645	1.960	2.576

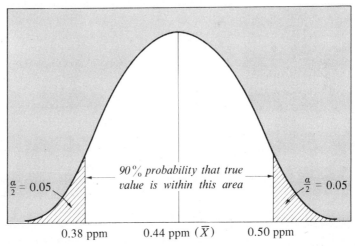

FIGURE 3-4. Graphical representation of the _t_ function for a mean of 0.44 ppm mercury, a standard deviation of 0.10 ppm mercury for ten results, and confidence limits of 90%.

| QUESTIONS AND PROBLEMS

(Answers to most even-numbered problems are in Appendix 5.)

Concepts and Definitions

1. Define the following terms using "the true value" in your definitions.
   a. Absolute error
   b. Relative error
   c. Accuracy (Do not use error in your definition.)
2. Define the following terms without using mathematical formulas or symbols.
   a. Precision (Do not use deviation in your definition.)
   b. Standard deviation
   c. Range
3. Explain the advantage of *relative* standard deviation over *absolute* standard deviation.
4. Explain why the median for $n = 4$ has a higher effectiveness than:
   a. The median for $n = 3$.
   b. The median for $n = 6$.
5. Explain why the median and the mean have the same effectiveness for $n = 2$.
6. Explain why the range is recommended over the standard deviation for $n = 3$ to 10, when the range is less effective than s.

Significant Figures

7. Add each of the following pairs of numbers and express the $[H^+]$ to the correct number of significant figures.
   a. $1 \times 10^{-10} M$ of HCl (a strong acid) + $1.0 \times 10^{-7} M$ H^+ from pure water
   b. $5.0 \times 10^{-5} M$ of HCl + $1.0 \times 10^{-7} M$ H^+ from pure water
   c. $4.0 \times 10^{-5} M$ of NaOH (a strong base) + $1.0 \times 10^{-7} M$ OH^- from pure water
8. Calculate the correct $[H^+]$ of pure water in each case below, assuming that K_w is known only to the number of significant figures given.
   a. For $K_w = 1.0 \times 10^{-14}$, $[H^+] = ?$
   b. For $K_w = 1.00 \times 10^{-14}$, $[H^+] = ?$
   c. For $K_w = 1.008 \times 10^{-14}$, $[H^+] = ?$
9. Calculate the pH of each solution below using the correct number of significant figures.
   a. $[H^+] = 1 \times 10^{-10}$
   b. $[H^+] = 1.40 M$

   c. $[OH^-] = 2.0 \times 10^{-11}$
10. Calculate the semiexponential value of K from the log of K for each case, using the correct number of significant figures.
    a. $\log K = 2.40$
    b. $\log K = 2.30$
    c. $\log K = 10._3$

Mean, Median, and Mode

11. The absolute value of the range for a titrimetric chloride analysis is usually 0.50%. For the % Cl values obtained below, decide whether the mean or the median should be used in each case and calculate it.
    a. 50.1%, 50.60%, 50.3%
    b. 50.0%, 50.3%, 50.8%
    c. 50.1%, 50.2%, 51.4%, 51.0%
12. No knowledge of the past precision is available for any of the samples below. Where a gross error is suspected, state what you would do first if you were in the laboratory. Then calculate the mean or median.
    a. 9.5%, 9.6%, 9.7%, 10.4%
    b. 9.5%, 9.6%, 9.7%, 11.4%
    c. 9.60% and 9.800%
13. Explain why the median is more effective for $n = 4$ than:
    a. The median where $n = 6$.
    b. The median where $n = 3$.
    c. The median where $n = 5$.
14. The following data were assembled by recording just one digit during a freshman chemistry measurement: 3, 3, 4, 4, 4, 4, 5, 5, 5, 6, 8, and 10 g.
    a. Calculate the mean, the median, and the mode.
    b. Decide if the data are skewed to the right or to the left, and then decide if the mean, the median, and the mode are located in the usual relative positions.
15. The following data were assembled by recording just one digit during a high school chemistry measurement: 1, 2, 3, 4, 4, 5, 5, 5, 5, 5, 6, 6, and 7 g.

a. Calculate the mean, the median, and the mode.

b. Decide if the data are skewed to the right or to the left, and then decide if the mean, the median, and the mode are located in the usual relative positions.

16. You are analyzing a chloride sample as your first laboratory experiment, using a method you have never used before. You obtain the following results for %Cl: 20.00, 21.00, and 20.10%. What should you report as the percentage chloride in the sample?

Precision and Confidence Limits

17. Calculate the standard deviation and the range for each of the following sets of results from the determination of lead in glazed pottery.

a. 0.100 ppm, 0.110 ppm, 0.120 ppm, and 0.150 ppm of Pb

b. 1.00 ppm, 1.10 ppm, 1.20 ppm, and 1.50 ppm of Pb

18. Calculate the relative values of the standard deviation and the range for each of the sets of results in Problem 17.

19. Calculate the standard deviation and the range for the following values of percentage chloride in a drug: 40.02, 40.11, 40.16, 40.18, 40.18, and 40.19%.

20. Calculate the 95% confidence limits for the data in Problem 19.

21. Given the following results from measuring the atomic weight of carbon—12.0097, 12.0095, 12.0111, 12.0102, 12.0080, 12.0112, 12.0101, 12.0210, 12.0118, 12.0106, and 12.0113—calculate the atomic weight using:

a. The mean and the absolute standard deviation.

b. The mean and its 95% confidence limits.

Testing for Gross Errors

22. Three high-precision pH measurements of gastric fluid give pH values of 1.802, 1.708, and 1.702. The laboratory notebook indicates that 1.802 was the first pH measurement and that it was taken as soon as the pH meter was activated for measurements, whereas the others were taken five minutes later. Assume the past precision is not known.

a. Assuming there is no time for additional measurements, decide whether any result should be rejected.

b. Calculate the mean or the median, whichever is appropriate.

c. Would the Q test have rejected any results in the first place?

23. Three high-precision pH measurements give pH values of 1.807, 1.910, and 1.800 for gastric fluid. The laboratory notebook indicates that 1.807 was the first pH measurement and that all measurements were taken when the pH meter was apparently functioning properly. Three additional pH measurements of 1.810, 1.805, and 1.812 are obtained. Assume the past precision is not known.

a. Decide whether any result of the original three should be rejected, first on the basis of determinate error and then on the basis of the Q test.

b. Decide what value to report.

24. Three high-precision pH measurements give pH values of 1.805, 1.700, and 1.710 for gastric fluid. The laboratory notebook indicates that 1.805 was the first pH measurement and that all measurements were taken when the pH meter was functioning properly. No time is available for additional measurements; however, past experience indicates the range is 0.02 pH units.

a. Decide whether any result should be rejected, using any criteria.

b. Decide what value to report.

25. Four measurements of serum glucose give results of 80.00, 80.42, 80.20, and 82.00 mg%. The laboratory notebook indicates that 80.00 mg% was the first measurement and that there was no apparent error in making any of the measurements. No time is available for additional measurements, and the past precision is not known.

a. Decide whether any result should be rejected, using any criteria.

b. Decide what value to report.

26. Five measurements of serum glucose give results of 80.00, 80.72, 80.20, 80.42, and 82.00 mg%. The laboratory notebook indicates that 80.00 mg% was the first

measurement and that there was no apparent error in making any of the measurements. Two additional measurements of 80.30 and 80.10 mg% are obtained. The past precision is not known.

a. Decide whether any result of the original five should be rejected, using any criteria.

b. Decide what value to report.

27. Six measurements of serum glucose give results of 80.00, 80.92, 80.42, 80.20, 80.30, and 82.00 mg%. The laboratory notebook indicates that 80.00 mg% was the first measurement and that there was no apparent error in making any of the measurements. Past experience indicates that the range is 1.2 mg%.

a. Decide whether any result should be rejected, using any criteria.

b. Decide what value to report.

Challenging Problems

28. The following results for percentage HCl were obtained by gravimetric analysis: 70.00%, 70.01%, and 71.00%. Assuming that the maximum deviation occurs from weighing 1.0000 g of silver chloride by difference, calculate the mean of the results after considering possible application of the Q test.

29. The calculation of the rejection quotient, $Q_{0.90}$, for $n = 3$ is done using the general formula of Dixon [*Anal. of Math. Statistics* **22**, 68 (1951)] for R_α:

$$R_\alpha = 0.500 + [(\sqrt{3}/2)\tan[\pi/3(0.500 - \alpha)]$$

where α is the fractional probability that a suspicious value is *larger* than R_α; for $Q_{0.90}$, α is thus 0.05.

a. Calculate $Q_{0.90}$ for $n = 3$ to three significant figures.

b. Calculate $Q_{0.99}$ for $n = 3$ to three significant figures.

30. Analyzing the same sample of fish for mercury, two FDA analysts find 0.495 and 0.497 ppm of mercury using the same method of analysis and the same equipment. Recall that the FDA tolerance for mercury is 0.5 ppm Hg. Apply the concepts of significant figures and decide whether the fish should be rejected as unsafe, whether the fish should be declared safe to eat, or whether more analyses should be done.

31. Without using the numerical values of effectiveness, predict whether the effectiveness of the median of a sample of six results should be the same, greater than, or smaller than the effectiveness of the median of a sample of three results. Then, rationalize the difference in the numerical values in Table 3-1.

32. Calculate the mean, the mode, and the median of the following data: 2, 3, 3, 3, 4, 4, 5, 6, 7, 7, 8, 8, 8, and 9. Rationalize the location of there parameters.

33. At the end point of an oxidation-reduction titration, the potential E in volts is calculated using a *weighted mean*:

$$E = \frac{n_a E_a^\circ + n_b E_b^\circ}{(n_a + n_b)}$$

Explain what *statistical* advantage there is in calculating E as a weighted mean rather than the arithmetic mean.

| NOTES

[1] D. E. Jones, *J. Chem. Ed.* **49**, 753 (1972).

[2] W. J. Blaedel, V. W. Meloche, and J. A. Ramsey, *J. Chem. Ed.* **28**, 643 (1951).

[3] R. B. Dean and W. J. Dixon, *Anal. Chem.* **23**, 636 (1951).

[4] L. H. Longley-Cook, *Statistical Problems and How to Solve Them*, Barnes & Noble, New York, 1970, pp. 44–47.

[5] D. H. Anderson, *Anal. Chem.* **44**, 2099 (1972).

[6] M. J. R. Healy, *Clin. Chem.* **25**, 675 (1979).

4 | Sampling and Preparation of the Sample

"What are you going to do, then?" I asked.
"To smoke," Holmes answered. "It is quite a
three pipe problem. . . ."

ARTHUR CONAN DOYLE
The Red-Headed League

Sampling and preparation of the sample do indeed constitute a "three pipe problem." The difficulty is that there is no one sampling procedure that is suitable for all samples. One reason for this is, of course, that a sample may have to be taken from a gas, a liquid, or a solid, but another equally important reason is the immense variety of samples and the incredible range of information that may be required. Finally, there is the discouraging maxim that if the sample is not chosen properly and prepared properly for analysis, all of the results may be meaningless because they will not represent the true nature of the material sampled.

In this chapter we will describe the analyst's procedures in the order in which the analyst does them: first, sampling from beginning to end, and then, the preparation of the sample for analysis.

4-1 | SAMPLING

Since the sampling method chosen and the size of sample taken in any analysis depend on the information asked for, we begin by discussing the latter.

Sampling and the Information Requested

Any of several different kinds of analytical information might be requested for a sample submitted for analysis. Recall from Section 1-1 that there are at least four unique areas of chemical measurement: measurement of physical and chemical properties, qualitative analysis, quantitative analysis, and diagnostic analysis. If a substance's physical or

61

chemical properties are to be measured, then the sampling method is not as important as the purity of the sample. A sample of the substance that is as pure as possible is chosen, and the sample is then purified by a technique such as distillation or recrystallization. Only a small amount of the purified sample is needed, unless a large number of measurements are to be made.

Suppose on the other hand that only a qualitative analysis is to be performed. In general a sample is taken that is as representative as possible of the average composition of the substance to be analyzed. This is done so that the minor constituents ($<1\%$) as well as the major constituent or constituents ($>1\%$) in the sample can be identified. In many cases the sample will not be purified, because the minor constituents are just as important as the major constituent(s). In some cases, however, only the main constituent must be identified. In these cases the sample may be purified to prevent interference by the minor constituents in the identification of the main constituent. The size of the sample of course will depend on the number of qualitative tests to be applied to the sample. If mainly optical methods of analysis (Sec. 1-1) are to be used, the sample can generally be quite small.

If a quantitative analysis or diagnostic analysis is to be performed, then a sample is taken that is as representative as possible of the substance to be analyzed. As in qualitative analysis, this is done so that the amounts of the major constituent(s) and minor constituents found in the sample can be stated to be representative of the substance to be analyzed. The size of the sample, of course, will depend on the number of each type of quantitative measurement performed on the sample and on the type of measurement (Table 1-1).

Sampling of Solids

The sampling of solids involves the greatest number of difficult sampling problems. Solid substances exist in so many shapes and sizes that it is difficult to decide how best to obtain a representative sample. To illustrate some of the problems involved, we will discuss the sampling of solids of large nonuniform particle size, solids of small uniform particle size, tablets, and contaminants from the surface of a solid.

Sampling of Large Nonuniform Particles

We will assume at this point that a large portion of the substance to be analyzed has been taken in a random fashion and brought to the laboratory for further treatment and that we are to select our sample from this portion. Solids consisting of large nonuniform particles must first be reduced to a small uniform particle size. This is accomplished by passing the substance through a disk pulverizer, grinding it in a ball mill, or grinding it with a mortar and pestle. A sieve may be used to ensure that the final sample is selected from a solid of uniform particle size. Care must be taken during the grinding and sieving so that the substance to be analyzed is not contaminated by dust, water, or other particles in the laboratory. Once the solid substance has been reduced to particles small enough to dissolve for analysis, the next problem is to reduce the amount of the solid to a small enough size for an analytical sample. This will be discussed next under the sampling of small particles.

Sampling of Small Particles

When a solid substance is judged to be composed of particles small enough to be dissolved for analysis without further grinding, the first step is to make sure that the particles are of uniform size. This is important because the inclusion of too many particles of a larger or smaller size may yield a sample that is not homogeneous and therefore not representative.

To achieve uniform particle size the particles are sieved, and only the particles passing through a certain size sieve are retained for selection of the sample. If there is too much of the substance to be analyzed left at this point, the amount may be reduced by forming a pile, dividing the pile into quadrants, and keeping opposite quadrants for the final sample.

The above method is, of course, a recommended *general* approach for all samples; however, many bottled powders, bottled crystals, and student laboratory samples are already homogeneous and may be sampled directly without any treatment. Still other bottled products may only need thorough mixing in the bottle to ensure homogeneity and a representative sample.

Sampling of Tablets

The sampling of pharmaceutical tablets or capsules poses different problems. One problem common to tablets is that the decomposition or other changes that have occurred in the tablets at the top of the container may not have occurred to the same extent in the tablets near the bottom of the container. It is therefore not correct to take one or two tablets from only one area in the bottle.

A good illustration of the way to sample some tablets is provided by the method used by the FDA to sample aspirin tablets. The main decomposition reaction of aspirin tablets in bottles that have been opened to moist air (or bottled improperly by the manufacturer) is the hydrolysis of acetylsalicylic acid (ASA):

$$\text{Acetylsalicylic acid} + H_2O \longrightarrow \text{Salicylic acid} + \text{Acetic acid}$$

The balanced reaction has been given in Section 1-1.

Naturally tablets at the top of an aspirin bottle will react to a greater extent with water than those at the bottom. The final sample should reflect the average composition of the aspirin, since the manufacturer cannot be held responsible for the handling of the aspirin after bottling. The FDA method calls for taking a representative sample of 20 tablets for analysis. The tablets are ground together to form a powder of uniform particle size, and a sample weight equivalent to the average weight of one tablet is taken for analysis.

The sample taken also influences the constituent determined. Since there is so much ASA present in an aspirin tablet, the most accurate indication of decomposition is the amount of salicylic acid or acetic acid formed. Since acetic acid is volatile, it will evaporate in varying degrees from the tablets—certainly to a larger degree from tablets at the top of the bottle. Therefore the determination of nonvolatile salicylic acid will be the more accurate indication of the degree of purity.

Sampling for Contaminants from the Surface of a Solid

Frequently it is necessary to test certain solids that release contaminants by leaching from the surface into a liquid or solid stored in them. A good example is contamination of foods and liquids by lead- and cadmium-based pigments used on pottery and ceramic ware. Analysis of the entire solid would be useless, since the problem is not how much lead or cadmium is coated on the pottery, but how much is leached off into the food or liquid stored therein.

The FDA solved this problem by adopting standard conditions for simulating the leaching of lead(II) and cadmium(II) into a standard liquid. The liquid selected was a dilute (4%) solution of acetic acid, similar in composition to table vinegar. This solution represents perhaps the most common degree of high food acidity, so that sampling using it represents the most possible contamination. The 4% acetic acid solution is allowed to remain in the pottery to be tested for a period of 24 hours. The 24-hour period of testing represents a practical compromise between the possible longer-term use by a consumer and the need for the FDA laboratory to evaluate pottery at a reasonable rate.

Sampling of Liquids

For liquids in general, the most difficult problem is *how* to sample the liquid. For body fluids, the most difficult problem is *when* to sample the fluid.

Liquids in General

If a liquid is pure, or at least homogeneous, it can be sampled with any appropriate volumetric device capable of opening and withdrawing a certain volume from the liquid to be analyzed. In the simplest case of a solution in the laboratory, a volumetric pipet (Sec. 6-3) can be inserted into the solution and suction exerted on the mouth of the pipet to withdraw a specific volume of liquid. Naturally the pipet should be clean and dry so as not to contaminate the liquid. Liquids in commercial containers or tanks may be sampled by other devices, such as a pipe inserted deep into the liquid.

If a liquid is not homogeneous or is a suspension, the sampling is more difficult. If the substance to be sampled can conveniently be shaken, then it may be that shaking is all that is necessary to obtain a homogeneous sample. For example, milk of magnesia may be sampled after thorough shaking to ensure that a pipetful of liquid will contain a representative amount of magnesium hydroxide. If the substance to be sampled cannot be mixed or shaken, then samples may have to be withdrawn from different levels of the liquid and a plot made of composition against depth.

Body Fluids

The sampling of blood and other body fluids must be done when the sample is representative of the state of the body to be tested. The ingestion of food, in particular, can vary the levels of chemicals in the blood so that analysis is not meaningful.

In this context it is interesting to consider the sampling of blood when testing for the body's response to glucose load. This is done in a *glucose tolerance test* in which the glucose level is measured over a period as long as six hours. To establish a base level of glucose, the blood is sampled before the patient has eaten any food in the morning. The patient drinks a flavored solution of glucose. Then blood samples are taken at exact time intervals, such as every hour for three hours or more. The glucose level at these times is plotted and compared to the normal glucose tolerance curve (lower curve, Fig. 4-1). For example, a rising trend in the glucose level after one hour (upper curve, Fig. 4-1) instead of a gradual lowering and leveling off indicates that the body is not being stimulated to consume the glucose. Such a symptom is a possible indication of diabetes.

Sampling of Air and Other Gases

The sampling of gases is a highly specialized area, so we will mention only one aspect— the general problems of sampling polluted air in cities. This has been described in detail by Warner [1].

The general requirements for a sampling method for polluted air are as follows:

1. Use of an accurate flow-measurement device to measure the total volume of air sampled.
2. Use of a sample collector to trap the desired pollutants. The collector is usually a filter and/or absorbing solution for which the efficiency must be determined under actual operating conditions, since air cannot usually be sampled with 100% efficiency.
3. Use of a pump that will ensure a constant flow of air through the collector.

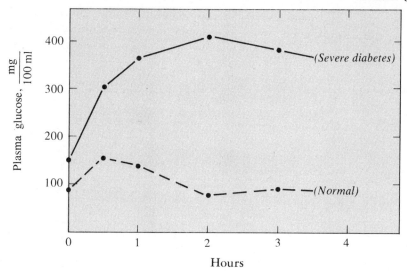

FIGURE 4-1. Glucose tolerance curves for normal individual (lower curve) and diabetic individual (upper curve).

The problems of sampling polluted air are quite complex and can only be alluded to in this chapter. For example, there are the problems of how to sample, where to sample, how long to sample, how many sampling sites to use in a large city, and what to sample.

Sampling of Gases

The problem of what to sample is of greatest interest. The most obvious constituents are the toxic gases that become a hazard when their concentrations reach certain levels. Some of these gases are carbon monoxide, sulfur dioxide, sulfur trioxide, nitrogen oxides, such as NO and NO_2, ozone (O_3), and certain organic aldehydes.

The analysis for carbon monoxide is important because it can affect driving efficiency on freeways; the government has set a level of 9 ppm as a maximum safe level. (Death occurs at 750 ppm when carbon monoxide prevents sufficient oxygen from reaching the brain via hemoglobin in the blood.) Carbon monoxide may be sampled by absorbing it in a solution of copper(I) chloride, with which it reacts to form a copper(I) complex ion. It may also be determined on a continuous basis by channeling it directly into an analyzer from the sample collector, omitting the absorption step.

The analysis for sulfur dioxide is even more important because of its toxic nature and because high-sulfur coal burning has for many years caused serious lung problems and even deaths. In the United States a monitoring system has been established for sulfur dioxide and other such gases in large cities, such as New York, Los Angeles, and Chicago, to sample, determine, and record the levels of these gases as a function of time. Keyed in with the analysis data are multistage alert plans that go into effect if the sulfur dioxide levels or other gas levels rise beyond safe limits.

Sampling of Particulates

In addition to gases, particulates (solids) suspended in the air are potentially dangerous and must also be sampled. Since there are many particles constantly settling because

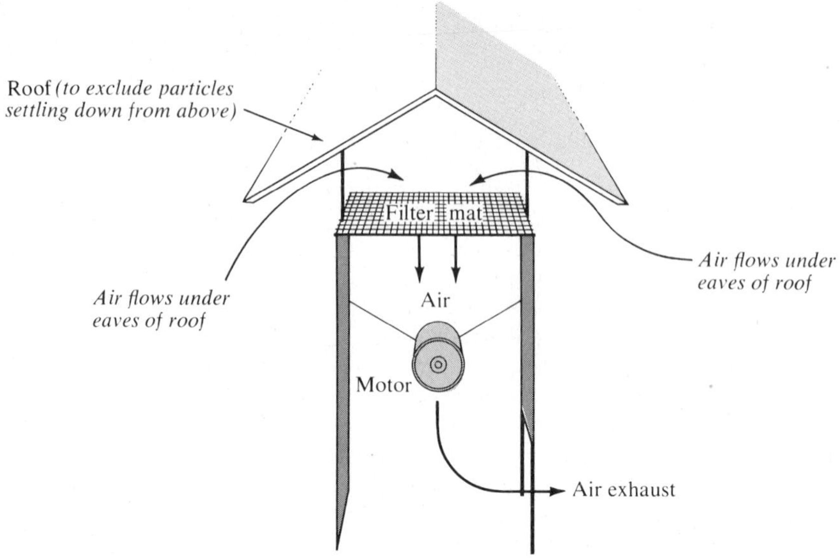

Figure 4-2. A schematic diagram of a high-volume air sampler showing filtration of suspended particulate matter from polluted air.

of gravity, there must be some way of separating particulates that are truly suspended from those that are not.

The air sampler used to obtain a representative sample of truly suspended particulates is shown schematically in Figure 4-2. It is based on the vacuum cleaner principle; air is sucked under the eaves of the roof, causing the air to change direction and excluding particles settling by gravity. Usually a standard 8 × 10-inch glass-fiber *filter mat* is employed to trap suspended particulate (aerosol). A normal sampling period is 24 hours, resulting in a typical particulate sample of about one-half gram [1]. Such particulates may consist of carbon soot, carcinogenic aromatic hydrocarbons, fly ash from power plant emission, iron oxide, silica (sand), and limestone from kilning.

4-2 | PREPARATION OF THE SAMPLE FOR MEASUREMENT

Once a sample has been properly chosen, it must be prepared for analysis. Some of the steps that may be necessary are drying, measuring, dissolving, and separating interfering substances.

Drying Samples Samples may contain water, and it must be decided whether it is necessary to dry the sample to obtain the proper analytical information. This decision will depend somewhat on the type of measurement to be made. In Section 1-1 four different areas were described: measurement of chemical and physical properties, qualitative analysis, quantitative analysis, and diagnostic analysis.

If a sample's physical or chemical properties are to be measured, it may have been automatically dried during purification. Since most substances should be as pure as possible for this kind of measurement, removal of water is generally essential. However, some samples may be solutions containing a constant amount of water, and therefore no water will be removed.

If a qualitative analysis is to be performed, it may not be necessary to remove any water as long as it does not interfere with the qualitative tests to be done. In fact, the properties of hydrates (compounds containing stoichiometric amounts of water) and other such substances might be altered by removing water.

If a quantitative analysis is desired, then drying of most inorganic samples is preferred, because the results can then be expressed as a percentage based on the dry sample rather than on a wet sample whose amount of water may vary. The drying of many organic and biological samples may be undesirable, because of decomposition of the sample or the fact that the sample is largely water. Inorganic and organic samples that need drying are usually dried in an oven at 100° to 110°C for one to two hours. Some temperature-sensitive samples are dried in a desiccator containing a chemical desiccant to absorb water without heating. In either case the percentage of water loss may be measured by weighing the sample before and after drying and dividing the weight loss by the weight of the sample before drying.

If a diagnostic analysis is to be performed, no drying is usually done because the sample is a body fluid, and the objective is to determine the concentration of the constituent in a certain volume of the sample.

Measuring the Amount of Sample to be Used

Solids and some liquid samples are usually measured by weighing on the analytical balance (Ch. 2) to obtain a known amount for analysis. Most analysts weigh out three or four samples of approximately the same weight, so that three or four results can be obtained to calculate an arithmetic mean (Sec. 3-2) which will then be reported. The results are usually calculated in terms of a weight percentage.

Weighing the Sample

$$\frac{\text{Weight of constituent determined}}{\text{Weight of sample}} (100) = \text{Wt}\% \text{ of constituent}$$

This is the type of calculation used in gravimetric analysis (Ch. 5) and in many volumetric analyses. Because the analytical balance is so accurate and precise, this is the most accurate way of measuring the amount of sample to be used.

Measuring the Volume of a Sample

Many liquid samples and some gas samples are measured by obtaining a known volume of the sample in a pipet (Ch. 6) or in a container whose volume is known. Liquids and solutions containing a known weight of a sample can be measured by using a pipet to take several *aliquots* (exact volume) for analysis, so that three or four results can be obtained. The results are calculated in terms of concentration, or on a weight/volume basis.

$$\frac{\text{Weight of constituent determined}}{\text{Volume of sample}} (100) = \text{wt/vol}\% \text{ of constituent}$$

If the weight is expressed in milligrams and the volume in liters, concentration in terms of ppm (Sec. 1-2) can be calculated by omitting the 100 multiplier. In general, measuring the volume of a sample is less accurate than weighing, so that the results using this type of measurement are less accurate than when the sample is weighed.

Some gases are measured in terms of a flow rate and time of flow rather than by volume. This is usually adequate for measuring the total volume of polluted air sampled, for example.

Dissolving or Diluting Samples

Since most analytical methods depend on measurements made on liquids or on solutions, a solid sample must be dissolved in an appropriate solvent. Solid samples are treated differently depending on whether they yield to dissolution by aqueous or nonaqueous type solvents.

Dissolution in Aqueous Type Solvents

Most inorganic compounds and some organic compounds can be dissolved in water or a mineral acid. If possible, water should be used as the solvent to avoid introducing anions of mineral acids that might cause interference. Many metals, oxides, carbonates, and even halides of silver(I), mercury(I), and lead(II) will not dissolve in water, however. Therefore it is necessary to employ an acid or mixture of acids to dissolve such compounds. Dilute nitric acid is a good general solvent for many metals and water-insoluble compounds because it forms soluble metal nitrates. It will not dissolve gold, platinum, antimony, certain sulfide compounds, and the halides of silver(I), mercury(I), and lead(II). Aqua regia, a mixture of nitric and hydrochloric acids, will dissolve gold, platinum, antimony, some sulfides, and the halides of lead(II) and mercury(I). Samples that fail to dissolve in the above acids can frequently be converted to a soluble form by high-temperature fusion using a flux of sodium carbonate, sodium peroxide (Na_2O_2), or a mixture thereof. The sample is ground together with the flux, and the mixture is heated to melting in a crucible to dissolve the sample. The melted material is allowed to cool and is then dissolved in water or an aqueous dilute acid to bring the sample into aqueous solution.

Dissolution in Nonaqueous Type Solvents

Most organic and biological compounds can be dissolved in organic solvents. It is difficult to generalize regarding the appropriate solvent for each class of compound. Usually, however, like compounds dissolve like compounds. For example, ketones will dissolve in acetone, alcohols dissolve in ethyl alcohol, and so on. If a compound has a polar functional group, such as the carbonyl group, it should dissolve in a hydrogen-bonding solvent, such as methyl alcohol or ethyl alcohol.

Separation of Inter-ferences

Once a sample has been dissolved in an appropriate solvent, the desired constituent may have to be separated from other constituents that may interfere in the measurement step. The nature of this interference will depend on which of the four general types of chemical measurements (Sec. 1-1) is to be made and on the specific method to be employed.

For example, suppose that a quantitative analysis is to be performed by precipitating the chloride ion from a solution. If silver(I) nitrate is to be added to precipitate the chloride as silver(I) chloride and if the bromide ion is also present, the bromide ion will interfere by also precipitating silver(I). In this case the bromide ion will first have to be separated by some separation technique. Some interfering constituents will not have to be separated if an adjustment in the composition of the solution is possible. For example, if ammonia is present when silver(I) ion is to be added to precipitate chloride ion, the acidity of the solution can be increased to convert the ammonia to the ammonium ion, thus preventing it from complexing the silver(I) ion as the diamminesilver(I) ion.

Various separation techniques can be employed: precipitation, electrodeposition, solvent extraction, chromatography, and so on. These techniques are elaborated on in Chapters 19, 20, and 21. In discussing the various methods of quantitative and qualitative analysis in Parts I and II we assume that a separation is not necessary. In many cases, however, separations are absolutely essential for an accurate analysis.

| QUESTIONS AND PROBLEMS

(Answers to even-numbered questions appear in Appendix 5.)

Sampling

1. What special problems are involved in sampling:
   a. Large nonuniform solid particles?
   b. Small solid particles?
   c. A nonhomogeneous liquid?
   d. Polluted air for particulate matter?
2. What differences in sampling difficulties might there be between aspirin tablets and aspirin capsules?
3. What difference is there between the treatment of a sample to be used for qualitative analysis and the treatment of a sample to be used for the measurement of physical properties?
4. How can a liquid that is not homogeneous be sampled if:
   a. It is contained in a one-pint bottle?
   b. It is contained in a large tank?
5. Explain the two different sampling requirements involved in the glucose tolerance test. One requirement should cover the first measurement and the other requirement should cover all succeeding measurements.
6. What difficulties are involved in sampling city air for air pollution as opposed to sampling laboratory air for natural gas pollution?

Preparation of the Sample

7. Explain the requirements for drying samples for qualitative analysis as opposed to those for drying samples for measurement of physical properties.
8. Why does drying a sample enable a more accurate determination of the weight percentage to be made?
9. Explain how by measuring the flow rate of a gas and the time it flows you can give an estimate of the volume of the gas taken for a sample.
10. Why is nitric acid a better solvent for lead and silver than hydrochloric acid is?

Challenging Question

11. Suggest a suitable solvent for each substance.
    a. Cholesterol, $C_{27}H_{45}OH$
    b. Glucose
    c. Glycine
    d. Vitamin A

| NOTES

[1] P. O. Warner, *Analysis of Air Pollutants*, Wiley-Interscience, New York, 1976, pp. 196–249.

5 Gravimetric Analysis

"... his [Holmes's] face showed rather the quiet and interested composure of the chemist who sees the crystals falling into position from his oversaturated solution."

Arthur Conan Doyle
The Valley of Fear, Chapter 2

In this chapter, we describe the use of the most fundamental measurement device, the analytical balance, in the determination of any species that can be quantitatively precipitated and isolated. We will assume that the proper sample has already been taken, as described in the previous chapter, so that the discussion can be limited to how to treat the sample to obtain a quantitative precipitation of the desired constituent and how to calculate the results. Because a theoretical understanding of precipitation is valuable, a section at the end of the chapter covers the use of the solubility product constant in solubility equilibria calculations.

5-1 | INTRODUCTION

Gravimetric analysis is a quantitative method in which the amount of the desired constituent is determined by isolating it in a known pure form and then weighing it. Most gravimetric methods involve the precipitation of an ion as an insoluble compound as, for instance:

$$\underset{\text{(as NaCl or HCl)}}{Cl^-} + \underset{(AgNO_3)}{Ag^+} \longrightarrow AgCl(s)$$

In the above reaction the chloride ion is determined by adding an excess of the silver(I) ion in the form of silver nitrate reagent. The silver(I) chloride precipitate is then filtered, dried, and weighed, and the amount of the chloride (or chloride compound) is calculated from the weight of the pure silver(I) chloride.

Although gravimetric analysis is not used as much as some newer methods of analysis, it is still important because of its inherent accuracy. Often, it can be used to check the accuracy of new and/or untested methods. It is also the method of choice for determining the concentration of certain titrants where high accuracy (four significant figures) is needed.

Major Advantages

The major advantages of gravimetric analysis are its inherent accuracy and the fact that it requires a single fundamental chemical precipitation reaction. The latter is in contrast to volumetric (titrimetric) analysis, which may require two or three chemical reactions. To illustrate, in the gravimetric determination of the sulfate ion, the only chemical reaction involved occurs with the addition of an excess of barium ion:

$$SO_4^{-2} + Ba^{+2} \longrightarrow BaSO_4(s)$$

However, if the same precipitation were used for a titration, an additional chemical reaction would be required to indicate the end point. Other advantages are that a gravimetric determination requires no accurately known standard solution (titrant) and that there is no need to prepare a series of known standards, such as in instrumental analysis for sulfate.

Comparing the two methods above in terms of accuracy, a substance can be weighed by difference on an ordinary analytical balance with an uncertainty of only ± 0.0002 g (Ch. 2), or for a 1-g sample, a relative uncertainty of 2 parts in 10,000 (0.02 pph). In contrast, the total uncertainty in reading a 50-ml buret twice is held to be 0.04 ml, or 0.2 pph for a 20-ml titration. On that basis a gravimetric method is at least ten times more accurate than a volumetric method.

Major Disadvantages

Gravimetric methods do have certain disadvantages; because of these most chemists prefer to use other analytical methods.

Although simple in theory, a gravimetric analysis is often very exacting and lengthy when carried out in the laboratory. The desired constituent is usually isolated by precipitating it from a solution. The precipitation process suffers from several inherent errors. Some of the desired constituent may be lost owing to the solubility of the precipitate, thus giving low results. On the other hand, if any impurities are present during the precipitation process, they may be carried down along with the desired constituent, thus giving high results. This latter phenomenon is known as *coprecipitation*.

Even when the precipitate is pure and no solubility losses occur, it must still be quantitatively transferred and filtered to separate it from the solution associated with it, and then it must be washed to free it from impurities dissolved in the solution. After this it must be carefully dried and finally weighed. All of these manipulations are tedious and time-consuming, and the loss of any of the precipitate at any point will result in errors.

Areas of Application

Gravimetric methods are most often used in the analysis of inorganic compounds and samples. Because of their great stability, these types of compounds can be dried for weighing without decomposition. Well-established gravimetric methods exist for the chloride ion, the sulfate ion, and a number of cations.

Gravimetric methods are sometimes used for the determination of inorganic substances in organic and biological samples but are seldom used for the determination of biological substances, since most of the latter cannot be dried without decomposition. A good example of an inorganic substance often determined in biological samples is the chloride ion. Another example is water, which is simply found by the loss in weight after drying at 110°C. The mineral content of biological samples may also be determined by the so-called *ashing procedure*. The organic matter is burned away and the nonvolatile inorganic residue is weighed to determine the total mineral content (Na, K, Ca, P, etc.).

Some specific examples of gravimetric methods in inorganic and pharmaceutical analysis will be discussed in Section 5-4.

5-2 | THE GRAVIMETRIC ANALYSIS METHOD

A typical gravimetric analysis method involves a number of steps which may be summarized as follows:

1. A representative sample must be obtained.
2. The sample must be weighed accurately and then dissolved; if it is a liquid, a known volume must be measured accurately using a pipet.
3. The conditions for quantitative and accurate precipitation must be achieved before the precipitation is carried out. This may include adjusting the pH of the solution, heating it, and possibly removing some substance that may interfere by coprecipitation with the constituent to be determined.
4. The appropriate precipitating reagent must be added to achieve quantitative precipitation. This may involve addition of an excess of the precipitating reagent, settling of the precipitate, and testing for complete precipitation by dropwise addition of more reagent.
5. The solution may be heated (digestion) to achieve filterable crystals and to cause impurities to dissolve.
6. The precipitate must be filtered, washed, and dried in the oven to achieve a precipitate of "constant weight."
7. The weight of the pure precipitate must be converted to the weight of the substance to be determined by multiplying it by the *gravimetric factor*. The percentage of the substance in the impure sample is then calculated.

These steps will be discussed in detail in the following subsections.

Obtaining a Representative Sample

To obtain a meaningful analytical result, you must choose a sample that represents the composition of the material being analyzed. If you do not, the results will probably not represent the true composition of the sample and your time will have been wasted. The problems of obtaining a truly representative sample were discussed in Chapter 4, and only a few practical words about laboratory samples will be added here.

Solid samples given to students in the laboratory are usually in a finely powdered form that needs no treatment before drying and weighing. Treat the sample carefully by putting it into a weighing bottle as soon as possible and covering it to protect it from the laboratory atmosphere. If possible, dry the sample ahead of time to avoid delays. When using the sample, guard against wasting any of it to make sure you will have enough if the analysis has to be repeated.

Liquid samples should be kept covered to prevent evaporation and contamination from the laboratory atmosphere. Be careful not to contaminate such samples with a pipet that is wet with another liquid.

Weighing and Dissolution of the Sample

Always write the exact weight or volume of the original sample in your notebook immediately after the measurement. This figure is important because the final result is almost always obtained from a calculation including the initial weight or volume of the sample. If there is any doubt at all as to the weight of the sample before or after analysis, the sample should be discarded and a fresh sample measured out.

For the gravimetric analysis of a solid sample, the experimental procedure will usually specify a solvent, such as water or a dilute acid. Always stir the sample so that it dissolves completely in the specified solvent. Consult the instructor if the sample does not dissolve completely. When dissolution seems complete, it may be necessary to wash down the sides of the beaker to ensure that all of the sample is in the body of the solution.

Preparation of the Solution for Precipitation

After dissolving the sample, certain adjustments may have to be made before or during addition of the precipitating reagent to prevent errors. Such adjustments may be the preliminary separation of potential interferences, the control of pH during precipitation, and the control of the temperature of the solution during precipitation.

Potential interferences may have to be separated either because they precipitate quantitatively along with the ion to be determined or because they *coprecipitate* with the ion to be determined. (Coprecipitation is said to occur when a normally soluble ion is carried down in varying amounts as part of the structure of the insoluble compound.) An example of the first type of interference: if the chloride ion is to be precipitated in the presence of the bromide ion, the silver nitrate reagent would precipitate the bromide ion quantitatively. An example of coprecipitation is when the ammonium ion is carried along during the precipitation of insoluble barium sulfate. To avoid errors, both the bromide and the ammonium ion in these examples would have to be removed prior to the precipitation.

The control of pH is important to achieve quantitative precipitation in most gravimetric methods. In some cases it is necessary to prevent decomposition of the precipitating reagent. For example, if the pH is not acidic when silver nitrate is added to precipitate chloride, some silver oxide may form:

$$2AgNO_3 + 2OH^- \longrightarrow Ag_2O(s) + H_2O + 2NO_3^-$$

Of course, the silver oxide will precipitate along with the silver chloride and cause high results. In other cases, an acidic pH is necessary to prevent precipitation of other ions along with the ion to be determined. For example, if the pH is not acidic when barium nitrate is added to precipitate sulfate in the presence of the carbonate ion, barium carbonate will precipitate. Here, an acidic pH changes the carbonate ion to bicarbonate ion, which is not precipitated by the barium ion.

In certain gravimetric methods, the pH must be controlled *very precisely*. For example, the precipitation of barium chromate is quantitative only at pH 5.7. At lower pH values losses occur from increased solubility of the precipitate; at higher pH values strontium chromate is coprecipitated with barium chromate.

In the precipitation of pharmaceuticals a preliminary separation of binders and/or lubricants is often necessary. For example, in the analysis of amobarbital tablets or similar substances (Sec. 5-4) the sample of the tablets is frequently washed with petroleum ether to remove any binder or lubricant in the tablets. The amobarbital or other medicinal agent is then isolated by extraction into chloroform or ether and ultimately weighed in its pure form.

Frequently, inorganic substances require a preliminary oxidation or reduction. In the determination of iron by precipitation of iron(III) hydroxide the solution is treated with bromine water or hydrogen peroxide to ensure that all the iron is oxidized to the $+3$ state.

In general, laboratory procedures give detailed instructions for preparing the solution for precipitation. Special samples may require special treatment not given in the procedure, however.

Precipitation of the Desired Constituent

Precipitation of the desired constituent is accomplished by adding a dilute solution of a reagent that reacts with the desired constituent to form an insoluble precipitate. A slight excess of this reagent is always added. This ensures that the desired ion is completely precipitated and the solubility of the precipitate is decreased by the common ion effect. The most important objective in such a procedure is to achieve an optimum *particle size* of the precipitated salt, which will be discussed next.

Particle Size of Precipitates

When carrying out a gravimetric determination it is important to have a precipitate that can be readily separated from a solution by filtration and that can be easily washed free of impurities. The particles must be large enough not to pass through the pores of the filter. There is an enormous range in the particle diameter of various precipitates, as can be seen from the particle diameter scale below.

Particle diameter scale (log)

Although there is really no sharp dividing line, but rather a broad intermediate region, between colloidal suspensions and crystalline particles, we will discuss precipitates in terms of these two categories. However, because colloidal suspensions coagulate into either curdy (granular) or gelatinous precipitates, we must begin our discussion by describing *three* different types of precipitates. These are crystalline (e.g., $BaSO_4$), curdy (e.g., $AgCl$), and gelatinous (e.g., $Fe(OH)_3$).

Types of Precipitates

1. *Crystalline precipitates* are recognized by the presence of many regularly shaped, discrete particles having smooth, shiny surfaces. A crystalline precipitate looks like dry sugar or salt. It is the most desirable of all precipitates, since it settles rapidly and is easy to filter and wash. The individual particles are large and compact. Errors caused by *occlusion* (entrapment within the crystal) and *adsorption* of lattice ions on the surface of the crystal can be minimized in most cases.

Coagulated colloidal suspensions

2. *Curdy (granular) precipitates* consist of a colloidal suspension that has been coagulated into a filterable colloid. In contrast to gelatinous precipitates, these retain very little water. These precipitates consist of small, irregularly shaped, individual

particles; the smooth surfaces and regular shape of a crystalline precipitate are absent. A curdy precipitate, like a crystalline precipitate, is easily filtered and washed, but the particles tend to be porous, which increases errors from adsorption.

3. *Gelatinous precipitates* form a sticky, jellylike mass that looks like jam or jelly, having no discrete particles and forming amorphous clumps. A gelatinous precipitate is not desirable, since it is difficult to filter and it entraps impurities that are impossible to wash out. Gelatinous precipitates are formed, for example, when the precipitate is highly insoluble or when concentrated solutions are mixed together.

Obviously any given insoluble salt will always be of one of the above types, but that does not mean the analyst cannot control conditions to improve particle size. For example, many crystalline precipitates are finely divided at first, but grow to larger crystals during the procedure. We will consider next the effect of experimental conditions on particle size.

Effect of Experimental Conditions on Particle Size
Many of the experimental conditions of gravimetric methods have a definite beneficial effect on particle size. In 1925, the German chemist P. P. von Weimarn showed that the *degree of relative supersaturation*, R, at the beginning of the precipitation process largely controls particle size [1]. R, which is also called the von Weimarn ratio, is defined as follows:

$$R = \frac{(Q - S)}{S} \qquad (5\text{-}1)$$

where Q represents the instantaneous molar concentration of the salt at the moment of mixing and S is the molar solubility of the salt at equilibrium.

According to von Weimarn, the first step in precipitation is the formation of a supersaturated solution. Thus $(Q - S)$ is an absolute measure of the amount of supersaturation and R is a relative measure of the amount of supersaturation. The second step involves *nucleation*, the combination of ions to form soluble *nuclei*, particles consisting of an aggregate of eight or more ions. The larger R is, the more nuclei are formed, and vice versa. The third and final step is the growth of the nuclei through a colloidal state to insoluble particles. As R is decreased, fewer and fewer new nuclei are formed so that established nuclei grow to larger crystals. With some salts, such as silver chloride, R cannot be decreased enough to prevent the formation of a colloidal suspension, so that these salts never precipitate in crystalline form.

Let us consider the precipitation of barium sulfate in the light of von Weimarn's ideas. If we mix a solution of barium chloride and a solution of sodium sulfate, we will observe the following net reaction:

$$Ba^{+2} + SO_4^{-2} \rightleftharpoons BaSO_4(s)$$

This reaction is slightly reversible; using the methods described in Section 5-3, we can calculate that S of barium sulfate is $1 \times 10^{-5} M$ to one significant figure. We will now consider two different situations involving the precipitation of barium sulfate for analytical purposes:

1. Precipitation of barium ion from 100 ml of $0.10M$ barium chloride by rapid addition of 100 ml of $0.10M$ sodium sulfate.

2. Precipitation of barium ion from 100 ml of $0.10M$ barium chloride by dropwise addition of 100 ml of $0.10M$ sodium sulfate.

In the first situation, we can calculate Q_1 as follows:

$$Q_1 = \frac{(0.10 \times 100)\text{mmole BaSO}_4}{200 \text{ ml final volume}} = 0.050M \text{ barium sulfate}$$

Substituting the value for Q_1 into 5-1, we obtain:

$$R_1 = \frac{0.050M - 1 \times 10^{-5}M}{1 \times 10^{-5}M} = 5 \times 10^3$$

In the second situation, we can calculate Q_2 after the first drop of $0.10M$ sodium sulfate has been added. To do this, we take one drop to be 0.05 ml and multiply this by the molarity of sodium sulfate to obtain the mmoles of barium sulfate formed by addition of one drop of sodium sulfate:

$$Q_2 = \frac{(0.10 \times 0.05)\text{mmole BaSO}_4}{100._{05} \text{ ml final volume}} = 5 \times 10^{-5}M \text{ barium sulfate}$$

(Note that the drop of sodium sulfate does not significantly change the volume of the solution.) Substituting the value for Q_2 into 5-1, we obtain:

$$R_2 = \frac{5 \times 10^{-5}M - 1 \times 10^{-5}M}{1 \times 10^{-5}M} = 4$$

Of course with succeeding drops, the volume will gradually increase, decreasing Q_2 and R_2. Nevertheless, R_2 will be much smaller than R_1 throughout most of the precipitation occurring in the second situation, resulting in a precipitate with larger crystals and therefore more readily filtered.

Recommenda- tions for Gravimetric Analysis

In light of the above calculations we can now make some recommendations for the most favorable conditions for gravimetric precipitations. We note first that to keep R small, 5-1 requires that Q be kept as small as possible and S be kept as large as practical. Q can be kept at a low value by using dilute sample solutions and dilute reagents. The use of dilute solutions also increases the solubility of any impurities present, thus reducing the chances of contamination of the precipitate.

The value of S may be increased by precipitating insoluble salts from hot solutions, thereby decreasing R and yielding larger crystals. The solubility of many salts can be increased judiciously by acidifying the solution to a point where quantitative precipitation is still obtained; hence, the pH in these cases can be adjusted carefully to a low value.

There are practical limits, however, to the above considerations. As the sample solution and reagent solution become more and more dilute, the final volume after mixing ultimately becomes too large. For example, it would be very awkward and time-consuming to manipulate several liters of solution just to obtain large crystals. Therefore, a practical working limit to the final volume is usually in the 200- to 500-ml range.

Also, in acidifying and heating solutions to increase S, one must be cautious not to increase the solubility to such an extent that the method no longer is quantitative. For example, the solubility of silver chloride increases from about $10^{-5}M$ at room temperature to $2 \times 10^{-3}M$ at 100°C, so that it would not be quantitatively precipitated at the boiling point of water.

Digestion of the Precipitate   In many procedures the precipitate, along with the supernatant solution (sometimes called the "mother liquor"), is allowed to "digest." Digestion means heating the precipitate along with the mother liquor below its boiling point for a period of time. The mixture is usually stirred occasionally during this time to encourage digestion. Digestion can accomplish several things:

1. A colloidal suspension, which is usually present with a curdy precipitate, is coagulated to a curdy precipitate. In addition, curdy precipitates form denser, more filterable particles during digestion.
2. Small crystalline particles, which tend to be more soluble than large crystalline particles, dissolve during digestion and are reprecipitated on the large crystalline particles. This increases the average particle size and makes the precipitate easier to filter.
3. Absorbed and entrapped impurities in crystalline precipitates tend to dissolve during digestion, resulting in purer crystalline particles.

The growth of a colloidal suspension into a curdy precipitate is not obvious, and to clarify it, it is necessary to describe the adsorption of ions on the surface of colloidal particles.

Adsorption and Precipitation of Colloidal Suspensions   The surface of a crystal lattice, precipitate, or colloidal (suspended) particle is usually contaminated with layers of two different kinds of ions, primary adsorbed ions and counter ions.

Primary Adsorbed Ions.   Closest to the surface of any particle are the primary adsorbed ions; these ions are always the same as the lattice ion that is present in excess. Thus for a silver halide precipitate, $AgX(s)$, an excess of silver(I) ion gives a $AgX:Ag^+(s)$ species, and an excess of X^- gives a $XAg:X^-(s)$ species. $AgX:Ag^+(s)$ symbolizes that a primary adsorbed cation is bonding to the surface by sharing a pair of electrons of a chloride ion at the lattice surface. $XAg:X^-(s)$ symbolizes that a primary adsorbed anion is bonding to the surface by sharing its own pair of electrons with a silver ion at the lattice surface. Even if there are other *potential* primary adsorbed cations or anions present, the lattice ion in excess will always form the strongest bond to the surface, just as both lattice ions do in the crystal or colloid.

Counter Ions.   The counter ion may be any ion of opposite charge to the primary adsorbed ion, because the two ions can only interact electrostatically. For a silver halide precipitate where silver(I) ion is the primary adsorbed ion and the nitrate is the counter ion, the symbol $AgX:Ag^+\cdot\cdot NO_3^-(s)$ is used. For the same precipitate where the halide ion is the primary adsorbed ion and the sodium ion is the counter ion, the symbol $XAg:X^-\cdot\cdot Na^+(s)$ is used.

If there are several ions present that are potential counter ions, the ion that forms the least soluble compound with the primary adsorbed ion will be the counter ion. For example, if both NO_3^- and ClO_3^- are present, the ClO_3^- will be the counter ion because the solubility of $AgClO_3$ is 10 g/100 ml while the solubility of $AgNO_3$ is 122 g/100 ml.

Precipitation of Chloride with Silver Nitrate.   When the chloride ion is precipitated initially using an excess of silver nitrate, it forms mainly a suspended white colloid.

At room temperature, the colloidal particles of $AgCl:Ag^+$ repel each other; even the nitrate counter ions do not reduce the repulsion enough for complete coagulation, because they are too far away. Digestion just below the boiling point enables coagulation to occur; the heat supplies energy to reduce repulsion and to effectively "shrink" the layers of primary ions and counter ions. This gives a curdy precipitate that is stable even at room temperature.

The digestion process may be summarized by the symbolic formation of a corner of a curd of silver chloride:

$$ClAgCl:Ag^+ \cdots\cdots NO_3^-$$
$$AgClAg \Updownarrow \qquad\qquad \xrightarrow{\text{digestion}} \qquad \begin{array}{l} ClAgCl:Ag^+ \cdot\cdot NO_3^- \\ AgClAg \\ ClAgCl:Ag^+ \cdot\cdot NO_3^- \\ \text{(corner of curd)} \end{array} \qquad (5\text{-}2)$$
$$ClAgCl:Ag^+ \cdots\cdots NO_3^-$$

Note that heating reduces the distance between each primary adsorbed ion and its counter ion, allowing the two primary adsorbed ions to approach more closely without repulsion (double arrow).

Postprecipita-
tion

A warning is necessary, however, against very long periods of digestion and against allowing precipitates, especially crystalline precipitates, to stand too long in contact with the mother liquor at room temperature. Certain impurities form supersaturated solutions which will precipitate slowly *after* the desired constituent. Precipitation that occurs as the solution stands is called *postprecipitation*. Barium sulfate (crystalline) is readily contaminated in this way, whereas silver chloride (curdy) is only slightly affected.

Filtration

Filtration is required to separate the precipitate quantitatively from its mother liquor and some impurities. The media commonly employed for filtration are

1. Sintered-glass filtering crucible
2. Gooch crucible with an asbestos mat
3. Filter paper

Each of these will be discussed in detail.

Sintered-
Glass Filtering
Crucibles

The sintered-glass filtering crucible is made entirely from glass (Fig. 5-1), having glass sides and a porous glass mat. The glass mats are made in different porosities, usually designated as coarse, medium, or fine. Suction must be employed when using these filters. The filters must be cleaned and then dried to constant weight in a drying oven before use.

These crucibles should be used only with a crystalline or granular type of precipitate. Gelatinous precipitates tend to seep through the filter when suction is applied; finely divided precipitates run through or plug up the pores in the filter.

After filtration and washing, the crucible containing the precipitate is dried in an oven at temperatures ranging from 110° to 250°C, depending on the nature of the precipitate. Sintered-glass crucibles should never be heated with a burner or to high temperatures, as they may crack or melt.

Gooch
Crucibles

The Gooch crucible is made of porcelain and has small round holes about the diameter of a pin in its base. To prepare it for use, a suspension of asbestos fibers in water

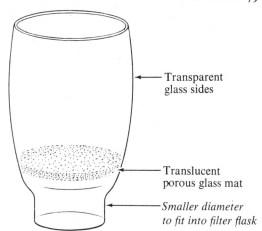

Transparent glass sides

Translucent porous glass mat

Smaller diameter to fit into filter flask

FIGURE 5-1. **A schematic diagram of a** sintered-glass filter crucible.

(asbestos soup) is poured into it and suction is applied. This forms a porous mat of asbestos 2–3 mm in thickness over the base. After the mat is washed with distilled water, the crucible is dried and ready for use. Although not as convenient as the sintered-glass crucible, the Gooch crucible must be used when the precipitate has to be heated to high temperatures, such as 800°–1000°C.

Filter Paper   Filter paper is the most widely used filtering medium. It is made in a variety of sizes and porosities. For quantitative analysis a so-called *ashless* filter paper is used. The minerals, or ash, are leached out from the paper by washing in a mixture of hydrochloric and hydrofluoric acids during the manufacturing process. The paper is then essentially pure cellulose and leaves a residue of less than 0.1 mg if it is ignited in a crucible.

The size of the filter paper to be employed is determined by the bulk of the precipitate, not by the volume of the solution to be filtered. The entire precipitate at the end of the filtration should fill the filter paper no more than half full; otherwise losses may occur by overfilling the filter.

In quantitative filtrations with filter paper, the filter paper is folded and placed in a long-stem funnel. (Techniques of filtration are discussed in the introduction to Exp. 3B, Part Four.) Suction is usually not employed in quantitative filtrations with filter paper, since the paper may burst or the precipitate may be pulled through the pores of the paper.

After the precipitate is filtered and washed, the paper containing it is folded and placed in a porcelain crucible. The paper and precipitate are carefully dried, and then the paper is slowly burned away by heating the crucible and paper in a gas flame or in an electric muffle furnace.

After the filter paper is completely burned away, the crucible and precipitate are cooled in a desiccator and then weighed. Not all precipitates may be treated in this fashion, since some chemical compounds are decomposed if heated to a high temperature with filter paper. Silver(I) chloride is one such precipitate. When heated to high temperatures with filter paper, it is partially reduced to free silver metal, leaving a precipitate of unknown composition. Barium(II) sulfate, on the other hand, is completely stable under this treatment.

Washing the Precipitate

During and after the filtration process the precipitate must be washed to free it from impurities. Washing is most effectively accomplished by stirring the precipitate in a beaker with the wash solution, then decanting the supernatant liquid through the filter. The wash solution is also used to quantitatively transfer the precipitate from the beaker to the filter. After the transfer is complete, the precipitate and the filter are again washed. A wash solution must accomplish several things. It must wash out all impurities from the precipitate but leave it unchanged in composition. It should introduce no insoluble product into the precipitate, and it should be completely volatile at the drying temperature of the precipitate.

The composition of a wash solution is determined by the chemical composition of the precipitate and the nature of the impurities associated with it. Pure water usually is not used, since it may cause peptization (form colloids) of the curdy type of precipitate. Since washing of this type of precipitate is so critical, we will discuss this next as an example of the importance of proper washing.

Washing Curdy Precipitates; Silver Chloride

When the chloride ion is precipitated using an excess of silver nitrate, the resulting silver chloride has a primary adsorbed layer of silver nitrate:

$$2\,Ag^+ + Cl^- + NO_3^- \longrightarrow AgCl:Ag^+ \cdot\cdot NO_3^- (s)$$

Of course, most of the excess silver nitrate remains in the solution merely in *physical* contact with the precipitate. The problem is to remove as much as possible of both the adsorbed silver nitrate and the silver nitrate in physical contact with the precipitate. We will consider two possibilities—washing the precipitate with pure water and washing it with an electrolyte solution.

Washing with Pure Water. If we wash the precipitate with pure water, the water will gradually dilute the nitrate counter ions away from the corresponding primary adsorbed ions. Consider the following symbolic representation of what will happen at the corner of a curd of silver chloride:

$$
\begin{array}{ccc}
\begin{array}{l}
ClAgCl:Ag^+ \cdot\cdot NO_3^- \\
AgClAg \\
ClAgCl:Ag^+ \cdot\cdot NO_3^-
\end{array}
& \xrightarrow[\text{wash}]{\text{1st}} &
\begin{array}{l}
ClAgCl:Ag^+ \cdots NO_3^- \\
AgClAg \\
ClAgCl:Ag^+ \cdots NO_3^-
\end{array}
\end{array}
\xrightarrow[\text{wash}]{\text{2nd}}
$$

$$
\begin{array}{l}
ClAgCl:Ag^+ \cdots\cdots NO_3^- \\
AgClAg \\
ClAgCl:Ag^+ \cdots\cdots NO_3^-
\end{array}
\tag{5-3}
$$

(Note that **5-3** is almost the reverse of **5-2**, the process of digestion.) After the second washing, the few counter ions left are too far away to prevent the positively charged primary adsorbed ions from repelling each other, causing a silver(I) ion to break away. This is followed by the breaking away of a silver chloride molecule as the first step in the process of peptization of the silver chloride to form colloidal (soluble) silver chloride.

This is shown symbolically for the corner of a curd of silver chloride (double arrow symbolizes repulsion):

$$ClAgCl:Ag^+ \cdots\cdots NO_3^-$$
$$AgClAg \updownarrow \qquad \xrightarrow{\text{peptization}}$$
$$ClAgCl:Ag^+ \cdots\cdots NO_3^-$$

$$Cl \text{-----}$$
$$AgClAg$$

$$ClAgCl:Ag^+ \cdots\cdots NO_3^- + \quad AgCl \quad + Ag^+ + NO_3^-$$
$$\text{(colloidal)}$$

(Note that the dashed line indicates where the silver chloride molecule has left the corner of the curd.) Thus water cannot be used as a wash solution.

Washing with an Electrolyte Solution. Although water is unsatisfactory as a wash, a dilute solution containing a certain type of electrolyte is very satisfactory. The main requirement for the electrolyte is that it be volatile; it is obvious that the electrolyte should be strongly ionized rather than weakly ionized. Of the several satisfactory electrolytes, nitric acid is often chosen. A dilute nitric acid solution fulfills the three functions of a wash solution for curdy precipitates:

1. It supplies a constant concentration of counter ion (nitrate ion) to prevent peptization.
2. It completely washes out the silver nitrate in physical contact with the precipitate.
3. It also washes off a little of the adsorbed silver nitrate left after digestion.

During the washing process, some of the electrolyte will be adsorbed on the surface of a curdy precipitate, such as silver chloride. If the primary adsorbed ion is a cation, then the cation of the electrolyte will be adsorbed as a primary adsorbed ion, and the anion of the electrolyte will serve as the counter ion. When the precipitate is dried, a volatile electrolyte will be vaporized:

$$AgCl:H^+ \cdot\cdot NO_3^- (s) \xrightarrow{110°C} AgCl(s) + HNO_3(g) \text{ or other } NO_x \text{ gases}$$

Unfortunately, a small amount of silver nitrate remains adsorbed and is not volatilized during drying at 110°C. Often, though, this is only of the order of one silver nitrate molecule for every 10^3 silver chloride molecules.

Drying the Final Precipitate   The final gravimetric precipitate must be in a pure, dry form of known composition for weighing. Most inorganic compounds are stable at high temperatures, and these precipitates are easily dried by placing the crucible containing them in an oven heated to 110°–250°C. They are kept in the oven until all moisture has evaporated, are placed in a desiccator to cool, and finally are weighed. Certain organic compounds, however, are decomposed even at temperatures of 100°C; hence, they cannot be heated to hasten the drying process. These are usually placed in a desiccator, perhaps even a vacuum desiccator, to remove the last traces of water.

Precipitates filtered through filter paper must be stable at temperatures as high as 1000°C. Filter paper cannot be dried to constant weight, since it begins to decompose at temperatures as low as $100°C$; it is eliminated completely from the precipitate by burning it away at a high temperature. This process is known as *igniting* the precipitate. Barium sulfate, used in the determination of sulfate ion, is filtered through ashless filter paper, then ignited, and the remaining precipitate of barium sulfate is then weighed. Precipitates such as ferric hydroxide frequently are ignited and then weighed as oxides.

Calculation of Results Using the Gravimetric Factor

The Gravimetric Factor

A gravimetric analysis gives the weight of a pure precipitated compound, not the weight of the ion or compound sought. To convert the weight of the precipitated compound to the weight of the species sought, the *gravimetric factor* is used. This may be defined as the stoichiometric ratio of the formula weight (form wt) of the species sought to the formula weight of the compound precipitated and weighed. Because *this ratio must reflect the stoichiometry of the reaction*, one or both of the formula weights may have to be multiplied by a number R.

The multiplication of the weight in grams of the pure precipitate by the gravimetric factor gives the weight in grams of the species sought.

$$\text{g precipitate} \times (R) \frac{\text{form wt species sought}}{\text{form wt precipitate}} = \text{g species sought}$$

Note that the *precipitate* term cancels out of the expression, making it easy to check whether the gravimetric factor is correct or must be inverted.

As an example, consider the analysis of N.F. pharmaceutical grade sodium phosphate by precipitation of the phosphate ion to form magnesium ammonium phosphate, $MgNH_4PO_4$, which is heated to form magnesium pyrophosphate, $Mg_2P_2O_7$, the compound actually weighed. The gravimetric factor for the calculation of the percentage of sodium phosphate (actually $\%Na_2HPO_4$) is

$$\frac{2 \,(\text{form wt } Na_2HPO_4)}{\text{form wt } Mg_2P_2O_7}$$

First, note that the formula weight of $Mg_2P_2O_7$ is in the denominator, because it is the formula weight of the precipitate. Second, note that it is necessary to multiply the formula weight of Na_2HPO_4 by two; this reflects the stoichiometry of the reaction, since two phosphates are equivalent to one pyrophosphate.

The precipitation of chloride ion from various chloride salts using silver nitrate also serves to illustrate the calculation of the gravimetric factor. One such reaction is

$$AlCl_3 + 3\,Ag^+(\text{excess}) = 3\,AgCl(s) + Al^{+3}$$

Note that the solid state of the silver chloride precipitate is indicated by the small *s* in parenthesis. Also note that for every one chloride ion *one* silver chloride is precipitated, but for every one aluminum ion *three* silver chloride molecules are precipitated. The example below illustrates these two relationships.

EXAMPLE: Calculate the gravimetric factors for **(a)** finding the grams of chloride (form wt 35.45 g/mole) in impure aluminum chloride (form wt 133.3 g/mole) by precipitating the chloride as silver chloride (form wt 143.3 g/mole), and **(b)** finding the grams of aluminum chloride in impure aluminum chloride by precipitating the chloride as silver chloride.

Solution to a: For finding the grams of chloride from silver chloride, the stoichiometric ratio of the formula weight of the species sought (Cl) to that of the compound precipitated is $1:1$. Thus the gravimetric factor is simply the formula weight of the chloride divided by that of the silver chloride. The calculation of the grams of chloride is

$$\text{g } \cancel{\text{AgCl precipitate}} \times \frac{35.45 \text{ g/mole Cl}}{143.3 \text{ g/mole } \cancel{\text{AgCl precipitate}}} = \text{g Cl}$$

Note that the *AgCl precipitate* term cancels out of the expression as a check on the correct form of the gravimetric factor.

Solution to b: For finding the grams of aluminum chloride from the silver chloride, the stoichiometric ratio of the formula weight of the species sought ($AlCl_3$) to that of the compound precipitated is $1:3$. This ratio equalizes the numbers of the *key atom*, the chloride, in the gravimetric factor. The calculation of the grams of aluminum chloride is

$$\text{g } \cancel{\text{AgCl precipitate}} \times \frac{133.3 \text{ g/mole } AlCl_3}{(3)\ 143.3 \text{ g/mole } \cancel{\text{AgCl precipitate}}} = \text{g } AlCl_3$$

Stoichiometric Ratios Note that the ratios of the formula weights in the two gravimetric factors in the previous example varied and depended on the formulas of the species sought. At first, it may seem contradictory that the gravimetric factors are different for the same analysis. However, note that in both cases the numbers of chloride atoms in the numerator and denominator of each gravimetric factor are equal. The chloride atom is the *key atom*, the common atom, in this case. If the numbers of the key atom are the same in the numerator and denominator, then the ratio of the formula weights will be stoichiometric and therefore correct.

Unfortunately, not all gravimetric factors involve a key atom. Consider the following reaction:

$$C_2O_4^{-2} + Ca^{+2}(\text{excess}) = CaC_2O_4(s) \xrightarrow{\text{heat}} CaCO_3(s) + CO(g)$$

The gravimetric factor for the calculation of grams of oxalate is

$$\frac{88.0 \text{ g/mole } C_2O_4^{-2}}{100.1 \text{ g/mole } CaCO_3}$$

In this reaction there is no key atom, because one of the carbon atoms and one of the oxygen atoms are lost during heating. Therefore, the gravimetric factor must be based on the $1:1$ stoichiometric ratio of the oxalate ion and the calcium ion in the precipitation reaction.

In most cases the key atom is a metal or nonmetal, such as Cl, Br, or S (but not O), that is common to both the species sought and the compound precipitated. Some typical examples are tabulated below:

Species sought	Compound precipitated	Gravimetric factor
S	$BaSO_4$	form wt S/form wt $BaSO_4$
P	$Mg_2P_2O_7$	2(form wt P)/form wt $Mg_2P_2O_7$
C	CO_2 (abs'd in NaOH)	form wt C/form wt CO_2
Fe	Fe_2O_3	2(form wt Fe)/form wt Fe_2O_3
$HgNH_2Cl$	HgS	form wt $HgNH_2Cl$/form wt HgS

Work the next self-test for more examples of gravimetric factor calculations.

SELF-TEST

Answers:

0.2473$_{83}$ (form wt Cl = 35.45)
0.3872$_{64}$
2.735$_{84}$

1.034$_{92}$

0.2729$_{15}$

0.1118$_{72}$

7. Calculation of Gravimetric Factors

Directions: Cover the answers in the left column before beginning the test. Calculate the gravimetric factor for each problem to four significant figures. Check your answers with those given to the left.

A. The chloride ion in calcium chloride, formula weight 110.99, can be analyzed by precipitating it as silver chloride. Calculate the gravimetric factor for:
   a. The percentage of chloride.
   b. The percentage of calcium chloride $CaCl_2$.

B. The purity of magnesium citrate, $MgHC_6H_5O_7 \cdot 5H_2O$, can be found by precipitating the magnesium as $MgNH_4PO_4$, formula weight 137.4, and weighing it as $Mg_2P_2O_7$, formula weight 222.6. Calculate the gravimetric factor for finding the percentage purity of the magnesium citrate, formula weight 304.5.

C. The manganese(II) ion can be analyzed by precipitating it as Mn_3O_4, formula weight 228.8. Calculate the gravimetric factor for reporting the purity as $\%Mn_2O_3$, formula weight 157.86.

D. Carbon-hydrogen compounds are analyzed by measuring the weights of carbon dioxide, formula weight 44.01, and water, formula weight 18.02, released during combustion.
   a. Calculate the gravimetric factor for converting the weight of carbon dioxide to the weight of carbon.
   b. Calculate the gravimetric factor for converting the weight of water to the weight of hydrogen.

Calculation of Results

Gravimetric analysis results may be expressed in terms of percentage composition, molarity, the number of halogen atoms present in a known molecule, or empirical formula.

Calculation of Percentage Composition. Once the gravimetric factor has been used to find the weight of the species sought, the percentage purity of the species sought can then be calculated. This is calculated as follows:

$$\frac{\text{g species sought}}{\text{g impure sample}} (100) = \% \text{ species sought}$$

Values for percentage may vary from zero to 100 or less, depending on whether or not the species sought is a compound. For example, if impure sodium chloride is analyzed and percentage sodium chloride is sought, it will vary from 0% to close to 100%. However, if percentage chloride is sought, it will vary only between 0% and 60.66%, the percentage of chloride in pure sodium chloride.

Gravimetric analyses are capable of high accuracy and precision for percentage composition. For example, the relative deviation between two chloride analyses from precipitation of silver chloride can be as low as 0.02 pph. That means that results of 50.10% and 50.12% chloride can be obtained by an experienced analyst. Other gravimetric methods of analysis are not quite as accurate or precise as the silver chloride precipitation. For example, the precision of the barium sulfate precipitation is not nearly as good as that of the silver chloride precipitation.

Calculation of Molarity. The molarity of a solution can be determined if the cation or the anion of the solute can be precipitated quantitatively. For example, hydrochloric acid titrant can be standardized to a high degree of accuracy and precision by precipitation.

$$H^+Cl^- + Ag^+(excess) = AgCl(s) + H^+$$

Since the moles of the silver chloride precipitate equal the moles of the acid, the molarity of the acid can be calculated by finding the moles of silver chloride precipitate obtained per measured volume of the acid.

$$M\ HCl = \frac{\text{mole HCl}}{\text{liter HCl}} = \frac{\text{g AgCl precipitate}/143.3\ \text{g/mole AgCl}}{\text{liters HCl}}$$

Since molarity is by definition also equal to mmoles/ml, it is also calculated as follows:

$$M\ HCl = \frac{\text{mmole HCl}}{\text{ml HCl}} = \frac{\text{mg AgCl precipitate}/143.3\ \text{mg/mmole AgCl}}{\text{ml HCl}}$$

Indirect Analysis of Halide Mixtures. Mixtures of two halide salts, such as potassium chloride and sodium chloride, cannot be analyzed *directly* by precipitation as silver chloride. They can be analyzed by *indirect analysis*, using one or two precipitations combined with a calculation involving one equation in one unknown or two equations in two unknowns.

If the mixture contains only two halide salts with the same halide anion (X^-) and no inert matter, then only one analysis is necessary. The calculation is done by setting up one equation in one unknown. The equation may be set up using as the left-hand term the sample weight, percentage of halide ion in the sample, or fraction of halide ion in the sample. For example, let us use the approach involving the familiar percentage of halide ion (%X) in a sample of NaX and KX. After precipitating AgX, we calculate the precentage of X^- in the sample as always. Then we set up one equation in one unknown, using x as the percentage of NaX and $(100 - x)$ as the percentage of KX. This equation can be set up with symbols and then with weights.

$$\%X = \%NaX\ (X/NaX) + \%KX\ (X/KX)$$
$$\%X = x\ (\text{form wt X/form wt NaX}) + (100 - x)\ (\text{form wt X/form wt KX})$$

If the mixture contains two halide salts and inert matter, then two analyses are necessary. One analysis may involve precipitation of AgX, but the other must involve precipitation of a salt other than AgX. The calculation is done by setting up two equations in two unknowns as above. The left-hand term in each equation is usually the weight of a precipitate or the weight of an anion.

Calculation of the Number of Halogens in a Molecule. In several organic reactions, molecules are reacted with halogens to give products substituted with an unknown number of halogen atoms. Many times the number of carbons, hydrogens, and oxygens in the product is known and only the number of halogens is unknown. Other times the number of hydrogens and halogens varies; the number of hydrogens decreases as the number of halogens increases.

Consider the following reaction:

$$C_4H_{12}N + yBr_2 \longrightarrow C_4H_{12}NBr_x$$

The number of bromines, x, substituted on the product is unknown, but a gravimetric analysis of the percentage of bromine will allow x to be calculated, as the following example demonstrates.

> **EXAMPLE:** The gravimetric analysis of 0.0962 g of *pure* $C_4H_{12}NBr_x$ yields 0.1730 g of silver bromide, formula weight 187.78. Find x.
>
> *Solution:* Calculate the percentage of bromine, formula weight 79.90, in the pure compound, and compare it with the percentage of bromine in all of the possible combinations of $C_4H_{12}NBr_x$.
>
> The percentage bromine in the pure compound is
>
> $$\frac{0.1730 \text{ g AgBr ppt} \times \dfrac{79.90 \text{ g/mole Br}}{187.78 \text{ g/mole AgBr}} (100)}{0.0962 \text{ g pure sample}} = 76.5_2\% \text{ Br}$$

Calculation of the percentage bromine for the various possible values of x gives: $C_4H_{12}NBr$, 51.8% Br; $C_4H_{12}NBr_2$, 68.4% Br; and $C_4H_{12}NBr_3$, 76.4% Br. It is obvious that the *theoretical* percentage of bromine in the latter compound matches quite closely the *experimental* percentage of bromine found in the analysis and that $x = 3$.

5-3 | AN EARLY LOOK AT SOLUBILITY PRODUCT CALCULATIONS

Solubility Product Constant

The solubility product constant, K_{sp}, is usually defined for a reaction in which the insoluble salt is the reactant and the ions are the products. It should be written to include water and energy (heat) as well; for example, for the dissolution of silver chloride the reaction is strictly

$$AgCl(s) + xH_2O + \text{energy} \rightleftharpoons Ag(OH_2)_4{}^+ + Cl(H_2O)_{x-4}^- \qquad (5\text{-}4)$$

This indicates the ions dissolve by bonding to water molecules. Secondly, **5-4** indicates that energy is required for silver chloride to dissolve, not enough of which is available at room temperature for a significant solubility. This is also indicated by the relative lengths of the two arrows, showing that the point of equilibrium lies on the left.

Equation **5-4** can be simplified as follows for the purposes of defining K_{sp}:

$$AgCl(s) = Ag^+ + Cl^- \qquad (5\text{-}5)$$

The K_{sp} expression can now be written and given a numerical value.

$$K_{sp} = [Ag^+][Cl^-] = 1.8 \times 10^{-10} M^2 \qquad (5\text{-}6)$$

Note that since the reactant is an insoluble salt, its concentration cannot be specified and so it is defined to be unity. Hence it does not appear in the above expression. Since the numerical value of K_{sp} is equal to the product of the molarities of the silver and chloride ions, it must have the units of M^2. Use of these units in calculations and in writing the ionic terms in the solubility product expression will be very helpful.

The value given for K_{sp} is at 25°C (room temperature). Since the dissolving of silver chloride requires energy **(5-4)**, raising the temperature increases the solubility and K_{sp}. Some typical values are

$$10°C \quad 4 \times 10^{-11} M^2$$
$$25°C \quad 1.8 \times 10^{-10} M^2$$
$$50°C \quad 1.3 \times 10^{-9} M^2$$

Formulation of the K_{sp} Expression

The formulation of a K_{sp} expression for a known salt, $A_n B_m$, is written

$$K_{sp} = [A^{+m}]^n [B^{-n}]^m = r \times 10^{-s} M^{(n+m)} \tag{5-7}$$

The molarity of each ion is always raised to a power equal to its subscript. The units of molarity following the numerical value of K_{sp} will always be the sum of all ions in the expression (see Appendix 1 for values with such units).

Formulating the K_{sp} expression for an *unknown* salt is done in one of two ways. Given the units of molarity, you can usually deduce a reasonable expression with the same value for $(n + m)$ in **5-7**. Consider AgN_3. Without units, you might incorrectly deduce that this is silver(III) ion with three N^{-1} anions. However the units of M^2 on the K_{sp} (Appendix 1) lead to the expression

$$K_{sp} = [Ag^+][N_3^-] = 2.9 \times 10^{-9} M^2$$

The other way to deduce the above expression involves consulting an inorganic reference for nitrogen compounds, which would indicate that N_3^- is the azide ion.

You should apply the above approaches to $Ca_3(PO_4)_2$ and $CaHPO_4 \cdot 2H_2O$, both of which precipitate on the surface of teeth as dental calculus (plaque).

Solubility in "Pure" Water

Theoretically, the step that limits whether or not a precipitation is quantitative is the step in which the precipitate is washed with pure water or a solution containing no common ions. It is useful to calculate S, the molar solubility of the insoluble salt, in such solutions, even though the presence of small amounts of ions renders the water wash less than "pure."

In "pure" water, the concentrations of all ions are represented by multiples of S, since they are unknown. This is done by writing a balanced equation using multiples of S as coefficients; for example, for $Ca_3(PO_4)_2$:

$$S Ca_3(PO_4)_2 \rightleftharpoons 3 S Ca^{+2} + 2 S PO_4^{-3}$$

The balanced equation automatically gives the correct symbolic concentrations to substitute into the K_{sp} expression.

$$K_{sp} = [3S]^3 [2S]^2 = 108 S^5$$

This equation is rearranged to give the relation between S and K_{sp}.

$$S = \sqrt[5]{\frac{K_{sp}}{108}} \tag{5-8}$$

Although a numerical value for S must still be obtained from **5-8**, the derivation is the most important part of the calculation; the rest is plain arithmetic.

It is important to be able to calculate any of the possible roots that may arise from relations such as **5-8**. Roots such as fifth or seventh roots cannot be calculated on inexpensive electronic calculators or on slide rules, but they can be calculated with logs. (See a math text for information on how to use logs. You can then use any inexpensive calculator to calculate any root.) In general, proceed as follows, for $S = \sqrt[n]{K_{sp}/x}$,

$$\log S = \frac{\log(K_{sp}/x)}{n}$$

$$S = \text{antilog}\left[\frac{\log(K_{sp}/x)}{n}\right] \qquad (5\text{-}9)$$

Effect of Inert Salts on Solubility in Pure Water

Inert salts such as $NaNO_3$, KNO_3, and $LiClO_4$ increase the ionic strength, μ, of a solution and the solubility of insoluble salts such as silver(I) chloride. (The ionic strength of inert salts of univalent cations and anions is the same as the total molarity of such salts; ionic strength for other inert salts may be calculated as shown in Section 1-1). The solubility increase is reflected by an increase in the value of K_{sp} for a given insoluble compound. For example, Appendix 1 lists values of K_{sp} for two different ionic strengths: $\mu = 0$ (pure water) and $\mu = 0.1$ (for example, a $0.1M$ solution of an inert salt such as KNO_3).

Calculation for Wash Water Solutions

The percentage of an insoluble compound that dissolves in a pure water wash (Sec. 5-2) may be calculated by deriving and using an equation such as **5-8**. If roughly *equal volumes of wash water and the unknown solution of the ion to be precipitated* are used, the percentage is calculated as follows:

$$\% \text{ dissolved} = \frac{\text{Molar solubility of compound in wash water}}{\text{Molarity of ion in unknown solution}}(100) \qquad (5\text{-}10)$$

No more than 0.1% should dissolve for a quantitative (99.9%) precipitation.

If the volumes are not roughly the same, then each molarity in **5-10** must be multiplied by the corresponding volume to convert molarities to millimoles:

$$\% \text{ dissolved} = \frac{(S \text{ in wash water})(\text{ml wash})}{(M \text{ ion in unknown})(\text{ml unknown})}(100)$$

EXAMPLE: The chloride ion is precipitated as silver chloride from 100 ml of a $0.0050M$ solution. The insoluble silver(I) chloride is washed with about 90 ml of water containing $0.03M$ nitric acid. Will the precipitation be quantitative?
Solution: Since the volumes of the unknown solution and the wash solution are about the same, **5-10** may be used to calculate the percentage of the chloride ion that dissolves in the wash solution. The solubility of silver chloride is

$$S = \sqrt{K_{sp(\mu=0)}} = \sqrt{1.8 \times 10^{-10}M^2} = 1.3 \times 10^{-5}M \text{ (pure water)}$$

Strictly speaking, the wash solution is not pure water since the nitric acid gives it an ionic strength of 0.03; however, the increase in the molar solubility is too small to be significant for the calculation of the percentage dissolved.

The percentage dissolved is calculated as follows:

$$\% \text{ dissolved} = \frac{1.3 \times 10^{-5}M \text{ chloride dissolved}}{5 \times 10^{-3}M \text{ chloride unknown}}(100) = 0.3\%$$

Since 0.3% is greater than 0.1%, the precipitation is not quantitative. The error could be reduced by washing with a smaller volume of water, concentrating the unknown solution, or lowering the solubility by lowering the temperature.

The Common Ion Effect

When an ion is precipitated for gravimetric analysis, the counter ion added to form an insoluble compound is usually added in approximately a 10% to 20% excess. The excess ion lowers the molar solubility; this effect is called the common ion effect. It is important to understand this effect and to be able to make calculations where it occurs.

The common ion effect acts to shift the point of equilibrium in any precipitation equilibrium to the left; for example, for silver ion being precipitated by chloride as common ion,

$$AgCl(s) \rightleftharpoons Ag^+ + Cl^-$$

pt. of equilibrium ← excess Cl^-
lies here

The excess chloride lowers the silver ion concentration, as well as the solubility of silver chloride, below that found in a saturated solution in pure water by precipitating more of the silver.

Since the excess chloride is added as HCl or KCl, the molar solubility of the silver chloride will be approximately equal to the molarity of the silver ion, rather than the molarity of the chloride ion. The K_{sp} expression can be solved for the molar solubility of silver chloride as follows:

$$S_{AgCl} = [Ag^+] = \frac{K_{sp}}{[Cl^-]} \cong \frac{K_{sp}}{M_{excess\ Cl^-}} \qquad (5\text{-}11)$$

Equation **5-11** contains an approximation in the right-hand term, because the equilibrium concentration of the chloride is, strictly speaking, the sum of the excess chloride ion plus the chloride resulting from a small amount of insoluble silver chloride dissolving. The latter is usually small compared to the excess chloride that has been added to the solution as HCl or KCl.

A more general equation than **5-11** can be written for calculating the solubility of any insoluble compound $M_x A_y$ in the presence of an excess of the anion A^{-x}:

$$S_{M_x A_y} = \frac{1}{x}[M^{+y}] = \frac{1}{x}\sqrt[x]{\frac{K_{sp}}{[A^{-x}]^y}} \cong \frac{1}{x}\sqrt[x]{\frac{K_{sp}}{[M_{excess\ A^{-x}}]^y}} \qquad (5\text{-}12)$$

This equation can be used to calculate the approximate molar solubility for any compound after precipitation while it is still in equilibrium with the excess common ion. (It cannot be used once the compound has been washed free of the common ion.) The following examples illustrate common ion calculations.

EXAMPLE: Silver chloride is precipitated from a solution to which an excess of 4 ml of 0.5M hydrochloric acid has been added. The final volume of the solution is 90–100 ml, so that a rough value of 100 ml can be used for the volume. What is the molar solubility of silver chloride, and is it less than that in pure water?
Solution: First, the final molarity of the excess chloride ion must be calculated.

$$M_{excess\ Cl} \cong \frac{4\ ml \times 0.5M}{100\ ml} \cong \frac{2\ mmole}{100\ ml} \cong 2 \times 10^{-2}M$$

Now the molar solubility of the silver chloride can be calculated, using the K_{sp} expression and the molarity of the excess chloride.

$$S_{AgCl} \cong [Ag^+] \cong \frac{K_{sp}}{M_{excess\ Cl^-}} \cong \frac{1.8 \times 10^{-10}M^2}{2 \times 10^{-2}M} \cong 9 \times 10^{-9}M$$

Since the molar solubility of silver chloride is pure water is $1.3 \times 10^{-5}M$, the common ion effect has lowered the concentration of silver chloride below that in pure water.

EXAMPLE: Calcium phosphate is precipitated from a solution to which $1 \times 10^{-3}M$ excess phosphate has been added. What is the molar solubility of calcium phosphate, and has it been lowered below its solubility in pure water?

Solution: Calculate the molar solubility using **5-12**, the K_{sp} expression for calcium phosphate, and the molarity of the excess phosphate.

$$S_{Ca_3(PO_4)_2} \cong \frac{1}{3}[Ca^{+2}] \cong \frac{1}{3}\sqrt[3]{\frac{K_{sp}}{[PO_4^{-3}]^2}} \cong \frac{1}{3}\sqrt[3]{\frac{1 \times 10^{-26}M^5}{(1 \times 10^{-3}M)^2}} \cong 7 \times 10^{-8}M$$

Since the solubility of calcium phosphate in pure water is of the order of $10^{-6}M$, the common ion effect has lowered the concentration of calcium phosphate below that in pure water.

SELF-TEST

8. Calculation of Solubility in Pure Water

Answers:

Directions: Cover the answers in the left column before beginning the test. Work one problem at a time, checking your answers with those to the left.

A. The precipitation of Ca^{+2} as $CaSO_4$ under certain conditions will be quantitatively complete if the molar solubility of $CaSO_4$ in pure water (used to wash the precipitate) is less than $4.8 \times 10^{-3}M$.

$S = 4.9 \times 10^{-3}M$

    a. Using $K_{sp} = 2.4 \times 10^{-5}M^2$, calculate the molar solubility of $CaSO_4$ in pure water.

No

    b. Is the precipitation quantitatively complete?

B. The precipitation of Pb^{+2} will be complete if the molar solubility of the lead(II) salt in pure water is less than $1 \times 10^{-3}M$. Given that $K_{sp} = 1.8 \times 10^{-14}M^2$ for $PbCrO_4$ and that $K_{sp} = 1.6 \times 10^{-5}M^3$ for $PbCl_2$:

$PbCrO_4$ sol. $= 1.3_4 \times 10^{-7}M$;
$PbCl_2$ sol. $= 1.5_9 \times 10^{-2}M$

    a. Calculate the molar solubilities of $PbCrO_4$ and $PbCl_2$ in pure water.

Only $PbCrO_4$

    b. Decide whether either compound is insoluble enough for complete precipitation of Pb^{+2}.

C. Milk of magnesia consists of suspended $Mg(OH)_2$ in pure water. Normally a solution of an alkali or alkaline earth metal hydroxide would be too basic to be safely ingested. Assuming such a suspension is a saturated solution:

$1.6_5 \times 10^{-4}M$

    a. Calculate the molar solubility of $Mg(OH)_2$, using $K_{sp} = 1.8 \times 10^{-11}M^3$.

$3.3 \times 10^{-4}M$

    b. Calculate the $[OH^-]$ of the solution.

Much less basic; safe

    c. Compare the $[OH^-]$ with the $[OH^-]$ of $0.10M$ NaOH as to basicity and safety.

D. Maalox antacid tablets consist of a maximum of 0.8 g of an insoluble base, $Al(OH)_3$, per tablet. Assuming you chewed one tablet to form a saturated solution of $Al(OH)_3$ in your mouth:

$3.6 \times 10^{-9}M$

Neglecting OH^- from water,
$\quad [OH^-] = 1.0_8 \times 10^{-8}M$; safe

a. Calculate the molar solubility of $Al(OH)_3$, using $K_{sp} = 4.6 \times 10^{-33}M^4$.
b. Calculate the $[OH^-]$ of the solution and comment on its safety.

5-4 | BIOLOGICAL AND PHARMACEUTICAL GRAVIMETRIC ANALYSIS

There are several gravimetric methods important in biological and pharmaceutical chemistry that deserve special attention. As an example of a biological gravimetric method we will discuss the determination of cholesterol. Thereafter, we will discuss some useful pharmaceutical gravimetric methods.

Determination of Cholesterol

Cholesterol is a steroid alcohol, or 3-sterol, with the formula $C_{27}H_{45}OH$. It can be precipitated by a high molecular weight organic saponin called digitonin. Digitonin has the formula $C_{55}H_{90}O_{29}$ and a formula weight of 1214. Cholesterol and digitonin react in a 1:1 molecular ratio to form an insoluble complex which apparently contains a molecule of water.

$$C_{27}H_{45}OH + C_{55}H_{90}O_{29} + H_2O \longrightarrow C_{27}H_{45}OH(C_{55}H_{90}O_{29})H_2O(s)$$

The precipitation may be used to isolate cholesterol for other measurements or for the gravimetric determination of cholesterol. The theoretical gravimetric factor for calculating the percentage of cholesterol is 0.2388. If esters of cholesterol must be converted to cholesterol before the precipitation, there is a small loss of cholesterol, which is corrected for by increasing the gravimetric factor to 0.243 [2]. Several investigators have found that an accurate analysis depends on the purity of the digitonin that is used [3]; a correction curve must be established for each sample of digitonin to obtain high accuracy.

The digitonin precipitation is quite specific for cholesterol among the steroids having a hydroxyl group. Even cholesterol esters are not precipitated, nor are sterols with a 3-hydroxyl group of a different steric arrangement than cholesterol. Unfortunately the precipitation is slow, so that it is recommended that it take place overnight [3].

Determination of Pharmaceuticals

Some pharmaceuticals can be determined gravimetrically by isolating the pure form of the organic medicinal agent without any chemical reaction. Others can be determined by conversion of the sodium salt to the acid form.

A partial list of pharmaceuticals determined without chemical reaction is given in Table 5-1. One such pharmaceutical is amobarbital, $C_{11}H_{17}N_2O_3H$, a weak acid. As long as it is the only pharmaceutical in the preparation, it can be isolated by extraction. However, before the amobarbital is extracted from solid preparations, the sample must be washed with petroleum ether to remove any inert binder or lubricant that might be extracted with the amobarbital. After the washing, the amobarbital is extracted with chloroform or ether. The extracting solvent is then evaporated and the pure dry residue is weighed. The weight gives the amount of amobarbital present in whatever amount of the sample was taken for analysis.

TABLE 5-1. Pharmaceuticals Determined Without Chemical Reaction

U.S.P. Pharmaceuticals	N.F. Pharmaceuticals
Amobarbital tablets	Amobarbital elixir
Aurothioglucose injection	Citrated caffeine
Caffeine and sodium benzoate	Ephedrine sulfate
Phenacetin tablets	Mephobarbital tablets
Sodium lauryl sulfate	Progesterone tablets

TABLE 5-2. Pharmaceuticals Determined by Conversion to the Acid Form

Pharmaceutical	Weighed as (Acid Form)
Sodium amobarbital, U.S.P.	$C_{11}H_{17}N_2O_3H$
Sodium butabarbital, N.F.	$C_{18}H_{19}O_2H$
Sodium fluorescein	$C_{20}H_{11}O_5H$
Sodium pentobarbital	$C_{11}H_{17}N_2O_3H$
Sodium phenobarbital	$C_{12}H_{11}H_2O_3H$

A partial list of pharmaceuticals determined by conversion to the acid form is given in Table 5-2. One typical pharmaceutical is sodium phenobarbital, $C_{12}H_{11}N_2O_3{}^-Na^+$. It is converted to phenobarbital (acid form) by adding a strong acid, such as hydrochloric acid, to it. After it has been washed to remove any binder or lubricant, it may also be extracted into ether or chloroform. (The sodium phenobarbital is not soluble in an organic solvent such as chloroform.) Evaporation of the ether or chloroform gives a pure dry solid, which can then be weighed to give an amount of phenobarbital equivalent to the sodium phenobarbital originally present in the pharmaceutical preparation. A gravimetric factor would have to be used to calculate the amount of sodium phenobarbital equivalent to the phenobarbital [4].

| QUESTIONS AND PROBLEMS

(Answers to most even-numbered problems are found in Appendix 5.)

Concepts and Definitions

1. List:
   a. Three major advantages of gravimetric methods.
   b. Three major disadvantages of gravimetric methods.
2. List the main steps in a gravimetric analysis.
3. What are a few of the preliminary adjustments that might have to be made to a dissolved sample before it can be precipitated?

4. Name the three different types of precipitates.
5. What is the purpose of digestion?
6. What are the advantages of:
   a. A sintered-glass crucible over a Gooch crucible?
   b. Filter paper over a sintered-glass crucible?
   c. A sintered-glass crucible over filter paper?
   d. A Gooch crucible over a sintered-glass crucible?

7. Define a gravimetric factor.
8. Explain why the gravimetric factor for the calculation of the percentage of chloride in impure aluminum chloride is calculated with the formula weight of silver(I) chloride when the gravimetric factor for the calculation of the percentage of aluminum chloride in impure aluminum chloride is calculated using a multiple of three times the formula weight of silver(I) chloride.
9. Why is it important to be able to calculate the solubility of an insoluble salt in the solution used to wash the precipitate of the salt?
10. Explain in detail how cholesterol is determined gravimetrically using digitonin.

Gravimetric Factor Calculations

11. Milk of magnesia can be analyzed by dissolving the magnesium hydroxide and then precipitating the magnesium(II) ion with 8-hydroxyquinoline to form $Mg[(C_9H_6NO)_2 \cdot 2H_2O]$. Calculate the gravimetric factor for reporting the percentage of magnesium hydroxide from the weight of the precipitate.
12. An iron-ore sample is analyzed by precipitation of $Fe(OH)_3$, form wt $= 314.03$, and ignition of the $Fe(OH)_3$ to Fe_2O_3, which is weighed. Calculate the gravimetric factor for finding the percentage of:
   a. Fe (at wt $= 55.85$)
   b. Fe_2O_3 (form wt $= 159.7$)
   c. Fe_3O_4 (form wt $= 231.54$)
   d. $2Fe_2O_3 \cdot 3H_2O$ (form wt $= 337.38$)
13. Carbon-hydrogen compounds are analyzed by measuring the weight of carbon dioxide (form wt $= 44.01$) and water (form wt $= 18.02$) released during combustion.
   a. Calculate the gravimetric factor for converting the weight of water to the weight of hydrogen.
   b. Calculate the gravimetric factor for converting the weight of carbon dioxide to the weight of carbon.
14. A certain fish containing mercury in the form of $(CH_3)_2Hg$ was analyzed and found to contain 3.5 ppm of mercury. Use the appropriate gravimetric factor to calculate the ppm of $(CH_3)_2Hg$ in the fish.

Calculation of Percentage Composition

15. The chloride ion in 0.9100 g of impure magnesium chloride (form wt $= 95.22$) is determined by precipitating it as 1.5000 g silver chloride (form wt $= 143.3$). Calculate:
   a. The percentage of chloride ion (at wt $= 35.45$) in the sample.
   b. The percentage of magnesium chloride in the sample.
16. Two samples of supposedly impure sodium chloride (form wt $= 58.44$) are analyzed for chloride ion (at wt $= 35.45$) by precipitating it as silver chloride (form wt $= 143.3$). Calculate the percentage of chloride ion in each sample and decide whether or not each sample is impure sodium chloride on the basis of the maximum percentage of chloride possible in pure sodium chloride.
   a. Weight of sample A $= 0.5100$ g; weight of resulting silver chloride $= 1.2330$ g.
   b. Weight of sample B $= 0.4800$ g; weight of resulting silver chloride $= 1.2330$ g.
17. The daily minimum requirement of Na^+ for the average healthy adult ranges between 250 and 500 mg of Na^+ (at wt $= 22.99$).
   a. Calculate the gravimetric factor for converting the weight of Na^+ to NaCl.
   b. Calculate the minimum requirement in mg of sodium chloride (form wt $= 58.44$).
   c. Calculate the minimum requirement in ounces of sodium chloride.
   d. Calculate the minimum requirement in eighths of a teaspoon, assuming one teaspoon $= 4.7$ g.
18. A 0.6000-g sample of sulfate (form wt $= 98.07$) is dried and found to contain 5.00% water. The sample yields 1.167 g of barium sulfate (form wt $= 233.4$). Calculate:
   a. The percentage of sulfur (at wt $= 32.06$) in the moist sample.
   b. The percentage of sulfur in the dry sample.
19. A 1.000-g sample of impure cholesterol yields 0.5000 g of the cholesterol-digitonin insoluble complex (form wt $= 1618.6$). Calculate the percentage of cholesterol (form wt $= 386.6$) in the sample.

Calculation of Molarity and Other Quantitites

20. A 25.00-ml volume of sodium chloride solution is analyzed by precipitating the chloride (at wt = 35.45) as 0.3740 g of silver chloride (form wt = 143.3). Calculate:
    a. The number of mmoles of silver chloride in the precipitate.
    b. The molarity of the sodium chloride solution.

21. Calculate the molarity of a hydrochloric acid solution, 50.00 ml of which gives 0.7510 g precipitate of silver chloride (form wt = 143.3).

22. Calculate the molarity of a magnesium chloride solution, 25.00 ml of which gives 0.7510 g precipitate of silver chloride (form wt = 143.3).

23. The FDA tolerance for mercury in food is 0.5 ppm (0.5 mg/kg food). A 120-g fish is found to contain 0.14 mg of mercury. Calculate:
    a. The ppm of mercury to see whether it exceeds FDA standards.
    b. The ppm of $(CH_3)_2Hg$ in the fish.

24. A 0.1500-g pure sample of an organic compound of the formula $C_6OCl_xH_{6-x}$ was decomposed to give 0.4040 g of silver(I) chloride. Calculate the numerical value of x and the correct molecular formula.

Washing and Digestion

25. Write equations for formation of a corner of a curd of the digested silver halide from:
    a. Adding excess $AgNO_3$ to NaBr.
    b. Adding excess $AgClO_4$ to NaCl.

26. Write equations for formation of a corner of a curd of digested silver bromide from:
    a. Adding excess NaBr to $AgNO_3$.
    b. Adding excess $AgClO_4$ to KBr.

27. State whether each solution below can be used to wash insoluble silver chloride after precipitation of sodium chloride with excess silver nitrate and also after precipitation of silver nitrate with excess sodium chloride.
    a. Dilute HNO_3
    b. Aqueous NH_4NO_3
    c. Dilute HCl
    d. Dilute H_2SO_4

28. Which of the following solutions would not be a good electrolyte for a wash solution for silver chloride? For those that would not, explain why not.
    a. Dilute HNO_3
    b. Dilute acetic acid
    c. Dilute HF
    d. Dilute H_2SO_4
    e. Dilute HNO_2
    f. Dilute HBr

Solubility Calculations

29. Using a log table (inside back cover) to find the root, calculate the molar solubility in pure water of each compound below. You must show the use of the logs even though you use a calculator.
    a. $Ca_3(PO_4)_2$; $K_{sp} = 1 \times 10^{-26}M^5$
    b. $Th(IO_3)_4$; $K_{sp} = 2.5 \times 10^{-15}M^5$
    c. $CoHg(SCN)_4$; $K_{sp} = 1.5 \times 10^{-6}M^6$
    d. $Fe_4[Fe(CN)_6]_3$; $K_{sp} = 3.0 \times 10^{-41}M^7$

30. Derive a relationship between the molar solubility of each of the following insoluble compounds in pure water and the K_{sp}. Then, using the K_{sp} value in Appendix 1, calculate a numerical value for S to the correct number of significant figures.
    a. Ag_2CrO_4
    b. Hg_2Cl_2
    c. $CaMg(CO_3)_2$
    d. $MgNH_4PO_4$

31. Assuming that each of the cations in the insoluble compounds in Problem 30 is precipitated from about 100 ml of a $0.001M$ solution, calculate the percentage of each dissolved in a 100-ml wash solution.

32. The chloride ion is precipitated from a solution to which an excess of 2.0 ml of $0.50M$ silver nitrate has been added. The final volume of the solution is about 100 ml.
    a. Calculate the molar solubility of silver chloride and the molarity of the chloride in the final solution.
    b. If the chloride in the original sample was $0.0010M$, is the precipitation of the chloride quantitative (99.9%)?

33. Magnesium(II) is precipitated from a solution to which hydroxide ion is added so that the

final pH is 10.3. Magnesium hydroxide precipitates.

a. Calculate the molar solubility of $Mg(OH)_2$, using $K_{sp} = 1.8 \times 10^{-11}M^3$.

b. If the magnesium(II) ion in the original sample was $0.01M$, is the precipitation quantitative (99.9%)?

Challenging Problems

34. An iron ore was incorrectly calculated to contain 10.0% of Fe_3O_4 instead of % Fe_2O_3. Calculate the % Fe_2O_3 without knowing the sample weight or any other measurements.

35. Calculate the molar solubility in pure water of each metal hydroxide below, assuming the water is neutral.

a. $Mg(OH)_2$; $K_{sp} = 1.8 \times 10^{-11}M^3$

b. $Fe(OH)_3$; $K_{sp} = 2.5 \times 10^{-39}M^4$

36. An organic compound containing only carbon and oxygen is found to contain 50.0% C and 50.0% O. Calculate:

a. Its empirical formula.

b. Its molecular formula, assuming a molecular weight of 289 ± 2.

37. Calculate the empirical formula of an inorganic compound containing 36.4% P, 37.6% O, 2.37% H, and 23.5% Ca. Also decide if the empirical formula is a reasonable molecular formula.

38. A mixture containing only sodium chloride (form wt = 58.44) and potassium chloride (form wt = 74.56) without any other substances present can be analyzed by a single gravimetric analysis. Calculate the percentage of sodium chloride in 191.44 mg of the mixture, if it yields 429.96 mg of silver chloride (form wt = 143.32).

39. A mixture containing sodium bromide (form wt = 102.9), potassium bromide (form wt = 119.0), and inert matter can be analyzed by two gravimetric analyses. Calculate the percentage of each present in 400.0 mg of the mixture, if it yields 563.34 mg of silver bromide (form wt = 187.78) in one analysis and 340.9 mg of sodium bromide plus potassium bromide (without inert matter) in the second analysis.

| NOTES

[1] P. P. von Weimarn, *Chem. Rev.* **2**, 217 (1925).

[2] R. Caminade, *Bull. Soc. Chim. Biol.* **4**, 601 (1922).

[3] J. J. Kabara in D. Glick, *Methods of Biochemical Analysis*, Vol. 10, Wiley-Interscience, New York, 1962, p. 270.

[4] G. L. Jenkins, A. M. Kenevel, and F. E. DiGangi, *Quantitative Pharmaceutical Chemistry*, McGraw-Hill, New York, 1967.

6 Fundamentals of Volumetric Methods of Analysis

"Let us have some fresh blood," Holmes said, . . .
drawing off the resulting drop of blood in a
chemical pipette. "Now I add this small quantity
of blood to a litre of water."

ARTHUR CONAN DOYLE
A Study in Scarlet, Chapter 1

A volumetric method of analysis is based on a titration in which the volume of a reagent (titrant) that reacts with a premeasured amount of a sample is measured. The sample may be measured by weighing it on the balance or by taking a known volume with a volumetric pipet (as Sherlock Holmes did with his "chemical pipette").

The fundamentals to be mastered include certain basic concepts, a number of calculations, and some laboratory techniques. All of these must of course be used together for a successful volumetric analysis, but we will discuss them in the order mentioned.

6-1 | BASIC CONCEPTS OF VOLUMETRIC ANALYSIS

To begin with, you must understand certain basic concepts of volumetric analysis. These are the standard solution, the primary standard, the equivalence point, the end point, the standardization process, and the notion of quantitative reaction. To illustrate these concepts the following general titrant reaction will be used.

$$tT + sS \longrightarrow tT_p + sS_p \qquad (6\text{-}1)$$

In this reaction, T and T_p are the reactant and product forms of the titrant, and S and S_p are the reactant and product forms of the sample, respectively.

The Titrant: A Standard Solution
In a titration, the sample S is measured into a flask, dissolved if necessary, and then titrated with a solution of the titrant T **(6-1)**. The volume of the titrant added to the flask is measured with a *buret* (Sec. 6-3). The solution of T usually reacts completely as

it is added to S. The point at which the reaction is complete is called the equivalence point. It is essential that the solution of T be a *standard solution*, for which there are several requirements:

1. *The concentration of such a solution must be known accurately, usually to four significant figures.* There are two ways of calculating the concentration. One is to add a known weight of a primary standard reagent (a reagent of known purity) to a volumetric flask (Sec. 6-3) and dilute to the known volume of the flask. The other way is to standardize a solution of approximate concentration by titrating it against a weighed amount of a primary standard reagent.
2. The solution should be stable for as long as the analysis and standardization require; that is, its concentration should not change significantly during this time.
3. The reaction of the titrant T with the sample S should be stoichiometric; that is, there should be a whole-number ratio between t and s in the reaction.
4. The reaction of the titrant T and the sample S should be rapid and quantitatively complete. Usually this means a 99.9% reaction.

Primary Standards

Preparing a standard solution depends on using the proper primary standard chemical reagent. For example, a standard solution of sodium hydroxide cannot be prepared by weighing sodium hydroxide into a volumetric flask, because the purity of the sodium hydroxide varies with time. If you know the requirements for a primary standard, you can readily decide whether a chemical meets such requirements.

The main requirements for a *primary standard chemical* are

1. The chemical must be of known composition and highly pure. It is desirable that it be a minimum of 99.9% pure. Since it is difficult to prepare an absolutely pure reagent, a purity of 99.99 + % is specified. (Sometimes a range, such as 99.95–100.05%, is given on bottles.) In addition to the purity, the label on the bottle should state that the reagent is a primary standard material.
2. The chemical should be stable at room temperatue and should dry to a constant weight and formula, in an oven if necessary. It should not absorb gases such as water vapor, carbon dioxide, or sulfur dioxide from the air.
3. To minimize the relative error from weighing (Sec. 2-2), the amount weighed should be relatively large. Thus a large formula and/or equivalent weight is desirable. Although this is not the most important requirement, it would be the deciding factor between two primary standards of otherwise equal properties.
4. The chemical should react rapidly and stoichiometrically with the titrant. The latter implies that the reaction be quantitative (99.9% complete).

These requirements make it clear why sodium hydroxide is not a primary standard material. It is not stable at room temperature, since it reacts with both water vapor and carbon dioxide. It is not available at initial 99.9% purity, and its formula weight is small.

Equivalence Point and End Point

The equivalence point is the *theoretical* point at which the amount of added titrant T is exactly equivalent to the amount of dissolved sample S. Assuming that the reaction is quantitative (99.9%), it is also the point at which the reaction is complete. Beyond this point, an insignificant amount ($\leqslant 0.1\%$) of the sample will react.

The end point is our perception of the equivalence point. Usually some physical property is continuously monitored, either visually or electrically, to indicate when the first excess of the titrant has been added beyond the equivalence point. An indicator that changes color is frequently used to detect the end point. A small portion of indicator is used so that the amount of titrant that reacts with the indicator is negligible. Under ideal conditions, the color change at the end point coincides with the equivalence point. Frequently there is a small error, but this can usually be compensated for by standardizing the titrant under the same conditions as the sample titration.

Standardiza-
tion
Unless a titrant can be prepared by weighing out a primary standard chemical, it must be standardized before or during use. This is done by titrating a weighed amount of a primary standard chemical.

As an example, let us consider the standardization of sodium hydroxide titrant. *Solid* sodium hydroxide is not a primary standard chemical. A *standard solution* of sodium hydroxide is an acceptable titrant if kept closed to contact with the air. It is made by diluting a saturated solution of sodium hydroxide to an approximate concentration in the 0.01–0.5M range. In this range, sodium hydroxide solutions are stable, provided they are not unduly exposed to carbon dioxide.

A primary standard acid is then weighed out for the standardization. The preferred primary standard acid is potassium acid phthalate, $KH[C_8H_4O_4]$, or just KHP. (The P^{-2} symbolizes the phthalate anion, $[C_8H_4O_4]^{-2}$.) The primary standard acid is then titrated with the sodium hydroxide to an end point.

$$KHP + NaOH \longrightarrow P^{-2} + H_2O + K^+ + Na^+$$

The precise concentration of the sodium hydroxide solution can then be calculated by the method described in the next section. The solution is now a standard solution, since its concentration is accurately known. Provided it is not contaminated by over-exposure to carbon dioxide, it can be used for weeks without a change occurring in the concentration of sodium hydroxide.

6-2 | CALCULATIONS OF VOLUMETRIC ANALYSIS

Definitions
A few definitions of concentration, which you should have mastered previously, are reviewed below. The mole-molarity system of calculations is discussed first, followed by the equivalent-normality system.

Symbols for
Volumes
At present there is a trend toward use of the symbol "L" instead of "l" for the liter. In deference to this trend, we will use the symbol dL for deciliter (0.1 L) and the symbol μL for microliter (10^{-6} L). However, we will retain the symbol ml for milliliter (10^{-3} L) because it is still commonly used.

The
Mole-Molarity
System
The mole, mmole, and often the μmole (Sec. 1-2) are used to measure quantities of chemical species. The operational definition of a mole is the formula weight of a species expressed in grams; the formula weight in turn is expressed in g/mole. Thus a mole of calcium carbonate contains 100.1 g, and the formula weight of calcium carbonate is 100.1 g/mole. It is also convenient to talk about a mole of an ion, so that we can say a mole of chloride ion contains 35.45 g.

The operational definition of a mmole is the formula weight of a species expressed in milligrams (mg); the formula weight in turn is expressed in mg/mmole. Thus a mmole of calcium carbonate contains 100.1 mg, and its formula weight is also 100.1 mg/mmole.

The molarity M of a solution can be expressed using moles and liters or millimoles and milliliters (ml).

$$M = \frac{\text{mole}}{\text{liter}} = \frac{\text{mmole}}{\text{ml}}$$

If the molarity of a solution is known and its volume is known, then the moles or millimoles of solute in the solution may be calculated.

$$\text{mole} = (M)(\text{liter}) \quad \text{or} \quad \text{mmole} = (M)(\text{ml})$$

The Equivalent- Normality System

Many chemical substances contain more than one ion per mole that reacts with a reagent. For example, sulfuric acid, H_2SO_4, may lose one or both of its ionizable hydrogens to a base, depending on the strength of the base. So one mole of sulfuric acid is *equivalent* to either one or two moles of a monoprotic acid. To correct the formula weight for this, the concept of *equivalent weight* is used.

The equivalent weight is calculated using the formula weight and a variable n, which depends on the chemical reaction of the substance in question. For acid-base reactions n is the number of moles of H^+ (or OH^-) neutralized per mole of acid (or base) in the reaction. For oxidation-reduction reactions n is the number of moles of electrons lost or gained per mole of reactant. For precipitation reactions n is the number of moles of univalent anion needed to precipitate the cation. For complex-forming reactions n is the number of moles of ligand needed to complex the metal ion. Regardless of the reaction, the calculation of the equivalent weight is the same.

$$\text{eq wt} = \frac{\text{form wt}}{n} \tag{6-2}$$

The calculation of equivalent weight will be demonstrated during the discussion of the calculation of titration results.

Analogous to the mole and the mmole, the equivalent (eq) and the milliequivalent (meq) are used to measure quantities in the equivalent system. The operational definition of an equivalent is the equivalent weight of a species expressed in grams; the equivalent weight is expressed in g/eq. Thus an equivalent of sulfuric acid contains 98.08/2 or 49.04 g, and the equivalent weight of sulfuric acid is 49.04 g/eq. We can also talk about an equivalent of an ion; for example, we say that an equivalent of hydrogen ion contains 1.008 g.

The operational definition of a milliequivalent is the equivalent weight of a species expressed in mg; the equivalent weight is expressed in mg/meq. Thus a milliequivalent of sulfuric acid contains 49.04 mg, and the equivalent weight of sulfuric acid is 49.04 mg/meq.

Often, the molarity of a solution must be corrected for the same reason that the formula weight is converted to an equivalent weight. When the molarity has been adjusted for the reaction of more than one ion per mole, it is called the normality, N. The adjustment is done mathematically using the variable n, which depends on the chemical reaction of the substance. For example, sulfuric acid usually loses 2 moles of H^+ per mole of sulfuric acid in acid-base reactions. Thus a $1M$ solution of sulfuric acid is actually $2M$ in H^+; its normality is therefore 2. To calculate normality the following relation is used:

$$N = (n)M \tag{6-3}$$

The value of n depends on the reaction, just as for the equivalent weight.

The normality of a solution can also be calculated directly from equivalents and liters or milliequivalents and milliliters.

$$N = \frac{\text{eq}}{\text{liter}} = \frac{\text{meq}}{\text{ml}} \tag{6-4}$$

If the normality of a solution is known and its volume is known, then the equivalents or milliequivalents of solute in the solution can be calculated from the following:

$$\text{eq} = (N)\,(\text{liter}) \qquad \text{or} \qquad \text{meq} = (N)\,(\text{ml})$$

Calculations with Molarity To calculate the results of a titration using molarity, you must know the volume and the molarity of the titrant as accurately as is necessary for the analysis. (Usually three or four significant figures are needed.) You must also know or obtain the balanced equation for the reaction occurring during the titration. If the reacting ratio of the titrant and the sample is one-to-one, then the calculation is straightforward. If not, a reaction ratio must be calculated.

Calculating the Titrant Concentration The first calculation necessary in most analyses is figuring the concentration of the titrant after standardization. Since the standardization can be done using either standard solution or a weighed amount of a primary standard, both types of calculations should be mastered. Let **6-1** represent any standardization reaction. If a standard solution (S) is used and a 1 : 1 reacting ratio exists, then the relation below can be used to calculate the molarity (M_T) of the titrant (T).

$$(\text{ml}_T)(M_T) = \text{mmole}_T = \text{mmole}_S = (\text{ml}_S)(M_S) \tag{6-5}$$

If the reacting ratio is not 1 : 1, then the mmoles of S will not equal the mmoles of T. The product of ml_S and M_S must be multiplied by the reacting ratio (t/s) from the balanced equation **(6-1)** to give the correct number of mmoles of T. The correct relation for calculating the molarity of T is then as follows:

$$(\text{ml}_T)(M_T) = \text{mmole}_T = (t/s)\,\text{mmole}_S = (t/s)(\text{ml}_S)(M_S) \tag{6-6}$$

EXAMPLE: Calculate the molarity of HCl titrant standardized against a standard sodium hydroxide solution and against a standard barium hydroxide solution. The reactions are as follows:

$$HCl + NaOH \longrightarrow H_2O + NaCl$$

$$2\,HCl + Ba(OH)_2 \longrightarrow 2\,H_2O + BaCl_2$$

Titration of 25.00 ml of $0.2000M$ NaOH requires 24.00 ml of HCl. Titration of 25.00 ml of $0.1000M$ Ba(OH)$_2$ requires 24.01 ml of HCl.

Solution: To calculate the molarity from the titration of NaOH, rearrange **6-5**:

$$M_T = M_{HCl} = \frac{(25.00 \text{ ml NaOH})(0.2000M \text{ NaOH})}{24.00 \text{ ml HCl}} = 0.2083_3 M$$

To calculate the molarity from the titration of $Ba(OH)_2$, rearrange **6-6** using $(t/s) = 2$:

$$M_T = M_{HCl} = \frac{(2)(25.00 \text{ ml } Ba(OH)_2)(0.1000M \text{ } Ba(OH)_2)}{24.01 \text{ ml HCl}} = 0.2082_4 M$$

Note that the calculated molarities agree closely even though the reactions are different.

If the standardization is done using a weighed amount of a primary standard, then the weight in mg and the formula weight in mg/mmole must be substituted for the volume and molarity of the standard. Since the weight in mg divided by the formula weight in mg/mmole gives mmoles of the standard, it follows from **6-6** that

$$(ml_T)(M_T) = mmole_T = \frac{(t/s) \text{ mg}_S}{(mg_S/mmole_S)} \tag{6-7}$$

EXAMPLE: Calculate the molarity of HCl titrant standardized against primary standard sodium carbonate powder. The reaction is as follows:

$$2HCl + Na_2CO_3 \longrightarrow H_2O + CO_2 + 2NaCl$$

Titration of 212.0 mg of sodium carbonate requires 22.00 ml of hydrochloric acid.
Solution: To calculate the molarity from the titration of sodium carbonate, whose formula weight is 106.0 mg/mmole, rearrange **6-7**:

$$M_T = M_{HCl} = \frac{(2)212.0 \text{ mg } Na_2CO_3}{(106.0 \text{ mg/mmole})(22.00 \text{ ml HCl})} = 0.1818M$$

Note that the reacting ratio is 2, which corrects for the fact that two mmoles (or moles) of hydrochloric acid titrant react with only one mmole (or mole) of sodium carbonate.

In oxidation-reduction standardizations the primary standard may be dissolved and then react in a different form from that in which it is weighed. To obtain the correct reacting ratio in such cases, it is recommended that the formula of the primary standard as it is weighed be written in the equation. As an example consider the standardization of iodine titrant against primary standard arsenic(III) oxide, As_2O_3. After it is dissolved, this oxide is converted to two moles of arsenious acid, H_3AsO_3, and it reacts with iodine as follows:

$$I_2 + H_3AsO_3 + H_2O \longrightarrow 2I^- + H_3AsO_4 + 2H^+$$

To obtain the correct reacting ratio, it is recommended that the reaction between arsenic(III) oxide and iodine be written as though it is the reaction occurring in solution.

$$2I_2 + As_2O_3 + 5H_2O \longrightarrow 2H_3AsO_4 + 4I^- + 4H^+ \tag{6-8}$$

EXAMPLE: Calculate the molarity of iodine titrant standardized against primary standard arsenic(III) oxide, according to **6-8**. Titration of 395.6 mg of arsenic(III) oxide requires 22.00 ml of iodine titrant.
Solution: To calculate the molarity of iodine from the titration of arsenic(III) oxide, whose formula weight is 197.8, rearrange **6-7**:

$$M_T = M_{I_2} = \frac{(2)395.6 \text{ mg } As_2O_3}{(197.8 \text{ mg/mmole})(22.00 \text{ ml } I_2)} = 0.1818M$$

Calculating
Percentage
Composition

After the concentration of the titrant has been calculated, the percentage composition of the sample may then be calculated. Equation **6-1** can be used to represent any reaction between a titrant and the substance sought, S. At the end point in the titration

$$(t/s)\text{mmole}_S = \text{mmole}_T = (\text{ml}_T)(M_T) \qquad (6\text{-}9)$$

The reacting ratio of t/s corrects for the difference in the number of mmoles.

The true weight of the substance sought is equal to $(\%S)(\text{mg}_{IS})/100$, the product of the percentage purity and mg_{IS}, the weight of the impure sample. Dividing the product of these by the formula weight in mg/mmole will then give mmoles of S. This quotient can then be substituted into **6-9** to give

$$\frac{(t/s)(\%S)(\text{mg}_{IS})/100}{\text{mg}_S/\text{mmole}_S} = \text{mmole}_T = (\text{ml}_T)(M_T) \qquad (6\text{-}10)$$

This is easily rearranged to calculate the percentage of the substance sought in the sample.

$$\%S = \frac{(\text{ml}_T)(M_T)(\text{mg}_S/\text{mmole}_S)\,100}{(t/s)\text{mg}_{IS}} \qquad (6\text{-}11)$$

Note that because of the rearrangement to obtain **6-11**, the reacting ratio appears in the denominator. If it seems more logical to you, you may use the equivalent form

$$\%S = \frac{(\text{ml}_T)(M_T)(s/t)(\text{mg}_S/\text{mmole}_S)\,100}{\text{mg}_{IS}} \qquad (6\text{-}12)$$

In either case, check your calculations by writing out **6-9** and **6-10** and satisfying yourself that the units are correct. The following examples will help.

EXAMPLE: The citric acid ($H_3C_6H_5O_7$) content of the pharmaceutical caffeine citrate, $H_3C_6H_5O_7(C_8H_{10}O_2N_4)$, can be determined by titration with standard NaOH:

$$3\,NaOH + H_3C_6H_5O_7(C_8H_{10}O_2N_4) \longrightarrow$$
$$C_6H_5O_7^{-3} + 3\,H_2O + 3\,Na^+ + C_8H_{10}O_2N_4$$

A 1.0000-g sample of N.F.-grade caffeine citrate requires 39.10 ml of 0.2000M NaOH for neutralization. Does the percentage of citric acid meet the N.F. requirement of 48–52%?
Solution: The formula weight of citric acid is 192.1 mg/mmole. To calculate the percentage purity, use either **6-11** or **6-12**.

$$\%H_3C_6H_5O_7 = \frac{(39.10 \text{ ml NaOH})(0.2000M \text{ NaOH})(192.1 \text{ mg/mmole})\,100}{(3)\,1000.0 \text{ mg}} = 50.05\%$$

The citric acid content of 50.05% does meet the N.F. requirement.

EXAMPLE: Calculate the percentage of arsenic (As) in an impure sample titrated with a standard iodine solution as follows:

$$I_2 + H_3AsO_3 + H_2O \longrightarrow 2I^- + H_3AsO_4 + 2H^+$$

Titration of 1.0000 g of impure sample requires 31.00 ml of 0.1000M iodine (I_2).
Solution: The formula weight of arsenic is the atomic weight of 74.92 mg/mmole. To calculate the percentage purity, note that the reacting ratio is 1 : 1 and need not be shown. Use either **6-11** or **6-12**.

$$\%As = \frac{(31.00 \text{ ml } I_2)(0.1000M \text{ } I_2)(74.92 \text{ mg/mmole})\,100}{(1)1000 \text{ mg}} = 23.22_5\%$$

Calculations To calculate the results of a titration using normality, you must know the volume and
with the normality of the titrant as accurately as is necessary for the analysis. (Usually
Normality three or four significant figures are needed.) You must decide what value n has and
use it in **6-2** to calculate the equivalent weight for the standardization of the titrant and
the analysis. Finally, the weight of the impure sample is used to calculate percentage,
or the weight of the pure sample is used to calculate the equivalent weight.

Calculating The first calculation necessary in most titrimetric analyses is the calculation of the nor-
the Titrant mality of the titrant after standardization. Let us use **6-1** to represent any standardiza-
Concentration tion reaction and assume that the primary standard (S) is in solution. At the end point
of the titration, by definition

$$\text{meq}_S = \text{meq}_T \quad \text{or} \quad \text{eq}_S = \text{eq}_T \tag{6-13}$$

If the primary standard is added in the form of a standard solution, then to calculate
normality **6-4** is rearranged to give meq $= (N)(\text{ml})$. This can be substituted into **6-13**
to give

$$(N_T)(\text{ml}_T) = (N_S)(\text{ml}_S) \tag{6-14}$$

Equation **6-14** is easily rearranged to calculate N_T, the normality of the titrant, from
N_S, the normality of the standard solution.

   If the primary standard is added in the form of a weighed amount, then you should
recognize that dividing the weight by the equivalent weight in mg/meq gives

$$\text{meq} = \frac{\text{mg}}{\text{mg/meq}} \tag{6-15}$$

Equations **6-14** and **6-15** can then be substituted into **6-13** to give

$$(\text{ml}_T)(N_T) = \frac{\text{mg}_S}{\text{mg}_S/\text{meq}_S} \tag{6-16}$$

Equation **6-16** is easily rearranged to calculate N_T; however, the equivalent weight of
the primary standard must first be calculated using **6-2**.

   *The calculation of the equivalent weight depends on the type of reaction and the value
of n in the reaction.* Consider the following example of an acid-base reaction:

$$\text{NaH}_3\text{P}_2\text{O}_7 + 2\,\text{NaOH} \longrightarrow \text{Na}_3\text{HP}_2\text{O}_7 + 2\,\text{H}_2\text{O} \tag{6-17}$$

The equivalent weight of monosodium pyrophosphate is equal to the formula weight
divided by two in the above reaction, because the number of moles of H^+ lost in the
reaction per mole of acid is two. If a reaction is not given, then you can usually assume
that all of the hydrogen ions are neutralized by base. For example, the equivalent
weight of sulfuric acid is generally 98.08/2 in acid-base reactions, because both hydrogen
ions are usually neutralized by common bases.

   Note that for both monosodium pyrophosphate and sulfuric acid the equivalent
weight is *smaller* than the formula weight, thus correcting for the fact that both lose
more than one proton per mole to base. It is obvious from **6-15** that the number of
meq of either acid would be twice the number of mmoles. (The number of mmoles of
H^+ neutralized by base is of course the same whether molarity or normality is used.)

   For other types of reactions n is calculated differently. For example, in oxidation-
reduction reactions n is the number of electrons gained or lost per mole of reactant.

This value can usually be obtained from a *half reaction* involving only the reactant in question. Consider the following half reaction involving oxidation of arsenious acid to arsenic acid:

$$H_3AsO_3 + H_2O \rightleftharpoons H_3AsO_4 + 2H^+ + 2e^- \qquad \text{(6-18)}$$

The value of n in the above reaction is two, so the equivalent weight of arsenious acid is 125.94/2.

Arsenic(III) oxide, a primary standard, can be weighed, dissolved to form arsenious acid, and oxidized to arsenic acid. Its equivalent weight can be obtained by writing the oxidation of arsenic(III) oxide directly to arsenic acid, as though it had occurred.

$$As_2O_3(s) + 5H_2O \rightleftharpoons 2H_3AsO_4 + 4H^+ + 4e^-$$

The value of n in the above reaction is four, so the equivalent weight of arsenic(III) oxide is 197.84/4.

EXAMPLE: Calculate the normality of an iodine titrant standardized by both of the following procedures: **(a)** titration of 25.00 ml of 0.1454N arsenious acid requires 10.00 ml of iodine titrant, and **(b)** titration of 395.6 mg of primary standard arsenic(III) oxide requires 22.00 ml of the same iodine titrant.

Solution to a: To calculate the normality of the iodine titrant, rearrange **6-14** and substitute the given values.

$$N_{I_2} = \frac{(25.00 \text{ ml } H_3AsO_3)(0.1454N \text{ } H_3AsO_3)}{10.00 \text{ ml } I_2} = 0.3635N$$

Solution to b: The equivalent weight of arsenic(III) oxide is 197.84/4 mg/meq. To calculate the normality of the iodine titrant, rearrange **6-16** and substitute the given values.

$$N_{I_2} = \frac{395.6 \text{ mg } As_2O_3}{(197.84/4 \text{ mg/meq})(22.00 \text{ ml } I_2)} = 0.3636N$$

Note that the standardizations give results that agree closely.

Calculating Percentage Composition After the concentration of the titrant has been calculated, the percentage composition of a sample may be calculated. Since **6-13** applies to the titration of any impure sample, as well as of pure samples, we can substitute the product $(ml_T)(N_T)$ for the meq of the titrant.

$$meq_S = (ml_T)(N_T) \qquad \text{(6-19)}$$

The true weight of the substance sought, mg_S, is equal to the product of the percentage purity, %S, and the weight of the impure sample, mg_{IS}:

$$mg_S = \frac{(\%S)(mg_{IS})}{100} \qquad \text{(6-20)}$$

The meq of S is equal to the quotient of the mg of S (right-hand side of **6-20**) and its equivalent weight. Substituting this quotient for the left-hand side of **6-19** gives

$$\frac{(\%S)(mg_{IS})/100}{mg_S/meq_S} = (ml_T)(N_T) \qquad \text{(6-21)}$$

Equation **6-21** is easily rearranged for the calculation of the percentage of S.

$$\%S = \frac{(ml_T)(N_T)(mg_S/meq_S)100}{mg_{IS}} \tag{6-22}$$

Equation **6-21** can also be rearranged to calculate any of the other unknown factors. For example, to calculate the equivalent weight of a pure unknown compound ($mg_{IS} = mg_S$), it rearranges to

$$mg_S/meq_S = \frac{mg_S}{(ml_T)(N_T)} \tag{6-23}$$

EXAMPLE: The purity of pharmaceutical-grade citric acid, $H_3C_6H_5O_7$, may be found by titration with standard sodium hydroxide.

$$3\,NaOH + H_3C_6H_5O_7 \longrightarrow C_6H_5O_7^{-3} + 3\,H_2O + 3\,Na^+$$

A 0.3000-g sample of U.S.P.-grade citric acid requires 46.60 ml of 0.1000N sodium hydroxide for neutralization. Does it meet the U.S.P. requirement of 99.5% purity?
Solution: The formula weight of citric acid is 192.1; its equivalent weight is 192.1/3, since three protons per molecule are neutralized. To calculate the percentage purity, use **6-22**.

$$\%H_3C_6H_5O_7 = \frac{(46.60\ ml\ NaOH)(0.1000N\ NaOH)(64.03\ mg/meq)\ 100}{300.0\ mg} = 99.46\%$$

The purity is just short of the requirement.

EXAMPLE: Calculate the percentage of arsenic (As) in an impure sample of arsenious acid titrated with iodine.

$$I_2 + H_3AsO_3 + H_2O \longrightarrow 2I^- + H_3AsO_4 + 2H^+$$

Titration of 1.0000 g of impure sample requires 31.00 ml of 0.0500N iodine.
Solution: The equivalent weight of arsenic is 74.92/2 mg/meq, since arsenious acid loses two electrons per molecule **(6-18)**. Use **6-22** to calculate the percentage purity.

$$\%As = \frac{(31.00\ ml\ I_2)(0.0500N\ I_2)(37.46\ mg/meq)\ 100}{1000.0\ mg} = 5.80_6\%$$

Clinical Calculations

Many clinical chemists now report results as milligrams per deciliter of liquid sample (mg/dL); this is used instead of the older term of milligram per cent (mg%), which, on a weight per volume basis, is equal to mg/100 ml of liquid sample. Also, electrolytes in blood are reported as meq/liter by clinical chemists.

To calculate results in mg/dL, the titrant concentration can be given in molarity, since most titration reactions involve a 1 : 1 reaction. An equation can be derived to calculate mg/dL by using the fact that the mg of a substance S may be calculated from the product $(ml_T)(M_T)(form\ wt_S)$. We begin by writing an equation relating the mg of substance S per ml of liquid sample to this product.

$$\frac{mg_S}{ml\ sample} = \frac{(ml_T)(M_T)(mg_S/mmole_S)}{ml\ sample}$$

To convert mg_S/ml to mg_S/dL (or $mg_S/100\,ml$), we simply multiply the right-hand side by a factor of 100:

$$\frac{mg_S}{\text{in } 100 \text{ ml sample}} = \frac{mg_S}{dL} = \frac{100(ml_T)(M_T)(mg_S/mmole_S)}{ml \text{ sample}} \qquad (6\text{-}24)$$

If the volume of the sample is given in μL, then the number of μL must be converted to ml by dividing by 1000.

> **EXAMPLE:** Calcium(II) ion can be titrated with ethylenediaminetetraacetic acid (EDTA) in a 1 : 1 reaction.
>
> $$Ca^{+2} + EDTA \longrightarrow Ca\text{-}EDTA$$
>
> Calculate the mg Ca^{+2}/dL of blood in a 100 μL sample of blood serum titrated with 2.00 ml of $1.00 \times 10^{-4}M$ EDTA. State whether the amount of Ca^{+2} is in the normal serum range, below it, or above it (Fig. 1-1).
> *Solution:* First, convert the volume of the sample to ml.
>
> $$\frac{100\ \mu L}{1000\ \mu L/ml} = 0.100 \text{ ml}$$
>
> Next, use **6-24** to calculate the mg/dL.
>
> $$\frac{mg\ Ca^{+2}}{dL} = \frac{100(2.00\ ml)(1.00 \times 10^{-4}M\ EDTA)(40.1\ mg\ Ca^{+2}/mmole\ Ca^{+2})}{0.100\ ml}$$
>
> $$= 8.02 \text{ mg/dL}$$
>
> The calculated value of 8.02 mg/dL is within the normal range of 7.9–10.5 mg Ca^{+2}/dL of blood serum (Fig. 1-1).

For calculating the results of an analysis in meq/liter, the concentration of the titrant is best expressed in molarity so that the mmoles of the substance can be calculated.

$$(ml_T)(M_T) = mmole_S \qquad (6\text{-}25)$$

For the purpose of calculating an electrolyte balance in blood, the number of meq of a substance is calculated from the number of mmoles by

$$meq_S = (\text{charge on ion})(mmole_S) \qquad (6\text{-}26)$$

By using this concept a check can be kept on the total number of meq of -1 charges and the total number of meq of $+1$ charges. (Obviously, the number of meq of $+1$ charges must equal that of -1 charges, but the number of mmoles of differently charged cations will not equal the number of mmoles of differently charged anions.) Once the number of meq of a substance is calculated, it is divided by the volume in liters to obtain meq/liter.

> **EXAMPLE:** Calculate the concentration of calcium(II) ion in meq/liter from the results of the analysis in the last example.
> *Solution:* First, calculate the mmole of Ca^{+2} using **6-25**.
>
> $$(2.00\ ml\ EDTA)(1.00 \times 10^{-4}M\ EDTA) = 2.00 \times 10^{-4} \text{ mmole } Ca^{+2}$$
>
> Next, use **6-26** to calculate the meq of Ca^{+2}.
>
> $$meq\ Ca^{+2} = (2)(2.00 \times 10^{-4}\ mmole) = 4.00 \times 10^{-4} \text{ meq}$$

Finally, calculate meq/liter.

$$\frac{\text{meq Ca}^{+2}}{\text{liter}} = \frac{4.00 \times 10^{-4} \text{ meq}}{1.00 \times 10^{-4} \text{ liter}} = 4.00 \text{ meq/liter}$$

6-3 | USING VOLUMETRIC GLASSWARE IN THE LABORATORY

Thus far in this chapter you have mastered some basic concepts of volumetric analysis and learned how to handle several types of calculations. Next you should become familiar with volumetric glassware and the techniques necessary to use it for a successful analysis.

The three basic pieces of volumetric glassware are the volumetric flask, the volumetric pipet, and the buret. Each of these items is manufactured to contain (TC) or to deliver (TD) a certain volume. Volumetric flasks are made to contain the volume written on them. Burets and volumetric pipets are made to deliver the stated volume by draining. Clinical (serological) pipets must be blown out to deliver the stated volume.

Because glassware is mass-produced, its volume cannot be assumed to be completely accurate. You must therefore understand something about the accuracy and precision of glassware to use it intelligently.

Accuracy and Precision There is a difference between the accuracy and the precision of glassware. The *accuracy* of a given piece of volumetric glassware may be expressed in terms of the maximum allowable error, or *tolerance*. The *precision* is expressed in terms of the uncertainty of a buret reading or the uncertainty of filling a flask or a pipet to the mark. By definition there is no precision of a single reading or a single filling, but it is proper to speak of the uncertainty (see Ch. 3).

Table 6-1 gives some of the National Bureau of Standards tolerances for volumetric glassware. The less expensive equipment found in instructional analytical laboratories does not meet these tolerances and may actually have errors twice as large as those in the table.

TABLE 6-1. Tolerances for Volumetric Glassware

Capacity, ml	Maximum error allowable, ml		
	Volumetric flasks	Volumetric pipets	Burets
5	—	0.01	0.01
10	—	0.02	0.02
25	0.03	0.03	0.03
50	0.05	0.05	0.05
100	0.08	0.08	0.10
500	0.15	—	—
1000	0.30	—	—

The Kimball brand Kimax, Class A, and the Corning brand Pyrex glassware conform to these specifications (National Bureau of Standards).

These tolerances are *absolute* values of the maximum allowable error. For example, the tolerance of 0.05 ml for a 50-ml buret means that the absolute error in the volume delivered can be as large as 0.05 ml. If a volume of 40 ml is used from the buret, the *relative* value of the maximum allowable error in parts per hundred (pph) will be

$$\frac{0.05 \text{ ml}}{40.00 \text{ ml}} (100) = 0.1_{25} \text{ pph (50-ml buret only)}$$

Since it is the most easily characterized, only the precision of a buret will be discussed. The uncertainty of a single 50-ml buret reading is somewhat subjective, but the *absolute* value of the uncertainty is thought to be ± 0.02 ml. Since a buret is always read twice (at the beginning and at the end of a titration), the *absolute* value of the total uncertainty may be as much as ± 0.04 ml. The absolute uncertainty is the same for any volume delivered from a 50-ml buret, but the *relative* uncertainty in parts per hundred or parts per thousand differs. The calculation of relative uncertainty for 10.00-ml and 40.00-ml volumes delivered from a 50-ml buret is as follows:

$$\frac{0.04 \text{ ml}}{10.00 \text{ ml}} (100) = 0.4_0 \text{ pph}$$

$$\frac{0.04 \text{ ml}}{40.00 \text{ ml}} (100) = 0.1_0 \text{ pph}$$

For the 10-ml volume the relative uncertainty is greater than the desirable relative precision of 0.1 to 0.2 pph for a volumetric method. It is obvious that a titration with a 50-ml buret should involve 35–40 ml of titrant for good precision and accuracy. A 10-ml buret is more proper for measuring volumes of 10 ml or less, because the 0.02-ml subdivisions on the buret permit a lower relative uncertainty in the reading.

Volumetric Flasks

Almost any size flask can be obtained for laboratory work; all are made to contain (TC) the exact volume of liquid when the bottom of the meniscus just touches the etched line across the neck. Most volumetric flasks employ ground-glass stoppers, but some are equipped with screw caps lined with polyethylene. If necessary, volumetric flasks may be cleaned by scrubbing with a dilute detergent solution.

Volumetric flasks are used to prepare standard solutions by exact dilution after a primary standard has been weighed out and dissolved. Another important use of the volumetric flask is for diluting samples to an exact volume prior to taking an aliquot for analysis.

A sample that requires heat for dissolving should be dissolved in a beaker and the resulting solution transferred quantitatively to a volumetric flask after cooling. (A volumetric flask should never be heated, since heating and cooling may change its volume.) The flask is filled to a point slightly below the etched line with distilled water (or other solvent) without contacting the ground-glass portion of the flask. A minute is allowed for draining from the upper portion of the neck, and then the flask is carefully filled to the mark by means of a long-barrelled medicine dropper or a pipet. The flask finally is stoppered, inverted, and agitated to mix thoroughly.

Volumetric flasks are not used for the storage of solutions. Solutions, especially alkaline ones, should be transferred immediately to plastic bottles after being made up in a volumetric flask.

Pipets and Their Use

The two types of pipets are the measuring pipet and the volumetric, or transfer, pipet. Both are shown in Figure 6-1. The measuring pipet is calibrated, but it does not deliver a given volume of liquid as accurately or as reproducibly as the buret or volumetric pipet. The volumetric pipet, not the measuring pipet, should be used when an aliquot must be taken from a standard solution (Table 6-1).

Cleaning the Pipet

If distilled water does not drain uniformly from the pipet but leaves water breaks or droplets of water adhering to the sides, the pipet must be cleaned. The following cleaning procedures are recommended in the order listed:

1. Fill with a hot dilute (2%) detergent solution and rotate the pipet to cover the inside thoroughly. Drain and rinse with distilled water.
2. Fill with a hot dilute (0.004M) alkaline (pH 12) solution of EDTA and soak for no more than 15 minutes. Drain, rinse with dilute acid, and rinse with distilled water.
3. Fill the pipet halfway *cautiously* with a hot (60°C) dichromate-sulfuric acid cleaning solution, using a rubber suction bulb, and rotate the pipet to wet the inside thoroughly. Return the cleaning solution to the storage bottle and rinse the pipet thoroughly with distilled water.

Use

In using the volumetric pipet, you should observe all of the following points.

1. *Use of rubber suction bulb to fill pipet.* Never fill a pipet using mouth suction; always use a rubber bulb to supply suction. This is especially important for concentrated acids, solutions of arsenic, and ammonia solutions.

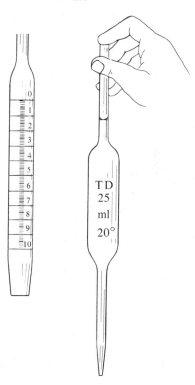

FIGURE 6-1. Measuring pipet (left) and volumetric pipet (right) showing meniscus.

2. *Rinsing.* Rinse the pipet with distilled water before using. Next rinse the pipet with the solution to be pipetted to avoid dilution by the water adhering to the inside of the pipet. Pour a small amount of the solution to be pipetted into a beaker and use this only to rinse the pipet. Never insert an unrinsed pipet into the container of solution. (If an alkaline solution is to be pipetted, it is preferable to take aliquots also from a beaker of the solution.) Rinse the pipet, not by filling completely, but by drawing in about one-fifth of the pipet's volume and twirling the pipet horizontally two or three times. Rinse above the mark by tipping the pipet slightly. Rinse at least twice in this manner with the solution to be pipetted. The pipet should then drain uniformly; if not, it needs cleaning or further rinsing.

3. *Filling.* Fill the pipet to about an inch above the etched line (place the fleshy part of a forefinger over the top of the pipet to stop the overflow). Then place the tip of the pipet against the inside of the vessel and rotate the pipet, allowing the solution to drain until the bottom of the meniscus just touches the etched line at eye level, as in Figure 6-1. There should be no air bubbles anywhere in the pipet.

4. *Carrying.* The pipet may be conveniently carried by tilting it slightly so that the solution flows back away from the tip slightly toward the other end.

5. *Draining.* Wipe the outside of the pipet tip with a tissue to remove any liquid before draining. Place the tip against the inside of the vessel into which the solution is to be transferred and allow the pipet to discharge. Keep the tip against the inside for 20 seconds after the pipet has emptied for complete drainage. Remove the pipet from the side of the container with a rotating motion to completely remove any of the drop on the tip. The small quantity of liquid remaining inside the tip is not to be blown out even though it appears to grow larger after a time. (The pipet has been calibrated to deliver (TD) a certain accurate volume.)

Storage The volumetric pipet should not remain unrinsed after use, especially after the transfer of alkaline solutions. It is good practice to fill the pipet with distilled water and to cap both ends with rubber bulbs from droppers. If this is not possible, the pipet should be thoroughly rinsed and stored in a rack or drawer where it will not easily be scratched or chipped.

Burets and Their Use Burets are made and calibrated to deliver variable volumes of liquid. The essential parts are shown in Figure 6-2. Most modern burets are equipped with a glass stopcock lubricated with hydrocarbon greases or with a Teflon plastic stopcock requiring no lubrication. Teflon stopcocks can be used for nonaqueous solvents and will not freeze even after long contact with basic solutions.

Care of the Buret Observe the following points.

1. *Lubrication of buret stopcock.* Remove all the old grease from the stopcock and stopcock hole (use a fine wire). Apply a thin uniform layer of grease (do not use silicone lubricant); use even less grease near the holes in the stopcock. Insert the stopcock and rotate several times. If too much grease has been applied, some of it will be forced from between the stopcock and barrel or may eventually work itself into the buret tip. If too little grease has been applied, the lubricant layer will not appear uniform and transparent.

2. *Cleaning the buret.* If distilled water does not drain uniformly from the buret but

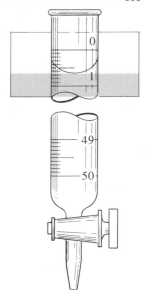

FIGURE 6-2. Essential parts of a buret, showing
the correct position of the meniscus illuminator
(dark portion just below meniscus).

leaves drops of water clinging to the sides, the buret must be cleaned. The following procedures are recommended in the order listed:

a. Use a hot, dilute (ca. 2%) detergent solution and scrub with a long-handled brush. Concentrated detergent may affect the glass and is difficult to rinse out completely. Rinse the detergent out with plenty of distilled water.

b. Soak for 10–15 minutes in a hot, dilute (0.004M, pH 12) solution of EDTA to remove metal ions. Rinse with dilute acid and then with distilled water.

c. Use a dichromate-sulfuric acid cleaning solution. Since a cold cleaning solution works slowly, the buret may require overnight soaking. Cleaning solution may disperse more stopcock lubricant than it removes and is properly drawn into the buret inverted, by suction; hence, it is inconvenient to use in a student laboratory. Cleaning solution should not be thrown away but returned to the storage bottle for reuse.

Use The lab instructor may check your buret technique during the standardization of sodium hydroxide against potassium acid phthalate or in the titration of acetic acid or potassium acid phthalate unknowns. In any titration the following points are important.

1. *Filling.* The buret must first be rinsed with titrant to remove water adhering to the inside of the buret. Do not rinse by filling the buret and draining, but by pouring about 10 ml of the titrant around the inside three times. Rotate the buret to wet the inside thoroughly; leave the stopcock open while rinsing and allow the buret to drain completely between rinses. Close the stopcock and fill the buret to at least an inch above the zero mark.

2. *Cleanliness.* During the rinsing, check whether the titrant drains uniformly from the buret. If it does, it is clean and may be used. If drops of titrant form on the inside after rinsing, the buret must be cleaned.

3. *Bubbles.* Check for bubbles in the tip of the buret by rapidly draining a milliliter or so of titrant after the buret is full. Check for bubbles in the entire buret when titrating with a dark solution, such as iodine or potassium permanganate.

4. *Reading the buret.* Allow the level of titrant to drain slowly to the zero mark. Using the meniscus illuminator shown in Figure 6-2, take an initial reading by estimating to 0.01 ml. The initial reading can be exactly 0.00 ml or a larger value. Record the initial reading. Bring the meniscus illuminator up so that the dark half is just below the meniscus; be sure your eye is at the same level as the meniscus to avoid parallax error. (Perception of the meniscus from an angle above or below causes the parallax error.)

5. *Titrating.* Fold a white index card or piece of paper and place under and behind the flask for a white background. Position the tip of the buret within the neck of the flask. Swirl the flask with the right hand and manipulate the stopcock with the left hand from behind the buret. This maintains a slight pressure on the stopcock and prevents leakage. (For more efficient stirring, use a magnetic stirrer and stirring bar.) Add titrant rapidly at first. As the color of the indicator changes more slowly, signaling the approach of the end point, add the titrant by drops.

6. *The end point.* Just before the end point, rinse down the sides of the flask with distilled water from a wash bottle. Split drops of titrant by allowing only a partial drop to form on the tip and washing it into the flask with distilled water. Since the buret can be read to 0.01 ml and since a drop is 0.05 ml, splitting drops is essential for accuracy. Allow a minute for drainage before the final reading.

7. *The "squirt" technique.* An alternative to dropwise titration and splitting drops just before the end point is the "squirt" technique. Hold the barrel of the stopcock steady with the left hand and quickly twist the stopcock 180° with the right hand. This will deliver a squirt of titrant of 0.01–0.05 ml.

8. *Vague end points.* If the color change at the end point is uncertain, a useful general method is to record the volume of titrant added for successive additions (0.01–0.05 ml), noting the color change with each addition. Usually, the point of maximum color change will be obvious after one or two further additions of titrant.

Storage   Burets filled with titrant should not be left standing, especially if they are filled with sodium hydroxide. To avoid freezing the stopcock, discard the titrant (do *not* return it to the original container) and rinse the buret several times with distilled water. It is good practice to fill the buret with distilled water after rinsing and cap it to keep dust out. If this is not possible, the buret should be stored upside down. Burets used for nonaqueous work may be rinsed with acetone and stored dry and upside down.

| QUESTIONS AND PROBLEMS

(Answers to even-numbered problems are found in Appendix 5.)

Definitions and Concepts

1. Define the following terms:
   a. Standard solution
   b. Primary standard
   c. Equivalence point and end point
   d. Molarity
   e. Normality

2. List at least four requirements for a primary standard material. Is it necessary that a primary standard be 100% pure?

3. Tell whether the following reagents are primary standard materials:
   a. Sodium hydroxide
   b. Hydrochloric acid

c. Potassium acid phthalate

d. Sulfuric acid

4. Explain the difference between a pipet that is stamped *TC* and one that is stamped *TD*.

5. What is the tolerance of a pipet or buret? Does it have anything to do with its accuracy? with its precision?

6. Describe the important steps that should be taken to locate an end point accurately with a buret, and then to read the buret properly at the end point.

Precipitation Titration Calculations

7. A 0.3000-g sample of impure magnesium chloride (form wt = 95.23) is titrated with 45.00 ml of 0.1000*M* silver nitrate according to the following equation:

$$MgCl_2 + 2AgNO_3 \longrightarrow$$
$$2AgCl(s) + Mg(NO_3)_2$$

a. Calculate the percentage chloride (at wt = 35.45) in the sample.

b. Calculate the percentage magnesium chloride in the sample.

8. A 8.5332-g portion of primary standard silver nitrate (form wt = 169.9) is weighed into a 500.0-ml volumetric flask. Calculate its molarity for titration of:

a. Sodium chloride (form wt = 58.44).

b. Calcium chloride (form wt = 110.9).

9. A 10.00-ml sample of sodium chloride solution is diluted to 50.00 ml. A 20.00-ml aliquot is then withdrawn and titrated with 3.923 ml of 0.0110*M* silver nitrate.

a. Calculate the molarity of the original sodium chloride solution.

b. Calculate the mg NaCl/dL in the original solution (form wt NaCl = 58.44).

10. A 25.00-ml sample of calcium chloride solution is diluted to 50.00 ml. A 20.00-ml aliquot is then withdrawn and titrated with 3.923 ml of 0.0110*M* silver nitrate.

a. Calculate the molarity of the original calcium chloride solution.

b. Calculate the mg $CaCl_2$/dL in the original solution (form wt $CaCl_2$ = 110.9).

11. A 1.000-g sample of impure aluminum chloride (form wt = 133.34) requires 44.00 ml of 0.1000*M* silver nitrate for complete precipitation of all the chloride. Calculate the percentage of aluminum chloride in the sample.

Acid-Base Calculations

12. Exactly 23.16 ml of sodium hydroxide is used to titrate a 600.0-mg sample of primary standard potassium acid phthalate (KHP); its formula weight is 204.2. Calculate the molarity (or normality) of the sodium hydroxide.

13. Exactly 27.08 ml of hydrochloric acid is required to titrate 25.00 ml of 0.1079*M* hydroxide. Calculate the molarity (or normality) of the hydrochloric acid.

14. Exactly 24.60 ml of hydrochloric acid is needed to neutralize a 278.0-mg sample of primary standard tris(hydroxymethyl)aminomethane (form wt = 121.1) according to the following reaction:

$$HCl + (CH_2OH)_3CNH_2 \longrightarrow$$
$$(CH_2OH)_3CNH_3{}^+Cl^-$$

Calculate the molarity (or normality) of the hydrochloric acid.

15. A 500.0-mg sample contains either impure H_3PO_4 or impure NaH_2PO_4. It is titrated with 21.00 ml of 0.1000*M* sodium hydroxide to the phenolphthalein end point to give entirely the HPO_4^{-2} ion.

a. Assuming the sample is H_3PO_4, calculate the %H_3PO_4 (form wt = 98.00).

b. Assuming the sample is NaH_2PO_4, calculate the %NaH_2PO_4 (form wt = 119.98).

16. Calculate the purity of a 500.0-mg sample of impure sodium carbonate requiring 22.00 ml of 0.1800*M* hydrochloric acid.

Oxidation-Reduction Calculations

17. A 93.0-mg sample of primary standard arsenic(III) oxide is dissolved according to the following equation:

$$As_2O_3 \longrightarrow 2H_3AsO_3$$

The resulting arsenious acid requires 18.40 ml of cerium(IV) for oxidation according to the following equation:

$$H_3AsO_3 + 2Ce^{IV} \longrightarrow 2H_3AsO_4 + 2Ce^{III}$$

Using the formula weight of 197.84 for As_2O_3, calculate either:
a. The molarity of the Ce^{IV}, or
b. The normality of the Ce^{IV}.

18. A 395.6-mg sample of primary standard arsenic(III) oxide is dissolved in 25.00 ml of acid, forming two molecules of H_3AsO_3 to one molecule of As_2O_3 (form wt = 197.84). Calculate either:
a. The molarity of H_3AsO_3, or
b. The normality of H_3AsO_3.

19. Calculate the normality (or molarity) of an iodine titrant, 24.10 ml of which is required to react with 25.00 ml of $0.0300N$ $(0.0150M)$ arsenious acid, H_3AsO_3.

20. An impure sample of 1.000 g of arsenious acid is titrated with 45.00 ml of $0.0800N$ $(0.0400M)$ iodine. Calculate:
a. The percentage of arsenious acid (form wt = 125.9).
b. The percentage of arsenic (at wt = 74.92).

Clinical Calculations

21. Calculate the mg/dL of calcium(II) ion in a 0.200-ml sample of blood serum titrated with 1.50 ml of $0.000100M$ EDTA. Is the concentration of calcium within, below, or above the normal range (Fig. 1-1)?

22. Calculate the mg/dL of calcium(II) ion in a 50-μL serum sample that requires 200 μL of $1.000 \times 10^{-4}M$ of EDTA for titration. Is the concentration of calcium within or outside the normal range (Fig. 1-1)?

23. Calculate the meq/liter chloride in a 100-μL serum sample that requires 2.00 ml of $0.00500M$ silver nitrate. If the average serum contains 104 meq/liter, is this sample above, at, or below this level?

Challenging Problems

24. A 10.00-ml aliquot of a sulfuric acid solution requires 28.16 ml of $0.1000M$ sodium hydroxide for titration. What volume of $0.1000M$ barium chloride will be required to titrate a second 10.00-ml aliquot of sulfuric acid to produce barium sulfate?

25. A 345.0-mg sample of a pure unknown monoprotic acid is dissolved and titrated with 27.40 ml of $0.1000M$ sodium hydroxide. Calculate the formula weight of the monoprotic acid.

26. The sum of the meq/liter of chloride and bicarbonate ions in blood serum should be within 10% of the meq/liter of the sodium ion. Check this by calculating the meq/liter of each from the following data:

$$Cl^- = 383 \text{ mg}\%$$
$$HCO_3^- = 17.7 \text{ mg}\%$$
$$Na^+ = 329 \text{ mg}\%$$

7 Introduction to Titrations

"Detection is, or ought to be, an exact science . . . ," said Holmes.

ARTHUR CONAN DOYLE
The Sign of the Four, Chapter 1

In the following chapters the analytical use of a number of different types of titration reactions will be discussed. Before you attempt to master these, you should be aware of some *general* concepts that apply to all titrations. In this chapter we will discuss the various types of titration equilibria, the quantitativeness of a specific titration, and the "exact science" of detection of the end point.

7-1 | TYPES OF TITRATION EQUILIBRIA AND TITRATION CURVES

In this section we describe the four main types of titration equilibria:

1. Acid-base equilibria
2. Precipitation equilibria
3. Complexation equilibria
4. Oxidation-reduction equilibria

In our discussion it is assumed that you have encountered each type of equilibrium in a previous course and therefore this is only a brief review.

In each type of equilibrium the concentration of a species is specified by the molarity, not the activity. Recall from Section 1-2 that activity is related to molarity by the activity coefficient, f:

$$\text{activity} = f[M^{+n}]$$

where $[M^{+n}]$ is the molarity of an ion with a charge of $+n$. The activity coefficient corrects for the interionic attraction among the ions and is usually less than one, which means that the activity of an ion is usually less than the molarity. The activity is a sort of molarity that has been corrected for interionic attraction. We will neglect this correction in our discussions to focus on the more important concept of equilibrium. The errors involved vary, ranging from less than 1% in many cases to more than 10% in a few cases. Conditions in analytical methods are always adjusted to compensate for these errors.

Acid-Base Equilibria and Titration Curves

We will first review the various types of acids and bases and their equilibria in water, starting with the ionization of water. Following this, we will present the titration curve for a strong acid–strong base titration, the fastest type of titration.

Ionization of Water

Since most acid-base analyses occur in aqueous solution, the ionization of water is always of prime importance when considering acid-base equilibria. This ionization is generally written using water as the reactant and the hydroxide and hydrogen (or H_3O^+) ions as the products.

$$H_2O = H^+ + OH^-$$

The equilibrium constant for this reaction is denoted as K_w, the ion product for water. It is defined as follows:

$$K_w = [H^+][OH^-] = 1.00 \times 10^{-14} M^2 \ (25°C)$$

Since the concentration of water is constant in dilute solutions, its concentration is not written in the definition of K_w; instead, the numerical value of the concentration ($55.5 M$) is included in K_w.

Ionization of Acids

Acids are classified as strong (completely ionized) or weak (less than 100% ionized). The common strong acids are hydrochloric, nitric, and perchloric acid. The first hydrogen of sulfuric acid is also completely ionized. When these acids are added to water the complete ionization is written as:

$$HCl \longrightarrow H^+ + Cl^-$$

$$H_2SO_4 \longrightarrow H^+ + HSO_4^-$$

Thus a hydrochloric acid solution, such as that found in gastric fluid, contains no HCl molecules, only H^+ and Cl^-. It is therefore impractical to measure an equilibrium constant for a strong acid; instead it is usually said that such constants approach infinity.

In constrast to strong acids most weak acids ionize 1% or less under the usual conditions. Typical weak acids are hydrofluoric acid, acetic acid, carbonic acid, and phosphoric acid. The ionization of acetic acid may be written as:

$$CH_3CO_2H = CH_3CO_2^- + H^+$$

Note that organic acids such as acetic acid are written so that the ionizable hydrogen is last in the formula rather than first. For this and other reasons such acids are sometimes written in an abbreviated form. For example, acetic acid is abbreviated as HOAc.

Carbonic acid, an important constituent in the blood's buffer system, is a diprotic acid; its ionization is written in two steps:

$$H_2CO_3 = H^+ + HCO_3^-$$

$$HCO_3^- = H^+ + CO_3^{-2}$$

The ionization constant for weak monoprotic acids is symbolized as K_a; ionization constants for diprotic acids may be symbolized as K_1 and K_2 or K_{a_1} and K_{a_2}. The ionization constants for acetic acid and carbonic acid are defined as follows:

$$K_a = \frac{[CH_3CO_2^-][H^+]}{[CH_3CO_2H]} = \frac{[H^+][OAc^-]}{[HOAc]} = 1.8 \times 10^{-5}$$

$$K_1 = \frac{[H^+][HCO_3^-]}{[H_2CO_3]} = 4.3 \times 10^{-7}$$

$$K_2 = \frac{[H^+][CO_3^{2-}]}{[HCO_3^-]} = 4.8 \times 10^{-11}$$

Note that the numerical value of K_2 for carbonic acid is much less than that of K_1. This is generally the case for all diprotic, triprotic, and other polyprotic acids.

Ionization of Bases

Bases are also classified as strong (completely ionized) or weak (less than 100% ionized). The common strong bases are sodium hydroxide, potassium hydroxide, and barium hydroxide. When these bases are added to water the complete ionization is written as:

$$NaOH \longrightarrow Na^+ + OH^-$$

$$Ba(OH)_2 \longrightarrow Ba^{+2} + 2OH^-$$

Note that, in contrast to sulfuric acid, *both* of the hydroxides of barium hydroxide are completely ionized.

A solution of sodium hydroxide thus contains no NaOH molecules, only Na^+ and OH^-. It is therefore impractical to measure an equilibrium constant for a strong base; again, such constants are usually said to approach infinity, just as for the strong acids.

In contrast to strong bases most weak bases ionize 1% or less under the usual conditions. The most common weak base is ammonia. It is correctly represented as NH_3, but is also written as NH_4OH. Actually very few molecules of NH_4OH exist in an ammonia solution, but most chemists use either formula. The ionization of ammonia can be written with either formula to give the same products:

$$NH_3 + H_2O = NH_4^+ + OH^-$$

$$NH_4OH = NH_4^+ + OH^-$$

Other common weak bases include any of the various organic amines, such as methylamine (CH_3NH_2), aniline ($C_6H_5NH_2$), and ethylenediamine ($NH_2C_2H_4NH_2$), and any anions of weak acids, such as acetate ion (in sodium acetate), fluoride ion (in sodium fluoride), and carbonate ion (in sodium carbonate). Ionization for such bases is written as follows:

$$CH_3NH_2 + H_2O = CH_3NH_3^+ + OH^-$$

$$CO_3^{-2} + H_2O = HCO_3^- + OH^-$$

(The carbonate ion can ionize further to carbonic acid but this reaction is not significant compared to the above reaction.)

The ionization constant for weak bases is symbolized as K_b; ionization constants for dibasic compounds are symbolized as K_1 and K_2, or K_{b_1} and K_{b_2}. The ionization constant for ammonia is defined as follows:

$$K_b = \frac{[NH_4^+][OH^-]}{[ammonia]} = 1.8 \times 10^{-5}$$

where [ammonia] may be written as either $[NH_3]$ or $[NH_4OH]$.

The ionization constants for ethylenediamine are defined as follows:

$$K_1 = \frac{[NH_2C_2H_4NH_3^+][OH^-]}{[NH_2C_2H_4NH_2]} = 1.28 \times 10^{-4}$$

$$K_2 = \frac{[NH_3C_2H_4NH_3^{+2}][OH^-]}{[NH_2C_2H_4NH_3^+]} = 2.0 \times 10^{-7}$$

Note that the numerical value of K_2 for ethylenediamine is much less than that of K_1. This is generally the case for all such bases.

The K_b for anions of weak acids is found by dividing K_w by the K_a for the conjugate acid. Thus for sodium acetate,

$$K_b = \frac{K_w}{K_a} = \frac{1.00 \times 10^{-14}}{1.8 \times 10^{-5}} = 5.6 \times 10^{-10} = \frac{[HOAc][OH^-]}{[OAc^-]}$$

The K_b for sodium carbonate is found by using the K_2 of carbonic acid; K_2 is used because the bicarbonate ion is the conjugate acid of the carbonate ion.

$$K_b = \frac{K_w}{K_2} = \frac{1.00 \times 10^{-14}}{4.8 \times 10^{-11}} = 2.1 \times 10^{-4} = \frac{[HCO_3^-][OH^-]}{[CO_3^{2-}]}$$

A pH Titration Curve

A titration curve can be constructed by plotting the concentration of one of the reactants or one of the products against the volume of titrant added. Because the change in concentration is so large, the negative log of the concentration is usually plotted against the volume of titrant. This is symbolized by putting a small "p" before the formula of the species whose negative log of concentration is being plotted. Three common examples are as follows:

1. Neutralization of HCl: $pH = -\log[H^+]$
2. Precipitation of Cl^-: $pCl = -\log[Cl^-]$
3. Complexation of Ca^{+2}: $pCa = -\log[Ca^{+2}]$

By inserting a pair of electrodes that are connected to a potentiometer into the sample solution, the pH, pCl, or pCa can be measured for each volume of titrant added. These readings can then be plotted to obtain a titration curve.

In general, readings for a titration curve are taken at the start of the titration (0 ml titrant or 0% reaction), at several points during the middle portion of the titration (10%, 40%, 50%, 60%, 90%, etc.), and then at a number of points close to the equivalence point region, also called the end point region. There is usually a large change in concentration in this region; e.g., for titration of HCl, there is a large change in pH in this region. Thus readings are taken at volumes of titrant representing such percent reactions as 99.0%, 99.5%, 99.9%, 100.0%, 101%, etc. The concentration at 100.0% will be the equivalence point, or theoretical end point.

Let us consider the titration of a strong acid, such as hydrochloric acid, with a strong base titrant, such as sodium hydroxide. The titration is conducted by following the change in pH on a pH meter in electrical contact with the solution by means of a pair of electrodes immersed in the hydrochloric acid. *To simplify matters we will neglect volume changes except at the equivalence point (theoretical end point).*

Suppose that 50 ml of $0.1M$ hydrochloric acid is the sample to be titrated. This is a strong acid which means it is completely ionized. Thus before any sodium hydroxide titrant is added, the pH is

$$pH = -\log[0.1M \; H^+] = -\log[1 \times 10^{-1}M \; H^+] = 1.0$$

Note that in Figure 7-1 the pH for 0 ml of titrant is 1.0. As the sodium hydroxide titrant is added, the $[H^+]$ will drop below $0.1M$ and the pH will increase. At 90% reaction, only 10% of the hydrogen ion is left and the pH is

$$pH = -\log(10\%)(0.1M) = -\log 0.01M = 2.0$$

After this point the pH begins to increase significantly. For example, at 99.0% reaction, only 1.0% of the original hydrogen ion is left and its concentration is $0.001M$. The pH at this point is 3.0. Then the concentration of the hydrogen ion and the pH change very drastically as the reaction passes through the end point region. As shown in Figure 7-1, the theoretical end point occurs at a pH of 7.0. The pH has this value because by definition the end point indicates the point at which the equilibrium concentration of the titrant equals that of the sample; that is, $[OH^-] = [H^+]$. The only

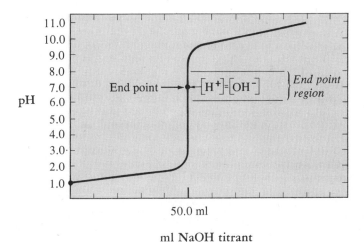

FIGURE 7-1. A titration curve showing the change in concentration of 50 ml of $0.1M$ HCl plotted as pH as the NaOH titrant is added. The end point occurs at pH = 7.0.

pH at which this can be true for mixtures of a strong acid and strong base is 7.0. This is readily calculated as follows:

$$K_w = [H^+][OH^-]$$
$$K_w = [H^+][H^+]$$
$$[H^+] = \sqrt{K_w} = 1.0 \times 10^{-7}$$

The region on either side of the end point pH of 7.0 (about three to four pH units in length) is called the end point region, because of the large pH change occurring per volume of titrant added. (About 0.1 ml of titrant is all that is required for this change to occur.)

When the end point is detected using electrodes and a pH meter, the titrant is usually added beyond the end point. In this region the concentration of hydrogen ion is so small that it does not change much. As an example, consider the change from 101% reaction with a pH of 11 to 150% reaction (75 ml of titrant). The pH is controlled at this point by the amount of sodium hydroxide added. The concentration of hydroxide present is now

$$\frac{(25 \text{ ml OH}^-)(0.1M \text{ OH}^-)}{50 + 25 \text{ total ml}} = \frac{2.5 \text{ mmole OH}^-}{75 \text{ ml}} = 0.033M \text{ OH}^-$$

The pOH of this solution is calculated as

$$pOH = -\log 0.033M = 1.5$$

The pH is therefore $14 - 1.5$, or 12.5. A change from pH 11 to pH 12.5 is, of course, a small change for such a large amount (25 ml) of titrant.

You have now seen how the course of a titration may be followed by plotting a titration curve. In particular you should note the sharp changes in concentration that occur at the end point *in favorable reactions*. Not all reactions are favorable; that is, not all reactions are necessarily quantitative and give a favorable end point. Later we consider how to predict whether or not a reaction is quantitative. If it is, then we can expect to find a favorable end point.

Precipitation Equilibria

In this section, we review the solubility product constant, then discuss some of the implications, and finally calculate a precipitation titration curve. You may wish to review the discussion of equilibria given at the end of Chapter 5.

Solubility Product Constants

The equilibrium reactions for the dissolving of two typical insoluble chloride salts are as follows:

$$PbCl_2(s) \rightleftharpoons Pb^{+2} + 2Cl^- \tag{7-1}$$
$$AgCl(s) \rightleftharpoons Ag^+ + Cl^- \tag{7-2}$$

The solubility product constant expressions for 7-1 and 7-2 are, respectively:

$$K_{sp} = [Pb^{+2}][Cl^-]^2 = 1.6 \times 10^{-5}M^3 \tag{7-3}$$
$$K_{sp} = [Ag^+][Cl^-] = 1.8 \times 10^{-10}M^2 \tag{7-4}$$

Note that 7-3 is the product of the molarity of the lead(II) ion and the square of the molarity of the chloride ion, so that its K_{sp} has units of M^3. For silver chloride the units for the K_{sp} in **7-4** are M^2, so that it is difficult to compare the solubilities of these salts using their K_{sp} values. Using the methods in Chapter 5, we calculate that the molar solubility of silver chloride is about $1 \times 10^{-5}M$, compared to $3 \times 10^{-2}M$ for lead chloride. Obviously silver(I) ion is the choice for a precipitation titration of the chloride ion.

Titration Curves Since precipitation titrations will be considered in greater detail in Chapter 8, we will comment here only on the calculation of a precipitation titration curve insofar as it is different from the calculation of an acid-base titration curve (Fig. 7-1). *To simplify calculations, we will neglect volume changes.*

Let us assume that we are titrating a 50-ml sample of $0.10M$ chloride ion with a $0.10M$ silver nitrate titrant, and that we are going to plot pCl against the percent precipitation, instead of against the volume of titrant. At the start of the titration, potentiometric measurement of pCl gives:

$$pCl = -\log[0.10M \ Cl^-] = 1.00$$

Figure 7-2 illustrates the increase in pCl as silver nitrate titrant is added. When 90% (0.90) of the chloride is precipitated, only 10% (0.10) chloride remains; *neglecting dilution*, pCl at this point is calculated:

$$pCl = -\log[(0.10)(0.10M \ Cl^-)] = 2.00$$

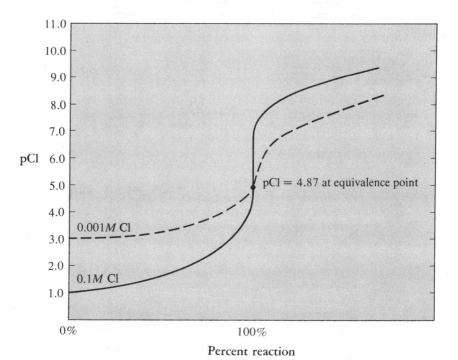

FIGURE 7-2. Titration curves of **0.1***M* chloride with **0.1***M* silver nitrate and **0.001***M* chloride with **0.0001***M* silver nitrate. Note that the pCl at the equivalence point is the same for both, but that the titration curve of the latter exhibits no steep portion in the region of the equivalence point.

At 99.9% (0.999) precipitation, only 0.1% (0.001) of the chloride remains; again *neglecting dilution*, pCl is calculated as follows:

$$pCl = -\log[(0.001)(0.10M\ Cl^-)] = 4.00$$

At the equivalence point, $[Cl^-] = [Ag^+]$, and

$$[Ag^+][Cl^-] = [Cl^-]^2 = K_{sp}$$
$$[Cl^-] = \sqrt{K_{sp}}$$

Inserting the value of K_{sp} for silver chloride at 25°C gives:

$$[Cl^-] = \sqrt{1.8 \times 10^{-10}M^2} = 1.3_4 \times 10^{-5}M$$

$$pCl = 4.87$$

At 100.1% reaction (when a fraction of 0.001 of titrant has been added after the equivalence point), an excess of 0.00010M silver(I) ion has been added, *neglecting dilution*. At this point, pCl is calculated as follows:

$$[Cl^-] = \frac{K_{sp}}{1.0 \times 10^{-4}M\ Ag^+} = 1.8 \times 10^{-6}M$$

$$pCl = 5.74$$

The rest of the titration curve is shown in Figure 7-2. Note that in the region of the equivalence point, pCl will undergo a relatively large change, especially for a sample as concentrated as 0.1M. With dilute samples, like 0.001M chloride, the pCl change in the equivalence point region is so small that an accurate end point cannot be obtained (Fig. 7-2).

Equilibria of Complexes   Recall that complex ions are formed by the reaction of a metal ion, M, with a ligand, L, where the ligand can be either a molecule or an anion. Complex ions are formed in steps:

$$M + L \rightleftharpoons ML$$
$$\vdots$$
$$ML_{(n-1)} + L \rightleftharpoons ML_n$$

where n is the coordination number of the metal ion, and ML_n is the usual form of the complex ion in the presence of an excess of the ligand.

Certain types of ligands complex with more than one of their atoms to more than one coordination position around the metal ion. This type of ligand is called a *multidentate* ligand, in contrast to ligands such as ammonia, which are *monodentate*, or *unidentate*, ligands. Such ligands will be discussed in more detail in Section 8-2, although one must be mentioned briefly here. This multidentate ligand is known commonly as EDTA (ethylenediaminetetraacetic acid) and is a *hexadentate* ligand. As such it forms 1:1 complex ions, called chelates, with most metal ions. It is used widely to titrate most +2 metal ions; for example, its titration reaction with calcium(II) ion is

$$Ca(H_2O)_6^{+2} + EDTA \longrightarrow Ca\text{-}EDTA + 6H_2O \qquad (7\text{-}5)$$

Stability Constants   The equilibrium constant for any type of complex ion is usually written in terms of the formation of the complex ion (like **7-5**) rather than in terms of its dissociation. So this constant is termed a *stability constant* or a *formation constant*. (Some authors write a dissociation reaction and then term the constant an instability constant.) For each step in the reaction of a metal with a ligand, there is a corresponding stability constant, $K_1, K_2, \ldots K_n$, where n is the coordination number.

The stability constants for $Ag(NH_3)_2{}^+$ are a simple example of stepwise constants. They are defined as folllows:

$$K_1 = \frac{[Ag(NH_3)^+]}{[Ag^+][NH_3]} = 2.3 \times 10^3$$

$$K_2 = \frac{[Ag(NH_3)_2{}^+]}{[Ag(NH_3)^+][NH_3]} = 6.0 \times 10^3$$

Note that K_2 is larger than K_1; if we titrate the silver(I) ion with ammonia, it is impossible to stop at the formation of $Ag(NH_3)^+$ since the relatively larger value of K_2 implies that $Ag(NH_3)^+$ will react preferentially with ammonia in comparison to Ag^+. In addition, it can be shown that for quantitative titration at the $0.01 M$ level, a stability constant must be at least 1×10^8; thus the reactions above will not be quantitative for silver ion.

Fortunately, the reactions of EDTA with metal ions are quantitative at appropriate pHs. For example, the stability constant for reaction **7-5** is

$$K_{\text{Ca-EDTA}} = \frac{[\text{Ca-EDTA}]}{[Ca^{+2}][\text{EDTA}]} = 5 \times 10^{10} \text{ (at pH} \geqslant 12)$$

The equilibrium involved is affected by pH, and this is handled by a special approach outlined in Section 8-2. All that need be said at this point is that the stability constants for EDTA complexation reactions are in general so large that most metals can be titrated over a specific pH range peculiar to the metal ion. Since in most cases titration curves are not used for EDTA titrations, none will be calculated here. The most important thing to remember is that EDTA stability constants must be at least 1×10^8 for quantitative titration at the $0.01 M$ level.

Oxidation-Reduction Equilibria and Titration Curves

We begin by reviewing the nature of an oxidation-reduction (redox) reaction, starting with oxidizing and reducing agents. Then we will present a brief discussion of the shape of a redox titration curve.

Oxidation-Reduction Equilibria

The reactants in an oxidation-reduction reaction can be examined in terms of half reactions and half-reaction equilibria. The species gaining electrons is the oxidizing agent or oxidant; the species losing electrons is the reducing reagent or reductant. Each is characterized not by an equilibrium constant, but by a *standard reduction potential*, symbolized as $E°$. This potential characterizes only half reactions written as reductions; thus the $E°$ values describe only the equilibria of a half reaction involving an oxidizing agent. We commonly state that the more positive the $E°$ value, the "stronger" the oxidizing agent, or the more readily it will gain electrons. Because the half reaction of a reducing agent is the reverse of that associated with the $E°$ values, we say that the more negative the $E°$ value, the stronger the reducing agent, or the more readily it will lose electrons.

For example, a good oxidizing agent (oxidant) is the cerium(IV) ion, which in sulfuric acid undergoes the half reaction and exhibits the standard reduction potential given below.

$$Ce(IV) + e^- \rightleftharpoons Ce^{+3} \qquad E^\circ = +1.44 \text{ V } (1N \text{ H}_2\text{SO}_4)$$

Since the cerium(IV) ion may exist in more than one form in acid solution, we symbolize it as Ce(IV) rather than as Ce^{+4}. Note that the E° of the above half reaction has a rather large positive value, indicating that cerium(IV) is a strong oxidizing agent.

As an example of a reducing agent, consider the iron(II) ion which undergoes the half reaction given below.

$$Fe^{+2} \rightleftharpoons Fe^{+3} + e^-$$

Because the above half reaction is written as an oxidation, we cannot give a standard reduction potential for it. However, if we reverse the direction of the half reaction, we can supply an E° value as follows:

$$Fe^{+3} + e^- \rightleftharpoons Fe^{+2} \qquad E^\circ = +0.771 \text{ V}$$

Since this E° has a moderate value, we can say that iron(II) is only a moderate reducing agent, not a strong reducing agent. However, since the E° of cerium(IV) is so much more positive than that of iron(II), we can predict that cerium(IV) will react spontaneously, and quantitatively, with iron(II):

$$Ce(IV) + Fe^{+2} \rightleftharpoons Ce^{+3} + Fe^{+3} \qquad \text{(7-6)}$$

Although the above reaction does reach equilibrium, the point of the equilibrium is far to the right, as indicated by the greater length of the arrow pointing to the right. This is also indicated by the difference in the E° values of $+0.66_9$ V. This difference is sometimes symbolized as E°_{rxn}, E°_{cell}, or the "cell emf." As long as the E°_{rxn} is at least $+0.354$ V, any redox reaction can be said to be quantitative (99.9% complete).

Redox Titration Curves A detailed calculation of all types of redox titration curves is beyond the scope of this text, but we will present a brief discussion of a titration curve for a reaction in which the oxidizing agent (titrant) gains just one electron and the reducing agent (sample) loses just one electron. The key points in such a curve are shown in Figure 7-3.

A redox titration curve is constructed by plotting potentiometer readings of E, the potential exhibited by a sample solution, in volts. This will be discussed more fully in Chapter 11, but E is related to the E° of the sample by the Nernst equation:

$$E = E^\circ + \frac{0.059 \text{ V}}{\text{no. of } e^-} \log \frac{[\text{Ox}]}{[\text{Red}]}$$

In our situation, the number of electrons is one, [Red] will be the concentration of the reduced form of the sample (a reducing agent), and [Ox] will be the concentration of the oxidized product.

As we start the titration, the concentration of the sample will be quite low and E will be less than E°, as you can deduce by inspecting the Nernst equation. When we have added enough oxidizing agent (titrant) to oxidize 50% of the sample, Red, to the product, Ox, then [Red] = [Ox] and $E = E^\circ$ (Fig. 7-3). If we were titrating Fe^{+2} with Ce(IV) as in **7-6**, then E would be $+0.771$ V.

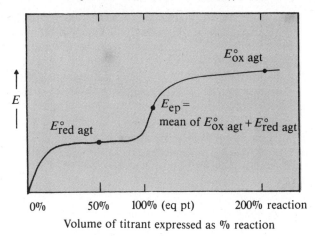

FIGURE 7-3.   Oxidation-reduction titration curve for a titration of a reducing agent sample that loses one electron with an oxidizing agent titrant that gains one electron.

After 50% titration, [Ox] will be greater than [Red], and E will be greater than $E°$ from inspection of the Nernst equation.   At the equivalence point, it can be shown by a derivation beyond the scope of this chapter that E is the mean of the respective $E°$ values:

$$E_{ep} = \frac{E°_{ox\ agt} + E°_{red\ agt}}{2} \tag{7-7}$$

Equation **7-7** applies only for titrations where both the oxidant and reductant half reactions involve only one electron.   Thus for titration of Fe^{+2} with Ce(IV), the potential at the equivalence point would be $+1.10_{55}$ V.   We can see that there must be a dramatic increase in E from $+0.771$ V to 1.10_{55} V in the equivalence point region; this is in fact shown on our general titration curve (Fig. 7-3.).

After the equivalence point, the ratio of [Ox] to [Red] increases very slowly and the Nernst equation predicts a very slow increase in E, again shown in Figure 7-3.   Ultimately, when an amount of oxidizing agent titrant equal to that at the equivalence point has been added, $E = E°$ of the titrant.   This is shown at 200% titration in Figure 7-3. This can be derived by applying the Nernst equation to the half reaction of the oxidizing agent; the derivation is left for an exercise for you.   For the titration of Fe^{+2} with Ce(IV), E at 200% titration would be $+1.44$ V.

7-2 | QUANTITATIVE ASPECTS OF SILVER(I) TITRATION OF CHLORIDE

Because the determination of the chloride ion in samples such as body fluids is so important and is performed so often, it is useful to use the titration of chloride with silver(I) ion to illustrate how to predict whether a titration will be quantitative.   We consider this reaction here from an equilibrium viewpoint only, but will present a detailed discussion of all silver(I) titration methods in the next chapter.

Two factors are important in defining a quantitative titration reaction. One factor is the equilibrium conditions involved (value of the equilibrium constant, etc.), and the other is the precision (and accuracy) of the buret used. For a 50-ml buret, the precision is of the order of 0.1 pph (0.1%). Therefore, the upper limit for a quantitative titration involving at least 40 ml of titrant measured in a 50-ml buret is $100.0\% - 0.1\%$, or 99.9% reaction. If a different size buret is used, or if appreciably smaller volumes are used with the 50-ml buret, then of course the upper limit will be different.

Silver(I) Titration of Chloride Ion

The two types of silver(I) titration methods for chloride are the direct titration and the back titration. The quantitativeness of each method will be considered separately because of the different equilibria involved.

Quantitative-ness of Direct Titration of Chloride

Two different approaches are used for estimating whether a precipitation titration reaction will be 99.9% complete or not: (1) calculate the percent reaction indirectly by calculating the percentage of chloride dissolved at the equivalence point, using K_{sp}; and (2) calculate the equilibrium constant, K_{rxn}, and compare it to the minimum theoretical value for K_{rxn}. We will outline both approaches.

A 99.9% reaction implies that the maximum percentage of chloride that dissolves at the equivalence point should be 0.1%. This percentage may be estimated by dividing the $[Cl^-]$ at the equivalence point by the initial concentration of chloride at the start of the titration. Since dilution of the sample is ignored, this is only an estimation. For example, let us calculate the percentage of chloride dissolved at the equivalence point in the titration of a sample of $0.10M$ chloride. At the equivalence point, the concentration of chloride is found by taking the square root of the K_{sp} of silver chloride (Sec. 7-1), so that the percentage dissolved is

$$\% \text{ dissolved} = \frac{\sqrt{K_{sp} \text{ of AgCl}}}{M \text{ of initial Cl}^-} (100) = \frac{1.3_4 \times 10^{-5}M}{0.10M} (100) = 0.013\%$$

Since this is less than 0.1%, the titration of $0.10M$ chloride is more than 99.9% complete and is quantitative.

In the second approach for estimating 99.9% reaction, we derive the minimum theoretical value for K_{rxn} for silver(I) titration of any -1 anion, such as chloride, and then compare the actual value of K_{rxn} with it. The equilibrium constant for any silver(I) titration of a -1 anion, X^-, is

$$K_{rxn} = \frac{1}{[Ag^+][X^-]} = \frac{1}{K_{sp}} \tag{7-8}$$

To calculate the minimum theoretical value for K_{rxn}, we recall that precipitation of X^- will be quantitative when $[X^-]$ is reduced to 0.1%, or a fraction of 0.001, of the initial value. Thus for a $0.10M$ sample:

$$\text{upper limit of } [X^-] = (0.001)(0.10M) = 1 \times 10^{-4}M$$

At the equivalence point, by definition $[Ag^+]$ must equal $[X^-]$. The minimum theoretical value of K_{rxn} is then calculated by substituting into **7-8**:

$$\text{min theo } K_{rxn} = \frac{1}{[1 \times 10^{-4}M \text{ Ag}^+][1 \times 10^{-4}M \text{ X}^-]} = 1 \times 10^8 \tag{7-9}$$

We generalize our derivation for any $1:1$ insoluble salt in the box below.

If the actual K_{rxn} for a $1:1$ insoluble salt is $\geqslant 1 \times 10^8$ (the minimum theoretical value), the precipitation titration will be quantitative (99.9%) at or above the $0.1M$ level.

Proceeding to the titration of $0.10M$ chloride, we use **7-8** to calculate the equilibrium constant for the proposed precipitation:

$$K_{rxn} = \frac{1}{1.8 \times 10^{-10}} = 5.5_5 \times 10^9$$

Note that the titration of chloride is quantitative. However, if we were to titrate $0.10M$ iodate, where the K_{sp} of $AgIO_3$ is 3.0×10^{-8}, the K_{rxn} would be 3.3×10^7, indicating a reaction that is not quantitative.

We might also point out that the calculation for chloride applies for any concentration *above* $0.10M$, since the minimum theoretical K_{rxn} will *decrease* below 1×10^8. If a chloride sample is significantly less concentrated than $0.10M$, then a new minimum theoretical K_{rxn} will have to be calculated. This is explored in the self-test at the end of this section.

Quantitativeness of Direction Titration of -2 Anions As an extension of the concepts discussed above for chloride or other -1 anions, let us briefly consider the quantitative aspects of a titration of a -2 anion, say B^{-2}, with silver(I) ion titrant. Again both approaches can be used. Suppose that we wanted to titrate $0.10M$ chromate, CrO_4^{-2}. At the equivalence point, the concentration of chromate is found by substituting the identity $[Ag^+] = 2[CrO_4^{-2}]$ into the K_{sp} expression and solving for $[CrO_4^{-2}]$:

$$K_{sp} = [Ag^+]^2[CrO_4^{-2}] = (2[CrO_4^{-2}])^2[CrO_4^{-2}]$$

$$[CrO_4^{-2}] = \sqrt[3]{\frac{K_{sp}}{4}} = \sqrt[3]{\frac{5 \times 10^{-12}M^3}{4}} = 1._{08} \times 10^{-4}M$$

We then calculate the percentage of chromate dissolved as before for chloride:

$$\% \text{ dissolved} = \frac{1._{08} \times 10^{-4}M \; CrO_4^{-2} \text{ eq pt}}{0.10M \text{ initial } CrO_4^{-2}} (100) = 0.1_1\%$$

Since this is slightly more than 0.1%, the titration of $0.10M$ chromate is less than 99.9% complete and is just short of being quantitative.

In the second approach for estimating 99.9% reaction, we derive the minimum theoretical value for K_{rxn} for silver(I) titration of any B^{-2} anion, such as chromate, and then compare the actual value of K_{rxn} with it. As in **7-8**, the equilibrium constant for the titration of any B^{-2} anion is

$$K_{rxn} = \frac{1}{[Ag^+]^2[B^{-2}]} = \frac{1}{K_{sp}} \tag{7-10}$$

At the equivalence point, the $[B^{-2}]$ will be reduced to $1 \times 10^{-4}M$, but the stoichiometry dictates that the $[Ag^+]$ must be twice that, or $2 \times 10^{-4}M$. The minimum theoretical value of K_{rxn} is then calculated by substituting into **7-10**:

$$\text{min theo } K_{rxn} = \frac{1}{[2 \times 10^{-4}M \, Ag^+]^2 [1 \times 10^{-4}M \, B^{-2}]} = 2.5 \times 10^{11} \quad \textbf{(7-11)}$$

We generalize for any 2:1 insoluble salt in the box below.

If the actual K_{rxn} for any 2:1 silver(I) salt, or any other 2:1 salt, is $\geqslant 2.5 \times 10^{11}$ (the minimum theoretical value), the precipitation titration will be quantitative (99.9%) at or above the $0.1M$ level.

Using **7-10**, the actual value of K_{rxn} for the titration of chromate is

$$K_{rxn} = \frac{1}{5 \times 10^{-12}} = 2 \times 10^{11}$$

Note that the titration of chromate just falls short of being quantitative at the $0.10M$ level; however, if the sample is significantly more concentrated, the titration undoubtedly will be quantitative. If the sample is $0.01M$, we can predict that the titration will not be quantitative, but, strictly speaking, we should calculate a new minimum theoretical K_{rxn}. This is explored in the self-test at the end of this section.

Quantitative-ness of Back Titration of Chloride

The Volhard back titration of chloride and other anions involves addition of a measured excess of silver nitrate to precipitate the anion, and then a back titration of the unreacted silver(I) nitrate with potassium thiocyanate, KSCN, titrant:

$$\underset{\text{(titrant)}}{KSCN} + \text{unreacted } Ag^+ \longrightarrow AgSCN(s) + K^+ \quad \textbf{(7-12)}$$

The limiting factor to the quantitativeness of the Volhard titration occurs near or at the end point in the back titration **(7-12)**. After reacting with all of the silver(I) ion, the thiocyanate ion can then react with any insoluble silver(I) salt, Ag_nB:

$$nSCN^- + Ag_nB(s) \rightleftharpoons nAgSCN(s) + B^{-n} \quad \textbf{(7-13)}$$

If the point of equilibrium lies on the right in **7-13**, too much potassium thiocyanate titrant will be added in the back titration and the results will be low.

One way of estimating whether the point of equilibrium does indeed lie on the right is to calculate the molar solubilities of both the AgSCN and Ag_nB as though each were in equilibrium alone in pure water. If AgSCN has a lower molar solubility than Ag_nB, then we can predict that the titration will be in error unless Ag_nB is removed from the equilibrium.

EXAMPLE: Decide whether there will be an error in the Volhard back titration of **(a)** chloride and **(b)** sulfite, given $K_{sp} = 1.5 \times 10^{-14}M^3$ for Ag_2SO_3. The molar solubility of AgSCN is the square root of the K_{sp} of $1.1 \times 10^{-12}M$; thus, $S = 1.0_5 \times 10^{-5}M$.

Solution to a: Given that S is the molar solubility of silver chloride, substitute the appropriate multiple of S into the K_{sp} expression and solve for S.

$$AgCl : K_{sp} = [S][S]$$

$$[S] = \sqrt{1.8 \times 10^{-10} M^2} = 1.3_4 \times 10^{-5} M$$

Since the solubility of silver thiocyanate is smaller than that of silver chloride, we predict an error in the titration from reaction of thiocyanate with silver chloride. (We could also have predicted this by comparing K_{sp}'s in this case because each has the same units of M^2.)

Solution to b: Given that S is the molar solubility of silver sulfite, substitute the appropriate multiple of S into the K_{sp} expression and solve.

$$Ag_2SO_3 : K_{sp} = [2S]^2[S]$$

$$[S] = \sqrt[3]{(1.5 \times 10^{-14} M^3)/4} = 1.5_{54} \times 10^{-5} M$$

Since the solubility of $1.0_5 \times 10^{-5} M$ of silver thiocyanate is smaller than that of silver sulfite, we predict again that the titration will be in error. (Note that we might have predicted just the opposite by comparing K_{sp}'s, but this comparison would have been wrong because each K_{sp} has different units.)

In spite of the potential error from the thiocyanate titrant reacting with an insoluble salt, such as silver chloride, the Volhard back titration method is still used for the determination of chloride and other such anions. The Volhard method is modified to avoid the error, as described in Chapter 8.

Another way of estimating whether the point of equilibrium lies on the right in **7-13** is to calculate a value for K_{rxn}, the equilibrium constant, as follows:

$$K_{rxn} = \frac{K_{sp} \text{ of } Ag_n B}{(K_{sp} \text{ of } AgSCN)^n} = \frac{[B^{-n}]}{[SCN^-]^n} \tag{7-14}$$

If K_{rxn} has a numerical value greater than one, we can be sure that a significant reaction of thiocyanate with the insoluble silver salt will occur and that the titration will be in error. For example, using **7-14**, the value of K_{rxn} for the thiocyanate reaction of silver sulfite is

$$K_{rxn} = \frac{1.5 \times 10^{-14} M^3}{(1.1 \times 10^{-12} M^2)^2} = 1.2 \times 10^{10} M^{-1}$$

Here, the fact that K_{rxn} is greater than one clearly indicates that thiocyanate will react with sulfite and cause an error.

This concludes the discussion of the quantitative aspect of titrations; you should now take the following self-test for further work on this subject.

SELF-TEST	**9. Deciding Whether a Reaction Is Quantitative**
Answers:	*Directions:* Cover the answers in the left column before beginning the test. Calculate the K_{rxn} for each problem to solve it. A. Decide whether or not a precipitation titration producing each of the following 1 : 1 insoluble salts will be quantitative at the $0.1M$ level. If not, suggest whether a change in concentration might achieve a quantitative reaction.

Quant ($K_{rxn} = 5.6 \times 10^9$)
Not quant ($K_{rxn} = 1 \times 10^7$)

a. $Ag^+ + Cl^- \rightarrow AgCl(s)$; $K_{sp} = 1.8 \times 10^{-10}$
b. $Pb^{+2} + SO_4^{-2} \rightarrow PbSO_4$; $K_{sp} = 1.0 \times 10^{-7}$

B. Decide whether or not a precipitation titration producing each of the following 2 : 1 insoluble salts will be quantitative at the $0.1M$ level.

Not quant ($K_{rxn} = 2.5 \times 10^{10}$)
Quant ($K_{rxn} = 1.0_3 \times 10^{15}$)

a. $2Ag^+ + C_2O_4^{-2} \rightarrow Ag_2C_2O_4(s)$; $K_{sp} = 4.0 \times 10^{-11}$
b. $2Ag^- + SeO_3^{-2} \rightarrow Ag_2SeO_3(s)$; $K_{sp} = 9.7 \times 10^{-16}$

C. Decide whether a precipitation titration producing each of the following insoluble salts at the $0.01M$ level will be quantitative. Calculate the theoretical K_{rxn}.

Not quant
Quant ($K_{rxn} > 1 \times 10^{10}$)
Not quant
Quant ($K_{rxn} > 2.5 \times 10^{14}$)

a. $Ag^+ + Cl^- \rightarrow AgCl(s)$; $K_{sp} = 1.8 \times 10^{-10}$
b. $Ag^+ + SCN^- \rightarrow AgSCN(s)$; $K_{sp} = 1.1 \times 10^{-12}$
c. $2Ag^+ + SO_3^{-2} \rightarrow Ag_2SO_3(s)$; $K_{sp} = 1.5 \times 10^{-14}$
d. $2Ag^+ + SeO_3^{-2} \rightarrow Ag_2SeO(s)$; $K_{sp} = 9.7 \times 10^{-16}$

D. Decide whether there will be an error in the Volhard back titration of each of the following anions if potassium thiocyanate is the titrant used. Use a K_{sp} of 1.1×10^{-12} M^2 for AgSCN.

Error
Error ($S = 6.2 \times 10^{-6}M$)
Error
$K_{rxn} = 1.6 \times 10^2$; error
$K_{rxn} = 8.0 \times 10^8$; error
$K_{rxn} = 9.8 \times 10^{15}$; error

a. Cl^-; K_{sp} of AgCl = $1.8 \times 10^{-10}M^2$
b. SeO_3^{-2}; K_{sp} of Ag_2SeO_3 = $9.7 \times 10^{-16}M^3$
c. PO_4^{-3}; K_{sp} of Ag_3PO_4 = $1.3 \times 10^{-20}M^4$

E. Calculate K_{rxn} using 7-14 for each thiocyanate-precipitate reaction in D to see whether there will be an error in the Volhard method.

7-3 | END POINT DETECTION METHODS

In this section we will illustrate some of the various methods of detection of the titration end point by considering specific methods used for precipitation titrations. We saw in Figure 7-1 that the concentration of the species being titrated changes throughout the titration, but especially at the equivalence point. End points are detected by taking advantage of the large change in concentration in the equivalence point region.

In general, the method for detecting the end point is chosen so that it is perceived as close as possible to the equivalence point; i.e., in the middle of the region where a sharp change in concentration of the species being titrated occurs. There are three types of end point detection that you should be aware of—visual methods, spectral methods, and electrical methods.

Visual Methods for End Point Detection In visual methods of end point detection the eye detects some type of color change or change in state that coincides as closely as possible with the true end point, or equivalence point. A chemical reagent, called an *indicator*, is usually added to achieve the color change. Before the end point the indicator exhibits one color; at the end point it exhibits the most pronounced color *change*; and after the end point it exhibits a different color than before. When the eye perceives the pronounced color change, the titration is stopped and the volume of titrant is recorded.

In precipitation titrations, visual end points are based on color changes as a result of precipitation, color changes as a result of adsorption, and color changes of soluble indicators. Each of these will be discussed.

Precipitation as an Indicator (Mohr Method)

In the Mohr method for the titration of chloride ion, the indicator is an anion that forms a colored insoluble silver(I) salt which precipitates only *after* the equivalence point has been reached. Thus an important requirement for the indicator anion is that the concentration of silver(I) ion present before or at the equivalence point not exceed the *minimum concentration* required to precipitate the colored insoluble silver(I) salt. One way to judge this is to calculate the silver(I) concentration at the equivalence point of the titration and to compare it with the calculated minimum concentration of silver(I) necessary for indicator precipitation.

Let us consider the choice of an indicator anion for the titration of chloride ion. Using **7-4**, the silver(I) concentration at the equivalence point for silver chloride is

$$[Ag^+]_{ep} = \sqrt{K_{sp}(AgCl)} = \sqrt{1.8 \times 10^{-10}M^2} = 1.3_4 \times 10^{-5}M \qquad (7\text{-}15)$$

Suppose that we choose the iodide ion as the indicator anion, since silver(I) iodide has a yellow color and is insoluble ($K_{sp} = 8.3 \times 10^{-17}M^2$). Since one recommended concentration for such an indicator anion is $0.0025M$, we can use that concentration and the K_{sp} to calculate the minimum concentration of silver(I) required to precipitate yellow silver(I) iodide:

$$[Ag^+]_{min} = \frac{8.3 \times 10^{-17}M^2}{0.0025M\ I^-} = 3.3 \times 10^{-14}M$$

Since the silver(I) concentration at the equivalence point is much greater than the above minimum concentration, silver ion will precipitate silver(I) iodide before the equivalence point and iodide will not be suitable as an indicator anion.

Now let us consider the use of chromate ion as the indicator; silver(I) chromate has a red color and is also insoluble ($K_{sp} = 1.1 \times 10^{-12}M^3$). Again using an indicator ion concentration of $0.0025M$ and the K_{sp} to calculate the minimum concentration of silver(I) required to precipitate Ag_2CrO_4, we obtain:

$$[Ag^+]_{min} = \sqrt{\frac{1.1 \times 10^{-12}M^3}{0.0025M\ CrO_4^{-2}}} = 2.1 \times 10^{-5}M \qquad (7\text{-}16)$$

In this case, the silver(I) concentration at the equivalence point for the titration of chloride is less than the above minimum concentration; thus chromate can be used as the indicator anion. When the silver nitrate titrant has reacted with all of the chloride ion, the indicator color change is

$$2Ag^+ + CrO_4^{-2} \longrightarrow Ag_2CrO_4(s)$$
$$\text{(yellow)} \qquad\qquad \text{(red ppt)}$$

In reality, the true end point is the first permanent reddening of the yellow color of the chromate ion caused by a small amount of suspended silver(I) chromate.

Although such a reaction is used specifically for the determination of the chloride ion, it illustrates a general principle of indicator action. After the titrant has reacted with the sample species, the first excess of the titrant reacts *similarly* with an indicator species of the same type as the sample.

Adsorption Indicator (Fajans Method)

In the adsorption indicator method (Fajans method) for chloride ion, the indicator is usually an anion that will be adsorbed on the surface of the precipitated silver(I) chloride at the equivalence point. Before the equivalence point, the indicator, usually a yellow color, remains in solution because it is repelled from the surface by the unprecipitated chloride ions adsorbed on the surface. Once all of the chloride ions have been precipitated, the indicator is adsorbed and exhibits a red color on the surface. This type of indicator action is rather unusual and does not illustrate any general principles useful for other types of titrations; hence it will be discussed in detail in Chapter 8.

Soluble Indicators in Precipitation Titrations (Volhard Method)

In the Volhard method for the back titration of chloride ion, the indicator is the iron(III) ion. When the first excess of potassium thiocyanate is added after precipitation of silver(I) thiocyanate (7-12), it forms the red Fe(SCN)$^{+2}$ ion. Since the iron(III) ion is a light yellow, the color change is from a light yellow to a red color. This reaction will also be discussed in more detail in Chapter 8.

Soluble Indicators in Other Types of Titrations

Most visual end points are based on the color change of a soluble indicator. Indicators for acid-base titrations, for example, are either weakly ionized acids or bases, both soluble. Indicators for complexation titrations are colored dyes that form colored complexes with the metal ion to be titrated. Indicators for oxidation-reduction titrations are usually colored oxidizing or reducing agents. Often the color change is simply

$$\text{Titrant + Indicator (color 1)} \longrightarrow \text{Indicator (color 2)}$$

Spectral Methods for End Points

In spectral methods of end point detection, a spectral change is followed by an instrument rather than the eye. The same types of indicators may be used, but the color change is measured on a spectrophotometer (see Ch. 12). The measurements are plotted against the volume of titrant added at each point. A titration curve such as that in Figure 7-1 is obtained. It is also possible to measure the fluorescence (see Ch. 14) of a fluorescent indicator and make the same kind of plot.

Electrical Methods for End Points

In the most common adaptation of the electrical method, a pair of electrodes and a potentiometer are used to measure the potential (voltage) of the solution as the titration is conducted. For acid-base titrations a special type of potentiometer called a pH meter is used. The so-called glass electrode is used to measure pH changes. Such a procedure is described in Chapter 16; the titration curve in Figure 7-2 is obtained by this means.

Other types of electrical methods for end point detection are based on measuring changes in conductance (conductometric titration) and on measuring the change in current passing through a cell in which the voltage applied across the electrodes is kept constant (amperometric titration).

| QUESTIONS AND PROBLEMS

(Answers to even-numbered problems are in Appendix 5.)

Titration Equilibria

1. Sketch a titration curve for the titration of 0.10M sodium hydroxide using 0.10M hydrochloric acid as the titrant. Plot pH versus percentage reaction for 0.0%, 90.00%, 99.00%, and 100.0% reaction.

2. Sketch a titration curve for the titration of 0.10M nitric acid using 0.10M sodium hydroxide as the titrant. Plot pH versus percentage reaction for 0.0%, 90.00%, 99.00%, and 100.0% reaction.

3. Sketch a titration curve for the titration of $0.01M$ sodium hydroxide using $0.010M$ hydrochloric acid as the titrant. Plot pH versus percentage reaction for 0.0%, 90.00%, 99.00%, and 100.0% reaction.

4. Sketch a titration curve for the titration of $0.0010M$ nitric acid using $0.0010M$ sodium hydroxide as the titrant. Plot pH versus percentage reaction for 0.0%, 90.00%, 99.00%, and 100.0% reaction.

5. Sketch a titration curve for the titration of $0.1M$ Fe^{+2} ($E° = +0.771$ V) in sulfuric acid using $0.1M$ Ce(IV) as the titrant. Locate the potentials at 50%, 100%, and 200% reaction, using $E° = 1.44$ V for Ce(IV).

6. Sketch a titration curve for the titration of $0.1M$ Fe^{+2} ($E° = +0.771$ V) in perchloric acid using $0.1M$ Ce(IV) as the titrant; use $E° = 1.70$ V for cerium(IV) in perchloric acid. Locate the potentials at 50%, 100%, and 200% titration.

7. Sketch a titration curve of pCl against percentage precipitation for the titration of 50 ml of $0.010M$ chloride with $0.010M$ silver nitrate; calculate pCl at the following points:
   a 0%
   b. 90% (neglect dilution)
   c. 99.9% (neglect dilution)
   d. 100.0%

8. Sketch a titration curve of pCl against ml of titrant for the titration of 50.00 ml of $0.100M$ chloride with $0.100M$ silver nitrate; calculate pCl at the following points:
   a. 0 ml titrant
   b. 45.0 ml titrant (neglect dilution)
   c. 49.50 ml titrant (neglect dilution)
   d. 49.95 ml titrant (neglect dilution)
   e. 50.00 ml titrant
   f. 50.05 ml titrant (neglect dilution)

9. Sketch a titration curve of pBr against ml of titrant for the titration of 50.00 ml of $0.100M$ bromide with $0.100M$ silver nitrate; calculate pBr at the following points. Neglect dilution at all points.
   a. 0 ml titrant
   b. 45.0 ml titrant
   c. 49.50 ml titrant
   d. 49.95 ml titrant
   e. 50.00 ml titrant
   f. 50.05 ml titrant

10. In the Volhard back titration method for chloride, 50 ml of $0.10M$ silver nitrate is added to a 40-ml sample of approximately $0.040M$ chloride ion. The unreacted silver(I) ion is back titrated with $0.100M$ potassium thiocyanate to the end point. Calculate the concentration of each ion specified and its negative log.
    a. $[Cl^-]$ and pCl before addition of $0.100M$ potassium thiocyanate
    b. $[Ag^+]$ and pAg before addition of $0.100M$ potassium thiocyanate
    c. $[Ag^+]$ and pAg at the equivalence point
    d. $[Ag^+]$ and pAg after 1 ml of $0.100M$ potassium thiocyanate has been added beyond the equivalence point (estimate the total volume to two significant figures)

11. In the Volhard back titration method for bromide, 50 ml of $0.10M$ silver nitrate is added to a 30-ml sample of approximately $0.045M$ bromide ion. The unreacted silver(I) ion is back titrated with $0.100M$ potassium thiocyanate to the end point. Calculate the concentration of each ion specified and its negative log.
    a. $[Br^-]$ and pBr before addition of $0.100M$ potassium thiocyanate
    b. $[Ag^+]$ and pAg before addition of $0.100M$ potassium thiocyanate
    c. $[Ag^+]$ and pAg at the equivalence point
    d. $[Ag^+]$ and pAg after 1 ml of $0.100M$ potassium thiocyanate has been added beyond the equivalence point (estimate the total volume to two significant figures).

Quantitative Reactions

12. Decide whether or not the following precipitation titrations will be quantitative at the $0.1M$ level. If not, suggest whether a change in concentration might achieve a quantitative reaction.
    a. $Ag^+ + N_3^- = AgN_3(s)$;
       $K_{sp} = 2.9 \times 10^{-9}$
    b. $Ag^+ + VO_3^- = AgVO_3(s)$;
       $K_{sp} = 5 \times 10^{-7}$
    c. $2Ag^+ + SO_3^{-2} = Ag_2SO_3(s)$;
       $K_{sp} = 1.5 \times 10^{-13}$
    d. $3Ag^+ + PO_4^{-3} = Ag_3PO_4(s)$;
       $K_{sp} = 1.3 \times 10^{-20}$

13. Decide whether the precipitation titrations in the preceding problem will be quantitative at the $0.01M$ level. Be sure to derive the minimum theoretical value for K_{rxn} for each type.

14. Decide whether the precipitation titration of each of the following concentrations of chloride will be quantitative, using either a calculation of a minimum theoretical K_{rxn} or the percentage of chloride dissolved at the equivalence point. In either case, ignore dilution.
    a. $0.2M$ chloride
    b. $0.02M$ chloride
    c. $0.014M$ chloride
    d. $0.013M$ chloride
    e. $0.012M$ chloride

15. Decide whether there will be an error in the Volhard back titration of each of the following anions if potassium thiocyanate is used as the back titrant. Use a K_{sp} of $1.1 \times 10^{-12}M^2$ for AgSCN. Unless other K_{sp}'s are given, check the appendix.
    a. Cl^-
    b. Br^-
    c. MoO_4^{-2} (for $Ag_2MoO_4(s)$, $K_{sp} = 2.8 \times 10^{-12}M^3$)
    d. S^{-2} (for $Ag_2S(s)$, $K_{sp} = 6 \times 10^{-50}M^3$)
    e. AsO_4^{-3} (for $Ag_3AsO_4(s)$, $K_{sp} = 1.1 \times 10^{-20}M^4$)
    f. $Fe(CN)_6^{-4}$ (for $Ag_4Fe(CN)_6(s)$, $K_{sp} = 1._6 \times 10^{-41}M^5$)

16. Assume that in the Volhard method, potassium thiocyanate back titrant is replaced by potassium carbonate, resulting in insoluble silver carbonate ($K_{sp} = 8.1 \times 10^{-12}M^3$). Decide whether there will be an error if such a Volhard method is used for each of the anions in the preceding problem.

End Point Detection Methods

17. Evaluate each of the anions below as a possible substitute for the chromate ion in the Mohr titration of the chloride ion. Use two criteria: there should be a definite color change (see a handbook), and the silver(I) ion present at the equivalence point should not precipitate a $0.0025M$ solution of the anion.
    a. $Br^- \rightarrow AgBr(s)$
    b. $S^{-2} \rightarrow Ag_2S(s)$; $K_{sp} = 6 \times 10^{-50}M^3$
    c. $AsO_4^{-3} \rightarrow Ag_3AsO_4(s)$; $K_{sp} = 1.1 \times 10^{-20}M^4$
    d. $PO_4^{-3} \rightarrow Ag_3PO_4(s)$
    e. $N_3^- \rightarrow AgN_3(s)$
    f. $Fe(CN)_6^{-4} \rightarrow Ag_4Fe(CN)_6(s)$; $K_{sp} = 1._6 \times 10^{-41}M^5$

18. It is desired to apply the Mohr method for the precipitation titration of the azide ion, precipitating $AgN_3(s)$. Evaluate each of the anions below as a possible indicator using two criteria: the insoluble silver salt should be colored (see a handbook), and the silver(I) ion present at the equivalence point should not precipitate a $0.0025M$ solution of the anion. Use a K_{sp} of $2.9 \times 10^{-9}M^2$ for $AgN_3(s)$.
    a. $CrO_4^{-2} \rightarrow Ag_2CrO_4(s)$
    b. $WO_4^{-2} \rightarrow Ag_2WO_4(s)$; $K_{sp} = 5.5 \times 10^{-12}M^3$
    c. $AsO_4^{-3} \rightarrow Ag_3AsO_4(s)$; $K_{sp} = 1.1 \times 10^{-20}M^4$
    d. $VO_3^{-1} \rightarrow AgVO_3(s)$; $K_{sp} = 5 \times 10^{-7}M^2$

19. The concentration of the chromate ion in the Mohr method is arbitrarily increased to $0.0070M$. Using appropriate calculations, decide whether the Mohr method can now be applied to the titration of the following ions:
    a. $Cl^- \rightarrow AgCl(s)$
    b. $SO_3^{-2} \rightarrow Ag_2SO_3(s)$; $K_{sp} = 1.5 \times 10^{-14}M^3$
    c. $SeO_3^{-2} \rightarrow Ag_2SeO_3(s)$; $K_{sp} = 9.7 \times 10^{-16}M^3$
    d. $PO_4^{-3} \rightarrow Ag_3PO_4(s)$

20. Compare the end point detection methods in the Mohr method and the Volhard method. Which is more generally applicable for anions other than chloride and why?

21. Suppose that potassium thiocyanate titrant were not available for the Volhard method, but that it was still desired to use a complex of iron(III) to form a color for the end point. Evaluate each of the anions below as a possible titrant in the back titration.
    a. Cl^-; forms light yellow $FeCl^{+2}$
    b. $C_2O_4^{-2}$; forms intense yellow $Fe(C_2O_4)^+$
    c. $C_6H_4OHCO_2^-$ (salicylate ion); forms violet $Fe(C_6H_4OHCO_2^-)^{+2}$

Challenging Problems

22. Calculate a minimum theoretical value for K_{rxn} for a precipitation reaction involving the ions below.

$$4\,Fe^{+3} + 3\,Fe(CN)_6^{-4} \longrightarrow$$
$$Fe_4[Fe(CN)_6]_3(s)$$

23. Evaluate the potential error at the equivalence point in the potassium thiocyanate back titration of the Volhard method for the determination of each of the following anions, by calculating K_{rxn} for the reaction of the insoluble silver(I) salt, $Ag_n B$, with the thiocyanate ion.

a. HVO_4^{-2}
b. S^{-2}; $K_{sp} = 6 \times 10^{-50} M^3$
c. $Fe(CN)_6^{-4}$; $K_{sp} = 1.6 \times 10^{-41} M^5$

24. The Volhard method is to be modified by replacing the potassium thiocyanate titrant with either sodium sulfite or sodium molybdate for the determination of phosphate as silver phosphate, $K_{sp} = 1.3 \times 10^{-20} M^4$. Using two different approaches, including calculation of K_{rxn}, show whether there will be an error at the equivalence point in the back titration to produce:

a. $Ag_2SO_3(s)$; $K_{sp} = 1.5 \times 10^{-14} M^3$
b. $Ag_2MoO_4(s)$; $K_{sp} = 2.8 \times 10^{-12} M^3$

8 | Precipitation and Complexation Titrations

In the previous chapter, the use of silver(I) ion to precipitate chloride ion and other ions was discussed as an example of the application of equilibrium *theory* in titrations. In this chapter, we take a broader look at the possible reactions of metal ions and anions, particularly the *practical* applications of such reactions. One class of reactions, of course, involves the use of metal ions to precipitate anions, and the other class involves the use of anions and similar species to complex metal ions. Unfortunately, there is no reagent that can be used to precipitate hemoglobin and nothing else.

Whether a metal ion is precipitated by an anion or complexed as a soluble complex ion depends on the type of anion involved and on its concentration. *Unidentate* anions, anions that occupy one coordination position around a metal ion, may precipitate certain +1 cations. For example, chloride ion precipitates silver(I) ion; however, an excess of chloride can also form soluble $AgCl_2^-$ ion. *Bidentate* anions, anions that occupy two coordination positions around a metal ion, may precipitate certain cations. For example, sulfate ion precipitates barium(II) ion as barium(II) sulfate. An excess of sulfate, however, can form soluble complex ions with certain other metal ions. Finally, *hexadentate* ligands, such as ethylenediaminetetraacetic acid (EDTA), are used for complexation titrations. We will show that EDTA is a ligand in two senses—it has negatively charged donor atoms that complex with a metal ion and it has neutral donor atoms that also complex with the metal.

8-1 | PRECIPITATION TITRATIONS

Most of the precipitation reactions used for gravimetric analyses are unsuitable for the titration of the anion or the cation involved. Two types of reactions are rapid enough and stoichiometric enough to be used for analysis—the titration of halide ions with silver(I) nitrate and the titration of sulfate ion with barium chloride. Back titration of silver(I) ion using a salt such as potassium thiocyanate (the Volhard method) is also possible. The titration of chloride ion is very important because of the need for chloride analysis of body fluids, water supplies, certain foods, and industrial products.

General Characteristics of Silver Nitrate Titrations

Direct vs. Back Titration

In the direct titration of chloride ion or other anions, silver nitrate titrant is added until the precipitation of the silver(I) salt is complete. There are three commonly used methods, each with different end point detection: the Mohr method, the adsorption indicator (Fajans) method, and the potentiometric method. Each of these methods depends on quantitative precipitation of the anion at the end point, as will be discussed below. The potentiometric method is more general because it does not require a visual indicator as do the Mohr and adsorption indicator methods. It can therefore be used in the determination of ions for which the visual indicators in the Mohr and adsorption indicator methods fail. However, because the potentiometric method usually requires that a titration curve, such as that in Figure 7-1, be plotted, it is rather tedious and slow unless instrumentation is used to plot the curve automatically. If such potentiometric instrumentation is not readily available, then the Mohr and adsorption indicator methods are faster and easier to organize.

In the back titration of chloride or other anions, a measured excess of silver nitrate is added to precipitate the silver(I) salt all at once. Then the unreacted silver nitrate is back titrated. This is the essence of the Volhard method. This method has an advantage over direct titration in that it can determine anions that are not quantitatively precipitated by direct titration. The common ion effect (Sec. 5-3) exerted by the excess silver(I) ion is sufficient to precipitate nearly any silver(I) salt of moderate insolubility. Another advantage of the Volhard method is that the end point is independent of the anion titrated, because the back titration with potassium thiocyanate of the unreacted silver(I) ion is always the same. Of course, using the back titration means that the Volhard method is a bit slower than direct titrations using visual indicators, but the difference in time spent is not great.

Quantitativeness of Silver Nitrate Titrations

Recall from Section 7-2 that the equilibrium constant, K_{rxn}, for precipitation titrations of any -1 anion, X^-, is calculated as follows:

$$K_{rxn} = \frac{1}{K_{sp}} = \frac{1}{[Ag^+][X^-]} \tag{8-1}$$

In Section 7-2, the concept of a *minimum theoretical value* of K_{rxn} was developed. By comparing the actual value of K_{rxn} calculated from **8-1** with the minimum theoretical values of K_{rxn} at different concentrations, you can decide whether or not a precipitation titration of X^- will be quantitative. The criteria are summarized in the box.

Precipitation titrations of X^- anions with $AgNO_3$ will be quantitative at the concentrations listed if:

for $0.1M\ X^-$, the actual K_{rxn} is $\geqslant 1 \times 10^8$,
for $0.01M\ X^-$, the actual K_{rxn} is $\geqslant 1 \times 10^{10}$.

As an example, for silver(I) chloride with a K_{sp} of 1.8×10^{-10}, the actual value of K_{rxn} is 5.6×10^9 and the titration of chloride at the $0.01M$ level is not quantitative. However, the titration of $0.01M$ levels of bromide or iodide is quantitative.

To evaluate the quantitativeness of the Volhard back titration is somewhat more complicated. Of course, the common ion effect of excess silver(I) ion reduces the solubility of most moderately insoluble silver(I) salts to a level where less than 0.1% of the anion is soluble (Sec. 7-2). For example, consider the determination of the bromate ion. The K_{sp} of silver(I) bromate is $5 \times 10^{-5}M^2$; this value is too large for bromate to be determined by direct titration. However, a concentrated sample, such as a $1.0M$ solution, of the bromate ion can be precipitated quantitatively under the right conditions. Suppose 50 ml of $0.60M$ silver nitrate is added to 5.0 ml of a $1.0M$ solution of the bromate ion. The amount of silver(I) ion remaining after precipitation is $(50 \times 0.60) - (5.0 \times 1.0) = 25$ mmoles, and the bromate concentration is

$$[BrO_3^-] = \frac{5 \times 10^{-5}M^2}{25 \text{ mmole } Ag^+/55 \text{ ml}} = 1._1 \times 10^{-4}M$$

Then the percentage of dissolved bromate ion can be calculated:

$$\% \text{ dissolved } BrO_3^- = \frac{1.1 \times 10^{-4}M}{5 \text{ mmole initial } BrO_3^-/55 \text{ ml}} (100) = 0.1_2\%$$

Since the percentage of dissolved bromate ion is approximately equal to the maximum allowable (0.1%) for a quantitative reaction (Sec. 7-2), the precipitation of bromate is considered quantitative under these conditions for the Volhard method. However, there is the possibility of reaction of the potassium thiocyanate back titrant with the precipitated silver(I) bromate. This will be considered later in this section in the discussion of the details of the Volhard method.

Precipitation of Chloride in Body Fluids

The determination of chloride ion in many body fluids is very important. Chloride ion is measured routinely in the blood serum as one indication of the electrolyte balance in the blood (Sec. 22-1). The mean serum level of chloride is about $0.1M$, so precipitation of chloride as silver chloride is quantitative.

Before chloride can be determined in most body fluids, any protein present must be removed from the fluid by precipitation and centrifugation. This prevents the side reaction of silver nitrate with the protein. If the protein is removed by adding nitric acid, the direct silver nitrate titration methods discussed in this section cannot be used since they require neutral or weakly acid pH. The Volhard (back titration) method can be used since it is carried out in nitric acid.

Chloride ion can also be determined by a coulometric method (Sec. 17-3) in which silver(I) ion is generated from an electrode and the end point is signaled by a sudden increase in current rather than by an indicator. There are also a few methods for chloride that are not based on precipitation of silver chloride. One of these is the titration of chloride with mercury(II) nitrate (Sec. 8-2). Another is the specific ion electrode measurement of "sweat chloride," used as a means of detecting cystic fibrosis in infants (Sec. 16-3).

Direct Titration of Halide Ions

There are three direct precipitation titration methods for halide ions: the potentiometric titration method, the Mohr method, and the adsorption indicator (Fajans) method. Since the end point detection places different limitations on each, they will be discussed separately, starting with an introduction to the potentiometric titration method.

The Potentiometric Titration Method

The potentiometric measurement of ion concentration is somewhat complicated and will be treated in detail in Chapter 16. However, the apparatus and procedure can be summarized simply enough to provide an introduction to this method. For potentiometric detection of the equivalence point, a meter (called a potentiometer) and two electrodes are needed. One electrode is called an *indicator electrode*; this responds to changes in concentration of either the sample ion or the titrant ion. The other electrode provides a necessary reference for the changes in potential occurring at the indicator electrode; it is called a *reference electrode*.

Since silver nitrate is usually the titrant for halide ions, the indicator electrode often chosen for their potentiometric titration is the silver indicator electrode. The reference electrode is usually a calomel electrode (Ch. 16). These electrodes are connected to an appropriate potentiometer which can be read in terms of $[Ag^+]$ or, more commonly, pAg. The electrodes are then inserted into a solution of the sample, and standard silver nitrate is added from a buret. After each increment of titrant has been added, a reading of pAg is taken from the scale of the potentiometer. This reading changes as a result of the change in the response of the indicator electrode to the decreasing concentration of whatever halide ion is being titrated.

The titration is continued past the equivalence point and pAg readings are taken at appropriate points. With the proper instrumentation, a potentiometric titration curve, such as those in Figure 8-1, can be plotted automatically as the titration proceeds. Otherwise, such curves are plotted manually. In either case, the equivalence point is at the midpoint of the sharp "break" in the titration curve. Note in Figure 8-1 that the break is largest for the titration of iodide ion, and smallest for chloride ion. All three halide ions can be titrated quantitatively with this method, without the use of an indicator.

The Mohr Method for Halide Ions

The Mohr method, published in 1856 [1], is still used for the titration of chloride, bromide, and cyanide ions. The general titration reaction for such anions, X^-, is as follows:

$$AgNO_3 + X^- \longrightarrow AgX(s) + NO_3^- \tag{8-2}$$

(The reaction of cyanide is not typical; it involves the formation of $Ag(CN)_2^-$.) Yellow sodium chromate is added at the beginning of the titration, and the approach of the end point is signaled by temporary flashes of red, or orange-red, silver(I) chromate. The end point is signaled by the first permanent appearance of insoluble red silver(I) chromate.

$$2AgNO_3 + CrO_4^{-2} \longrightarrow Ag_2CrO_4(s) + 2NO_3^-$$

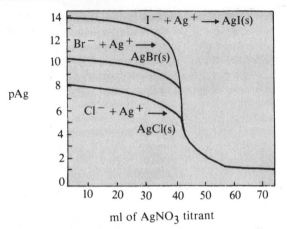

FIGURE 8-1. Potentiometric titration curves for titrations of iodide, bromide, and chloride ions in separate solutions.

The silver(I) chromate is more soluble than silver(I) chloride, so it does not precipitate until the equivalence point is reached. Unfortunately, the end point is somewhat difficult to see over the white silver(I) chloride precipitate and the yellow sodium chromate; a slight excess of titrant is usually necessary to obtain enough silver(I) chromate to see the color change. If the silver nitrate titrant is standardized against primary standard potassium chloride using the same conditions, this error may cancel out. However, it is best to run a blank and subtract this from each titration; the correction is usually in the 0.05–0.15 ml range.

The concentration of chromate is critical. If not enough is present, the end point is late; if too much is present, the end point is premature. The correct concentration can be estimated using the solubility product expression of silver(I) chromate and the calculated concentration of silver(I) at the equivalence point. At this point, from **7-15**, $[Ag^+] = [Cl^-] = 1.3_4 \times 10^{-5} M$. Substituting this value into the silver(I) chromate expression we have:

$$[CrO_4^{-2}] = \frac{K_{sp}}{[Ag^+]^2} = \frac{1.1 \times 10^{-12}}{[1.3_4 \times 10^{-5} M]^2} = 6.1 \times 10^{-3} M$$

Theoretically, therefore, the correct chromate concentration for a color change to occur exactly at the equivalence point is $0.006 M$. Experimentally, $0.006 M$ is slightly too high for a distinct end point in some cases. The color of a $0.006 M$ chromate solution is a bit too deeply yellow to allow one to see the red silver(I) chromate at the end point. As an answer to this difficulty some authors recommend a solution as low as $0.0025 M$ and others one as low as $0.005 M$ chromate, depending on the conditions.

The pH is also critical in this method; the pH should be in the range between 6.5 and 10.3 [2]. Below pH 6.5 the acidity of the solution dissolves too much silver(I) chromate. The reactions are as follows:

$$Ag_2CrO_4(s) = 2Ag^+ + CrO_4^{-2} + H^+ = HCrO_4^-$$

$$2HCrO_4^- = \underset{\text{(dichromate)}}{Cr_2O_7^{-2}} + H_2O$$

As the pH decreases, the ratio of CrO_4^{-2} to $HCrO_4^{-}$ decreases. At pH 7 $[CrO_4^{-2}]/[HCrO_4^{-}] = 3/1$, but at pH 6 this ratio drops to 0.3/1. Above pH 10.3 silver(I) begins to coprecipitate as silver(I) hydroxide. Thus, the pH cannot be acid because of the indicator, or too basic because of a side reaction of the solvent.

Applications: Theoretically, the Mohr method should be satisfactory for the determination of any anion that forms a more insoluble salt than silver chromate. In the previous chapter, we used **7-16** to calculate $[Ag^+]_{min}$, the minimum concentration of silver(I) to precipitate silver chromate. Using this, we summarize in the box the theoretical limitation of the Mohr method for both of the recommended minimum concentrations of 0.0025M and 0.005M chromate.

The Mohr method theoretically should be feasible for the determination of anions at which:

using 0.0025M CrO_4^{-2}, the $[Ag^+]$ at the equivalence point is less than a $[Ag^+]_{min}$ of $2.1 \times 10^{-5}M$, or

using 0.0050M CrO_4^{-2}, the $[Ag^+]$ at the equivalence point is less than a $[Ag^+]_{min}$ of $1.4_8 \times 10^{-5}M$.

In practice, other factors limit the application of the Mohr method. For instance, highly colored silver(I) salts prevent the perception of the indicator color change. As an example, arsenate cannot be titrated because silver(I) arsenate has an intense red color; in fact, it can be used as an indicator in place of silver(I) chromate. Another factor limiting the application is the adsorption of chromate or other anions on the surface of the precipitate.

Of the common anions, chloride, bromide, and cyanide are the principal ones whose titrations are feasible using the Mohr method. This includes the anions in hydrochloric acid and hydrobromic acid, which of course must first be neutralized, as well as alkali metal chlorides, bromides, and cyanides. (Other metal ions tend to precipitate the chromate ion.) Although titrations of both iodide and thiocyanate are theoretically feasible, silver salts of these adsorb the chromate ion, resulting in vague end points. In addition, silver iodide is yellow, and its color obscures the end point.

The Adsorption Indicator (Fajans) Method

In 1924 Fajans and his coworkers [3] observed that certain organic dyes changed colors when adsorbed on the surface of insoluble silver halide salts. The fluorescein family of dyes was found to work especially well for the titration of the halide ions. The indicator action involved the fluorescein anion Fl^- ionizing from the acid form, HFl, of the fluorescein molecule.

The titration reaction for the chloride ion is the same as given in **8-2** for the Mohr titration method. The *overall* reaction at the end point is as follows:

$$Ag^+ + AgCl(s) + Fl^- \longrightarrow AgCl:Ag^+\cdot\cdot Fl^-(s) \qquad \textbf{(8-3)}$$
(excess   (yellow)   (red)
titrant)

Note that the fluorescein anion is yellow before the end point while it is in solution. At the end point it adsorbs on the surface of the silver chloride precipitate and changes color to a pink or red. Apparently the adsorption of the anion changes the type of light absorbed by the anion.

Because the color change occurs on the surface of the precipitate, there must be enough surface for a reasonable amount of the indicator to be adsorbed. Silver halide precipitates are known to coagulate, which reduces the amount of available surface. Therefore, dextrin or polyethylene glycol is usually added to keep the precipitate in the colloidal form and prevent coagulation. Coagulation is also retarded by performing the titration quickly and avoiding excessive stirring. Dilute samples of halide ions limit the amount of surface available because the amount of precipitate is small. The most suitable concentrations are in the $0.005-0.025M$ range.

Because the fluorescein dyes are weakly ionized acids, too high an acidity reduces the concentration of the Fl^- anion to a point where the pink-red color is too faint to be seen. When fluorescein itself is used, the pH must be above 7; dichlorofluorescein is less affected by acidity and can be used above pH 4.

Some of the fluorescein dyes are not suitable for the titration of chloride because their color changes occur before the equivalence point. At pH 7, dichlorofluorescein cannot be used for chloride for this reason. It apparently displaces the chloride ion from the surface of silver chloride somewhat before the equivalence point. In the equation below, adsorption on surface silver ions is shown symbolically.

$$ClAg:Cl^- \cdot\cdot Na^+(s) + Fl^- \longrightarrow ClAg:Fl^- \cdot\cdot Na^+(s) + Cl^-$$

Lowering the pH to 4 reduces the concentration of the anion to a point where chloride can be titrated accurately.

One other limitation is that these titrations cannot be carried out in sunlight or strong room light because of photochemical decomposition, which blackens the precipitate and interferes with the color change.

Applications: Theoretically, the adsorption indicator method should be satisfactory for the determination of any anion that is not displaced from the surface of its insoluble silver(I) salt by the indicator anion. Of course, it is useful to be able to titrate at low pH and thereby avoid neutralizing acidic samples. We have summarized in the box the minimum pH and the ions that are usually titrated with the three main adsorption indicators.

For each adsorption indicator below, the Fajans method is theoretically feasible for the ions given at the minimum pH listed.

Fluorescein (pH $\geqslant$ 7):   usually Cl^-; also feasible for Br^-, I^-, and SCN^-.

Dichlorofluorescein (pH $\geqslant$ 4):   usually Cl^-; also feasible for Br^-, I^-, and SCN^-.

Eosin (pH $\geqslant$ 2):   usually Br^-, I^-, and SCN^-.

The silver(I) ion can also be titrated with a standard solution of potassium chloride, using a positively charged dye such as methyl violet. This has an advantage over the Mohr method, which is impractical for the determination of silver(I) because silver chromate does not dissolve completely at the equivalence point. Neither the adsorption indicator method nor the Mohr method appears to be satisfactory for the titration of chloride in blood serum or urine because of the interference of protein and other organic materials [4].

Back Titration of Halide and Other Anions

In contrast to the three direct titration methods for halide ions, there is essentially only one back titration method for halide ions—the Volhard method. Because of the back titration approach used in the Volhard method, it can be used for many more ions than just the halide ions, as will be shown below.

The Volhard Method

The Volhard method, published in 1874 [5], is widely used because it permits halide and other anions to be titrated in acid solution. In this method a measured excess of standard silver(I) nitrate is added to an acid solution of B^{-n}, the sample anion.

$$nAgNO_3(excess) + B^{-n} \longrightarrow Ag_nB(s) + nNO_3^- \tag{8-4}$$

Note that B^{-n} can represent a -1 ion such as Cl^-, a -2 ion such as $C_2O_4^{-2}$, or even a -3 ion such as AsO_4^{-3}. The unreacted silver nitrate is then back titrated with standard thiocyanate.

$$KSCN + \text{unreacted } AgNO_3 \longrightarrow AgSCN(s) + K^+ + NO_3^- \tag{8-5}$$

At the start of the back titration, $0.02M$ iron(III) is added to the solution. At the equivalence point, the first excess of thiocyanate complexes with the yellow iron(III) to form the red thiocyanatoiron(III) complex ion.

$$KSCN + Fe(III) \longrightarrow Fe(SCN)^{+2} + K^+$$

Quantitativeness of the Back Titration. Since silver(I) thiocyanate is fairly insoluble $(K_{sp} = 1.1 \times 10^{-12} M^2)$, some of the Ag_nB salts formed in **8-4** are more soluble than it. Thus the thiocyanate ion added in **8-5** can react significantly with the insoluble salt as follows:

$$nSCN^- + Ag_nB(s) \rightleftharpoons nAgSCN(s) + B^{-n} \tag{8-6}$$

If this occurs, too much thiocyanate titrant is added in the back titration and the results for chloride will be low. There are two ways to estimate whether or not the point of equilibrium in **8-6** will lie on the right (Sec. 7-2). The first is to compare the calculated molar solubilities of both AgSCN and Ag_nB in pure water. The other is to calculate K_{rxn}, the equilibrium constant for **8-6**. K_{rxn} is defined as:

$$K_{rxn} = \frac{K_{sp} \text{ of } Ag_nB}{(K_{sp} \text{ of } AgSCN)^n} = \frac{[B^{-n}]}{[SCN^-]^n} \tag{8-7}$$

If K_{rxn} is greater than one, there are two modifications of the Volhard method used to avoid the error. One modification is to filter the Ag_nB precipitate before back titrating; this is slow and subject to possible errors from adsorption of some of the excess silver(I) ion. The other modification is to add a little liquid nitrobenzene before the back titration to coat the precipitated Ag_nB and prevent the equilibrium in **8-6** from being established. It is claimed that the nitrobenzene only slows down the reaction and does not prevent it [6], so the titration should be conducted as rapidly as possible.

Applications: The Volhard method has different applications depending on whether either of the two modifications is used. The theoretical limitations of each procedure are summarized in the box.

The Volhard modification in which $Ag_n B$ is not removed or coated with nitrobenzene should theoretically be feasible for the determination of anions whose silver(I) salts have solubilities in pure water significantly less than that of AgSCN ($1 \times 10^{-6}M$) or whose K_{rxn} is significantly less than 1.0.

The Volhard modifications in which $Ag_n B$ is removed or coated with nitrobenzene should theoretically be feasible for the determination of any anion whose silver(I) salt can be precipitated quantitatively.

To determine the chloride ion, a Volhard modification involving either filtration of precipitated silver chloride or coating with liquid nitrobenzene must be used. The comparison of molar solubilities has already been made in the example in Section 7-2. K_{rxn} for the reaction between silver chloride and thiocyanate is calculated:

$$K_{rxn} = \frac{K_{sp} \text{ of AgCl}}{K_{sp} \text{ of AgSCN}} = \frac{1.8 \times 10^{-10}M^2}{1.1 \times 10^{-12}M^2} = 1.6_{36} \times 10^2$$

(Note that the M^2 on each K_{sp} cancels out of the expression, indicating that the setup is correct.) We see that K_{rxn} is indeed greater than 1.0, implying the need for filtering or nitrobenzene.

It can be shown that the Volhard method can be used without filtering or nitrobenzene for the determination of bromide, iodide, thiocyanate, and arsenate. A number of other anions can be determined after filtering or adding nitrobenzene. Although acid is usually added at the start of the Volhard method, its addition can be delayed if necessary. For example, anions of weak acids, such as oxalate and carbonate, may be precipitated in neutral solution, filtered away from the unreacted silver nitrate, and then dissolved in acid and the resulting silver(I) ion titrated with potassium thiocyanate.

Because the digestion of protein requires nitric acid, chloride ion in blood serum and urine can be analyzed after digestion in this acid, using the Volhard method [4].

Volhard Calculations. The results of a Volhard titration may be calculated by subtracting either mmoles or meq. The only complication arises if an anion with a charge greater than -1 is to be determined.

$$B^{-n} + nAgNO_3 \longrightarrow Ag_n B(s) + NO_3^-$$

If mmoles are subtracted and n in the above equation is not one, then the number of mmoles of silver nitrate reacting with B^{-n} must be corrected to agree with the stoichiometry through division by n. In general, the number of mmoles of any anion determined by the Volhard method can be calculated as follows:

$$\text{mmole anion} = \frac{(M \text{ AgNO}_3)(\text{ml}) - (M \text{ KSCN})(\text{ml})}{n} \tag{8-8}$$

Using meq, the stoichiometry is automatically corrected for by using the proper equivalent weight for the B^{-n} anion or its salt. Thus, the meq of any anion determined by the Volhard method can be calculated as follows:

$$\text{meq anion} = (N\,AgNO_3)(ml) - (N\,KSCN)(ml) \qquad \text{(8-9)}$$

The following example illustrates Volhard calculations with both equations above.

EXAMPLE: Calculate the percentage of sodium oxalate (form wt = 134.0) in a 0.5100-g impure sample of $Na_2C_2O_4$. The sample requires 50.00 ml of $0.1111M(N)$ silver nitrate for precipitation of $C_2O_4^{-2}$ as $Ag_2C_2O_4$ and 11.00 ml of $0.1050M(N)$ potassium thiocyanate for back titration.

Solution using mmoles: First calculate the number of mmoles of sodium oxalate from the mmoles of silver nitrate used to precipitate it, using **8-8**.

$$\text{mmole } Na_2C_2O_4 = \frac{(0.1111M)(50.00\text{ ml}) - (0.1050M)(11.00\text{ ml})}{2}$$

$$= \frac{(5.555 - 1.155)\text{ mmole } AgNO_3}{2} = 2.200\text{ mmole}$$

Then calculate the percentage.

$$\% Na_2C_2O_4 = \frac{(2.200\text{ mmole } Na_2C_2O_4)(134.0\text{ mg/mmole})(100)}{510.0\text{ mg}} = 57.80_{39}\%$$

Solution using meq: First calculate the number of meq of sodium oxalate, using **8-9**.

$$\text{meq } Na_2C_2O_4 = (0.1111N)(50.00\text{ ml}) - (0.1050N)(11.00\text{ ml}) = 4.400\text{ meq}$$

Then calculate the percentage.

$$\% Na_2C_2O_4 = \frac{(4.400\text{ meq } Na_2C_2O_4)(67.00\text{ meq/mmole})(100)}{510.0\text{ mg}} = 57.80_{39}\%$$

Adsorption Indicator Method for Sulfate

There is also an adsorption indicator method for the titration of sulfate ion that is quick and reasonably accurate when interfering cations are removed by ion-exchange [7]. The titration is carried out at pH 3.5 in a solvent of approximately 50% water and 50% methyl alcohol.

Alizarin red S, a dye, is used as the indicator. It is yellow in solution but forms a pink complex when it is adsorbed on the surface of the barium sulfate in the presence of a slight excess of barium(II) ion. Using AR^- as the symbol for the indicator anion, the reaction at the end point is as follows:

$$Ba^{+2} + BaSO_4(s) + AR^- \longrightarrow BaSO_4:Ba^{+2}\cdots2AR^-(s)$$

(excess   (yellow)   (pink)

titrant)

Cations such as aluminum(III), iron(III), and potassium(I) interfere seriously, but they can be removed by an ion-exchange column (Ch. 21) before the titration. Chloride, bromide, and perchlorate anions cause only a small error. Nitrate causes a large positive error.

SELF-TEST	**10. Calculation of Precipitation Titration Results**

Answers:

Directions: Cover the answers in the left column before beginning the test. Compare your answers with those given at the left of the test.

A. Calculate the percentage bromide in a 1.0000-g sample analyzed by the Volhard method. 50.00 ml of 0.1000M silver nitrate is added; after precipitation the unreacted silver nitrate is back titrated with 15.00 ml of 0.0800M potassium thiocyanate. Use 79.9 mg/mmole for the formula weight of bromide.

30.36%

B. Using either the molarity or normality system of calculation in Section 6-2 (see **6-2** for equivalent weight), calculate the percentage purity of each compound analyzed by direct titration.

a. A 1.0000-g sample of impure magnesium chloride (form wt = 95.22) requires 35.00 ml of 0.1000M silver nitrate in the Mohr method.

16.66%

b. A 1.000-g sample of impure aluminum chloride (form wt = 133.34 mg/mmole) requires 44.00 ml of 0.1000M silver nitrate in the adsorption indicator method.

19.56%

C. A 1.0000-g sample of chloride is analyzed by the Mohr method; it requires 31.00 ml of 0.1100M silver nitrate for titration. Another 1.0000-g sample is analyzed by the adsorption indicator method using dichlorofluorescein at pH 7; it requires 31.50 ml of 0.1050M silver nitrate.

12.09% (Mohr); 11.73% (adsorption)

a. Calculate the percentage chloride (form wt = 35.453 mg/mmole) found in the sample by both methods.

The pH should be 4 for titration of of chloride using dichlorofluorescein.

b. If there is a difference in the percentage chloride, suggest which of the two methods is the more accurate, and why.

D. A pure compound is either barium chloride (form wt = 208.25) or barium chloride dihydrate ($BaCl_2 \cdot 2H_2O$, form wt = 244.27). Titration of 732.8 mg of the pure compound requires 37.50 ml of 0.1600M silver nitrate. Which compound is it?

$BaCl_2 \cdot 2H_2O$

8-2 | COMPLEXATION TITRATIONS

The ligands discussed in the previous section were mainly *unidentate* ligands which *precipitated* the metal ion as an insoluble salt or as a special type of insoluble 1 : 1 co-ordination (complex) compound. These precipitation reactions are exceptions to the usual situation in which a *soluble* complex ion, or coordination compound, is produced by the reaction of a metal ion with one or more ligands. The advantage of precipitation is that it terminates the reaction of the metal ion with the ligand at a *definite number* of ligands. Furthermore, the reactions are quantitative because the precipitation favors this. The formation of complex ions, in contrast, does not necessarily possess either of these advantages. This is especially true in the case of unidentate ligands. This can be best understood by considering both complex ion equilibria and their equilibrium constants (stability constants).

Complex Ion Equilibria

We will treat these equilibria in terms of the reaction of a metal ion, M, with one or more ligands, L. Such reactions occur in stepwise fashion with one ligand at a time reacting with the metal ion. The equilibrium reaction is usually described by writing the metal and the ligand as reactants and the complex ion or compound (ML_n) as the product.

$$M + L = ML$$

$$ML + L = ML_2$$

$$\vdots$$

$$ML_{(n-1)} + L = ML_n$$

Depending on the conditions, the metal complexes with one, two, or more ligands. It complexes with some maximum number, n, of ligands. That number is characteristic of the particular ligand complexing with it. The value of n varies from two, as in $Ag(NH_3)_2^+$, to four, as in $Cu(NH_3)_4^{+2}$, or to six, as in $Cu(H_2O)_6^{+2}$.

Chelates

Certain ligands use more than one atom to occupy more than one coordination position of the metal ion. This type of ligand is known as a multidentate ligand, in contrast to ligands such as ammonia, which are *monodentate*, or *unidentate*, ligands. The complex ion formed is known as a chelate because so-called chelate rings are formed between the metal ion and the ligand.

One of the simplest of the multidentate ligands is ethylenediamine, $NH_2C_2H_4NH_2$. It can form a complex with both of the nitrogen atoms at either end. A common chelate formed by it is $Cu(NH_2C_2H_4NH_2)_2^{+2}$, or $Cu(en)_2^{+2}$, where *en* is an abbreviation for ethylenediamine. The structure of this chelate is depicted as:

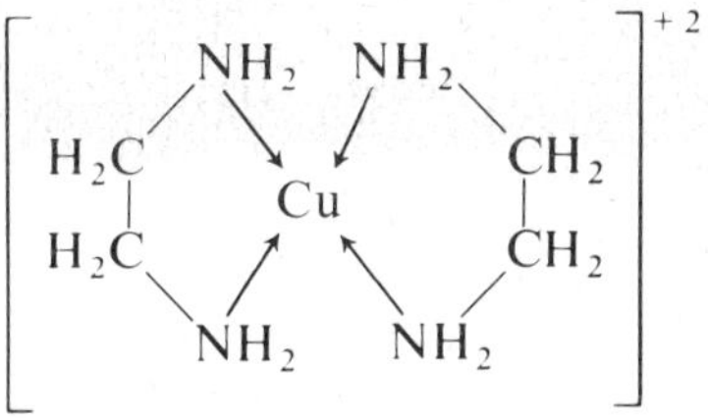

Note that each of the ethylenediamine molecules forms a five-membered ring with the copper(II) ion. The fact that ethylenediamine is able to complex with two of its nitrogen atoms rather than one, as in ammonia, makes its chelate with copper(II) more stable that the $Cu(NH_3)_2^{+2}$ complex ion.

One of the most useful multidentate ligands is known commonly as EDTA (ethylene-diaminetetraacetic acid); this ligand is discussed in more detail later in this section. At this point it is important to know that it is a hexadentate ligand and can form 1 : 1 chelates with most metal ions. Such chelates are very stable, so EDTA can be used for the titration of numerous metal ions. For example, it can be used to titrate calcium(II) ions in water.

$$Ca(H_2O)_6^{+2} + EDTA = Ca\text{-}EDTA + 6H_2O \tag{8-10}$$

Note that calcium(II) ion exists as a hydrated metal ion in water and that when EDTA reacts with it all of the water molecules are displaced by the EDTA in forming the Ca-EDTA chelate.

Stability Constants

The equilibrium constant for complexes and chelates is usually written for the formation of the complex or chelate rather than for the dissociation of either. The constant is termed the *stability constant* or the formation constant. (If the reaction is written as a dissociation, the constant is termed the instability constant.) For each step in the reaction of a metal M with a ligand L there is a corresponding stepwise stability constant, $K_1, K_2, \ldots, K_n$, where n is the maximum number of Ls coordinating to M.

The constants for $Cu(NH_3)_4^{+2}$ are a good example of the stepwise stability constants. They are defined as follows:

$$K_1 = \frac{[Cu(NH_3)^{+2}]}{[Cu^{+2}][NH_3]} = 1.3 \times 10^4 \qquad K_2 = \frac{[Cu(NH_3)_2^{+2}]}{[Cu(NH_3)^{+2}][NH_3]} = 3 \times 10^3$$

$$K_3 = \frac{[Cu(NH_3)_3^{+2}]}{[Cu(NH_3)_2^{+2}][NH_3]} = 8 \times 10^2 \qquad K_4 = \frac{[Cu(NH_3)_4^{+2}]}{[Cu(NH_3)_3^{+2}][NH_3]} = 1.3 \times 10^2$$

Note that the four stability constants describe the equilibria involved, starting with the reaction of Cu^{+2} with the first ammonia ligand and ending with the fourth and final ammonia ligand reacting to form $Cu(NH_3)_4^{+2}$. This is obviously a very complicated equilibrium situation; handling it mathematically is beyond the scope of this book.

The 1:1 chelates are not difficult to handle because only one stability constant is involved. The very important reactions of EDTA with various metals can readily be understood because only one step and one constant are involved. Thus a "conditional" stability constant for the calcium-EDTA reaction **(8-10)** at a constant pH is defined as follows:

$$K_{\text{M-EDTA}} = \frac{[\text{Ca-EDTA}]}{[Ca^{+2}][\text{EDTA}]} = 5 \times 10^{10}$$

The conditional stability constant for any 1:1 metal-EDTA chelate is the same as the equilibrium constant for the titration reaction of the metal and EDTA. There is no need to derive an equation to calculate the equilibrium constant.

Mercury(II) Titration of Chloride and Bromide

Because the stability constants for reactions involving unidentate ligands are usually not large, there are very few complexation titrations based on reactions of unidentate ligands with metal ions. The major exceptions are titrations involving the reactions of mercury(II), usually used as a titrant in the form of mercury(II) nitrate. Two of the more important ions titrated with mercury(II) are the chloride ion and the bromide ion.

Titration of Chloride in Blood or Urine

Although mercury(II) ion can complex with three or more chloride ions, the first two stability constants are much larger than the others:

$$K_1 = 2 \times 10^5 \text{ for } HgCl^+$$

$$K_2 = 3 \times 10^7 \text{ for } HgCl_2$$

$$K_3 = 1.4 \times 10^1 \text{ for } HgCl_3^-, \text{ etc.}$$

Because of this difference, mercury(II) ion is used as a titrant for chloride, with the reaction stopping at the step in which *soluble*, unionized mercury(II) chloride is formed.

$$Hg(NO_3)_2 + 2Cl^- \longrightarrow HgCl_2 + 2NO_3^-$$

The titration can be performed in the presence of high concentrations of nitric acid and is used for chloride concentrations as low as $5 \times 10^{-4} M$. A number of organic ligands can be added as indicators to form a colored chelate, with the first excess of mercury(II) beyond the equivalence point. Diphenylcarbazone, $(C_6H_5)_2CON_2H_2N_2$, is used in one method to form a violet-colored complex with mercury(II) ion to indicate the end point.

$$Hg(NO_3)_2 + (C_6H_5)_2CON_2H_2N_2 \longrightarrow Hg[(C_6H_5)_2CON_2H_2N_2]^{+2} + 2NO_3^-$$
$$\text{(violet)}$$

To determine chloride in blood or urine, the protein must first be precipitated to avoid errors in the titration.

Titration of Bromide Ion

The above approach can also be used for the determination of the bromide ion. Again, the stepwise stability constants K_1 and K_2 are much larger than K_3 for the complexation of mercury(II) with bromide ion. In the case of bromide, an excess of standard mercury(II) perchlorate is added to it, as in the Volhard method.

$$Hg(ClO_4)_2 + 2Br^- \longrightarrow HgBr_2 + 2ClO_4^-$$
$$\text{(excess)}$$

Note that mercury(II) bromide is *soluble*; the reaction terminates at the $HgBr_2$ stage rather than at the $HgBr^+$ or $HgBr_3^-$ stages because of the great stability of this coordination compound. The unreacted mercury(II) ion is then back titrated using a standard solution of thiocyanate.

$$Hg(ClO_4)_2 + 2KSCN \longrightarrow Hg(SCN)_2 + 2KClO_4$$
$$\text{(titrant)}$$

Titrations with EDTA

Because of the disadvantages of most of the unidentate ligands, *hexadentate* ligands have been employed to react with metal ions for analysis. These ligands form rings with metal ions in which as many as six of the coordination positions around the metal ion

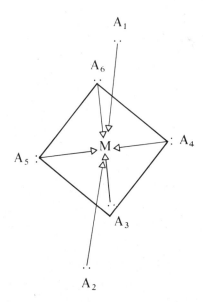

are occupied by six different atoms of the *same* ligand. If we use A_1 through A_6 for the six atoms, the positions they occupy can be pictured as shown. This is the well-known *octahedral* structure in which four of the atoms (A_3–A_6) are positioned in a plane around the central metal M, atom A_1 is above the plane, and atom A_2 is below the plane. We will next discuss in detail the most important of such hexadentate ligands, ethylene-diaminetetraacetic acid (EDTA).

The Chemistry of EDTA

Ethylenediaminetetraacetic acid (EDTA) is a tetraprotic acid. Its structure is depicted as:

$$\left[\begin{array}{c} H\!-\!OOCCH_2 \\ \\ H\!-\!OOCCH_2 \end{array} \!\!\!\! \underset{N-CH_2CH_2-N}{} \!\!\!\! \begin{array}{c} CH_2COO\!-\!H \\ \\ CH_2COO\!-\!H \end{array} \right]$$

$$EDTA\ (=H_4Y)$$

Each of the protons outside the brackets is lost when its adjoining oxygen bonds to a metal ion through the pair of electrons freed by the loss of the proton. EDTA is thus a tetraprotic acid which can be represented as H_4Y, where Y^{-4} represents the anion inside the brackets. As many as four of the above oxygen atoms can occupy as many as four of the octahedral positions around a metal ion. The other two positions are occupied by the nitrogen atoms, which also bond to the metal via their electron pairs. Because the resulting complex ion consists of ring structures, it is called a *chelate*. The general formula for any +2 metal–EDTA chelate is MY^{-2}. A simplified representation of the structure of such a chelate may be drawn by showing only the four oxygens and two nitrogens of EDTA as follows:

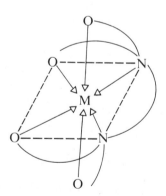

Note that the metal ion is completely surrounded by the EDTA; it is apparent why such chelates are so stable and can be used for quantitative analysis.

The usual form of EDTA used in the laboratory is the disodium salt Na_2H_2Y. When this form is used as the titrant the titration reaction is as follows:

$$Na_2H_2Y + M^{+2} \longrightarrow MY^{-2} + 2H^+ + 2Na^+ \qquad \textbf{(8-11)}$$

TABLE 8-1. Stability Constants of Metal-EDTA Chelates

Metal ion	log K_{MY}	Metal ion	log K_{MY}
Fe^{+3}	25.1	Co^{+2}	16.3
Th^{+4}	23.3	Al^{+3}	16.1
Cr^{+3}	23	Ce^{+3}	16.0
Bi^{+3}	22.8	La^{+3}	15.4
VO^{+2}	18.8	Mn^{+2}	14.0
Cu^{+2}	18.8	Ca^{+2}	10.7
Ni^{+2}	18.6	Mg^{+2}	8.7
Pb^{+2}	18.0	Sr^{+2}	8.6
Cd^{+2}	16.5	Ba^{+2}	7.8
Zn^{+2}	16.5		

Note that this reaction produces hydrogen ion; near the equivalence point it is possible that this would prevent the reaction from being quantitative by breaking apart the MY^{-2} chelate to form HY^{-3}, H_2Y^{-2}, and so on. Therefore a buffer is used for the titration of many metals; for the titration of calcium(II) or magnesium(II), the solution is generally buffered at a pH of 10.

A large number of metal ions may be determined by the kind of direct titration described by **8-11**. The reason for this is that they form very stable 1:1 chelates with EDTA. Their stability constants, K_{MY} (Sec. 7-1), are listed in Table 8-1. All of these constants are large enough that any metal can be titrated quantitatively with EDTA.

The calcium(II)-EDTA chelate has an important medical use in treating patients with lead poisoning. When it is administered to such a person, the lead(II) displaces the calcium(II) from the calcium(II)-EDTA because the lead(II)-EDTA chelate is more stable.

$$\text{Body-Pb(II)} + CaY^{-2} \longrightarrow PbY^{-2} + \text{Body-Ca(II)}$$

(The stability constant for PbY^{-2} is $10^{18.0}$, whereas that for CaY^{-2} is only $10^{10.7}$.) The lead(II) can no longer adhere to the body surfaces and is readily washed out of the body as PbY^{-2}. The EDTA is administered as the calcium chelate to prevent it from reacting with calcium(II) already present in the body.

The indicators used for EDTA titrations are mainly weakly acidic organic dyes. The first of these was Eriochrome Black T (EBT), reported in 1948 [8]. As an indicator, EBT functions by forming a colored metal complex, $M\text{-}EBT^-$, at the start of the titration. As long as some metal remains unchelated by EDTA, the solution remains the color of the $M\text{-}EBT^-$ complex. At the equivalence point, the EDTA removes the metal ion from the $M\text{-}EBT^-$ complex by chelating it, and the solution changes color.

$$Na_2H_2Y + M\text{-}EBT \longrightarrow HEBT^{-2} + MY^{-2} + 2Na^+ + H^+ \qquad \textbf{(8-12)}$$

(titrant)　(color before　(color at
　　　　　end point)　end point)

Formerly, the disodium salt of EDTA was not available in pure enough form for use as a primary standard. There is now available a 99.0% minimum purity A.C.S. grade which can be used as a primary standard for analyses that can tolerate 1% error.

Calculation of Titration Results The concentration of EDTA titrant may be expressed in molarity or in terms of the *titer* of EDTA for a certain ion or compound. The titer of EDTA for any species is defined as the weight of that species reacting with or equivalent to a unit volume of EDTA. Usually this is defined specifically as follows:

$$\text{EDTA titer} = \frac{\text{mg of species}}{1 \text{ ml EDTA}} = \frac{\text{g of species}}{1 \text{ liter of EDTA}}$$

The titer may be calculated from the molarity of the EDTA, but is most often calculated from the standardization data. For example, if EDTA is standardized against calcium carbonate, its titer may be calculated in terms of mg of calcium carbonate reacting (as Ca^{+2}) with one ml of EDTA.

EXAMPLE: Exactly 0.1001 g of pure calcium carbonate (form wt = 100.1 mg/mmole) is dissolved in 100.0 ml of water. A 10.0-ml aliquot is titrated with 9.00 ml of EDTA. Calculate the molarity of the EDTA and its titer for calcium carbonate.
Solution: First calculate the concentration of the calcium carbonate.

$$\text{mg CaCO}_3/\text{ml} = 100.1 \text{ mg}/100.0 \text{ ml} = 1.001 \text{ mg CaCO}_3/\text{ml}$$

$$M \text{ of CaCO}_3 = \frac{1.001 \text{ mg CaCO}_3/\text{ml}}{100.1} = 0.0100M$$

Then calculate the molarity of the EDTA.

$$M \text{ of EDTA} = \frac{(10.0 \text{ ml CaCO}_3)(0.0100M \text{ CaCO}_3)}{9.00 \text{ ml EDTA}} = 0.0111M$$

Finally, calculate its titer by either method below.

$$\text{EDTA titer} = \frac{(10.00 \text{ ml CaCO}_3)(1.001 \text{ mg CaCO}_3/\text{ml})}{9.00 \text{ ml EDTA}} = \frac{1.11_2 \text{ mg CaCO}_3}{\text{ml EDTA}}$$

or

$$\text{EDTA titer} = \frac{(10.00 \text{ ml CaCO}_3)(0.0100M \text{ CaCO}_3)(100.1 \text{ mg/mmole})}{9.00 \text{ ml EDTA}} = \frac{1.11_2 \text{ mg CaCO}_3}{\text{ml EDTA}}$$

The most common EDTA *direct titration* is the water hardness determination (see below). This consists of titrating the *sum* of calcium(II) and magnesium(II) ions in one titration. The results are calculated as parts per million (ppm) of calcium carbonate since this is the major compound causing water hardness. One definition of ppm of any species is the number of mg of that species dissolved in one liter. Thus, for calcium carbonate:

$$\text{ppm CaCO}_3 = \frac{\text{mg CaCO}_3}{\text{liter water}}$$

EXAMPLE: A 50.00-ml water sample from the Los Angeles River requires 12.00 ml of 0.0100M EDTA. Calculate the hardness of this sample as ppm calcium carbonate (form wt = 100.1 mg/mmole).
Solution: First calculate the mg of calcium carbonate from the titration data, recognizing that mmoles of $CaCO_3$ equal mmoles of EDTA.

$$\text{mg CaCO}_3 = (12.00 \text{ ml EDTA})(0.0100M \text{ EDTA})(100.1 \text{ mg CaCO}_3/\text{mmole})$$
$$= 12.01 \text{ mg}$$

Then calculate ppm $CaCO_3$, after converting 50.00 ml to 0.0500 liter of water.

$$\text{ppm } CaCO_3 = \frac{12.01 \text{ mg } CaCO_3}{0.0500 \text{ liter water}} = \frac{240.2 \text{ mg } CaCO_3}{\text{liter water}} = 240.2 \text{ ppm}$$

This is a very "hard" water, since most city water supplies are below 100 ppm.

The most common EDTA *back titration* methods involve the determination of iron(III), aluminum(III), and titanium(IV). Few indicators function in the presence of these metal ions, so an excess of EDTA is added to chelate these metals. Then a standard solution of zinc(II) is used to back titrate the unreacted EDTA, using an indicator such as EBT. Since the above metal ions are already chelated by the EDTA, they do not interfere with the action of the indicator. The calculations for these back titrations are similar to the calculations for the Volhard method. The number of mmoles of the zinc(II) titrant must be subtracted from the number of mmoles of standard EDTA added at the beginning to obtain the number of mmoles of iron(III), and so on. The following self-test includes such a calculation.

SELF-TEST | **11. Calculation of EDTA Titration Results**

Answers:

Directions: Cover the answers in the left column before beginning the test. Compare your answers with those given at the left.

A. A volume of 20.00 ml of EDTA is required to titrate 25.00 ml of standard 0.0100M calcium carbonate (form wt = 100.1 mg/mmole).

0.0125M
1.25 mg/ml
0.500 mg/ml

 a. Calculate the molarity of the EDTA.
 b. Calculate the titer of the EDTA for calcium carbonate.
 c. Calculate the titer of the EDTA for calcium(II) ion.

B. Titration of a 50.00-ml Detroit water sample requires 5.02 ml of 0.0100M EDTA.

100.$_5$ ppm

 a. Calculate the hardness of the water as ppm calcium carbonate (form wt = 100.1 mg/mmole).

0.00100M

 b. Calculate the total molarity of calcium(II) and magnesium(II) ions in the hard water.

C. Titration of a 25.00-ml Des Moines (Iowa) water sample requires 7.50 ml of 0.0100M EDTA.

300.$_3$ ppm

 a. Calculate the hardness of the water as ppm calcium carbonate

120.$_3$ ppm

 b. Calculate the hardness of the water as ppm calcium ion (form wt = 40.1 mg/mmole).

Des Moines water is harder.

 c. Compare the hardness of the Detroit and Des Moines water supplies.

D. Back titration of a 50.00-ml Albuquerque water sample for iron(III) requires 20.80 ml of 0.0110M zinc(II) after the addition of 25.00 ml of 0.0100M EDTA. Calculate the ppm iron (form wt = 55.85 mg/mmole).

23.6$_8$ ppm

The Effect of Acid on Quantitative Reaction

We have so far discussed the chemistry of EDTA titrations only enough to permit us to calculate EDTA titration results. Before we discuss EDTA methods in depth, we must emphasize that the pH of the solution can permit or prevent a quantitative metal-EDTA reaction. Recall from **8-11** that H^+ is a product of the EDTA titration, so too low a pH will reverse the reaction near the end point. You should read this section for whatever level of understanding your instructor recommends: a general grasp of the effect of pH, or a specific mastery of the calculations presented. Then go on to the following discussion of EDTA methods in depth.

We begin by recalling that EDTA is a weakly ionized tetraprotic acid (H_4Y). (At this point, you may want to preview the ionization of polyprotic acids in the next chapter.) As such it has four ionization constants, K_1, K_2, K_3, and K_4. The four ionization steps and the values of the corresponding ionization constants are as follows:

$$H_4Y = H^+ + H_3Y^- \qquad K_1 = 10^{-2.00}$$
$$H_3Y^- = H^+ + H_2Y^{-2} \qquad K_2 = 10^{-2.67}$$
$$H_2Y^{-2} = H^+ + HY^{-3} \qquad K_3 = 10^{-6.16}$$
$$HY^{-3} = H^+ + Y^{-4} \qquad K_4 = 10^{-10.26}$$

Note that we express the ionization constants here in full exponential form instead of semiexponential form. This is done to simplify our calculations later on.

The important species of EDTA is the Y^{-4} anion; all of the stability constants of EDTA (Table 8-1) are defined in terms of this anion. Thus, for the reaction of a $+2$ metal ion with EDTA, the stability constant is defined as follows:

$$K_{MY} = \frac{[MY^{-2}]}{[M^{+2}][Y^{-4}]} \qquad (8\text{-}13)$$

If all of the EDTA is not in the form of the Y^{-4} anion, then K_{MY} does not give adequate information on the equilibrium. We can decide whether this is true by calculating the fraction of EDTA in the Y^{-4} form. This fraction is called α_4, and it is defined as follows:

$$\alpha_4 = \frac{[Y^{-4}]}{[EDTA]} \qquad (8\text{-}14)$$

where [EDTA] is the total concentration of all forms of EDTA, including the Y^{-4} form. The mathematical relation needed to calculate α_4 can be derived using the definitions of the four ionization constants.

$$\alpha_4 = \frac{K_1K_2K_3K_4}{[H^+]^4 + [H^+]^3K_1 + [H^+]^2K_1K_2 + [H^+]K_1K_2K_3 + K_1K_2K_3K_4} \qquad (8\text{-}15)$$

An alternative approach for the calculation of α_4 is to use the equation below to calculate $1/\alpha_4$ and then invert this value.

$$\frac{1}{\alpha_4} = \frac{[H^+]^4}{K_1K_2K_3K_4} + \frac{[H^+]^3}{K_2K_3K_4} + \frac{[H^+]^2}{K_3K_4} + \frac{[H^+]}{K_4} + 1$$

The variation of α_4 with pH is estimated in Figure 8-2. You will notice that at pH 12 and above it has a value close to unity. Unfortunately, most EDTA titrations are

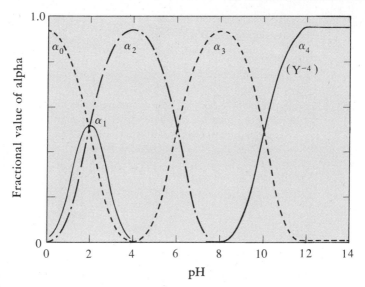

FIGURE 8-2. A plot of the fraction of EDTA present in various forms at different pH values. The left-hand portion of each curve approaches, but does not become, zero. The subscript on each α value refers to the charge on the EDTA species; thus α_4 refers to the fraction of EDTA present in the Y^{-4} form.

conducted at pH 10 and below, so it is important to be able to calculate a value for α_4 at any given pH. This calculation can usually be simplified by using only the significant terms (generally two) in the denominator of **8-15**, as the following example will show.

EXAMPLE: Calculate α_4 for EDTA at pH 10.0 and indicate what fraction of EDTA is present as Y^{-4} at this pH.

Solution: First calculate the several products of K_1, K_2, K_3, and K_4 for the numerator and denominator of **8-15**. They are

$$K_1 K_2 K_3 K_4 = 10^{-21.09}$$

$$K_1 K_2 K_3 = 10^{-10.83}$$

$$K_1 K_2 = 10^{-4.67}$$

Then substitute these values into **8-15** with the $[H^+]$.

$$\alpha_4 = \frac{10^{-21.09}}{[10^{-10}]^4 + [10^{-10}]^3 10^{-2.00} + [10^{-10}]^2 10^{-4.67} + [10^{-10}] 10^{-10.83} + 10^{-21.09}}$$

Next decide which of the terms in the denominator are significant. Looking at Figure 8-2, we see that Y^{-4} and HY^{-3} are the important species present at pH 10, and since the last two terms in the denominator reflect these two species, we evaluate these first. We find that they are indeed the two largest terms, and we proceed to add them up.

$$[10^{-10}][10^{-10.83}] = 10^{-20.83} = 1.48 \times 10^{-21}$$

$$10^{-21.09} = 8.1 \times 10^{-22} = 0.81 \times 10^{-21}$$

$$\text{sum} = \overline{2.29 \times 10^{-21}} = 10^{-20.64}$$

Finally, we calculate α_4.

$$\alpha_4 = \frac{10^{-21.09}}{10^{-20.64}} = 10^{-0.45} = 3.5 \times 10^{-1}$$

Note that this is the value given in Table 8-2.

Equation **8-15** can be used to calculate values for α_4 for all the pHs at which EDTA titrations are conducted. These values are listed in Table 8-2.

Now that values for α_4 are known, we can use them to calculate the effect of pH on the stability constant. First we rearrange **8-14**.

$$[Y^{-4}] = \alpha_4[EDTA] \tag{8-16}$$

Then we substitute the right side of **8-16** for $[Y^{-4}]$ in **8-13**.

$$K_{MY} = \frac{[MY^{-2}]}{[M^{+2}][EDTA]\alpha_4} \tag{8-17}$$

A careful look at this equation reveals that it is close to being an equilibrium constant expression for the reaction of all forms of EDTA with a metal ion.

$$EDTA + M^{+2} \longrightarrow MY^{-2} \quad (+xH^+)$$

Rearranging **8-17** gives the desired expression for the above reaction.

$$\alpha_4 K_{MY} = K_{M\text{-EDTA}} = \frac{[MY^{-2}]}{[M^{+2}][EDTA]} \tag{8-18}$$

The product of α_4 and K_{MY} yields a new stability constant, $K_{M\text{-EDTA}}$, which is a *conditional stability constant*, or conditional formation constant. It is constant only if the pH is constant during the titration. Since buffers are used in most EDTA titrations, this is

TABLE 8-2. Values of α_4 for EDTA at Different pHs

pH	α_4	
	Semiexponential	Exponential value
2.0	3.7×10^{-14}	$10^{-13.44}$
3.0	2.5×10^{-11}	$10^{-10.60}$
4.0	3.3×10^{-9}	$10^{-8.48}$
5.0	3.5×10^{-7}	$10^{-6.45}$
6.0	2.2×10^{-5}	$10^{-4.66}$
7.0	4.8×10^{-4}	$10^{-3.33}$
8.0	5.1×10^{-3}	$10^{-2.29}$
9.0	5.1×10^{-2}	$10^{-1.29}$
10.0	0.35	$10^{-0.45}$
11.0	0.85	$10^{-0.07}$
12.0	0.98	$10^{-0.01}$
13.0	1.0	$10^{-0.00}$

a very useful constant. It is more useful than K_{MY}, because it can be used for accurate equilibrium calculations and to decide if a reaction is quantitative at a given pH.

The criterion used in Section 7-2 to decide whether a reaction is quantitative is as follows:

> If the *actual* conditional stability constant is $\geqslant$ the theoretical minimum value of $K_{M\text{-EDTA}}$, then direct EDTA titration of the metal ion will be quantitative.

The minimum theoretical value of $K_{M\text{-EDTA}}$ varies with concentration. A few values follow.

Concn	Min theo $K_{M\text{-EDTA}}$
$0.01M$	1×10^8
$0.001M$	1×10^9
$0.0001M$	1×10^{10}

Since most EDTA titrations are conducted at the $0.01M$ level, the actual value of the conditional stability constant should be compared with the theoretical minimum of 1×10^8. The most acidic pH at which an EDTA titration can be conducted can also be calculated using the above values and a rearranged form of **8-18**.

$$\min \alpha_4 = \frac{\text{min theo } K_{M\text{-EDTA}}}{K_{MY}} \tag{8-19}$$

Suppose that we wish to know the most acidic pH at which a particular metal ion can be titrated at the $0.01M$ level. We divide the value of K_{MY} (Table 8-1) into 1×10^8 to obtain a minimum value for α_4. Then we select the pH value closest to it in Table 8-2. This will be the most acidic pH at which the titration can be conducted and still be quantitative. Of course, the titration will still be quantitative at less acidic pHs, but other metal ions may interfere at these pHs. The more acidic the pH, the fewer the metal ions that will interfere by reacting with EDTA.

EXAMPLE: It is proposed to titrate $0.01M$ levels of magnesium(II) at a pH of 9. Will the titration be quantitative? If not, what is the most acidic pH at which the titration will be quantitative?
Solution: We see from Table 8-2 that α_4 is 0.051 or $10^{-1.29}$. We then use **8-18** to calculate the conditional stability constant at pH 9 and insert the value of $10^{8.7}$ for K_{MgY} as follows:

$$K_{M\text{-EDTA}} = (10^{-1.29})(10^{8.7}) = 10^{7.4}$$

Since this is less than the minimum theoretical value of 10^8, the titration will not be quantitative.

The most acidic pH at which the titration will be quantitative is found by solving for the minimum value of α_4 using **8-19**.

$$\min \alpha_4 = \frac{1 \times 10^8}{10^{8.7}} = 10^{-0.7}$$

We see in Table 8-2 that the above value is closest to the value of α_4 of $10^{-0.45}$ at pH 10. Therefore, the most acidic pH at which this titration will be quantitative is pH 10.

The Effect of a Second Metal Ion
To titrate one metal ion, say M_1, in the presence of a second metal ion, M_2, it is necessary that the ratio of their stability constants (Table 8-1) be as follows:

$$\frac{K_{M_1Y}}{K_{M_2Y}} \geq \frac{10^6}{1}$$

It is also necessary to reduce the pH so that after M_1 has been titrated quantitatively, M_2 does not react appreciably with EDTA before the indicator color change for M_1. The safest thing to do is to adjust the pH so that the conditional stability constant for M_2-EDTA is 10^2 or less.

The Effect of Other Ligands
If there are other ligands, such as ammonia, present in a solution, they will reduce the ability of EDTA to chelate the metal ion. Their effect on the conditional stability constant can be calculated by means of the parameter β.

$$\beta = \frac{[M^{+2}]}{\text{total } M \text{ of } M^{II} \text{ not chelated by EDTA}}$$

A numerical value for β for any ML_4 species is calculated as follows:

$$\frac{1}{\beta} = 1 + K_1[L] + K_1K_2[L]^2 + K_1K_2K_3[L]^3 + K_1K_2K_3K_4[L]^4 \qquad \textbf{(8-20)}$$

where the Ks are the stepwise stability constants and $[L]$ is the molarity of the free ligand. The value of β is then multiplied by the conditional stability constant to obtain a new conditional stability constant, which is valid only for that value of $[L]$.

EDTA Methods
Of all of the EDTA methods, the titrations of calcium(II) and magnesium(II) ions are the most common. We first discuss the titration of these ions separately, and then together, as in the water hardness titration.

Titration of Magnesium(II)
Magnesium is usually titrated at a pH of 10; at higher pHs, such as pH 12, it precipitates as magnesium hydroxide. Eriochrome Black T may be used as the indicator, but because it is somewhat unstable, Calmagite is often preferred. Both indicators exhibit the same colors; that is, both form a pink-red magnesium complex, so we will discuss the color change at the end point in terms of Eriochrome Black T.

Since Eriochrome Black T is a triprotic acid, we symbolize it as H_3EBT. Its first hydrogen is strongly ionized, but $pK_2 = 6.9$ and $pK_3 = 11.5$. Using these constants the predominant forms over the entire pH range can be deduced.

pH	Predominant form
1–6.9	Purple-red H_2EBT^-
6.9–11.5	Blue $HEBT^{-2}$
11.5–14	Orange EBT^{-3}

The color change at the titration end point for magnesium(II) at the usual pH of 10 is as follows:

$$MgEBT^- + Na_2H_2Y \longrightarrow MgY^{-2} + HEBT^{-2} + H^+ \qquad \textbf{(8-21)}$$

$$\text{(pink-red)} \quad \text{(titrant)} \qquad\qquad \text{(blue)}$$

It is obvious that a distinct color change for this reaction can occur theoretically only in the pH range of 6.9–11.5. (In practice, the color change is found to be indefinite until the pH is above 8.) Below pH 6.9 the color change would be pink-red to purple-red; above pH 11.5 the change would be pink-red to orange. Both of these changes are difficult to see.

Because magnesium(II) forms only a moderately stable EDTA chelate, the most acidic pH at which it can be titrated is close to pH 10 (see the example on p. 157). Therefore, the titration of magnesium(II) is limited to a pH range of 10–11 by the indicator pH range, its own EDTA chelate stability, and the fact that it precipitates at pH 12. An ammonia buffer is usually used to maintain the pH around 10. Fortunately, magnesium(II) does not form complex ions with ammonia, so its conditional stability constant is not affected by the ammonia buffer.

| Titration of Calcium(II) | The EDTA chelate of calcium(II) is more stable than that of magnesium(II); K_{CaY} is $10^{10.7}$ (Table 8-1). The most acidic pH at which it can be titrated is calculated to be pH 8.0, using the method of the example on p. 157. However, it is usually titrated in a pH range of 10–13 because of the pH requirements of the available indicators. |

Because the CaEBT$^-$ complex is unstable, calcium(II) cannot be titrated *alone* using Eriochrome Black T or Calmagite. It can be titrated if a small amount of magnesium(II) is present to form the stable MgEBT$^-$ complex. Therefore, a small amount of magnesium(II) is added to the titrant or a small amount of magnesium-EDTA is added to the solution. In either case, the pink-red MgEBT$^-$ complex immediately forms. After the calcium(II) has been titrated, the reaction in **8-21** occurs to give a color change at the equivalence point.

Calcium(II) can also be titrated at pH 13 using Calcein or Calcein Blue indicators. The calcium chelate of these indicators exhibits a *fluorescent* yellow-green color; the indicators alone are a *nonfluorescent* light brown. The color change at the end point is therefore the disappearance of the fluorescent yellow-green color and the appearance of the light brown color. The titration can also be carried out in the dark, using an ultraviolet lamp to cause the indicator to fluoresce yellow-green (see Ch. 14 for a discussion of fluorescence). At the end point the fluorescence ceases and the solution is dark.

Determination of Water Hardness — EDTA can be used to titrate the *total* of calcium(II) and magnesium(II) in hard water. The results are usually calculated as parts per million (ppm) of calcium carbonate (see the example on p. 152) rather than the total molarity of both ions. Such an analysis is very important, since the amount of calcium and magnesium carbonate residue left behind by water that has evaporated affects the operation of equipment ranging from large industrial boilers to home hot-water heaters to steam irons.

The usual water hardness determination is conducted by titration at a pH of 10; an ammonia buffer is used to keep the pH at this value. Either Eriochrome Black T or Calmagite may be used as the indicator. After the magnesium(II) and calcium(II) are chelated, the MgEBT$^-$ reacts with EDTA **(8-21)** to provide the color change from pink-red to blue.

Small amounts of iron(III), aluminum(III), and copper(II) in the hard water are *potential* interferences. These ions react irreversibly with Eriochrome Black T and Calmagite, forming stable pink-red complexes. If this is not avoided, the water hardness end point is difficult or impossible to perceive, since the color does not change to blue. The interference of small amounts of iron(III) and aluminum(III) can be avoided by adding the ammonia buffer *before* the indicator is added. This complexes these cations as hydroxo (OH$^-$) complexes, so that they cannot react with the indicator. The addition of the cyanide ion, or hydroxylamine, prevents the interference of copper(II) by reduction to copper(I). These reagents also prevent the interference of small or large amounts of iron(III).

Titration of
+3 and +4
Cations

Many +3 and +4 cations cannot be titrated in the basic solution needed for Eriochrome Black T and Calmagite because they precipitate as insoluble hydroxides. Such titrations can be accomplished using xylenol orange indicator, which forms red complexes and is yellow in its uncomplexed form.

Thorium(IV) and bismuth(III), which form two of the most stable EDTA chelates (Table 8-1), can be titrated in acid solution at any pH in the range of 1.5–3.0. The color change at the end point for the titration of bismuth(III) is as follows (XO = xylenol orange):

$$\text{Bi}^{III}(\text{XO}) + \text{Na}_2\text{H}_2\text{Y} \longrightarrow \text{BiY}^- + \text{XO} + 2\text{H}^+ + 2\text{Na}^+ \quad \textbf{(8-22)}$$

$$\text{(red)} \qquad \text{(titrant)} \qquad\qquad \text{(yellow)}$$

The color change is the same for the direct titration of thorium(IV). Theoretically, any metal ion with a stability constant of $10^{15.44}$ or less will have a conditional stability constant of 10^2 or less at pH 2 (see Table 8-2 for the value of α_4) and should not interfere in the titration of these two metal ions at that pH. Thus thorium(IV) and bismuth(III) can be titrated in the presence of lanthanum(III), calcium(II), and so on (Table 8-1).

Zirconium(IV) and iron(III) cannot be determined by direct titration using xylenol orange, but they can be determined by back titration. An excess of standard EDTA is added to either of these ions at a pH of 1–2. The unreacted EDTA is then back titrated with a standard solution of bismuth(III).

$$\text{H}_2\text{Y}^{-2} + \text{Bi}^{III} + \text{FeY} \longrightarrow \text{BiY}^- + 2\text{H}^+ + \text{FeY}$$

$$\text{(unreacted EDTA)} \quad \text{(titrant)}$$

Even though the iron(III)-EDTA, or zirconium(IV)-EDTA, is present at the end point, it does not react with the bismuth(III) because the iron(III)-EDTA chelate is more stable than the bismuth(III)-EDTA chelate (Table 8-1).

Certain +2 cations, such as lead(II) and cadmium(II), can also be titrated using xylenol orange at higher pHs, such as pH 5.

| QUESTIONS AND PROBLEMS

(Answers to even-numbered problems are found in Appendix 5.)

Concepts and Definitions

1. Write the titration reaction and the indicator reaction for the titration of potassium bromide with standard silver nitrate using each of the following:
   a. The Mohr method
   b. The adsorption indicator method
   c. The Volhard method
2. Suggest whether each of the three methods in Problem 1 can be used for titration of silver(I) ion with standard potassium chloride as titrant. If any method cannot be used, give a reason.

3. What advantage does the Volhard method have over the two direct titration methods for the determination of an anion whose insoluble silver(I) salt has a K_{sp} value appreciably less than 1×10^{-8}?
4. Define each term and give an example.
   a. Unidentate ligand
   b. Hexadentate ligand
   c. Bidentate ligand
5. Explain why EDTA titrations are usually conducted in the presence of a buffer.
6. Explain why EDTA is superior to a unidentate ligand for the titration of metal ions.

Precipitation Titration Equilibria

7. For silver(I) bromide, $K_{sp} = 4.9 \times 10^{-13}$.
   For the titration of $0.10M$ bromide with
   $0.10M$ silver nitrate, calculate the following,
   neglecting dilution:
   a. pAg at 99.9% reaction
   b. pAg at the equivalence point
   c. pAg at 100.1% reaction

8. Decide whether each of the anions below can
   be titrated quantitatively by direct titration
   with silver nitrate. Calculate a numerical
   value for K_{rxn}.
   a. $0.10M$ IO_3^- ($0.10M$ $AgNO_3$)
   b. $0.010M$ Cl^- ($0.010M$ $AgNO_3$)
   c. $0.0010M$ SCN^- ($0.0010M$ $AgNO_3$)
   d. $0.00010M$ Br^- ($0.00010M$ $AgNO_3$)

9. Decide whether each of the anions in
   Problem 8 can be precipitated quantitatively
   by adding 50.0 ml of the specified silver
   nitrate solution to 20 ml of the specified
   solution of the anion (the first step of the
   Volhard method).

10. Evaluate each of the anions below as a
    possible substitute for the chromate ion in the
    Mohr titration of the chloride ion. Use two
    criteria: there should be a definite color
    change (check a handbook), and the silver(I)
    ion present at the equivalence point should
    not precipitate a $0.0025M$ solution of the
    anion.
    a. $Br^- \rightarrow AgBr(s)$
    b. $S^{-2} \rightarrow Ag_2S(s)$; $K_{sp} = 6 \times 10^{-50}M^3$
    c. $AsO_4^{-3} \rightarrow Ag_3AsO_4(s)$;
       $K_{sp} = 1.1 \times 10^{-20}M^4$
    d. $PO_4^{-3} \rightarrow Ag_3PO_4(s)$
    e. $N_3^- \rightarrow AgN_3(s)$
    f. $Fe(CN)_6^{-4} \rightarrow Ag_4Fe(CN)_6(s)$;
       $K_{sp} = 1._6 \times 10^{-41}M^5$

11. It is desired to apply the Mohr method for the
    precipitation titration of the azide ion,
    precipitating $AgN_3(s)$. Evaluate each of the
    anions below as a possible indicator using
    two criteria: the insoluble silver salt formed
    should be colored (check a handbook), and
    the silver(I) ion present at the equivalence
    point should not precipitate a $0.0025M$
    solution of the anion. Use a K_{sp} of
    $2.9 \times 10^{-9}M^2$ for $AgN_3(s)$.
    a. $CrO_4^{-2} \rightarrow Ag_2CrO_4(s)$

b. $WO_4^{-2} \rightarrow Ag_2WO_4(s)$;
   $K_{sp} = 5.5 \times 10^{-12}M^3$
c. $AsO_4^{-3} \rightarrow Ag_3AsO_4(s)$;
   $K_{sp} = 1.1 \times 10^{-20}M^4$
d. $VO_3^{-1} \rightarrow AgVO_3(s)$;
   $K_{sp} = 5 \times 10^{-7}M^2$

12. The concentration of the chromate ion used
    in the Mohr method is arbitrarily increased to
    $0.0070M$. Using appropriate calculations,
    decide whether the Mohr method can now be
    applied to the titration of the following ions:
    a. $Cl^- \rightarrow AgCl(s)$
    b. $SO_3^{-2} \rightarrow Ag_2SO_3(s)$;
       $K_{sp} = 1.5 \times 10^{-14}M^3$
    c. $SeO_3^{-2} \rightarrow Ag_2SeO_3(s)$;
       $K_{sp} = 9.7 \times 10^{-16}M^3$
    d. $PO_4^{-3} \rightarrow Ag_3PO_4(s)$

13. Decide whether there will be an error in the
    Volhard back titration of each of the following
    anions if potassium thiocyanate is used as the
    back titrant without filtration or addition of
    nitrobenzene. Use a K_{sp} of $1.1 \times 10^{-12}M^2$
    for AgSCN; other K_{sp}s are in Appendix 1 if
    not given below.
    a. Cl^-
    b. Br^-
    c. MoO_4^{-2} (for $Ag_2MoO_4(s)$,
       $K_{sp} = 2.8 \times 10^{-12}M^3$)
    d. AsO_4^{-3} (for $Ag_3AsO_4(s)$,
       $K_{sp} = 1.1 \times 10^{-20}M^4$)
    e. $Fe(CN)_6^{-3}$ (for $Ag_4Fe(CN)_6(s)$,
       $K_{sp} = 1._6 \times 10^{-41}M^5$)

Calculation of Precipitation Titration Results

14. Calculate the percentage iodide in a 1.000-g
    sample to which is added 50.00 ml of $0.1000M$
    silver nitrate. The unreacted silver nitrate is
    back titrated with 16.00 ml of $0.0800M$
    potassium thiocyanate.

15. A 0.5000-g sample of chloride requires 15.50
    ml of $0.1100M$ silver nitrate when titrated by
    the Mohr method. A 1.0000-g portion of the
    same sample requires 31.50 ml of $0.1050M$
    silver nitrate using dichlorofluorescein at
    pH 7. Calculate the percentage chloride
    results from both methods and rationalize any
    differences.

16. Calculate the percentage of sodium carbonate
    (form wt = 106.0) in a 0.5100-g impure

sample of Na_2CO_3. The sample requires 50.00 ml of $0.1111M$ (N) silver nitrate for precipitation of CO_3^{-2} as Ag_2CO_3, and 11.00 ml of $0.1050M$ (N) potassium thiocyanate for back titration.

17. Calculate the percentage of potassium arsenate (form wt $= 256.2$) in a 0.6200-g impure sample of K_3AsO_4. The sample requires 50.00 ml of $0.1111M$ (N) silver nitrate for precipitation of AsO_4^{-3} as Ag_3AsO_4, and 12.00 ml of $0.1010M$ (N) potassium thiocyanate for back titration.

Complexation Titrations

18. Calculate values of α_4 at the following pHs:
    a. pH $= 7.5$
    b. pH $= 8.5$
    c. pH $= 9.5$
    d. pH $= 10.5$

19. It is proposed to titrate $0.01M$ calcium(II) ion at each of the following pHs. Will each titration be quantitative?
    a. pH $= 8.0$
    b. pH $= 7.5$
    c. pH $= 8.5$

20. It is proposed to titrate $1 \times 10^{-4}M$ calcium(II) ion in urine samples. Is there a pH or a pH range at which the titration will be quantitative? What is the pH or pH range?

21. Titration of a 50.00-ml hard water sample requires 4.20 ml of $0.0100M$ EDTA. Calculate the ppm calcium(II) carbonate in the water.

22. A 21.00-ml volume of EDTA titrant is standardized against 25.00 ml of standard $0.0110M$ calcium carbonate. A 75.00-ml hard water sample requires 29.50 ml of the EDTA titrant for a total hardness (mg $CaCO_3$ per liter) analysis. Calculate both the ppm calcium (mg Ca/liter) and the ppm calcium carbonate in the hard water sample.

23. Calcium may be titrated in the presence of magnesium by raising the pH above 12 and precipitating magnesium hydroxide. Calcein indicator is used in place of EBT at this pH. A 50.00-ml aliquot of a hard water sample is adjusted to pH 12 with base, precipitating

magnesium hydroxide. After addition of Calcein, 11.00 ml of $0.0110M$ EDTA are required for titration. A second 50.00-ml aliquot is adjusted to pH 10 and, after addition of EBT indicator, is titrated with 14.50 ml of $0.0120M$ EDTA. Perform the following calculations for this hard water sample:
    a. Molarity of calcium
    b. Molarity of magnesium plus calcium
    c. Molarity of magnesium

24. Given the log of each stability constant for $Cu(NH_3)_4^{+2}$ below, calculate the new conditional stability constant for the titration of copper(II) ion with EDTA at pH 9 in $0.1M$ ammonia. Decide whether titration will be quantitative.

$$\log K_1 = 4.1 \qquad \log K_2 = 3.5$$
$$\log K_3 = 2.9 \qquad \log K_4 = 2.1$$

Challenging Problems

25. What is the most acidic pH at which Cu^{+2} can be titrated with EDTA at:
    a. The $0.01M$ level?
    b. The $1 \times 10^{-5}M$ level?
    c. The $5 \times 10^{-7}M$ level?

26. By comparing the solubilities of the silver(I) salts formed with the solubility in pure water of silver(I) chromate and silver(I) thiocyanate, state for each anion below whether it can be determined by the Mohr method and whether nitrobenzene must or need not be used in the Volhard determination of the anion.
    a. Br^-
    b. IO_3^-
    c. CO_3^{-2}
    d. $Co(CN)_6^{-3}$
       ($K_{sp} = 8.0 \times 10^{-21}M^4$ for the silver salt)

27. A mixture containing only sodium chloride (form wt $= 58.44$) and potassium chloride (form wt $= 74.56$) without any other substances present can be analyzed by a single direct titration with silver nitrate titrant. Calculate the percentage of sodium chloride in 191.44 mg of the mixture if 30.00 ml of $0.1000M$ silver nitrate are required for complete precipitation of the chloride ion.

28. Calculate the mg/dL of calcium(II) ion in a 0.200-ml blood sample that is titrated with 2.50 ml of 0.000100M EDTA. Is the concentration of calcium(II) within, below, or above the normal range (Fig. 1-1)?

| **NOTES**

[1] F. Mohr, *Annalen der Chemie und Pharmacie* **97**, 335 (1856).
[2] R. Belcher, A. M. G. MacDonald, and E. Parry, *Anal. Chim. Acta* **16**, 524 (1957).
[3] K. Fajans and H. Wolff, *Z. Anorg. Allg. Chem.* **137**, 221 (1924).
[4] R. L. Searcy, *Diagnostic Biochemistry*, McGraw-Hill, New York, 1969, p. 156.
[5] J. Volhard, *J. Prakt. Chem.* **9**, 217 (1874).
[6] H. F. Walton, *Principles and Methods of Chemical Analysis*, Prentice-Hall, Englewood Cliffs, N.J., 1964, p. 375.
[7] J. S. Fritz and M. Q. Freeland, *Anal. Chem.* **26**, 1593 (1954).
[8] G. Schwarzenback and W. Biedermann, *Helv. Chim. Acta* **31**, 678 (1948).

9 | Elementary Acid-Base Equilibria

"Interesting though *elementary*," said Holmes.

ARTHUR CONAN DOYLE
The Hound of the Baskervilles, Chapter 1

In this chapter your goal should be to review and master the fundamentals of acid-base equilibria so that you will be able to comprehend the equilibrium aspects of acid-base titrations presented in the next chapter. We will not present rigorous methods of calculation but rather useful approximations for the purpose of dealing with acid-base titration calculations. You should be familiar with the Bronsted acid-base theory and the terms conjugate acid and conjugate base; these are very important in acid-base equilibria.

9-1 | THE IONIZATION OF WATER AND pH

In this section the emphasis is placed on the definition of pH and pOH and on the importance of the ionization of water in defining pH. First it is necessary to consider the ionization of water when it is the solvent.

Ionization of Water

Any proton-containing solvent such as water undergoes an autoprotolysis, or self-ionization. This can be written in two ways:

$$2H_2O = H_3O^+ + OH^- \quad \text{or} \quad H_2O = H^+ + OH^-$$

In either case we understand that the hydrogen ion is solvated by one or more water molecules acting as Bronsted bases toward the proton.

We can speak accurately of the activity (Sec. 7-1) of the hydrogen ion, a_{H^+}, or the molarity of the hydrogen ion, $[H^+]$, in pure water. The same is true for the hydroxide ion. Thus, in pure water,

$$a_{H^+} = [H^+] \quad \text{and} \quad a_{OH^-} = [OH^-]$$

However, when ionic compounds are dissolved in the water, the equalities above are not valid. To simplify, we will use molarities rather than activities. Using this approximation, the equilibrium constant, K_w, for the ionization of water is defined as follows:

$$K_w = [H^+][OH^-] = [H_3O^+][OH^-] \tag{9-1}$$

Although a value of 1.0×10^{-14} is commonly used for K_w, this value is valid only at 25°C. This is because the ionization is endothermic; that is,

$$2H_2O + heat = H_3O^+ + OH^- \tag{9-2}$$

Thus, as the temperature is increased, the degree of ionization and the hydrogen ion concentration of water increase. Some values of K_w for other temperatures are given in Table 9-1.

TABLE 9-1. Values of K_w at Other Temperatures

Temperature, °C	K_w	pK_w
0	1.1×10^{-15}	14.94
10	2.9×10^{-15}	14.53
20	6.8×10^{-15}	14.17
25	1.0×10^{-14}	14.00
30	1.47×10^{-14}	13.83
37 (body temp.)	2.5×10^{-14}	13.60
40	2.9×10^{-14}	13.53
60	9.6×10^{-14}	13.02

Note that K_w gradually *increases* as the temperature increases. Even at body temperature, which is only 12°C above room temperature, K_w is significantly larger than it is at room temperature. Also note that the negative log of K_w, pK_w, gradually *decreases* as the temperature increases. Either constant may be used to calculate pH, as will be shown next.

The pH Scale

Strictly speaking, pH is defined in terms of the activity of the hydrogen ion.

$$pH = -\log a_{H^+}$$

The same is true for pOH. Since we are assuming that molarity is approximately the same as activity, we will use the following approximate definitions of pH and pOH:

$$pH = -\log[H^+]$$

$$pOH = -\log[OH^-]$$

It is also true that

$$pK_w = pH + pOH \tag{9-3}$$

At 25°C, where the $pK_w = 14.00$,

$$pH = 14.00 - pOH$$

$$pOH = 14.00 - pH$$

The *usual* pH scale ranges from the pH of a $1M$ H^+ solution to the pH of a $1M$ OH^- solution. The pH of $1M$ $H^+ = 0.0$. The pOH of a $1M$ OH^- solution is 0.0; the pH of this solution is 14.0 at 25°C **(9-3)**. Obviously it is possible to have a negative pH if the $[H^+]$ is larger than $1.0M$, just as it is possible to have a pH larger than 14 if the $[H^+]$ is smaller than $1.0 \times 10^{-14}M$.

A *neutral solution* at any temperature is one in which $[H^+] = [OH^-]$, or the pH = pOH. Substituting this equality into **9-1**, we obtain for any temperature:

$$K_w = [H^+][OH^-] = [H^+][H^+] = [H^+]^2$$
$$[H^+] = \sqrt{K_w} \tag{9-4}$$

or

$$pH = \tfrac{1}{2}pK_w \tag{9-5}$$

Thus a neutral solution at 25°C is one in which the $[H^+] = 1.0 \times 10^{-7}$, and the pH = 7.00.

It is interesting to compare the pH scale at 25°C with that at the body temperature of 37°C. At 37°C, the pH scale would again encompass the range from a pH of $1M$ H^+ to the pH of $1M$ OH^-. At this temperature, the pH of $1M$ H^+ would still be 0.0. The pH of $1M$ OH^- would not be 14.0, however. It is true that the pOH of $1M$ OH^- would still be 0.0; however, using **9-3** and substituting the appropriate value of pK_w from Table 9-1,

$$pH_{37°C} = pK_{w37°C} - pOH = 13.60 - 0 = 13.60$$

A neutral solution (NS) at 37°C does not have a pH of 7.00. Using **9-5**, the pH of such a solution is as follows:

$$pH = \tfrac{1}{2}pK_w = \tfrac{1}{2}(13.60) = 6.80(NS)$$

From these values we can compile the side-by-side comparison in Table 9-2.

Note in Table 9-2 that the pH of the stomach has the same value at both temperatures; the pH of $0.02M$ H^+ from stomach hydrochloric acid is the same for each. This is true by definition for any acid solution at body temperature. In the basic pH range there is a difference. A blood pH of 7.4 is basic by only 0.4 pH units at 25°C; however, at body temperature a blood pH of 7.4 is basic by 0.6 pH units. Thus, at 37°C a blood pH of 7.4 is a more basic solution than at room temperature by 0.2 pH unit. The pOH of a pH 7.4 blood sample at 37°C is

$$pOH_{37°C} = pK_{w37°C} - pH = 13.60 - 7.4 = 6.2$$

whereas at room temperature the pOH is $14.00 - 7.4 = 6.6$.

TABLE 9-2. pH Scales at 25° and 37°C (Body Temperature)

Solution	pH at 25°C	pH at 37°C (body)
1M H$^+$	0.0	0.0
0.02M HCl in stomach	1.7	1.7
		6.80(NS)
Neutral: [H$^+$] = [OH$^-$]	7.00(NS)	
A blood pH of 7.4	7.4	7.4
		13.60
1M OH$^-$	14.00	

9-2 | CALCULATION OF pH OF STRONG AND WEAK ACIDS AND BASES

Strong acids and bases are defined as those acids and bases that are completely ionized in water. Weak acids and bases are defined as those acids and bases that are quite a bit less than 100% ionized. Keeping that in mind, we will next demonstrate how the pH of such solutions is calculated.

The pH of Strong Acids and Bases

Recall that strong acids and bases have no equilibrium constants because they are 100% ionized to either protons and anions or cations and hydroxide ions. We will discuss strong acids first.

Calculations for Strong Acids

The common strong acids are nitric acid, hydrochloric acid, perchloric acid (HClO$_4$), and the first hydrogen of sulfuric acid. The second hydrogen of sulfuric acid has a K_2 value of 1×10^{-2}, so it is not a strong acid.

To calculate the pH of any strong acid, simply assume the molarity of the hydrogen ion is the same as the molarity of the strong acid and take its negative log. The correct number of significant figures should be used; recall from Section 3-1 that a pH value should have the same number of significant figures to the right of the decimal point as the total number of significant figures of the [H$^+$]. The following examples illustrate these points.

> **EXAMPLE:** Calculate the pH of a stomach fluid containing 0.020M hydrochloric acid.
> *Solution:* Assume that the acid is 100% ionized and that the hydrogen ion concentration equals that of the hydrochloric acid. Write the [H$^+$] in exponential form.
>
> $$[H^+] = 0.020M = 2.0 \times 10^{-2}M$$
>
> Calculate the negative log of the [H$^+$] using two significant figures to the right of the decimal point in the pH.
>
> $$pH = -\log(2.0 \times 10^{-2}) = 2 - 0.3010 = 1.69_9 \text{ or } 1.70$$

Note that the two is an exact number because it comes from an exact exponent; it has no effect on the number of significant figures in the pH because significant figure rules apply only to measured numbers, not to exact numbers. The log of 2.0 was given to four digits because it was read from the four-place log table inside the back cover. The correct number of significant figures in the pH was obtained by rounding off after subtraction.

EXAMPLE: Calculate the pH of a $0.100M$ solution of H_2SO_4, sulfuric acid ($K_1 =$ infinity; $K_2 = 1.0 \times 10^{-2}$).
Solution: To obtain an answer valid to one significant figure to the right of the decimal point, assume the first hydrogen is 100% ionized and neglect ionization of the HSO_4^- ion to H^+ and SO_4^{-2}. Write the $[H^+]$ in exponential form.

$$[H^+] \cong 0.1M \cong 1 \times 10^{-1}M$$

Calculate the pH using one significant figure to the right of the decimal point.

$$pH = -\log(1 \times 10^{-1}M) = 1 - 0.0000 = 1.0$$

To obtain a more exact answer, substitute the $0.1M$ H^+ from the first ionization into the second ionization constant expression, using $x = [SO_4^{-2}]$.

$$K_2 = 1.0 \times 10^{-2} = \frac{[H^+][SO_4^{-2}]}{[HSO_4^-]} = \frac{[0.100 + x][x]}{[0.100 - x]}$$

Using the solution to the quadratic equation (see equation **9-36**), $x = 8.4 \times 10^{-3}M$ and the exact $[H^+] = 0.108_4M$. The pH is 0.965.

Calculation for Strong Bases
The common strong bases are sodium hydroxide, potassium hydroxide, and barium hydroxide. Unlike sulfuric acid, both the hydroxide ions of barium hydroxide are 100% ionized.

To calculate the pH of any strong base, simply assume the molarity of the hydroxide is the same as the molarity of the strong bases sodium hydroxide and potassium hydroxide. (The molarity of the hydroxide ion is twice the molarity of barium hydroxide.) Then calculate the pOH. Finally, use **9-3** to calculate the pH. As for strong acids, the correct number of significant figures should be used. The following examples illustrate the calculation of pH from a given concentration and the calculation of the $[OH^-]$ from a given pH.

EXAMPLE: Calculate the pH of a $0.10M$ $Ba(OH)_2$ solution.
Solution: Assume that $0.010M$ barium hydroxide is 100% ionized to $0.020M$ OH^- and $0.010M$ Ba^{+2}. Write the $[OH^-]$ in exponential form.

$$[OH^-] = 0.020M = 2.0 \times 10^{-2}M$$

Calculate the negative log of the $[OH^-]$ using two significant figures to the right of the decimal point in the pOH.

$$pOH = -\log(2.0 \times 10^{-2}M) = 2 - 0.3010 = 1.70 \text{ or } 1.69_9$$

Because the two is an exact number, it has no effect on the number of significant figures in the pOH. The log of 2.0 is given to four digits because it was taken from the four-place log table inside the back cover of this text. The correct number of significant figures was obtain by rounding off after subtaction. Now **9-3** is used to find the pH.

$$pH = pK_w - pOH = 14.00 - 1.70 = 12.30$$

EXAMPLE: The pH of a certain blood sample at 37°C is 7.4. Calculate (a) the $[H^+]$ of this blood sample and (b) the $[OH^-]$ of this sample.

Solution to a: Substitute the pH value of 7.4 into the definition for pH and solve for the $[H^+]$.

$$pH = 7.4 = -\log[H^+]$$

$$-7.4 = \log[H^+]$$

$$+0.6 - 8 = \log[H^+]$$

Now take the antilog of both sides of the equation.

$$4 \times 10^{-8} M = [H^+]$$

Solution to b: Use **9-3** to calculate the pOH at 37° from the pH.

$$pOH_{37°C} = pK_{w37°C} - pH$$

$$pOH_{37°C} = 13.60 - 7.4 = 6.8$$

$$pOH = 6.8 = -\log[OH^-]$$

$$+0.2 - 7 = -\log[OH^-]$$

$$1._6 \times 10^{-7} = [OH^-]$$

Calculations for Weak Acids

In contrast to strong acids, most weak acids ionize 1% or less. The definition of ionization constants (K_a) for weak monoprotic and polyprotic acids was given in Section 7-1. The K_a values for a number of weak acids are found in Appendix 1. Some references give the pK_a values; therefore, you should be able to calculate the K_a from the pK_a. This is done by simply solving for the K_a from the definition of the pK_a $(= -\log K_a)$. For example, the pK_a of acetic acid is 4.74. The K_a is found as follows:

$$pK_a = 4.74 = -\log K_a$$

$$-4.74 = \log K_a$$

$$+0.26 - 5 = \log K_a$$

$$1.8 \times 10^{-5} = K_a$$

To calculate the pH of a solution of a pure weak acid, HA, simply write the definition of the ionization constant.

$$K_a = \frac{[H^+][A^-]}{[HA]} \tag{9-6}$$

and assume that $[H^+] = [A^-]$. To use the approximate method, assume that $[HA]$ is approximately equal to the initial molarity of HA. This means that the ionization of HA to H^+ and A^- is neglected. To calculate the approximate $[H^+]$, use the following equation:

$$[H^+] \cong \sqrt{M_{HA}(K_a)} \tag{9-7}$$

where M_{HA} is the initial molarity of HA in the solution before ionization. To check this approximation, calculate the % error, using the following equation:

$$\% \text{ error} = \frac{[H^+]}{M_{HA}}(100) \tag{9-8}$$

If the % error is less than 1%, the approximate $[H^+]$ is valid to two or less significant figures. The following examples illustrate the use of 9-6 and 9-8.

EXAMPLE: The pain-killing ingredient in aspirin tablets is acetylsalicylic acid (ASA), $C_8H_7O_2CO_2H$. This is a weak monoprotic acid with $K_a = 3.3 \times 10^{-4}$. Assume two tablets, each containing 315 mg of ASA, are dissolved in 100 ml of aqueous body fluid. (a) Calculate the pH of this solution. (b) If this fluid were the stomach gastric fluid, would the ASA make the stomach significantly more acid?

Solution to a: First calculate the molarity of ASA, using a formula weight of 180.

$$M = \frac{\text{mmole ASA}}{100 \text{ ml}} = \frac{630 \text{ mg ASA}/180 \text{ mg/mmole}}{100 \text{ ml}} = 0.0350M \text{ ASA}$$

Next, use the initial molarity of 0.0350M ASA to set up an *exact* equation for calculation of the $[H^+]$. Assume that $[H^+] = [C_8H_7O_2CO_2^-]$, the equilibrium concentration of the anion of ASA. Equation 9-6 becomes

$$K_a = \frac{[H^+][H^+]}{0.0350M - [H^+]}$$

where the terms in the denominator represent the initial amount of ASA minus the amount ionized, or the amount of un-ionized ASA at equilibrium.

Then we neglect the ionization of ASA to $[H^+]$ and obtain the following *approximate* equation:

$$K_a \cong \frac{[H^+]^2}{0.0350M}$$

Substituting the value of K_a and rearranging, we obtain

$$H^+ \cong \sqrt{0.0350M(3.3 \times 10^{-4})} \cong 3.4 \times 10^{-3}M$$

Using 9-8, the % error is calculated.

$$\% \text{ error} = \frac{3.4 \times 10^{-3}M}{0.035M \text{ ASA}} (100) = 9.7\%$$

Since the % error is more than 1%, we cannot report the $[H^+]$ to two significant figures. However, the % error is less than 10% so we can estimate the $[H^+]$ to one significant figure.

$$[H^+] = 3._4 \times 10^{-3}M$$

$$pH = 2.4_7 \text{ or } 2.5$$

Solution to b: Since the $[H^+]$ of the stomach is about $2 \times 10^{-2}M$, the $3 \times 10^{-3}M$ hydrogen ion produced by ionization of ASA will not significantly affect the H^+ of the stomach. In fact, the acid in the stomach gastric fluid represses the ionization of ASA to its ions.

In addition to organic acids like ASA, certain types of nitrogen salts can act as acids in the Bronsted sense. Examples are $NH_4^+Cl^-$, ammonium chloride; $NH_3OH^+Cl^-$, hydroxylammonium chloride; and $CH_3NH_3^+Cl^-$, methylammonium chloride. The K_as for these acids are calculated from the K_b of the corresponding bases: NH_3, ammonia; NH_2OH, hydroxylamine; and CH_3NH_2, methylamine, respectively.

$$K_a = \frac{K_w}{K_b} \qquad \qquad (9-9)$$

The following example illustrates the calculation of the $[H^+]$ of such acids.

EXAMPLE: The blood plasma can contain as high as $0.040M$ ammonium ion.   Assuming there are no other acids or bases present, calculate the pH of such a solution, given that the pK_b of the conjugate base NH_3 is 4.74.

Solution: First calculate the K_b of NH_3 from the pK_b in a fashion similar to the method used for acetic acid at the beginning of the discussion of weak acids.

$$pK_b = 4.74 = -\log K_b$$

$$+0.26 - 5 = \log K_b$$

$$1.8 \times 10^{-5} = K_b$$

Now use **9-9** to calculate the K_a of $NH_4{}^+$.

$$K_a = \frac{1.00 \times 10^{-14}}{1.8 \times 10^{-5}} = 5.5_5 \times 10^{-10}$$

Write the ionization of $NH_4{}^+$.

$$NH_4{}^+ = H^+ + NH_3$$

The ionization constant for this acid is defined as follows:

$$K_a = 5.5_5 \times 10^{-10} = \frac{[H^+][NH_3]}{[NH_4{}^+]}$$

This is similar to the general definition for any K_a given in **9-6**, except that NH_3 is uncharged rather than being an anion.   Therefore **9-7** can be used to calculate the approximate $[H^+]$.   Substituting the value of $0.040M$ for initial concentration, M_{HA}, and the value of $5.5_5 \times 10^{-10}$ for K_a in **9-7** gives the following:

$$[H^+] \cong \sqrt{0.040M\ NH_4{}^+(5.5_5 \times 10^{-10})} \cong 4.7_1 \times 10^{-6}M$$

Using **9-8**, the % error is calculated.

$$\%\ error = \frac{4.7_1 \times 10^{-6}M}{0.040M\ NH_4{}^+}(100) \doteq 0.01\%$$

Since the % error is less than 1%, we can report the $[H^+]$ to two significant figures.

$$[H^+] = 4.7_1 \times 10^{-6}M \text{ or } 4.7 \times 10^{-6}M$$

$$pH = 5.32_7 \text{ or } 5.33$$

Calculations for Weak Bases   Most weak bases ionize 1% or less, in contrast to strong bases.   The definition of the ionization constant (K_b) for weak bases was given in Section 7-1, and typical values are given in the appendix.   The pK_b value is also listed in some references; this can be converted to a K_b value in the same manner as in the last example.

To calculate the pH or pOH of a pure weak base, B, first write its ionization reaction with water.

$$B + H_2O = BH^+ + OH^-$$

Then write the definition of the ionization constant.

$$K_b = \frac{[BH^+][OH^-]}{[B]} \qquad\qquad (9\text{-}10)$$

To use an approximate method for calculating pH, first assume that $[BH^+] = [OH^-]$. Then assume that $[B]$ is approximately equal to the initial molarity of B.   This means that we neglect the ionization of B to BH^+ and OH^-. To calculate the approximate $[OH^-]$, use the following equation:

$$[OH^-] \cong \sqrt{M_B(K_b)} \qquad (9\text{-}11)$$

where M_B is the initial molarity of B in the solution before ionization. To check this approximation, calculate the % error as follows:

$$\% \text{ error} = \frac{[OH^-]}{M_B}(100) \qquad (9\text{-}12)$$

If the % error is less than 1%, the approximate $[OH^-]$ is valid to two or less significant figures. An error of 1%–10% implies that the $[OH^-]$ may be estimated to only one significant figure. The following examples illustrate the use of 9-11 and 9-12.

EXAMPLE: Calculate the pH of a $0.10M$ solution of ammonia. Ammonia is a weak base whose $K_b = 1.8 \times 10^{-5}$ and whose ionization may be written as

$$NH_3 + H_2O = NH_4^+ + OH^- \qquad (9\text{-}13)$$

or as

$$NH_4OH = NH_4^+ + OH^- \qquad (9\text{-}14)$$

Solution: Using 9-13 to represent the ionization of ammonia, we define K_b in a manner similar to 9-10.

$$K_b = \frac{[NH_4^+][OH^-]}{[NH_3]} \qquad (9\text{-}15)$$

Next assume that $[NH_4^+] = [OH^-]$ and substitute this equality into 9-15.

$$K_b = \frac{[OH^-][OH^-]}{0.10M - [OH^-]} \qquad (9\text{-}16)$$

The terms in the denominator represent the initial amount of ammonia minus the amount ionized, or the amount of un-ionized ammonia at equilibrium.

Then we neglect the ionization of ammonia to OH^- and obtain from 9-16 the approximate equation.

$$K_b \cong \frac{[OH^-]^2}{0.10M}$$

Substituting the value of K_b and rearranging, we obtain the following:

$$[OH^-] \cong \sqrt{0.10M(1.8 \times 10^{-5})} \cong 1.3_4 \times 10^{-3}M$$

Using 9-12, the % error is calculated as follows:

$$\% \text{ error} = \frac{1.3 \times 10^{-3}M}{0.10M}(100) = 1.3\%$$

Since the % error is essentially 1%, we can report the $[OH^-]$ to two significant figures.

$$[OH^-] = 1.3 \times 10^{-3}M$$

$$pOH = 2.87$$

$$pH = 14.00 - 2.87 = 11.13$$

In addition to organic nitrogen bases and ammonia, certain types of anions can act as bases in the Bronsted sense. These are the anions of weak acids, such as the acetate ion, the fluoride ion, the anion of ASA, and the benzoate ion, $C_6H_5CO_2^-$ (the anion of benzoic acid). The K_bs for these bases are calculated from the K_a of the corresponding acids: acetic acid, hydrofluoric acid, ASA, and benzoic acid, respectively.

$$K_b = \frac{K_w}{K_a} \tag{9-17}$$

EXAMPLE: Calculate the pH of a $0.10M$ solution of sodium acetate, $CH_3CO_2^-Na^+$. The K_a of acetic acid is 1.8×10^{-5}.
Solution: First calculate K_b of sodium acetate using **9-17**.

$$K_b = \frac{1.00 \times 10^{-14}}{1.8 \times 10^{-5}} = 5.5_5 \times 10^{-10}$$

Write the ionization of sodium acetate omitting the sodium ion, since it does not affect the pH.

$$CH_3CO_2^- + H_2O = OH^- + CH_3CO_2H$$

The ionization constant for this base is defined as follows:

$$K_b = 5.5_5 \times 10^{-10} = \frac{[OH^-][CH_3CO_2H]}{[CH_3CO_2^-]}$$

This is similar to the general definition for any K_b given in **9-10**, except that the acetate ion is charged. Therefore, **9-11** can be used to calculate the approximate $[OH^-]$. Substituting the value of $0.10M$ for the initial concentration, M_B, and the value of $5.5_5 \times 10^{-10}$ for K_b in **9-11** gives the following:

$$[OH^-] \cong \sqrt{0.10M\ CH_3CO_2^-(5.5_5 \times 10^{-10})} \cong 7.4_5 \times 10^{-6}M$$

Using **9-12**, the % error is calculated as follows:

$$\% \text{ error} = \frac{7.5 \times 10^{-6}M}{0.10M}(100) = 7.5 \times 10^{-3}\%$$

Since the % error is less than 1% we can report the $[OH^-]$ to two significant figures.

$$[OH^-] = 7.4_5 \times 10^{-6}M$$

$$pOH = 5.13$$

$$pH = 8.87$$

SELF-TEST | **12. Strong and Weak Acids and Bases**

Answers:

Directions: Cover the answers in the left column before beginning the test. Check your answers with those given to the left of the self-test.

A. Calculate the pH of the following strong acids or bases:.

1.60   a. Gastric fluid containing $0.025M$ HCl

1.602   b. Gastric fluid containing $0.0250M$ HCl

11.30   c. A $0.0010M$ $Ba(OH)_2$ solution

0.1M

0.108M (see example on p. 168)

1.1$_7$ × 10^{-8}

2.1 × 10^{-7}

5.46$_5$ (% error in
approximation = 0.34%)
4.99 (% error in
approximation = 1%)

B. Calculate the pH of 0.100M sulfuric acid
($K_2 = 1.0 \times 10^{-2}$):
a. To one significant figure.
b. To three significant figures.
C. Calculate the K_a for each of the following:
a. A pK_a of 7.93 for hemoglobin (H-Hb).
b. A pK_a of 6.68 for oxyhemoglobin (H-HbO$_2$).
D. Calculate the pH of each of the solutions below to see
whether they will make blood more acid or more basic;
also calculate the % error.
a. A 0.0010M hemoglobin (H-Hb) solution
(pK_a = 7.93)
b. A 0.0010M oxyhemoglobin (H-HbO$_2$) solution
(pK_a = 6.68)

9-3 | MIXTURES AND SIMPLE BUFFERS

In the preceding section we discussed the calculation of pH of solutions of *pure* acids
and *pure* bases. No mixtures were treated. However, a mixture of a weak acid and its
anion, or a mixture of a weak base and its cation, is a very common occurrence. Such
mixtures are called *conjugate acid-base pairs*, according to the Bronsted acid-base
theory. They are made by mixing an acid with a salt containing a common anion or by
mixing a base with a salt containing a common cation. Frequently mixtures of such
pairs are used as *buffers* to control the pH. We will first discuss the calculation of the
pH of such mixtures and then their use as buffers.

**Calculation
of pH**
First, you should recognize that a mixture of an acid and its conjugate anion is not as
acidic as a solution of the pure acid alone. This may be understood by considering the
ionization equilibrium of any weak acid HA.

$$HA \rightleftharpoons H^+ + A^-$$

point of
equilibrium ← excess
anion

In a solution of a pure weak acid at equilibrium, the anion and the hydrogen ion are
present in nearly equal concentrations. In a mixture of acid plus its conjugate anion,
the large excess of the anion displaces the point of the equilibrium farther to the left.
This lowers the hydrogen ion concentration as compared to that of a solution of the
pure weak acid alone. For actual comparison, the [H^+] of a 0.10M solution of acetic
acid alone is 1.3 × 10^{-3}M, whereas the [H^+] of an equimolar mixture of acetic acid and
the acetate ion is 1.8 × 10^{-5}M.

Similarly, a mixture of a base and its conjugate cation is not as basic as a solution of
the pure base alone. Consider the following ionization equilibrium of any weak base B:

$$B + H_2O \rightleftharpoons B^+ + OH^-$$

point of
equilibrium ← excess
cation

In a solution of a pure weak base at equilibrium, the cation and the hydroxide ion are
present in nearly equal concentrations. In a mixture of the base plus its conjugate

cation, the large excess of the cation displaces the point of equilibrium farther to the left. This lowers the hydroxide ion concentration as compared to that of a solution of the pure weak base alone. For actual comparison, the $[OH^-]$ of a $0.10M$ solution of ammonia alone is $1.3 \times 10^{-3}M$, whereas the $[OH^-]$ of an equimolar mixture of ammonia and ammonium chloride is $1.8 \times 10^{-5}M$.

The pH of Acid-Anion Mixtures
To calculate the pH of a mixture of a weak acid, HA, and a salt, NaA, containing its conjugate anion A^-, we start by writing the definition of the ionization constant.

$$K_a = \frac{[H^+][A^-]}{[HA]} \tag{9-18}$$

To make an exact calculation, we must recognize that there are two sources of A^-—the salt NaA and the HA molecules that ionize to H^+ and A^-. Thus,

$$[A^-] = M_{A^-} + [H^+] \tag{9-19}$$

where M_{A^-} is the initial molarity of A^- from the salt NaA. To calculate the pH from 9-18, we substitute the right-hand side of 9-19 for $[A^-]$.

$$K_a = \frac{[H^+](M_{A^-} + [H^+])}{M_{HA} - [H^+]} \tag{9-20}$$

Here M_{HA} is again the initial molarity of HA before ionization. To make an approximate calculation, we assume that the $[H^+]$ can be neglected in both the numerator and denominator of 9-20 and solve for $[H^+]$.

$$[H^+] \cong K_a \frac{M_{HA}}{M_{A^-}} \cong K_a \frac{\text{mmole HA}}{\text{mmole A}^-} \tag{9-21}$$

Note that because 9-21 involves a ratio of the molarity of HA to the molarity of A^-, the volume terms in a molarity definition such as mmole/ml cancel out of the ratio. Thus the pH can be calculated without knowing the volume of the mixture.

To check the approximations in 9-21, calculate the % error using the following equation:

$$\% \text{ error} = \frac{[H^+]}{M}(100) \tag{9-22}$$

where M is the molarity of either HA or the salt NaA. If the % error is less than 1%, the approximate $[H^+]$ is valid to two or less significant figures. If it is between 1% and 10%, the approximate $[H^+]$ may be estimated to one significant figure. The following example illustrates all of the above concepts.

EXAMPLE: Calculate the pH of a solution prepared by mixing 20 ml of $0.30M$ acetic acid and 10 ml of $0.30M$ sodium acetate, $CH_3CO_2^-Na^+$. The K_a of acetic acid is 1.8×10^{-5}.

Solution: First calculate the molarity of the acetic acid and sodium acetate.

$$M_{CH_3CO_2H} = \frac{\text{mmole}}{\text{final vol}} = \frac{(20\text{ ml})(0.30M)}{20 + 10\text{ ml}} = \frac{6.0\text{ mmole}}{30\text{ ml}} = 0.20M$$

$$M_{CH_3CO_2^-Na^+} = \frac{\text{mmole}}{\text{final vol}} = \frac{(10\text{ ml})(0.30M)}{20 + 10\text{ ml}} = \frac{3.0\text{ mmole}}{30\text{ ml}} = 0.10M$$

Next substitute these numbers into **9-21** to obtain an approximation.

$$[H^+] \cong 1.8 \times 10^{-5} \frac{6.0 \text{ mmole CH}_3\text{CO}_2\text{H}/30 \text{ ml}}{3.0 \text{ mmole CH}_3\text{CO}_2^-\text{Na}^+/30 \text{ ml}} \cong 3.6 \times 10^{-5} M$$

Note that 30 ml final volume appears in both the numerator and denominator and does not need to appear in the calculation. To check the approximation the % error is calculated, using **9-22** twice.

$$\% \text{ error} = \frac{3.6 \times 10^{-5} M}{0.20 M \text{ CH}_3\text{CO}_2\text{H}} (100) = 1.8 \times 10^{-2} \%$$

$$\% \text{ error} = \frac{3.6 \times 10^{-5} M}{0.10 M \text{ CH}_3\text{CO}_2^-\text{Na}^+} (100) = 3.6 \times 10^{-2} \%$$

Since the % error in both cases is less than 1%, we can report the $[H^+]$ to two significant figures.

$$[H^+] = 3.6 \times 10^{-5} M$$

$$pH = 4.44$$

The pH of Base-Cation Mixtures

To calculate the pH of a mixture of a weak base, B, and a salt, BH^+Cl^-, containing its conjugate cation BH^+, we start by writing the definition of the ionization constant.

$$K_b = \frac{[BH^+][OH^-]}{[B]} \qquad (9\text{-}23)$$

To make an exact calculation, we must note that there are two sources of BH^+—the salt BH^+Cl^- and the molecules of B that ionize to BH^+ and OH^-. Thus

$$[BH^+] = M_{BH^+} + [OH^-] \qquad (9\text{-}24)$$

where M_{BH^+} is the initial molarity of BH^+ from the salt BHCl. To calculate the pH from **9-23**, we substitute the right-hand side of **9-24** for $[BH^+]$.

$$K_a = \frac{(M_{BH^+} + [OH^-])[OH^-]}{M_B - [OH^-]} \qquad (9\text{-}25)$$

Here M_B is again the initial molarity of B before ionization. To make an approximate calculation, we assume that the $[OH^-]$ can be neglected in both terms in the numerator and denominator of **9-25** and solve for $[OH^-]$.

$$[OH^-] \cong K_b \frac{M_B}{M_{BH^+}} \cong K_b \frac{\text{mmole B}}{\text{mmole } BH^+} \qquad (9\text{-}26)$$

Note that the volume terms in a molarity definition such as mmole/ml cancel out of the ratio.

To check the approximation in **9-26**, calculate the % error using the following equation:

$$\% \text{ error} = \frac{[OH^-]}{M} (100) \qquad (9\text{-}27)$$

where M is the molarity of either B or the salt BHCl. If the % error is less than 1%, the approximate $[OH^-]$ is valid to two or less significant figures. The following example illustrates the above.

EXAMPLE: Calculate the pH of a solution prepared by mixing 5.0 ml of $0.50M$ ammonia with 20 ml of $0.25M$ ammonium chloride. The K_b of ammonia is 1.8×10^{-5}.
Solution: First calculate the molarity of the ammonia and ammonium ion.

$$M_{NH_3} = \frac{mmole}{final\ vol} = \frac{(5.0\ ml)(0.50M)}{20 + 5\ ml} = \frac{2.5\ mmole}{25\ ml} = 0.10M$$

$$M_{NH_4^+} = \frac{mmole}{final\ vol} = \frac{(20\ ml)(0.25M)}{20 + 5\ ml} = \frac{5.0\ mmole}{25\ ml} = 0.20M$$

Next substitute these numbers into **9-26** to make an approximation.

$$[OH^-] \cong 1.8 \times 10^{-5} \frac{2.5\ mmole\ NH_3/25\ ml}{5.0\ mmole\ NH_4^+/25\ ml} \cong 9.0 \times 10^{-6}M$$

Note that the 25 ml final volume appears in both the numerator and denominator and does not need to be used in the calculation. To check the approximation, the % error is calculated, using **9-27** twice.

$$\%\ error = \frac{9.0 \times 10^{-6}M}{0.1M\ NH_3}(100) = 9.0 \times 10^{-3}\%$$

$$\%\ error = \frac{9.0 \times 10^{-6}M}{0.20M\ NH_4^+}(100) = 4.5 \times 10^{-3}\%$$

Since the % error in both cases is less than 1%, we can report the $[OH^-]$ to two significant figures.

$$[OH^-] = 9.0 \times 10^{-6}M$$
$$pOH = 5.05$$
$$pH = 8.95$$

An interesting observation can be made after making another calculation for the ammonia-ammonium ion system of the preceding example. If we calculate the pH of a mixture of 5.0 mmole of ammonia and 5.0 mmole of ammonium chloride, we obtain a pH of 9.26. Thus even though the amount of weak base has been *doubled*, the pH has increased only about 0.3 pH units! The same small change in pH is also noted with similar changes in the amount of weak acid from the conditions of the example on p. 175. This ability of mixtures of conjugate acids and conjugate bases to resist the change in pH has led to their use as *buffers*; this will be discussed next.

Buffers A buffer is a mixture of a conjugate acid and conjugate base that will resist a change in pH from:

1. Addition of a strong or weak acid,
2. Addition of a strong or weak base, or
3. Dilution with pure water.

The buffer action is the result of the conversion of all other acids or bases to two species in equilibrium with each other, as in these examples:

$$HA \rightleftharpoons A^- + H^+ \qquad (9\text{-}28)$$
$$\text{(acid)} \qquad \text{(base)}$$

$$B + H_2O \rightleftharpoons BH^+ + OH^- \qquad (9\text{-}29)$$
$$\text{(base)} \qquad \text{(acid)}$$

The $[H^+]$ or $[OH^-]$ is thus controlled by the relative amounts of acid and base present in the equilibrium.

Note that a buffer *resists* a pH change; it does *not* keep the pH constant to two significant figures to the right of the decimal point. Generally a buffer only maintains a pH constant to the digit(s) to the left of the decimal point; it cannot prevent small pH changes of 0.01–0.2 units. For example, buffers in the blood maintain the pH within the range of 7.3–7.5, but do not hold the pH at exactly pH 7.40.

Simple buffers consist of a *relatively high* concentration of a weak acid, such as acetic acid, and its conjugate base, the acetate ion, or *high* concentrations of a weak base, such as ammonia, and its conjugate acid, the ammonium ion. The buffering action destroys added acid, producing more of the conjugate weak acid and leaving less of the conjugate base. Only two species remain after buffering action occurs, and a *single equilibrium*, such as that in **9-28** or **9-29**, *is in control of the pH*. The buffer must have a relatively high concentration of both the conjugate acid and conjugate base so that neither is used up by the added acid or base. If either is used up, the pH will be affected by more than one equilibrium.

The addition of a strong acid to a buffer mixture of the type in **9-28** results in the following reaction:

$$H^+ \quad + \quad A^- \quad \longrightarrow \quad HA$$
$$\text{strong acid} \quad \text{buffer base} \qquad \text{buffer acid}$$

Addition of a strong base to the same buffer mixture involves this reaction:

$$OH^- \quad + \quad HA \quad \longrightarrow \quad A^-$$
$$\text{strong base} \quad \text{buffer acid} \qquad \text{buffer base}$$

Note that in each case the strong acid or base is neutralized completely, producing a weak acid or base, which is part of the single equilibrium in **9-28**. The addition of strong acid or base to a buffer mixture of the type in **9-29** involves the same type of reactions, except that B is the buffer base and BH^+ is the buffer acid.

The dilution of a buffer with a reasonable volume of pure water should not cause a significant change in the pH of the buffer. However, if the buffer is diluted to a concentration of about $10^{-4}M$ or lower, then the hydrogen ion from the ionization of water becomes significant and the buffer pH will change.

Strong Acids or Bases as Buffers In one sense, strong acids and strong bases are not buffers, but in another sense they are. For example, the pH of a 1.0M solution of hydrochloric acid would not change significantly from the addition of a small amount of any strong acid. So we can say that a fairly concentrated solution of strong acid or strong base acts as a buffer toward the addition of smaller amounts of strong acid or strong base.

Calculation
of pH
Changes
during
Buffering

In making a calculation of the pH of a buffer mixture, you should use the K_a or K_b equilibrium expression of the weak acid or weak base of the buffer. Begin by calculating the total mmoles of strong acid or strong base added. Then calculate the total mmoles of buffer acid and buffer base. Then subtract the mmoles of strong acid or base from the mmoles of the buffer base or acid. The reaction will also produce an equivalent amount of the buffer acid (from a strong acid) or the buffer base (from a strong base), and this must be added to the amount of the buffer acid, or buffer base, present at the start. After these calculations are finished, the mmoles of buffer acid and buffer base are substituted into the K_a or K_b equilibrium expression to calculate the $[H^+]$.

EXAMPLE: A $0.30M$ acetic acid–$0.30M$ sodium acetate buffer has a pH of 4.74. Calculate the change in pH when 10 ml of $0.050M$ sulfuric acid is added to 40 ml of the buffer. The K_a is 1.8×10^{-5}.

Solution: The $0.050M$ sulfuric acid furnishes $0.10M$ H^+. The number of mmoles of H^+ (total acid) added is calculated.

$$(10 \text{ ml})(0.10M \text{ H}^+) = 1.0 \text{ mmole H}^+$$

The number of mmoles of acetic acid and sodium acetate present at the start in 40 ml of the buffer is calculated.

$$(40 \text{ ml})(0.30M) = 12 \text{ mmole each of } CH_3CO_2H \text{ and } CH_3CO_2^-Na^+$$

Next write the reaction between the buffer base and the strong acid (H^+), and underneath write the amount of each at the start, the amount of change in each, and the amount of each present at equilibrium.

	$CH_3CO_2^-$	+	H^+	$\rightarrow$	CH_3CO_2H
Start:	12 mmole		1.0 mmole		12 mmole
Change:	-1.0 mmole		-1.0 mmole		$+1.0$ mmole
Equilibrium:	11 mmole		0 mmole		13 mmole

Next substitute these numbers into the approximate equilibrium constant expression **(9-21)**.

$$[H^+] \cong 1.8 \times 10^{-5} \frac{13 \text{ mmole } CH_3CO_2H/50 \text{ ml}}{11 \text{ mmole } CH_3CO_2^-/50 \text{ ml}} \cong 2.1 \times 10^{-5} M$$

Note that the final volume of 50 ml appears in both the numerator and denominator and cancels out of the expression. To check the approximation, the % error is calculated using **9-22**. Since the largest error involves sodium acetate, only that error is calculated.

$$\% \text{ error} = \frac{2.1 \times 10^{-5} M \text{ H}^+}{0.22M \text{ } CH_3CO_2^-Na^+} (100) = 9.5 \times 10^{-3}\%$$

Since the % error is less than 1%, we can report the $[H^+]$ to two significant figures.

$$[H^+] = 2.1 \times 10^{-5} M$$

$$pH = 4.68$$

Choosing the Proper Buffer

To choose the proper buffer, follow this general rule:

> For optimum buffer action, use a buffer pair in which the pK_a of the *buffer acid* is as close as possible to the specified pH.

The reason this rule is valid is that a buffer has its maximum effect when the ratio of the amount of buffer acid to that of buffer base is $1:1$. Consider the following logarithmic equation relating pH to the concentrations of buffer acid and buffer base:

$$pH = pK_a - \log \frac{M \text{ of HA}}{M \text{ of A}^-} \qquad (9\text{-}30)$$

If the ratio of the molarity of HA to the molarity of A^- is $1:1$, then the pK_a of the buffer acid will equal the pH of the buffer pair. You can see by inspection that addition of a given amount of a strong acid or a strong base causes the smallest change in pH at a $1:1$ ratio compared to any other ratio, such as $2:1$, $1:2$, and so on. The following example illustrates this point.

EXAMPLE: Addition of 1.0 mmole of strong acid in 10 ml of solution to 40 ml of $0.30M$ acetic acid–$0.30M$ sodium acetate buffer results in a change of pH from 4.74 to 4.68, or 0.06 pH units (see last example). Calculate the change in pH when 1.0 mmole of strong acid in 10 ml of solution is added to 40 ml of a $0.03M$ acetic acid–$0.15M$ sodium acetate buffer, which has an initial pH of 4.44. (See example on p. 175 for a similar calculation.) *Solution:* First calculate the number of mmoles of acetic acid and sodium acetate.

$$(40 \text{ ml})(0.30M) = 12 \text{ mmole of CH}_3\text{CO}_2\text{H}$$

$$(40 \text{ ml})(0.15M) = 6.0 \text{ mmole of CH}_3\text{CO}_2{}^-\text{Na}^+$$

Next write the reaction between the buffer base and the strong acid (H^+). Underneath write the amount of each at the start, the amount of change in each, and the amount of each present at equilibrium.

	$CH_3CO_2{}^-$	+	H^+	$\rightarrow$	CH_3CO_2H
Start:	6.0 mmole		1.0 mmole		12 mmole
Change:	−1.0 mmole		−1.0 mmole		+1.0 mmole
Equilibrium:	5.0 mmole		0 mmole		13 mmole

Next substitute these numbers into the approximate equilibrium constant expression **(9-21)**.

$$[H^+] = 1.8 \times 10^{-5} \frac{13 \text{ mmole CH}_3\text{CO}_2\text{H}/50 \text{ ml}}{5.0 \text{ mmole CH}_3\text{CO}_2{}^-/50 \text{ ml}} = 4.6_8 \times 10^{-5}M$$

Checking the approximation using **9-22** gives a % error that is less than 1%, so the $[H^+]$ can be reported to two significant figures.

$$[H^+] = 4.7 \times 10^{-5}M$$

$$pH = 4.33$$

The pH change from the addition of 1.0 mmole of strong acid is from 4.44 to 4.33, or 0.11 pH units. This is a *larger change* than the 0.06 pH change when the same amount of strong acid was added to a $0.30M$ acetic acid–$0.30M$ sodium acetate buffer. The two situations can be summarized as follows:

M acid–M base	Ratio, acid : base	Initial pH	pH of buffer +1 mmole H⁺	pH change
$0.30M$–$0.30M$	$1:1$	4.74	4.68	-0.06
$0.30M$–$0.15M$	$2:1$	4.44	4.33	-0.11

Calculation of Buffer Acid-Base Ratio

After the buffer acid whose pK_a is closest to the specified pH has been chosen, it is usually necessary to calculate the ratio of the buffer acid to the buffer base to achieve the exact value of the specified pH. The pK_a will usually be within ± 0.3 units of the pH, but it will seldom match the pH to two digits to the right of the decimal point. The buffer ratio can be calculated by using a rearranged form of **9-30**.

$$\log \frac{M \text{ of HA}}{M \text{ of A}^-} = pK_a - pH \qquad (9\text{-}31)$$

The following example illustrates the use of this equation.

EXAMPLE: It is desired to prepare a buffer for a pH of 3.48. The following acids are available: acetic acid ($K_a = 1.8 \times 10^{-5}$), formic acid ($K_a = 1.70 \times 10^{-4}$), and nitrous acid ($K_a = 5.0 \times 10^{-4}$). Which should be used for optimum buffer action? What ratio of buffer acid to buffer base should be used?

Solution: First calculate the pK_a of each of the acids to see which is closest to the specified pH of 3.48. The pK_as are as follows:

$$\text{acetic acid, } pK_a = 4.74$$

$$\text{formic acid, } pK_a = 3.77$$

$$\text{nitrous acid, } pK_a = 3.30$$

Since the pK_a of nitrous acid is only 0.18 units less than the pH of 3.48, nitrous acid should be chosen.

To calculate the ratio of buffer acid to buffer base, use **9-31**.

$$\log \frac{M \text{ of HNO}_2}{M \text{ of Na}^+\text{NO}_2^-} = 3.30 - 3.48 = -0.18$$

Multiplying both sides of the equation by -1 and then inverting the log term to eliminate the negative sign, we obtain the following:

$$\log \frac{M \text{ of Na}^+\text{NO}_2^-}{M \text{ of HNO}_2} = 0.18$$

$$\frac{M \text{ of Na}^+\text{NO}_2^-}{M \text{ of HNO}_2} = \frac{1.5}{1}$$

Thus we need 1.5 times as much of the buffer base, the nitrite ion, as of the buffer acid. For example, a $0.15M$ sodium nitrite–$0.10M$ nitrous acid solution would make a suitable buffer, provided that this is sufficient to neutralize the strong acid or base to be added.

Although the buffer ratio should be as close as possible to $1:1$, there are special situations, such as in the blood buffers, where a ratio of $20:1$ still provides effective buffering. This will be discussed in the next section.

9-4 | POLYPROTIC ACIDS AND PHYSIOLOGICAL BUFFERS

In the preceding two sections we discussed the equilibria of monoprotic acids and simple buffers made from monoprotic acids and their salts. Polyprotic acids exhibit more complicated equilibria so we have waited until the end to discuss some of these. The discussion will not be exhaustive; only a few key differences will be discussed as an *introduction* to the equilibria involved.

Two of the most important polyprotic acids in chemistry and in the body are carbonic acid, H_2CO_3, and phosphoric acid, H_3PO_4. Each of these acids ionizes in steps and each step has an ionization constant with a subscript that refers to the number of the step involved.

Since carbonic acid is really hydrated carbon dioxide, $CO_2(aq)$, its first ionization step can be written either of the following two ways:

$$H_2CO_3 = H^+ + HCO_3^-$$
$$CO_2(aq) + H_2O = H^+ + HCO_3^-$$

Note that the products are the same in either case. The equilibrium expression for K_1, the first ionization constant, is therefore defined in two ways.

$$K_1 = 4.3 \times 10^{-7} = \frac{[H^+][HCO_3^-]}{[H_2CO_3]} = \frac{[H^+][HCO_3^-]}{[CO_2(aq)]}$$

However, the second step is only defined in one way, since it is the ionization of the bicarbonate ion to hydrogen ion and carbonate ion.

$$K_2 = 4.8 \times 10^{-11} = \frac{[H^+][CO_3^{-2}]}{[HCO_3^-]}$$

As for most polyprotic acids, the value of K_1 is much larger than that of K_2.

Phosphoric acid ionizes in three steps, which are defined as follows:

$$K_1 = 7.5 \times 10^{-3} = \frac{[H^+][H_2PO_4^-]}{[H_3PO_4]}$$

$$K_2 = 6.2 \times 10^{-8} = \frac{[H^+][HPO_4^{-2}]}{[H_2PO_4^-]}$$

$$K_3 = 4.8 \times 10^{-13} = \frac{[H^+][PO_4^{-3}]}{[HPO_4^{-2}]}$$

Phosphoric acid is more typical of polyprotic acids than carbonic acid is because the first ionization constant is fairly large, that is, of the order of 10^{-3} to 10^{-2}. The second ionization constant of phosphoric acid is about 10^{-5} smaller than the first, while for carbonic acid the second constant is 10^{-4} smaller.

Calculation of the pH of a Solution of a Pure Polyprotic Acid To calculate the pH of a solution containing only a polyprotic acid, you should first recognize that K_1 for almost all of the common polyprotic acids is significantly larger than K_2 and K_3. This means that K_2 and K_3 can be neglected when considering the ionization of H_2CO_3 or H_3PO_4. In other words, less than a 1% error is incurred by neglecting the ionization of HCO_3^- in a solution of H_2CO_3 (or by neglecting the ionization of $H_2PO_4^-$ in a solution of H_3PO_4). For a solution of H_2CO_3 alone, the following equality can be written:

$$[H^+] = [HCO_3^-] \qquad (9\text{-}32)$$

Similarly, for a solution of phosphoric acid alone, this equality is valid:

$$[H^+] = [H_2PO_4^-] \qquad (9\text{-}33)$$

To calculate the pH of such solutions, the procedure for weak monoprotic acids outlined in Section 9-2 is used. The ionization of the polyprotic acid to H^+ and the corresponding anion is neglected, and an equation for the approximate $[H^+]$ analogous to **9-7** is written.

$$[H^+] \cong \sqrt{M_{H_nA}(K_1)} \qquad (9\text{-}34)$$

Here M_{H_nA} is the initial molarity of polyprotic acid H_nA in the solution before ionization. To check this approximation, calculate the % error using the following equation:

$$\% \text{ error} = \frac{[H^+]}{M_{H_nA}}(100) \qquad (9\text{-}35)$$

If the % error is less than 1%, the approximate $[H^+]$ is valid to two or less significant figures. If the % error is less than 10%, the $[H^+]$ may be estimated to one significant figure. The next example illustrates the use of the above equations.

EXAMPLE: Rapid breathing changes the carbon dioxide concentration of a hypo-thetical unbuffered bloodstream in the lungs to $0.010M$ $CO_2(aq)$. Assuming no other acidic species are present, calculate the pH of this bloodstream.
Solution: First, you should recognize that the predominate ionization is that of carbon dioxide to hydrogen ion and bicarbonate ion.

$$CO_2(aq) + H_2O = H^+ + HCO_3^- \qquad K_1 = 4.3 \times 10^{-7}$$

This is essentially the first ionization step of carbonic acid, whose K_1 value is given above. Since K_1 is much larger than K_2 for this acid, the equality in **9-32** is valid, and we can use **9-34** to calculate the approximate $[H^+]$.

$$[H^+] \cong \sqrt{0.010M\ CO_2(4.3 \times 10^{-7})} \cong 6.5_6 \times 10^{-5}M$$

Checking the approximation by using **9-35** gives the following:

$$\% \text{ error} = \frac{6.6 \times 10^{-5}M\ H^+}{0.010M\ CO_2}(100) = 0.66\%$$

Since this is less than 1% error, the $[H^+]$ can be expressed to two significant figures.

$$[H^+] = 6.6 \times 10^{-5}M$$

$$pH = 4.18$$

(The pH of normal, buffered blood is between 7.3 and 7.5.)

The above example is not typical of all polyprotic acids. Many polyprotic acids, such as phosphoric acid, have large K_1 values and the % error for most concentrations is more than 1%. In such cases the solution to the quadratic equation must be used to calculate the exact value of the $[H^+]$.

$$[H^+] = \frac{-(K_1) \pm \sqrt{(K_1)^2 - 4(K_1)(M \text{ of acid})}}{2} \qquad (9\text{-}36)$$

A few polyprotic acids, such as H_2SO_4 and $H_4P_2O_7$, are strong acids with respect to the ionization of the first proton. The pH of such acids is calculated exactly by using the K_2 ionization (see the example on p. 168).

Calculation of the pH of Acid Salts of Polyprotic Acids

Acid salts of polyprotic acids are *amphiprotic*; that is, they can ionize to give up H^+ or they can act as bases. Examples of such salts are $NaHCO_3$, sodium bicarbonate; KH_2PO_4, potassium dihydrogen phosphate; Na_2HPO_4, sodium hydrogen phosphate; $KHC_8H_4O_4$, potassium acid phthalate; and KHC_2O_4, potassium hydrogen oxalate.

Consider the reactions of a general acid salt, NaHA, the acid salt of the general diprotic acid, H_2A. The anion of this salt can ionize to give up a proton to water.

$$HA^- = H^+ + A^{-2} \qquad (9\text{-}37)$$

It can also act as a base and remove small amounts of H^+ from the solution.

$$HA^- + H^+ = H_2A \qquad (9\text{-}38)$$

At the same time water is also ionizing.

$$H_2O = H^+ + OH^- \qquad (9\text{-}39)$$

Using a lengthy derivation, an *exact equation* relating the $[H^+]$ to $[HA^-]$, the molarity of the salt containing the amphiprotic anion, can be obtained.

$$[H^+] = \sqrt{\frac{K_2[HA^-] + K_w}{1 + [HA^-]/K_1}} \qquad (9\text{-}40)$$

Equation **9-40** can be derived by first writing an equation summing up the contributions to the $[H^+]$ from **9-37**, **9-39**, and **9-38**.

$$[H^+] = \underset{(9\text{-}37)}{H^+} + \underset{(9\text{-}39)}{H^+} - \underset{(9\text{-}38)}{H^+}$$

Then we substitute equalities from each equation for the H^+ in each.

$$[H^+] = \underset{(9\text{-}37)}{[HA^-]} + \underset{(9\text{-}39)}{[OH^-]} - \underset{(9\text{-}38)}{[H_2A]}$$

Next we substitute identities for each term from the three corresponding equilibrium expressions.

$$[H^+] = \frac{K_2[HA^-]}{[H^+]} + \frac{K_w}{[H^+]} - \frac{[H^+][HA^-]}{K_1}$$

Rearrangement of the above equation leads to **9-40**.

Exact equation **9-40** can be simplified by two approximations. In the first we assume that K_w is less than 10% of the $K_2[HA^-]$ term in the numerator and neglect K_w. This is always true for acid salts where K_2 is about 10^{-8}. For salts where K_2 is 10^{-11} or smaller, such as $NaHCO_3$, it is true for solutions more concentrated than $0.01 M$. Neglecting K_w in **9-40** gives the following:

$$[H^+] \cong \sqrt{\frac{K_2[HA^-]}{1 + [HA^-]/K_1}} \qquad (9\text{-}41)$$

In the second approximation, we assume that in the denominator the value of one is less than 10% of the $[HA^-]/K_1$ term. This gives the following:

$$[H^+] \approx \sqrt{\frac{K_2[HA^-]}{[HA^-]/K_1}} \approx \sqrt{K_1 K_2} \qquad (9\text{-}42)$$

The double approximation of **9-42** is valid only if $[HA^-]$, the concentration of the acid salt, is at least five times the value of K_1. This should be checked before using **9-42** by making the following calculation:

$$\frac{[HA^-]}{K_1} \geqslant \frac{5}{1} \tag{9-43}$$

If **9-43** is true, then the $[HA^-]/K_1$ term in **9-41** is large enough that dropping the one will only result in an error of 9.5% or less.

Note that the double approximation of **9-42** implies that the $[H^+]$ and the pH of acid salts are *independent of concentration*. Compared to results with other types of acids and bases, this appears to be unreasonable. Intuitively, we insist that increasing the concentration of an acid salt should change the $[H^+]$. A look at **9-37** and **9-38** reminds us that acid salts have a double action in solution, so their tendency to act as bases counteracts their tendency to act as acids. In fact, these salts act as a *type of buffer*. Recall that a buffer resists changes in pH; with that in mind, **9-42** does not seem unreasonable. The following examples illustrate the above ideas.

EXAMPLE: Calculate the pH of $0.020M$ and $0.20M$ solutions of $NaHCO_3$, sodium bicarbonate. For H_2CO_3, $K_1 = 4.3 \times 10^{-7}$ and $K_2 = 4.8 \times 10^{-11}$.
Solution: First check to see if the approximation converting **9-41** to **9-42** is valid for these cases by using **9-43**.

$$\text{For } 0.020M \text{ NaHCO}_3: \quad \frac{0.020M \text{ HCO}_3^-}{4.3 \times 10^{-7}} = \frac{4.4 \times 10^4}{1}$$

$$\text{For } 0.20M \text{ NaHCO}_3: \quad \frac{0.20M \text{ HCO}_3^-}{4.3 \times 10^{-7}} = \frac{4.4 \times 10^5}{1}$$

In each case, the $[HA^-]/K_1$ ratio from **9-43** is far greater than $10:1$, so **9-42** can be used to calculate $[H^+]$.

$$[H^+] \approx \sqrt{(4.3 \times 10^{-7})(4.8 \times 10^{-11})} \approx 4.5_4 \times 10^{-8} M$$

$$pH = 7.34$$

Since the $[H^+]$ is independent of concentration, the pH of $0.020M$ and of $0.20M$ sodium bicaronate is the same, 7.34.

EXAMPLE: Calculate the pH of $0.015M$ KH_2PO_4, potassium dihydrogen phosphate. For H_3PO_4, $K_1 = 7.5 \times 10^{-3}$ and $K_2 = 6.2 \times 10^{-8}$.
Solution: First check to see if the approximation converting **9-41** to **9-42** is valid here by using **9-43**.

$$\frac{0.015M \text{ H}_2PO_4^-}{7.5 \times 10^{-3}} = \frac{2}{1}$$

Since the $[HA^-]/K_1$ ratio is much less than $5:1$, **9-42** is not valid and **9-41** will have to be used.

$$[H^+] \cong \sqrt{\frac{6.2 \times 10^{-8}[0.015M \text{ H}_2PO_4^-]}{1 + [0.015M \text{ H}_2PO_4^-]/7.5 \times 10^{-3}}}$$

$$[H^+] \cong \sqrt{\frac{9.3 \times 10^{-10}}{1 + 2.0}} \cong 1.7_6 \times 10^{-5} M$$

$$pH = 4.75$$

Calculation of the $[H^+]$ using **9-42** in this case would give $2.1_6 \times 10^{-5}M$, a value which is 23% high.

EXAMPLE: It is desired to buffer a solution near pH 9.6. Would any of the following acid salts be satisfactory: KH_2PO_4, $NaHCO_3$, or Na_2HPO_4?
Solution: From the preceding two examples, we know that the pH of $NaHCO_3$ is 7.34 and the pH of KH_2PO_4 is 4.75. Since neither of these is satisfactory, we will use **9-42** to calculate the approximate pH of Na_2HPO_4. Since this is a -2 anion, K_2 and K_3 must be used instead of K_1 and K_2, which are only used for -1 anions.

$$[H^+] \approx \sqrt{(6.2 \times 10^{-8})(4.8 \times 10^{-13})} \approx 1.7_2 \times 10^{-10}M$$

$$pH = 9.76$$

Since the calculated pH is close to pH 9.6, the Na_2HPO_4 can be used as a buffer. Any proposed concentration should be checked using **9-43** to see if **9-42** is valid for that concentration.

Mixtures and Physiological Buffers

Thus far we have discussed solutions in which only one species is present—a polyprotic acid alone or an acid salt of a polyprotic acid. Now we will discuss mixtures of a polyprotic acid and an acid salt or of an acid salt and its salt. Examples of such mixtures are H_2CO_3 with $NaHCO_3$ and $NaHCO_3$ with Na_2CO_3. Such mixtures also function as buffers.

Calculation of pH of Mixtures

In a mixture of a general diprotic acid H_2A and an acid salt NaHA, the H_2A functions as the acid and the HA^- anion functions as the base. The presence of an excess of the HA^- anion serves to repress the ionization of the diprotic acid.

$$H_2A \; = \; H^+ + HA^-$$

point of equilibrium $\longleftarrow$ excess anion

This lowers the hydrogen ion concentration of the solution as compared to that of a solution of the pure diprotic acid alone. At the same time, the solution is more acidic than a solution of the acid salt, NaHA, alone. The following example illustrates this.

EXAMPLE: Compare the pH of a $0.10M$ solution of H_2CO_3 with that of a mixture of $0.10M$ H_2CO_3 and $0.10M$ $NaHCO_3$, and with that of a solution of $0.10M$ $NaHCO_3$ alone.
Solution: First we calculate the $[H^+]$ of $0.10M$ H_2CO_3 using **9-34**.

$$[H^+] = \sqrt{0.10M\ H_2CO_3(4.3 \times 10^{-7})} \cong 2.0_7 \times 10^{-4}M$$

Since **9-35** gives the % error as only 0.2%, the $[H^+]$ may be expressed to two significant figures.

$$[H^+] = 2.1 \times 10^{-4}M$$

$$pH = 3.68$$

Next we use **9-21** to approximate the $[H^+]$ of the $0.10M$ H_2CO_3–$0.10M$ $NaHCO_3$ mixture.

$$[H^+] \cong 4.3 \times 10^{-7} \frac{0.10M\ H_2CO_3}{0.10M\ HCO_3^-} \cong 4.3 \times 10^{-7}M$$

Use of **9-22** shows that the error is negligible in this calculation, so the $[H^+]$ may be reported to two significant figures.

$$[H^+] = 4.3 \times 10^{-7}M$$

$$pH = 6.37$$

From the example on p. 185, we see that a solution of $0.10M$ $NaHCO_3$ has a $[H^+]$ of $4.5 \times 10^{-8}M$ and a pH of 7.34. We can summarize the acidities of the three solutions in table form.

Solution	$[H^+]$	pH
$0.10M$ H_2CO_3	$2.1 \times 10^{-4}M$	3.68
$0.10M$ H_2CO_3 + $0.10M$ $NaHCO_3$	$4.3 \times 10^{-7}M$	6.37
$0.10M$ $NaHCO_3$	$4.5 \times 10^{-8}M$	7.34

We see that a solution of a diprotic acid alone is more acidic than a solution of a diprotic acid plus the conjugate acid salt, which in turn is more acidic than a solution of the conjugate acid salt alone.

Mixtures of a general acid salt, NaHA, with a general salt, Na_2A, can be calculated in a manner similar to that used in the last example for mixtures of H_2A with NaHA, except that K_2 must be used.

EXAMPLE: Calculate the pH of a mixture of $0.10M$ $NaHCO_3$ and $0.10M$ Na_2CO_3.
Solution: Use **9-21** to approximate $[H^+]$, substituting K_2 of H_2CO_3 for the ionization constant.

$$[H^+] \cong 4.8 \times 10^{-11} \frac{0.10M\ HCO_3^-}{0.10M\ CO_3^{-2}} \cong 4.8 \times 10^{-11}M$$

Use of **9-22** gives a % error of 4.8×10^{-8}%, so the $[H^+]$ can be expressed to two significant figures.

$$[H^+] = 4.8 \times 10^{-11}M$$

$$pH = 10.32$$

Physiological Buffers

The living tissues of the body are very sensitive to changes in the surrounding fluids; in particular, they are very sensitive to pH changes. Because of this, buffers exist in body fluids to protect the tissues against excess acid buildup, or *acidosis*, and against excess base buildup, or *alkalosis*. Buffers are very important both in the blood and in the muscles. A normal adult is reported to have sufficient buffer in the blood to neutralize about 0.150 mole (150 mmoles) of hydrogen ion [1]. The buffering bases in the muscles have the capacity to neutralize five times as much acid.

There are three main buffer systems in the blood. They are the $CO_2(aq)/HCO_3^-$ buffer pair, the $H_2PO_4^-/HPO_4^{-2}$ buffer pair, and certain protein pairs including hemoglobins. The hydrated carbon dioxide/bicarbonate ion buffer pair is most important in buffering alveolar (lung) blood; it reacts with acids or bases as follows:

$$HCO_3^- + H^+ \underset{\text{excess base}}{\overset{\text{excess acid}}{\rightleftharpoons}} CO_2(aq) + H_2O$$

$$\text{buffer base} \qquad\qquad \text{buffer acid}$$

Recall that we said at the beginning of this section that H_2CO_3 was really hydrated carbon dioxide, $CO_2(aq)$.

The $H_2PO_4^-/HPO_4^{-2}$ buffer pair has a pK_a of 7.2, which is quite close to the normal blood pH of 7.4. This pair reacts with acids or bases as follows:

$$HPO_4^{-2} + H^+ \underset{\text{excess base}}{\overset{\text{excess acid}}{\rightleftharpoons}} H_2PO_4^-$$

$$\text{buffer base} \qquad\qquad\qquad \text{buffer acid}$$

The buffer pairs of the protein function in a similar way.

In Section 9-3 it was stated that the maximum buffer action is obtained when the pK_a of the buffer acid is as close as possible to the specified pH. It is interesting that this is not true for the $CO_2(aq)/HCO_3^-$ buffer pair, and yet it functions as an effective buffer. The reason is that the concentration of the bicarbonate ion in the blood is much higher than that of the hydrated carbon dioxide. The example below gives you a typical situation.

> **EXAMPLE:** Calculate the buffer ratio for the carbon dioxide/bicarbonate ion buffer system in alveolar (lung) blood whose pH is 7.40.
>
> *Solution:* Following the method of the example on p. 181, we first calculate the pK_a from the K_1 of the carbonic acid system. Several authors [1,2] use a value of 8.0×10^{-7} for K_1 under body conditions. The pK_a would then be 6.10.
>
> To calculate the ratio of bicarbonate to hydrated carbon dioxide, we use a modified form of **9-31**.
>
> $$\log\frac{[HCO_3^-]}{[CO_2(aq)]} = pH - pK_a = 7.40 - 6.10 = 1.30$$
>
> $$\frac{[HCO_3^-]}{[CO_2(aq)]} = \frac{20}{1}$$
>
> This ratio will vary somewhat depending on the actual blood pH.

The reactions that occur in the alveolar blood as oxygen combines with hemoglobin (HHb) and forms the more acidic oxyhemoglobin (HHb-O_2) are a good illustration of how a 20:1 buffer ratio such as found in the last example is maintained. First inhaled oxygen combines with hemoglobin in the lung blood to form the more acidic oxyhemoglobin.

$$HHb + O_2 = HHb\text{-}O_2$$

The oxyhemoglobin ionizes, raising the $[H^+]$.

$$HHb\text{-}O_2 = H^+ + Hb\text{-}O_2^-$$

To remove the excess H^+, bicarbonate ion migrates to the region and reacts with the excess H^+.

$$HCO_3^- + H^+ \longrightarrow H_2CO_3$$

(A shift of chloride ion away from this region maintains a balance in the positively and negatively charged ions.) To prevent a buildup in carbonic acid, which is really hydrated carbon dioxide, it is rapidly decomposed by the enzyme carbonic anhydrase.

$$H_2CO_3(CO_2\text{-}H_2O) = CO_2(g) + H_2O$$

The gaseous carbon dioxide, $CO_2(g)$, diffuses into the lungs and is exhaled. This keeps the buffer ratio constant, near 20:1 (or whatever ratio exists at the pH of the blood).

| QUESTIONS AND PROBLEMS

(Answers to even-numbered problems are in Appendix 5.)

Ionization of Water

1. Explain why water undergoes *self-ionization*.
2. Give a mathematical equation that describes a neutral solution at *any* temperature.
3. Explain why the ionization of water is endothermic.
4. Calculate the pH of a neutral solution at:
   a. 37°C.
   b. 0°C.
   c. 60°C.
5. Construct a pH scale for the temperatures of 0° and 60°C. Locate the pH of $1M$ H^+, $1M$ OH^-, neutral solution, and the blood pH of 7.4.

Strong and Weak Acids and Bases

6. Calculate the pH of the following:
   a. Gastric fluid containing $0.020M$ HCl
   b. Gastric fluid (after being mixed with food in the stomach) containing $1.5 \times 10^{-4}M$ HCl
   c. Rationalize the difference between the two pH values above
7. Calculate the pH of the following:
   a. $0.002M$ sodium hydroxide
   b. $0.00200M$ sodium hydroxide
   c. $0.0020M$ barium hydroxide
8. Calculate the pH of $0.200M$ sulfuric acid ($K_2 = 1.0 \times 10^{-2}$):
   a. To one significant figure.
   b. To three significant figures.
9. Calculate the pH of a $0.10M$ solution of phenol, C_6H_5OH, a weak monoprotic acid with a K_a of 1.4×10^{-10}. Also calculate the % error in your approximation.
10. Calculate the $[OH^-]$ of a $0.100M$ solution of sodium benzoate, $C_6H_5CO_2^-Na^+$. The K_a of benzoic acid is 6.3×10^{-5}.

Mixtures and Simple Buffers

11. Calculate the pH of a solution prepared by mixing 10 ml of $0.30M$ acetic acid and 20 ml of $0.30M$ sodium acetate.

12. Calculate the pH of a solution prepared by mixing 25 ml of $0.30M$ ammonia with 5.0 ml of $0.15M$ ammonium chloride.
13. A $0.20M$ acetic acid–$0.20M$ sodium acetate buffer has a pH of 4.74. Calculate the change in pH when 10 ml of $0.10M$ hydrochloric acid is added to 40 ml of the buffer. The K_a of acetic acid is 1.8×10^{-5}.
14. Choose the proper pair of acid and base and calculate the ratio of the acid to the base necessary to achieve the following pHs:
   a. 4.60
   b. 9.26
   c. 9.10
15. According to recent research [R. C. Plumb, *J. Chem. Ed.* **49**, 179 (1972)], pain is produced when cells are ruptured, releasing their acidic buffers into the nerve areas so that the pH of the nerve endings falls below 6.2. What would the $[H^+]$ of a cell solution have to be to cause pain?

Polyprotic Acids

16. Calculate the pH of a solution containing only $0.0050M$ carbon dioxide.
17. Calculate the pH of the following solutions of phosphoric acid:
   a. $5.0M$ H_3PO_4
   b. $0.10M$ H_3PO_4
18. Calculate the pH of the following acid salt solutions:
   a. $0.30M$ $NaHCO_3$
   b. $0.010M$ KH_2PO_4
19. Select a suitable buffer for a solution near pH 9.70.

Challenging Problems

20. Calculate the pH of a $1.0 \times 10^{-5}M$ solution of the monoprotic acid phenol whose $K_a = 1.4 \times 10^{-10}$. Be sure your answer is reasonable.
21. Show that the double approximation of **9-42** is valid with exactly 9.5% error if the ratio in **9-43** is exactly 5 : 1.

22. Suppose that 10 mmoles of hydrogen ion are to be absorbed by the hydrated carbon dioxide/bicarbonate buffer pair with only a 0.5% change in the blood pH of 7.40. What number of mmoles of each would have to be present at the start to limit the change to 0.5%?

| NOTES

[1] R. A. Day and A. L. Underwood, *Quantitative Analysis*, 3rd ed., Prentice-Hall, Englewood Cliffs, N.J., 1974, p. 100.

[2] H. W. Davenport, *The ABC of Acid-Base Chemistry*, University of Chicago Press, Chicago, 1969, pp. 44–49.

10 Acid-Base Titrations

"Well, I gave my mind a thorough rest by
plunging into a *chemical analysis*," Holmes said.

ARTHUR CONAN DOYLE
The Sign of the Four, Chapter 10

In the preceding chapter, we reviewed the essential acid-base equilibria necessary for understanding the use of acid-base reactions in analysis. You are now prepared for "plunging" into the subject of acid-base titrations. Acid-base titrations are so convenient and so accurate that scientists in fields other than chemistry employ them for many different purposes. Chemists are able to titrate hundreds of inorganic and organic acids and bases in both aqueous solutions and nonaqueous solvents. Both direct and indirect acid-base titration methods are used widely in chemistry, biology, pharmacy, medical technology, etc. We will discuss primarily titrations in water, but we will devote some space to nonaqueous titrations at the end of the chapter.

Recall from Section 6-1 that an acid or base titrant must first be standardized against a primary standard chemical. An indicator or pH meter is used to locate the end point in the standardization as well as in the actual analysis. There are various types of titrations that must be understood before one attempts to apply any acid-base titration method.

10-1 | CHOOSING AND STANDARDIZING A TITRANT

In theory, any strong acid or strong base can be used as a titrant. The reason for this is that most reactions involving a strong acid or a strong base are quantitative. The only limitation is the reaction of the anion of the weak acid, or the cation of the weak base, with the solvent in the region of the end point. Since we will be discussing only titrations in water in the first part of the chapter, we will deal with the limits that the above reactions place on aqueous titrations.

Limits to Titrations in Water

In establishing the limits to aqueous titrations, we will rely on the concept of quantitative reactions established in Section 7-2. There a quantitative reaction was defined as a 99.9% reaction; that is, no more than 0.1% of the reactants are left at equilibrium.

Consider the reaction of any acid, HA, with sodium hydroxide titrant and the reaction of any base, BOH, with hydrochloric acid titrant:

$$\text{HA} + \text{OH}^- \rightleftharpoons \text{A}^- + \text{H}_2\text{O}$$
$$\text{(NaOH)}$$

$$\text{BOH} + \text{H}^+ \rightleftharpoons \text{B}^+ + \text{H}_2\text{O}$$
$$\text{(HCl)}$$

In the end point region, the anion, A^-, of the weak acid or the cation, B^+, of the weak base may react with water to produce more than 0.1% of their reactants (HA or BOH) at equilibrium. In their reactions with water, A^- behaves as a base and has a corresponding K_b, and B^+ behaves as an acid and has a corresponding K_a (Sec. 9-2). In this discussion, we will develop an approach for dealing with the reaction of A^- with water in the end point region that will also apply to the reaction of B^+ with water. Before doing this, we should state that the equilibrium constant K_b for the reaction of A^- with water at the end point is defined as follows:

$$K_b = \frac{K_w}{K_a} = \frac{[\text{HA}][\text{OH}^-]}{[\text{A}^-]} \tag{10-1}$$

To calculate the *maximum theoretical value* of K that corresponds to 0.1% or less reaction of A^- at the end point, we must first assume that $[\text{A}^-]$ will have a typical value of $0.1M$. Then the limiting value of $[\text{HA}]$ will be 0.1% of that, or $1 \times 10^{-4}M$. Since HA and OH^- are produced in equal quantities by the reaction of A^- with water, $[\text{OH}^-]$ will also equal $1 \times 10^{-4}M$. The maximum theoretical value of K_b is then calculated.

$$\text{max theo } K_b = \frac{[1 \times 10^{-4}M \text{ HA}][1 \times 10^{-4}M \text{ OH}^-]}{[0.1M \text{ A}^-]} = 1 \times 10^{-7} \tag{10-2}$$

By comparing it to the actual value of K_b for a specific anion, you can decide whether a sodium hydroxide titration will be quantitative at the end point. The guideline is summarized in the box below.

If the actual K_b for the anion-water reaction at the end point is $\leqslant 1 \times 10^{-7}$ (maximum theoretical value), the sodium hydroxide titration of the weak acid will be quantitative at the $0.1M$ level.

For example, consider the sodium hydroxide titration of phenol, whose K_a is 1.6×10^{-10} and whose formula is $\text{C}_6\text{H}_5\text{OH}$. The titration reaction is as follows:

$$\text{NaOH} + \text{C}_6\text{H}_5\text{OH} \rightleftharpoons \text{C}_6\text{H}_5\text{O}^- + \text{Na}^+ + \text{H}_2\text{O}$$

Note that the $C_6H_5O^-$ anion is produced at the end point; we assume its concentration is a typical $0.1M$. Using **10-1**, the actual value for the equilibrium constant for its reaction with water is calculated.

$$K_b = \frac{K_w}{K_a} = \frac{1.0 \times 10^{-14}}{1.6 \times 10^{-10}} = 6.3 \times 10^{-5}$$

This is larger than 1×10^{-7}, so the titration will not be quantitative at the $0.1M$ level. Near the end point the reaction of the $C_6H_5O^-$ anion with water will regenerate phenol in amounts such that 99.9% of the phenol can never be neutralized to the anion. The fundamental reason for this is that phenol is too weak an acid in water for quantitative neutralization. This is also the reason phenol is used medicinally as a disinfectant. It is just acidic enough to disinfect by killing microorganisms, but not too acid at low concentrations to be harmful to mammals.

The same reasoning can be applied to the titration of a weak base with a strong acid to give the guideline summarized in the box below.

If the actual K_a for the cation-water reaction at the end point is $\leqslant 1 \times 10^{-7}$ (maximum theoretical value), the titration of a weak base with a strong acid will be quantitative at the $0.1M$ level.

Another way of stating the guidelines in the boxes is that an acid can be titrated quantitatively at the $0.1M$ level only if its K_a is $\geqslant 1 \times 10^{-7}$ and a base can be titrated quantitatively at the $0.1M$ only if its K_b is $\geqslant 1 \times 10^{-7}$. However, using the guidelines as stated in terms of the ion reacting with water at the end point is more general. You should work Problem 11 at the end of this chapter to gain an understanding of these guidelines.

The only way to avoid the limitations that water imposes on the above titrations is to use nonaqueous solvents; this will be discussed at the end of this chapter.

Choosing and Standardizing Strong Acid Titrants

In theory, hydrochloric acid (HCl), nitric acid (HNO_3), and perchloric acid ($HClO_4$) are all equally useful as strong acid titrants. Nitric acid has an advantage in that it forms almost no insoluble salts to obscure end points and disadvantages in that it oxidizes many species and is frequently photodecomposed to a brown product. For these reasons hydrochloric acid is usually chosen; however, it forms insoluble salts in the presence of silver(I), lead(II), and mercury(I) ions, so it would not be used in the presence of these cations. Perchloric acid has none of the above disadvantages and can be used as a $0.1M$ titrant without problems. None of these strong acids are obtained commercially as primary standards, although constant-boiling hydrochloric acid, being of known concentration, can be prepared; all must be standardized using primary standards.

Primary Standard Bases for Acid Titrants

Acid titrants may be standardized using standardized sodium hydroxide (this standardization is described below); this reaction has a sharper end point than any of the titrations involving the primary standards discussed below. The only disadvantage is that any error made in standardizing the sodium hydroxide is incorporated into the standardization of the acid titrant.

The primary standards for strong acid titrants must meet all the requirements for a primary standard (Sec. 6-1). The two common primary standard bases are the following:

$$\text{Sodium carbonate, } K_{b_2} = 2.0 \times 10^{-8}$$

$$\text{THAM, Tris(hydroxymethyl)aminomethane, } K_b = 1.26 \times 10^{-6}$$

Sodium carbonate, Na_2CO_3, is the classic primary standard; it is available in pure form, except for a trace of sodium bicarbonate which is converted to sodium carbonate by drying at 300°C [1]. When it is titrated to carbonic acid, its formula weight of 106.00 must be divided by two to obtain an equivalent weight (Sec. 6-2). The end point is not sharp unless the carbonic acid is boiled off as carbon dioxide, a tedious process.

THAM, tris(hydroxymethyl)aminomethane, $(CH_2OH)_3CNH_2$, is now preferred to sodium carbonate, but it has a low formula weight (121.1) and its basicity is not that much stronger than that of sodium carbonate. Thus the end point is fairly sharp but not excellent.

Gravimetric standardization of hydrochloric acid by precipitation as silver chloride (Ch. 5) is, of course, the most accurate but the slowest standardization.

Choosing and Standardizing Strong Base Titrants

Sodium hydroxide is the most common strong base titrant. As discussed in Section 6-1, sodium hydroxide is not a primary standard, but a solution of sodium hydroxide, once standardized and protected from CO_2, is a stable titrant. It is usually made from a saturated solution that has been allowed to stand to permit precipitation of sodium carbonate impurity. Once such a solution is diluted, it is stable provided it is not exposed to carbon dioxide for too long a time. If it is, the following reaction becomes significant:

$$2OH^- + CO_2 \longrightarrow CO_3^{-2} + H_2O$$

This reaction, of course, lowers the concentration of the sodium hydroxide.

Primary Standard Acids for Sodium Hydroxide

To standardize sodium hydroxide, any of a number of primary standard weak acids may be used. They must all meet the requirements for a primary standard (Sec. 6-1). Some primary standard acids are the following:

$$\text{Sulfamic acid, } K_a = 1 \times 10^{-1}$$

$$\text{Benzoic acid, } K_a = 6.3 \times 10^{-5}$$

$$\text{Potassium acid phthalate, } K_a = 3.9 \times 10^{-6}$$

Sulfamic acid, HSO_3NH_2, is thought by some authors [1] to be the best primary standard for the standardization of strong bases. It is close to being a strong acid, is anhydrous and nonhygroscopic, and is stable up to 130°C. Unfortunately, its formula weight of 97.09 is not as large as is desirable for a primary standard.

Benzoic acid, $C_6H_5CO_2H$, is not nearly as strongly ionized as sulfamic acid but is commonly available. It is stable, anhydrous, and nonhygroscopic. Its formula weight of 122.12 is also not as large as those of some other primary standards, but it is satisfactory. Its low solubility in water (0.3 g/100 ml) and high solubility in organic solvents make it more suitable as a primary standard for nonaqueous titrations.

Potassium acid phthalate, $KHC_8H_4O_4$ or KHP, is weakly ionized but is generally preferred for standardization of sodium hydroxide. Not only is it stable at 110°C, anhydrous, and nonhygroscopic, but it has a large formula weight of 204.22. This, of course, reduces the effect of weighing errors, because it requires that a larger amount be weighed to obtain the usual 4 mmoles of primary standard for standardization of $0.1M$ sodium hydroxide.

Potassium hydroxide may also be employed as a titrant; it is standardized in the same manner that sodium hydroxide is, using the same primary standards.

10-2 | TITRATION CURVE OF A PRIMARY STANDARD WEAK ACID

The standardization of any titrant using a primary standard is best understood by calculating a titration curve. The process is very similar for both the standardization of a strong base using a weak acid and the standardization of a strong acid using a weak base. We will calculate the titration curve for the standardization of sodium hydroxide using a primary standard weak acid. You should then be able to use the same approach for calculating the curve for the standardization of hydrochloric acid using a primary standard weak base.

Calculations Involving a Molecular Weak Acid, HA

Of the three primary standard acids discussed in Section 10-1, both benzoic acid and sulfamic acid are of the molecular type HA; it is this type we will use for our calculations. (Potassium acid phthalate will be discussed briefly later.)

At the start of the titration, the solution in the flask consists of a weighed amount of the primary standard, HA, in a known volume of water. Thus the initial molarity of HA, M_{HA}, is known and the $[H^+]$ can be calculated to a good approximation using equation **9-7**:

$$[H^+] \cong \sqrt{M_{HA}(K_a)} \tag{10-3}$$

After sodium hydroxide has been added, a certain concentration of A^- will always be present because of neutralization; this results in a buffer system. From the discussion in Section 9-3, we write the following:

$$[H^+] \cong K_a \frac{M_{HA}}{M_{A^-}} \cong K_a \frac{\text{mmole HA}}{\text{mmole A}^-} \tag{10-4}$$

At the equivalence point, only the base, A^-, remains. The numerical value of its K_b is calculated from the quotient K_w/K_a (see **10-1**). From the discussion in Section 9-2, we calculate the $[OH^-]$ using the following equation:

$$[OH^-] \cong \sqrt{M_{A^-}(K_b)} \tag{10-5}$$

If the titration is carried beyond the equivalence point, the pH of the solution depends essentially on the concentration of added sodium hydroxide titrant:

$$[OH^-] = \frac{\text{mmole xs NaOH}}{\text{total volume, ml}} \tag{10-6}$$

In the following example, we use the standardization of sodium hydroxide using benzoic acid to illustrate a typical titration.

EXAMPLE: Calculate the pH for the standardization of approximately $0.06M$ sodium hydroxide using 366.36 mg/100 ml of primary standard benzoic acid, with $K_a = 6.3 \times 10^{-5}$ and formula weight of 122.12, at the following points: 0 ml, 10 ml, 25 ml, 50.0 ml, and 60 ml of sodium hydroxide (Fig. 10-1).

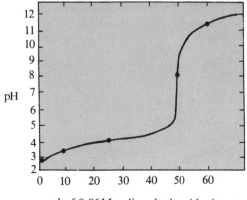

ml of 0.06M sodium hydroxide titrant

FIGURE 10-1. **Calculated titration curve for 0.0300M benzoic acid with 0.06M sodium hydroxide titrant.**

Solution at 0 ml: First we calculate the initial molarity of the benzoic acid, using HA as the symbol for benzoic acid.

$$M_{HA} = \frac{366.36 \text{ mg}/122.12 \text{ mg/mmole}}{100 \text{ ml}} = 0.0300M$$

Then we use **10-3** to calculate the [H$^+$].

$$[H^+] \cong \sqrt{0.0300M(6.3 \times 10^{-5})} = 1.3_{75} \times 10^{-3}M$$

$$pH = -\log[1.3_{75} \times 10^{-3}M] = 2.86_2 \quad (\text{see Fig. 10-1})$$

Solution at 10 ml: At this point there will be 0.6 mmole of benzoate anion, A$^-$, and 3.0 − 0.6 or 2.4 mmoles of benzoic acid at equilibrium. This is a buffer system for which the total volume can be ignored, since it is the same for A$^-$ and HA. Using **10-4**, we obtain the following:

$$[H^+] \cong (6.3 \times 10^{-5}) \frac{2.4 \text{ mmole HA}}{0.6 \text{ mmole A}^-} \cong 2._{52} \times 10^{-4}M$$

$$pH = 3.6 \quad (\text{see Fig. 10-1})$$

Solution at 25 ml: Addition of 25 × 0.06 mmoles of sodium hydroxide produces 1.5 mmoles of benzoate anion, A$^-$, and leaves 3.0 − 1.5 or 1.5 mmoles of benzoic acid at equilibrium. This is also a buffer system. Using **10-4**, we obtain the following:

$$[H^+] \cong (6.3 \times 10^{-5}) \frac{1.5 \text{ mmole HA}}{1.5 \text{ mmole A}^-} = 6.3 \times 10^{-5}M$$

$$pH = 4.20 \quad (\text{see Fig. 10-1})$$

Solution at 50.0 ml: This is the equivalence point, since 50 × 0.06 mmoles of sodium hydroxide has been added to 3.0 mmoles of benzoic acid. This leaves a 3.0 mmole/150 ml

$= 0.020M$ solution of benzoate anion, A^-, at equilibrium. Since benzoate anion is a base, we can use **10-5** to calculate the $[OH^-]$ and then the pH.

$$K_b = \frac{K_w}{K_a} = 1.5_9 \times 10^{-10}$$

$$[OH^-] \cong \sqrt{1.5_9 \times 10^{-10}(0.020M)} = 1.7_8 \times 10^{-6}M$$

$$pOH = 5.74_9$$

$$pH = 8.25_1 \qquad \text{(see Fig. 10-1)}$$

Solution at 60 ml: This is an excess of 10 ml of $0.06M$ sodium hydroxide past the equivalence point. Since sodium hydroxide is a much stronger base than the benzoate anion, we can calculate the pH essentially by finding the concentration of the hydroxide ion from the added sodium hydroxide, using **10-6**.

$$[OH^-] = \frac{10 \times 0.06 \text{ mmole OH}^-}{100 + 60 \text{ ml}} = 3._{75} \times 10^{-3}M$$

$$pOH = 2.4_{26}$$

$$pH = 11.5_{74} \qquad \text{(see Fig. 10-1)}$$

Calculation of molarity of NaOH: Finally, we complete the calculations by using the methods of Chapter 6 to calculate the molarity of the sodium hydroxide:

$$M_{NaOH} = \frac{366.36 \text{ mg}/122.12 \text{ mg/mmole}}{50.0 \text{ ml}} = 0.0600M$$

Any of the weak acids discussed later in this chapter can be treated as above to calculate a titration curve such as that in Figure 10-1. There are a few differences if the weak acid is the salt of a diprotic acid; those will be illustrated next.

Calculations Involving a Salt of a Diprotic Acid (KHP)

The widely used primary standard potassium acid phthalate (Sec. 10-1) is the salt of the diprotic acid phthalic acid, H_2P. Therefore, potassium acid phthalate, or KHP, is a monoprotic acid whose ionization constant of 3.9×10^{-6} is, of course, equal to the K_2 of phthalic acid.

The titration curve for KHP is very similar to that in Figure 10-1, with the exceptions of the pH at 0 ml of titrant and the pH at the equivalence point. At 0 ml of titrant, the HP^- ion can behave as an acid or a base (Sec. 9-4). Its $[H^+]$ can be calculated using a double approximation **(9-42)**:

$$[H^+] \cong \sqrt{K_1 K_2}$$

which is valid since **9-43** holds. Of course, the values used for K_1 and K_2 are those of phthalic acid.

The calculation of the pH at the equivalence point for KHP makes use of the concepts leading to **10-5**, except that K_b is calculated from K_2 of phthalic acid. Thus **10-5** is modified as follows:

$$[OH^-] \cong \sqrt{M_{P^{-2}}(K_w/K_2)}$$

You should test your comprehension of the above material by working Problems 14, 15, and 16 at the end of this chapter.

10-3 | ACID-BASE INDICATORS

Acid-base indicators function by changing color just after the equivalence point of a titration; this color change is called the *end point*. The end point is most often detected visually (Sec. 7-3). Most acid-base indicators are organic dye molecules which are either acids or bases. We will discuss the acid-base nature of indicators before showing how they are used.

Acid-Base Nature of Indicators   Indicators may be monoprotic (HIn) or diprotic (H_2In) acids. The acid form of an indicator is usually colored; when it loses a proton, the resulting anion (In^-), or base form of the indicator, exhibits a different color. The two forms exist in equilibrium with one another just as with any weak monoprotic acid (Sec. 9-2).

$$HIn \quad = H^+ + \quad In^- \tag{10-7}$$

(color *A*)   (color *B*)

A monoprotic acid indicator has an ionization constant, K_{in}, that is analogous to the K_a of a weak monoprotic acid. The ionization constant expression is

$$K_{in} = \frac{[H^+][In^-]}{[HIn]} \tag{10-8}$$

A diprotic acid indicator, H_2In, ionizes in two steps, giving the HIn^- and In^{-2} forms, each of which may exhibit a different color than the H_2In form.

$$H_2In \quad = H^+ + \quad HIn^- = H^+ + \quad In^{-2} \tag{10-9}$$

(color *A*)   (color *B*)   (color *C*)

Such an indicator would have two ionization constants, just as a diprotic acid does (Sec. 9-4).

A good example of a monoprotic acid indicator is *p*-nitrophenol, which ionizes to the *p*-nitrophenoxide anion as follows:

$$O_2N-\!\!\bigcirc\!\!-OH = H^+ + O_2N-\!\!\bigcirc\!\!-O^-$$

($O_2NC_6H_5OH$)   ($O_2NC_6H_5O^-$)

In dilute solution the acid form of this indicator is colorless; the base form, however, is yellow. The ionization constant for *p*-nitrophenol has a value of 1×10^{-7}.

$$K_{in} = 1 \times 10^{-7} = \frac{[H^+][O_2NC_6H_5O^-]}{[O_2NC_6H_5OH]} \tag{10-10}$$

This indicator exhibits a *pH transition range* of 6.2–7.5. Below pH 6.2, it exists in the colorless acid form; slightly above pH 6.2, a *faint yellow* color develops because of the presence of a significant fraction of the yellow phenoxide anion form. In the middle of this transition range (pH 7.0), a *definite yellow* color is observed. It can be seen by rearranging **10-10** that in the middle of the transition range for *p*-nitrophenol the ratio of the base form to the acid form will be 1 : 1.

$$\frac{[O_2NC_6H_5O^-]}{[O_2NC_6H_5OH]} = \frac{K_{in}}{[H^+]} = \frac{1 \times 10^{-7}}{1 \times 10^{-7}M\ H^+} = \frac{1}{1}$$

The *definite yellow* color observed at this point is the result of a mixture of equal amounts of the *intense yellow* phenoxide anion form and the *colorless* acid form of *p*-nitrophenol. Finally, above a pH of 7.5, an *intense* yellow color is observed as the indicator is completely converted to the phenoxide anion. These changes can be summarized in the *bar graph* form below.

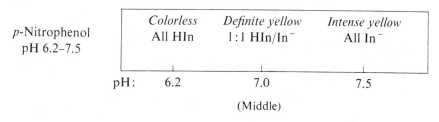

p-Nitrophenol
pH 6.2–7.5

| | Colorless | Definite yellow | Intense yellow |
| | All HIn | 1:1 HIn/In$^-$ | All In$^-$ |

pH: 6.2 7.0 7.5

(Middle)

An example of a diprotic acid indicator is the thymol blue molecule; it has two pH transition ranges (Table 10-1). Using the same terminology as in **10-9**, the colors can be ascribed to the following forms:

Red: H_2In form
Yellow: HIn^- form
Blue: In^{-2} form

The color changes in the two pH transition ranges are summarized using the two bar graphs below (recall that red *and* yellow together appear orange).

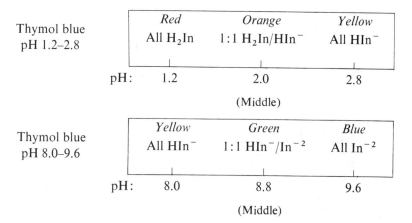

Thymol blue
pH 1.2–2.8

| Red | Orange | Yellow |
| All H_2In | 1:1 H_2In/HIn^- | All HIn^- |

pH: 1.2 2.0 2.8

(Middle)

Thymol blue
pH 8.0–9.6

| Yellow | Green | Blue |
| All HIn^- | 1:1 HIn^-/In^{-2} | All In^{-2} |

pH: 8.0 8.8 9.6

(Middle)

From these observations for thymol blue and *p*-nitrophenol, we can make the following generalization for all indicators.

In the middle of the pH transition range, an indicator exhibits a color usually consisting of equal amounts of the color of the acid form and the color of the base form.

TABLE 10-1.   Some Acid-Base Indicators

Common Name	Transition Range (pH)	Color Change Acid	Color Change Base
Crystal violet	0.1–1.5	yellow	blue
Thymol blue	1.2–2.8	red	yellow
Methyl yellow	2.9–4.0	red	yellow
Methyl orange	3.1–4.4	red	yellow
Bromcresol green	3.8–5.4	yellow	blue
Methyl red	4.2–6.3	red	yellow
Chlorophenol red	4.8–6.4	yellow	red
Bromothymol blue	6.0–7.6	yellow	blue
Phenol red	6.4–8.0	yellow	red
Neutral red	6.8–8.0	red	yellow-orange
Cresol purple	7.4–9.0	yellow	purple
Thymol blue	8.0–9.6	yellow	blue
Phenolphthalein	8.0–9.7	colorless	red
Thymolphthalein	9.3–10.5	colorless	blue
Alizarin yellow	10.1–12.0	colorless	violet

Selecting and Using Indicators   In Section 7-3, the selection of indicators in general was discussed. Essentially, the idea is to choose an indicator that undergoes a distinct color change at the equivalence point (true end point) of an acid-base titration. A color change, rather than the appearance of a particular color or shade of a color, is preferred, because it is easier for the eye to perceive.

After the preceding discussion of the acidic nature and color changes of indicators, we can conclude that the *greatest* color change must occur in the *middle half* of the pH transition range, from 25%–75% color change. This is shown in the bar graph below.

Acid color	(Slight color change)	Greatest color change Acid + Base colors	(Slight color change)	Base color
0% Base form	25% Base form	50% Base form	75% Base form	100% Base form
100% Acid form	75% Acid form	50% Acid form	25% Acid form	0% Acid form

It is in the middle half of the transition range that we observe the inversion in the ratio of the base to acid form, going from 25% base/75% acid to 75% base/25% acid. Stated in terms of **10-8**, the inversion starts at the following ratio of indicator ionization constant to $[H^+]$:

$$\frac{K_{in}}{[H^+]} = \frac{[In^-]}{[HIn]} = \frac{25\% \text{ Base}}{75\% \text{ Acid}} = \frac{1}{3}$$

The inversion ends at the following ratio:

$$\frac{K_{in}}{[H^+]} = \frac{[In^-]}{[HIn]} = \frac{75\% \text{ Base}}{25\% \text{ Acid}} = \frac{3}{1}$$

This means the $[H^+]$ has decreased to almost one-tenth of its value at the start of the middle half of the transition range.

To choose an indicator, we must know two things: the pH transition range of the indicator (Table 10-1) and the pH at the equivalence point, or at the steepest part of the titration curve. We then use the following generalization.

Choose an indicator whose middle half of the pH transition range encompasses the pH at the equivalence point, or the pH at the steepest part of the titration curve.

Finding the pH at the Equivalence Point

The pH at the equivalence point can either be estimated by calculation or found by performing a pH titration and plotting a titration curve. The former method is faster, but the latter method gives a better picture of the pH change. The latter method may reveal that several indicators are suitable.

The estimation of the pH at the equivalence point by calculation depends on the type of reaction involved. For the titration of strong acid with strong base, or vice versa, we showed in Section 7-1 that at the end point:

$$[H^+] = [OH^-]$$

$$[H^+] = \sqrt{K_w}$$

$$pH = \frac{pK_w}{2} = 7.00 \ (25°C) \tag{10-11}$$

EXAMPLE: Choose an indicator for the titration of 50 ml of $0.10M$ HCl with $0.10M$ NaOH.

Solution: Regardless of the concentrations, the pH at the end point of a titration of a strong acid with a strong base will always be 7.00 **(10-11)**. There are three indicators in Table 10-1 whose transition ranges include pH 7.0; they are, with the *middle half* of their transition ranges, as follows:

Bromothymol blue	pH 6.4–7.2
Phenol red	pH 6.8–7.6
Neutral red	pH 7.1–7.7

Both bromothymol blue and phenol red are suitable indicators, because the middle half of their transition ranges includes the pH of 7.0. Theoretically, neutral red would be rejected as an indicator, because the middle half of its range does not include pH 7.0. A glance at

Figure 7-1 will indicate that the experimental pH change at the end point is broad enough to allow neutral red to be used. (Other indicators whose middle half of the transition range does not include pH 7.0 can also be used; see the self-test at the end of this section.)

For the titration of a weak acid with a strong base, the pH at the end point is controlled entirely by the concentration of the anion of the weak acid. For example, consider the titration of $0.20M$ acetic acid with $0.20M$ sodium hydroxide titrant. The titration reaction is

$$CH_3CO_2H + NaOH \longrightarrow CH_3CO_2^- + H_2O + Na^+$$
$$\text{(acetic acid)} \qquad\qquad\qquad \text{(acetate ion)}$$

At the end point, the acetate ion produced acts as a Bronsted base and reacts with water to make the solution slightly basic.

$$CH_3CO_2^- + H_2O = OH^- + CH_3CO_2H \qquad\qquad (10\text{-}12)$$

All that is needed is to calculate the pH of the solution using the methods in Section 9-2 and to consult Table 10-1.

EXAMPLE: Choose an indicator for the titration of $0.2M$ acetic acid with $0.2M$ sodium hydroxide titrant.
Solution: At the end point of the titration, the concentration of the acetate ion will be one half that of the acetic acid because of dilution. It is now necessary to calculate the $[H^+]$ of a $0.1M$ solution of acetate, which ionizes according to **10-12**. The K_b of the acetate ion is equal to K_w/K_a of acetic acid, and the calculation is done following the approach in the example on p. 173, where the pH was calculated to be 8.87. There are three indicators in Table 10-1 whose transition ranges include pH 8.87; they are, with the middle half of their transition ranges, as follows:

Cresol purple	pH 7.8–8.6
Thymol blue	pH 8.4–9.2
Phenolphthalein	pH 8.4–9.2

Obviously cresol purple is unsatisfactory, because the pH of 8.87 at the equivalence point falls outside the middle half of its transition range. Both thymol blue and phenophthalein are satisfactory, because pH 8.87 falls inside the middle half of their transition ranges (see Fig. 10-2).

If a pH titration curve is available, the steepest part of the curve should be used to locate the equivalence point pH, and the middle half of the indicator transition range should include this pH.

EXAMPLE: Choose an indicator for the titration of $0.2M$ acetic acid, the pH titration curve of which is shown in Figure 10-2.
Solution: The middle of the steepest part of the titration curve in Figure 10-2 is approximately pH 8.9 (see dot). The same three indicators considered in the last example can be considered here: cresol purple (pH 7.4–9.0), thymol blue (pH 8.0–9.6), and phenolphthalein (pH 8.0–9.7). Although the transition range of each does occur on the steepest part of the curve, it can be seen from Figure 10-2 that the middle half of the range of cresol purple does not encompass the middle of the steepest part of the titration curve (see dot). The middle half of the range for phenolphthalein (and tor thymol blue, not shown) does, however. Either of the latter two would be a much better choice for an indicator than cresol purple.

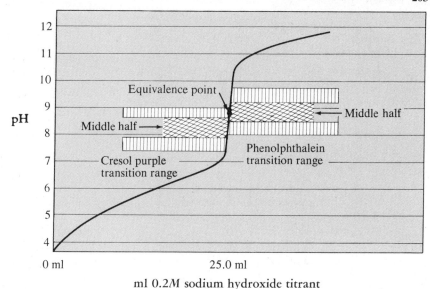

FIGURE 10-2. Titration curve for 25 ml of 0.2*M* acetic acid with 0.2*M* sodium hydroxide.

SELF-TEST	**13. Choosing Indicators**

Answers:

Directions: Cover the answers in the left column before beginning the test. Check your answers with those given at the left.

A. The indicator methyl red has a pH transition range of 4.2–6.3.

    a. Calculate half of the difference between these two pH values, carrying a nonsignificant figure along if necessary.

1.0_5

    b. Since the middle point of this transition range must have a quarter of the range on either side, calculate a quarter of the range.

0.5_3

    c. Subtract a quarter of the range from 6.3 and add a quarter of the range to 4.2 to obtain the middle half of the range.

4.7–5.8

B. Repeat the process in Problem A for the indicator bromcresol green which has a pH transition range of 3.8–5.4.

0.8; 0.4; 4.2–5.0

C. Decide whether either, both, or neither of the above indicators are suitable for equivalence points with the following calculated pH values:

Methyl red

    a. pH = 5.5

Both

    b. pH = 4.8

(over)

Neither

Both

Bromcresol green

Methyl red, bromothymol blue,
and phenolphthalein

c. pH = 4.0

d. pH = 5.0

e. pH = 4.3

D. The pH at the equivalence point in the titration of $0.1M$
HCl with $0.1M$ NaOH changes sharply from 4.0 to 10.5.
Suggest several indicators from Table 10-1 that are suitable.

10-4 | AQUEOUS TITRATION METHODS

In this section we will discuss the application of acid-base titration methods to the
analysis of various types of samples in water only. Nonaqueous titration methods
will be discussed in the next section.

The types of samples that can be analyzed by acid-base methods are simple one-
component samples, simple two-component mixtures, and complex mixtures. Both
direct and back titrations may be employed in the analysis. A back titration is one in
which an excess of a standard solution is added to react with all of the acid or base; the
remaining standard solution is then back titrated with a second titrant.

**Analysis of
Simple
One-
Component
Samples**

A one-component sample is either a solid or liquid sample containing just one acid or
base, either strong or weak. Because the analysis of strong acids or strong bases is
relatively simple compared to the analysis of weak acids or weak bases, the former will
be discussed first.

**Titration of
Strong Acids
or Bases**

A one-component strong acid sample is simply a solution of hydrochloric acid, nitric
acid, sulfuric acid, or perchloric acid in water. A one-component strong base sample is
just a solution of sodium hydroxide, potassium hydroxide, or barium hydroxide in
water. Sodium hydroxide is the usual titrant for strong acids, and a typical titration
reaction is

$$\text{NaOH} + \quad \text{HCl} \quad \longrightarrow \quad \text{Na}^+ + \text{Cl}^- + \text{H}_2\text{O} \quad \textbf{(10-13)}$$

(titrant)   (hydrochloric acid)

Note that in **10-13** there is no Bronsted acid or base produced; neither the sodium ion
nor the chloride ion reacts with water to produce hydrogen ion or hydroxide ion. The
only source of either of the latter at the titration end point is from the ionization of
water; therefore,

$$[\text{H}^+] = [\text{OH}^-] \quad \textbf{(10-14)}$$

$$[\text{H}^+] = \sqrt{K_w} \quad \textbf{(10-15)}$$

At room temperature, the $[\text{H}^+]$ will always be $1.0 \times 10^{-7}M$ at the equivalence point
for all strong acids. However, the steepest part of the titration curve will vary in length
depending on the initial concentration (see Fig. 10-3). Because of this, several indicators
may be used with equal accuracy to locate the end point. Note in Figure 10-3 that although the transition range of bromothymol blue falls in the middle of the steepest part
of the titration curve of $0.1M$ hydrochloric acid, both methyl red and phenolphthalein
indicators can also be used. Neither of the latter indicators is satisfactory for the
titration of $0.001M$ strong acid, however. Therefore, if the concentration of the strong

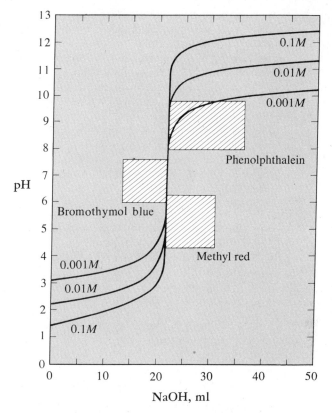

FIGURE 10-3. **Titration curves of different concentrations of strong acid with the corresponding concentration of sodium hydroxide, showing the feasibility of using three different indicators.**

acid is not known, it is safest to choose an indicator for which the middle half of the transition range will include pH 7.00.

The titration of most strong bases is just the reverse of the titration of strong acids. As long as the base is soluble, it is titrated directly with a strong acid titrant, such as hydrochloric acid. A titration reaction for a typical strong base, barium hydroxide, is

$$Ba(OH)_2 \quad + \; 2HCl \quad \longrightarrow \quad H_2O + Ba^{+2} + 2Cl^- \qquad \textbf{(10-16)}$$

(barium hydroxide) (titrant)

Note that there is no Bronsted acid or base produced in **10-16**, just as there was none in **10-13**. The barium(II) ion, like the sodium ion, does not react with water.

Cations in Groups IA and IIA of the periodic table do not react with water, in contrast to other cations, such as Al^{+3} which reacts slightly to give small amounts of $Al(OH)^{+2}$ and H^+.

At the end point in the titration of any strong base, the only source of hydrogen ion or hydroxide ion is the ionization of water. Equations **10-14** and **10-15** give the $[H^+]$ and the pH, which is 7.00 at the equivalence point. The same choice of indicators is available for the titration of strong bases at $0.1M$ concentration or at different concentrations such as are shown in Figure 10-3 for acids.

Certain insoluble, medically important "strong" bases, such as magnesium hydroxide and aluminum hydroxide, cannot be analyzed by direct titration because they form white suspensions and obscure the indicator color change. Magnesium hydroxide is used in a liquid suspension as milk of magnesia and is frequently mixed with aluminum hydroxide in solid antacids, such as Maalox. Such insoluble bases are analyzed by a *back titration* procedure instead of a direct titration, using standard solutions of hydrochloric acid and sodium hydroxide. For example, milk of magnesia can be analyzed for magnesium hydroxide content by the following steps. A measured excess of standard hydrochloric acid is added to the milk of magnesia.

$$2\,HCl(excess) + Mg(OH)_2(s) \longrightarrow 2\,H_2O + Mg^{+2} + 2Cl^-$$

Then the unreacted acid is back titrated with standard sodium hydroxide.

$$HCl + NaOH \longrightarrow H_2O + Na^+ + Cl^-$$
$$\text{(titrant)}$$

To calculate the percentage of magnesium hydroxide, the amount of $Mg(OH)_2$ in mmoles or meq is calculated. If the molarity system of calculation (Sec. 6-2) is used, calculate mmoles.

$$\text{mmole } Mg(OH)_2 = 1/2[(M\ HCl)(ml\ HCl) - (M\ NaOH)(ml\ NaOH)]$$

If the normality calculation system (Sec. 6-2) is used, calculate meq.

$$\text{meq } Mg(OH)_2 = (N\ HCl)(ml\ HCl) - (N\ NaOH)(ml\ NaOH)$$

Weak Monoprotic Acids and Bases

A one-component weak acid sample may be any one of a number of solutions or solids, such as the official pharmaceuticals listed in Table 10-2. These include both inorganic and organic pharmaceuticals in tablet, solid, suspension, or liquid form. One-component weak base samples are similar; typical weak bases are ammonia, methylamine (CH_3NH_2), and salts of weak acids.

The usual titrant for weak acids is sodium hydroxide; a typical titration reaction is the neutralization of acetic acid.

$$NaOH + CH_3CO_2H \longrightarrow H_2O + CH_3CO_2^- + Na^+ \qquad \textbf{(10-17)}$$

Note that in **10-17** there is a Bronsted base produced—the acetate ion. At the equivalence point it reacts with water as follows:

$$CH_3CO_2^- + H_2O = OH^- + CH_3CO_2H$$

This makes the solution slightly basic. For example, if $0.2M$ acetic acid is titrated with $0.2M$ sodium hydroxide, the pH of the resulting $0.1M$ acetate ion at the end point is 8.87 (see the example on p. 173 and the titration curve in Fig. 10-2). Other weak acids can be treated similarly.

TABLE 10-2. Acidic Pharmaceuticals Determined by Titration

Pharmaceutical	Formula (Organic acidic hydrogen written last)
Acetic acid, U.S.P.	CH_3CO_2H
Benzoic acid, U.S.P.	$C_6H_5CO_2H$
Boric acid, N.F.	H_3BO_3
Citric acid, U.S.P.	$C_6H_5O_7H_3$
Citrated caffeine, N.F.	Caffeine plus $C_6H_5O_7H_3$
Glutamic acid hydrochloride, N.F.	$C_5H_9NO_4 \cdot HCl$
Niacin, N.F.	$C_6H_4NO_2H$
Phenobarbital, U.S.P.	$C_{12}H_{11}N_2O_3H$
Potassium bitartrate, N.F.	$KHC_4H_4O_6$
Saccharin, U.S.P.	$C_7H_4N_2OSH$
Salicylic acid, U.S.P.	$C_7H_4O_3H_2$
Tartaric acid, N.F.	$C_6H_4O_6H_2$

EXAMPLE: Phenobarbital ($C_{12}H_{11}N_2O_3H$) is an important pharmaceutical, a weak acid ($K_a = 5 \times 10^{-8}$) that can barely be titrated in water. Calculate the pH at the end point in the titration of a $0.4M$ solution with $0.40M$ sodium hydroxide.

Solution: First calculate the K_b of the $C_{12}H_{11}N_2O_3^-$ anion at the end point.

$$K_b = \frac{K_w}{K_a} = \frac{1.00 \times 10^{-14}}{5 \times 10^{-8}} = 2 \times 10^{-7}$$

Then calculate the $[OH^-]$ at the end point, following the method of the example on p. 173.

$$[OH^-] \cong \sqrt{M\, C_{12}H_{11}N_2O_3^-(K_b)} \cong \sqrt{0.2M(2 \times 10^{-7})} \cong 2 \times 10^{-4}M$$

The % error in the approximation is calculated as follows:

$$\% \text{ error} = \frac{2 \times 10^{-4}M\, OH^-}{0.2M}(100) = 0.1\%$$

Since the % error is less than 1%, the $[OH^-]$ *may* theoretically be reported to two significant figures; however, only one significant figure is justified because the base concentration was given to only $0.2M$.

$$[OH^-] = 2 \times 10^{-4}M$$

$$pH = 10.3$$

Since the pH at the end point is quite basic, none of the indicators in Table 10-1 are perfectly suitable. Neither thymolphthalein nor alizarin yellow indicators are satisfactory, since the middle half of each of their ranges is outside the calculated pH.

In Section 10-1 it was stated that titrations of $0.1M$ weak acids are quantitative only if the K_a is 1×10^{-7} or greater. It is also true that many titration curves for $0.1M$ weak acids do not exhibit a reasonably sharp break at the equivalence point so the latter cannot be located graphically. Figure 10-4 illustrates this point. Note that the titration curves for weak acids with a K_a greater than or equal to 1×10^{-7} exhibit steep portions so that the equivalence point can be located accurately. An acid with a smaller K_a,

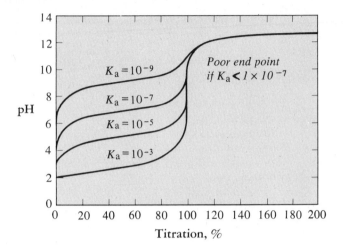

FIGURE 10-4. Theoretical titration curves of 0.1M solutions of various weak acids each having the K_a values shown with 0.1M sodium hydroxide.

such as 1×10^{-9}, exhibits no steep portion but merely a "wiggle" near the equivalence point. Of course, these titration curves apply only to 0.1M concentrations; at lower concentrations, such as 0.01M, steep portions of the titration curve are observed only for weak acids with a K_a as low as 1×10^{-6}. Then the titration reaction is not quantitative if the K_a is less than 1×10^{-6}.

An example of a typical method for a single weak acid component is the analysis of vinegar for acetic acid. Commercial vinegar is distilled from the raw materials used in its manufacture; of the acids, only acetic acid (boiling point 118°) distills over. Other possibly acidic materials are left behind because they boil at higher temperatures. Thus the vinegar contains only one acid component. Because the vinegar is relatively concentrated (5% or 1M), either it is diluted before analysis or a very small sample is taken. The calculation of the concentration of such samples is usually done in terms of weight percent acetic acid, which is generally expressed in grams of acetic acid per milliliter of solution. See the example below.

EXAMPLE: Calculate the weight percent of acetic acid in 5.00 ml of vinegar that requires 40.00 ml of 0.1000M sodium hydroxide. Use 60.0 mg/mmole as the formula weight of acetic acid, CH_3CO_2H.

Solution: First calculate the number of mg of CH_3CO_2H.

mg CH_3CO_2H = (40.00 ml NaOH)(0.1000M NaOH)(60.0 mg/mmole CH_3CO_2H)

mg CH_3CO_2H = 240 mg

Convert the mg to g.

$$\frac{240 \text{ mg}}{1000} = 0.240 \text{ g } CH_3CO_2H$$

Then calculate the weight percent in terms of g CH_3CO_2H/ml.

$$Wt\% \ CH_3CO_2H = \frac{0.240 \ g \ CH_3CO_2H}{5.00 \ ml}(100) = 4.80\%$$

Strictly speaking, 5.00 ml of water should be converted to grams of water, but since 5.00 ml of water is also 5.00 g of water, it is not necessary.

Another example of a typical analysis for a single weak acid component is the titration of acid salts of diprotic acids. These salts behave as monoprotic acids, since they lose but one hydrogen per formula weight. Typical salts are potassium acid oxalate (KHC_2O_4), potassium acid tartrate ($KHC_4H_4O_6$), and potassium acid phthalate ($KHC_8H_4O_4$). The titration reaction with the latter is

$$NaOH + KHC_8H_4O_4 \longrightarrow H_2O + Na^+ + K^+ + C_8H_4O_4^{-2}$$
(titrant)

The titration of most weak bases is similar to the titration of weak acids. They are titrated directly with a strong acid titrant, such as hydrochloric acid. A typical titration reaction for ammonia can be written with NH_3 or with NH_4OH as follows:

$$NH_3 + HCl \longrightarrow NH_4^+ + Cl^- \qquad \textbf{(10-18)}$$
(titrant)

$$NH_4OH + HCl \longrightarrow NH_4^+ + H_2O + Cl^-$$

Note that in **10-18** a Bronsted acid is produced; this is the ammonium ion. At the equivalence point this is the only acidic species present; it reacts with water as follows:

$$NH_4^+ + H_2O \rightleftharpoons H_3O^+ + NH_3 \qquad \textbf{(10-19)}$$

This means the solution is slightly acidic, the pH being controlled by the concentration of the ammonium ion present. See the following example.

EXAMPLE: Calculate the pH at the end point in the titration of $0.20M$ ammonia with $0.20M$ hydrochloric acid. Choose an indicator.

Solution: First calculate the K_a of the ammonium ion in **10-19** using the general equation **9-9**.

$$K_a = \frac{K_w}{K_b} = \frac{1.00 \times 10^{-14}}{1.8 \times 10^{-5}} = 5.5_5 \times 10^{-10}$$

Then, using the methods of the example on p. 171, calculate the approximate $[H^+]$. Recall that the approximate relation of **9-7** is valid for any weak acid, such as the ammonium ion.

$$[H^+] \cong \sqrt{M \ of \ NH_4^+(K_a)} \cong \sqrt{0.10M(5.5_5 \times 10^{-10})} \cong 7.4_5 \times 10^{-6}M$$

Using **9-8**, calculate the % error.

$$\% \ error = \frac{7.4 \times 10^{-6}M}{0.10M}(100) = 7.4 \times 10^{-3}\%$$

Since the % error is less than 1%, we can report the $[H^+]$ to two significant figures.

$$[H^+] = 7.4_5 \times 10^{-6} M$$

$$pH = 5.13$$

Because the middle half of the transition range of methyl red indicator is 4.7–5.8, this indicator is satisfactory. Bromcresol green is not satisfactory, since the middle half of its range is 4.2–5.0.

The titration curve for $0.20M$ ammonia using $0.20M$ hydrochloric acid titrant is shown in Figure 10-5. The middle of the steepest part of the titration curve is at approximately pH 5.1 (see dot). There are three indicators whose transition ranges include pH 5.1.

Indicator	Middle half of pH transition range
Bromcresol green	4.2–5.0
Methyl red	4.7–5.8
Chlorophenol red	5.2–6.0

Although the transition range of each indicator does enclose the steepest part of the curve, it can be seen in Figure 10-5 that the *middle half* of the range of bromcresol green (as well as that of chlorophenol red, not shown) does not include the middle of the steepest part of the curve. That of methyl red indicator does, making this a better choice than either of the other two indicators.

In Section 10-1 we stated that the titration of $0.1M$ solutions of weak bases having a K_b of 1×10^{-7} or larger is quantitative. It is likewise true that such titration curves do

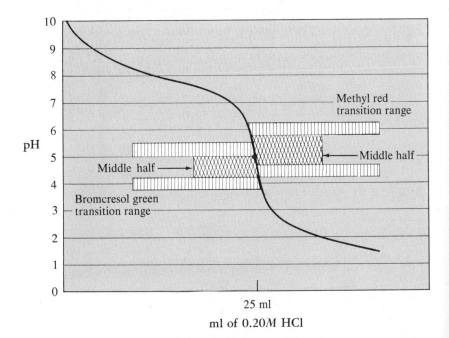

FIGURE 10-5. Titration curve for 25 ml of $0.20M$ ammonia with $0.20M$ hydrochloric acid.

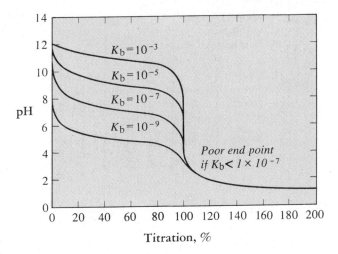

FIGURE 10-6. Theoretical titration curves of $0.1M$ solutions of various weak bases having the K_b values shown with $0.1M$ hydrochloric acid.

exhibit reasonably sharp breaks at the equivalence point. Figure 10-6 illustrates this situation for $0.1M$ solutions. Note that the titration curves for weak bases with values of K_b from 1×10^{-3} through 1×10^{-7} exhibit steep portions so that the equivalence point can be located accurately. A base with a smaller K_b, such as 1×10^{-9}, exhibits no steep portion, just a "wiggle." At lower concentrations, such as $0.01M$, a steep portion is observed at the equivalence point only if the K_b is 1×10^{-6} or larger. Then the titration reaction is not quantitative if the K_b is less than 1×10^{-6}.

An example of a typical method for a single weak base component is the determination of ammonia in cleaning products, such as household ammonia solutions. Many such solutions contain only water, ammonia, coloring matter, and perhaps a detergent. Equation **10-18** would then be the only acid-base reaction that could occur for such a sample. The calculation of the concentration of such samples is usually done in terms of weight percent ammonia, which is usually expressed in terms of grams of ammonia per milliliter of solution. See the example below.

EXAMPLE: Calculate the weight percent of ammonia in 5.00 ml of an ammonia solution that requires 30.00 ml of $0.1000M$ hydrochloric acid. Use 17.0 mg/mmole as the formula weight for NH_3.

Solution: First calculate the number of mg of NH_3.

$$mg\ NH_3 = (30.00\ ml\ HCl)(0.1000M\ HCl)(17.0\ mg/mmole\ NH_3)$$

$$mg\ NH_3 = 51.0\ mg$$

Next convert mg NH_3 to grams.

$$\frac{51.0\ mg\ NH_3}{1000} = 0.0510\ g\ NH_3$$

Then calculate the weight percent NH_3 in terms of g NH_3/ml.

$$wt\ \%\ NH_3 = \frac{0.0510\ g\ NH_3}{5.00\ ml}(100) = 1.02\%\ NH_3$$

A second example of an analysis for a single weak base component is the *direct* titration of sodium bicarbonate or potassium bicarbonate in tablet (pharmaceutical) form. The titration reaction is

$$HCl + NaHCO_3 \longrightarrow H_2CO_3 + Na^+ + Cl^-$$

A number of other weakly basic pharmaceuticals cannot be titrated directly but can be determined by back titration. For example, the weak nitrogen base morpholine, C_4H_9ON, is determined as follows:

$$C_4H_9ON + HCl(xs) \longrightarrow C_4H_9ONH^+ + Cl^-$$

The unreacted acid is then back titrated with standard sodium hydroxide.

$$HCl + NaOH \longrightarrow H_2O + Na^+ + Cl^-$$

An example of the calculations involved appears below.

EXAMPLE: Calculate the percentage of morpholine (form wt = 87.1) in a 1.0000-g sample of morpholine pharmaceutical that was neutralized with 50.00 ml of 0.2000M hydrochloric acid and back titrated with 20.00 ml of 0.1000M sodium hydroxide.
Solution: First calculate the number of mmoles of morpholine by subtracting the number of mmoles of NaOH from the number of mmoles of excess HCl added.

mmole C_4H_9ON = (50.00 ml HCl)(0.2000M HCl) − (20.00 ml NaOH)(0.1000M NaOH)

mmole C_4H_9ON = 10.00 − 2.00 = 8.00 mmole

$$\% \ C_4H_9ON = \frac{(8.00 \ \text{mmole})(87.1 \ \text{mg/mmole})(100)}{1000 \ \text{mg}} = 69.6_8 \%$$

Analysis of Polyprotic Acids and Bases and Two-Component Samples

A two-component sample is either a solid or liquid sample containing two acids or two bases, both strong, both weak, or one strong and one weak. Because there is a difference in the magnitudes of K_1, K_2, and so on, of polyprotic acids, the titration of a polyprotic acid is much like the titration of a two-component mixture. We will therefore discuss the titration of polyprotic acids first to establish the principles that can also be used for the titration of two-component mixtures.

Titration of Polyprotic Acid One-Component Samples

The stepwise ionization of polyprotic acids has already been discussed in Section 9-4. Because there is a difference in the magnitudes of K_1, K_2, and so on, the first hydrogen may be totally neutralized before the second hydrogen is neutralized, or both hydrogens may be neutralized essentially together. The general rules are given in the box.

If K_1/K_2 is $\geqslant 10^4$, the titration curve of a polyprotic acid will exhibit an end point for the first hydrogen. (If K_2 is $\geqslant 1 \times 10^{-7}$, an end point for the second hydrogen will also be observed.) If K_1/K_2 is $< 10^4$, the titration curve will not exhibit a sharp enough "break," or steep portion, for the first end point to be located accurately.

To illustrate the general rules, let us consider the following acids:

maleic acid $K_1 = 1.2 \times 10^{-2}$ $K_2 = 6.0 \times 10^{-7}$
carbonic acid $K_1 = 4.3 \times 10^{-7}$ $K_2 = 4.8 \times 10^{-11}$
phosphoric acid $K_1 = 7.5 \times 10^{-3}$ $K_2 = 6.2 \times 10^{-8}$ $K_3 = 4.8 \times 10^{-13}$

The ratio K_1/K_2 for maleic acid is

$$\frac{K_1}{K_2} = \frac{1.2 \times 10^{-2}}{6.0 \times 10^{-7}} = 2.0 \times 10^4$$

Since K_1/K_2 is greater than 10^4, the titration curve for maleic acid will exhibit an end point for the neutralization of the first hydrogen.

$$\text{NaOH} + \text{C}_2\text{H}_2(\text{CO}_2\text{H})_2 \longrightarrow \text{C}_2\text{H}_2(\text{CO}_2\text{H})\text{CO}_2^- + \text{Na}^+ + \text{H}_2\text{O}$$
(titrant)   maleic acid

The value of K_2 for maleic acid is greater than 1×10^{-7}, so the titration curve will also exhibit an end point for the neutralization of the second hydrogen.

$$\text{NaOH} + \text{C}_2\text{H}_2(\text{CO}_2\text{H})\text{CO}_2^- \longrightarrow \text{C}_2\text{H}_2(\text{CO}_2)^{-2} + \text{Na}^+ + \text{H}_2\text{O}$$
(titrant)   (acid maleate anion)   (maleate anion)

The ratio of K_1/K_2 for carbonic acid is only slightly less than 10^4, so the titration curve for carbonic acid should exhibit an end point for the neutralization of the first hydrogen.

$$\text{NaOH} + \text{H}_2\text{CO}_3 \longrightarrow \text{HCO}_3^- + \text{Na}^+ + \text{H}_2\text{O}$$

The value of K_2 for carbonic acid is less than 1×10^{-7}, so the titration cannot be carried accurately beyond the neutralization to the bicarbonate ion above.

The titration of phosphoric acid is like titrations of two diprotic acids, since K_1 must be compared to K_2 for the first end point and K_2 must be compared to K_3 for the second end point. The ratio of K_1 to K_2 for phosphoric acid is as follows:

$$\frac{K_1}{K_2} = \frac{7.5 \times 10^{-3}}{6.2 \times 10^{-8}} = 1.2 \times 10^5$$

Since this is larger than 10^4, the titration curve will exhibit an end point for the neutralization of the first hydrogen.

$$\text{NaOH} + \text{H}_3\text{PO}_4 \longrightarrow \text{H}_2\text{PO}_4^- + \text{Na}^+ + \text{H}_2\text{O}$$

This is shown in Figure 10-7. Since the only species present is the dihydrogen phosphate anion, its pH can be calculated by using **9-42** (see the example on p. 185).

$$[\text{H}^+] \approx \sqrt{K_1 K_2} \approx 2.1_6 \times 10^{-5} M$$

$$\text{pH} = 4.75$$

Because phosphoric acid is a triprotic acid, we must also decide whether the titration curve will exhibit a second end point by calculating the ratio K_2/K_3.

$$\frac{K_2}{K_3} = \frac{6.2 \times 10^{-8}}{4.8 \times 10^{-13}} = 1.3 \times 10^5$$

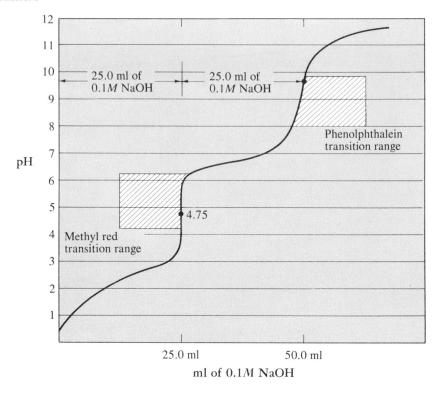

FIGURE 10-7. Titration of 0.1M phosphoric acid with 0.1M sodium hydroxide.

This verifies that the titration curve shown in Figure 10-7 exhibits an end point for the neutralization of the second hydrogen.

$$NaOH + H_2PO_4^- \longrightarrow HPO_4^{-2} + Na^+ + H_2O$$

Since the only species present at the second end point is the HPO_4^{-2} ion, the pH can be calculated in the same manner as at the first end point (see the example on p. 186).

$$[H^+] \approx \sqrt{K_2 K_3} \approx 1.7 \times 10^{-10}$$

$$pH = 9.76$$

The K_3 of phosphoric acid is less than 1×10^{-7}, so titration of the HPO_4^{-2} ion cannot be carried out accurately.

To calculate the amount of phosphoric acid present in a sample it is desirable to use the volume of titrant at the first end point, because the titration curve is steeper there than at the second end point (see Fig. 10-7). However, if a different acid is also present, whichever end point requires the smallest volume of titrant will give the only accurate results. Because a second acid may be titrated with H_3PO_4 or with $H_2PO_4{}^-$, two situations, summarized below, are possible.

Relative size of K_a of second acid	Relation of buret readings at end points	To calculate $\% H_3PO_4$, use
$K_a \geqslant K_1$	$ml_{1st} > (ml_{2nd} - ml_{1st})$	$(ml_{2nd} - ml_{1st})$
$K_1 > K_a \geqslant K_2$	$(ml_{2nd} - ml_{1st}) > ml_{1st}$	ml_{1st}

Note that the buret reading at the second end point (ml_{2nd}) is actually the total volume of titrant added for the neutralization of H_3PO_4 and $H_2PO_4{}^-$. To calculate the volume required for neutralization of $H_2PO_4{}^-$ alone, the ml of titrant used at the first end point must be subtracted from that used at the second end point. The calculation is clarified by considering the following example.

EXAMPLE: A 1.0000-g sample of phosphoric acid mixed with an unknown mono-protic acid is titrated with $0.1000M$ sodium hydroxide. The buret reading at the first end point is 30.00 ml, and the buret reading at the second end point is 55.00 ml. Calculate the percentage of phosphoric acid present. The formula weight of phosphoric acid is 98.0.

Solution: First decide which end point represents only the phosphoric acid by comparing the volumes.

$$ml_{1st} \text{ for neut. of } H_3PO_4 = 30.00 \text{ ml}$$

$$(ml_{2nd} - ml_{1st}) \text{ for neut. of } H_2PO_4{}^- = 55.00 - 30.00 = 25.00 \text{ ml}$$

Since the volume for the neutralization of $H_2PO_4{}^-$ is smaller, this volume represents only the titration of phosphoric acid. The monoprotic acid is obviously being titrated along with H_3PO_4 to the first end point (contrast this with Fig. 10-7). Now calculate the percentage of phosphoric acid.

$$\% H_3PO_4 = \frac{(25.00 \text{ ml NaOH})(0.1000M)(98.0 \text{ mg/mmole})(100)}{1000.0 \text{ mg}} = 24.5\%$$

In conclusion, we can say that the number of mmoles of the weak monoprotic acid is $(5.00 \text{ ml})(0.1000M)$, or 0.500 mmole. Unless the molecular weight is known, nothing further can be calculated.

Titration of Boric Acid

Although boric acid is a triprotic acid, all three hydrogens are very weakly ionized. The value of 5.9×10^{-10} for K_1 is less than 1×10^{-7}, so even the first hydrogen cannot be titrated quantitatively. (The fact that boric acid is a very weakly ionized acid makes it ideal for use in eye wash solutions. For example, a $0.1M$ solution is acidic enough at $[H^+] = 8 \times 10^{-6}M$ for medicinal purposes, but not acidic enough to harm the cornea of eye.)

Fortunately, boric acid forms stable complexes with organic glycols; such complexes are acidic enough to be titrated quantitatively. For example, glycerine reacts with boric acid as follows:

$$H_3BO_3 + C_3H_5OH \longrightarrow (HO)_2B\underset{OCH_2}{\overset{OCH-CH_2OH}{|}} + H_2O \quad (10\text{-}20)$$

(boric acid) (glycerine)

The K_1 value of boric acid is increased by the reaction to a value larger than 1×10^{-7} [2]; the actual value depends on the amount of complexing glycol added. After the complex is formed, it can be titrated as a monoprotic acid to a phenolphthalein end point [3].

$$(HO)_2B{=}O_2C_3H_5OH + NaOH \longrightarrow [O(HO)B{=}O_2C_3H_5OH]^- + Na^+ + H_2O$$
$$\text{(complex)} \qquad \text{(titrant)}$$

$$(10\text{-}21)$$

The remaining hydrogen is still too weak to be titrated, so only one hydrogen can be titrated quantitatively from the complex.

Titration of
Polybasic
Base One-
Component
Samples

The most important polyfunctional base is sodium carbonate. Its two ionization constants are not listed in the appendix, but can be calculated as K_b for any anion that is calculated by using **9-17**. Thus K_1 for the carbonate ion is calculated by inserting the ionization constant for the bicarbonate ion, its conjugate acid, into **9-17**.

$$K_1 = \frac{K_w}{K_{HCO_3^-}} = \frac{1 \times 10^{-14}}{4.8 \times 10^{-11}} = 2.1 \times 10^{-4}$$

The value of K_2 for the carbonate ion is similarly calculated from K_w and the ionization constant for carbonic acid.

$$K_2 = \frac{K_w}{K_{H_2CO_3}} = \frac{1 \times 10^{-14}}{4.3 \times 10^{-7}} = 2.3 \times 10^{-8}$$

The same general rules for the stepwise neutralization of polyprotic acids apply to the stepwise neutralization of polyfunctional bases. Thus if K_1/K_2 for such bases is at least 10^4, the titration curve will exhibit an end point for the reaction of one proton of titrant per formula weight of the base. For sodium carbonate, the ratio K_1/K_2 is

$$\frac{K_1}{K_2} = \frac{2.1 \times 10^{-4}}{2.3 \times 10^{-8}} = 9.1 \times 10^3$$

Since this is almost 10^4, the titration curve will exhibit an end point for the neutralization of the carbonate ion to bicarbonate.

$$CO_3^{-2} + HCl \longrightarrow HCO_3^- + H_2O$$

This is shown in Figure 10-8. Since the only species present at the end point is the bicarbonate ion, the pH can be calculated by using **9-42** (see the example on p. 185) and *the ionization constants for carbonic acid.*

$$[H^+] \approx \sqrt{K_1 K_2} \approx \sqrt{(4.3 \times 10^{-7})(4.8 \times 10^{-11})} \approx 4.5 \times 10^{-8} M$$

$$pH = 7.34$$

The K_2 of the carbonate ion is slightly less than 1×10^{-7}, so the titration of $0.1 M$ solutions should be just short of quantitative. Indeed, the color change at the methyl orange end point (Fig. 10-8) is notorious for being difficult to perceive. Usually a pH 4 buffer is prepared and the color of methyl orange in the titration flask is matched with the color of methyl orange in the pH 4 buffer.

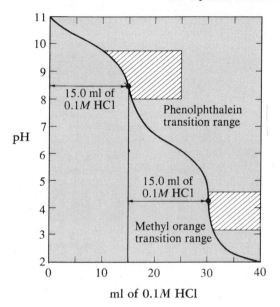

FIGURE 10-8. Titration of 1.5 mmoles of sodium carbonate with 0.1M hydrochloric acid.

To calculate the amount of sodium carbonate present in a sample, it is still relatively desirable to use the volume of titrant at the second end point rather than at the first end point, which is not as steep. If a different base is also present, whichever end point requires the smallest volume of titrant will give the only accurate results. Because the second base may be a strong base such as sodium hydroxide or a weak base such as sodium bicarbonate, two situations are possible. They are summarized below.

Strength of second base	Relation of buret readings at end points	To calculate % Na_2CO_3 use
Strong (NaOH, etc.)	$ml_{1st} > (ml_{2nd} - ml_{1st})$	$(ml_{2nd} - ml_{1st})$
Weak (NaHCO$_3$)	$(ml_{2nd} - ml_{1st}) > ml_{1st}$	ml_{1st}

Note that the buret reading at the second end point (ml_{2nd}) is actually the total volume of titrant added for neutralization to H_2CO_3. To calculate the volume required for neutralization of HCO_3^- to H_2CO_3, the ml of titrant used at the first end point must be subtracted from the buret reading at the second end point. See the example below.

EXAMPLE: A 1.0000-g sample of sodium carbonate mixed with an unknown base (monofunctional) is titrated with 0.1000M hydrochloric acid. The buret reading at the first end point is 25.00 ml; the reading at the second end point is 40.00 ml. Calculate the percentage of sodium carbonate present. Its formula weight is 106.0.

Solution: First decide which end point represents only the sodium carbonate by comparing the end point volumes.

$$ml_{1st} \text{ for neut. of } CO_3^{-2} = 25.00 \text{ ml}$$

$$(ml_{2nd} - ml_{1st}) \text{ for neut. of } HCO_3^- = 40.00 - 25.00 = 15.00 \text{ ml}$$

Because the volume for the neutralization of the CO_3^{-2} is smaller, this volume must represent the titration of only sodium carbonate. The unknown base is obviously being titrated with the CO_3^{-2} to the first end point (contrast this to Fig. 10-8). Now calculate the percentage.

$$\% \, Na_2CO_3 = \frac{(15.00 \, ml \, HCl)(0.1000M \, HCl)(106.0 \, mg/mmole)(100)}{1000.0 \, mg} = 15.90\%$$

The number of mmoles of the unknown monofunctional base is (10.00 ml) (0.1000M), or 1.000 mmole. Unless the molecular weight is known, nothing else can be calculated.

Analysis for Two Components in a Sample

Two-component mixtures may consist of two monoprotic acids or two monofunctional bases, or they may be more complex samples, such as a mixture of a monoprotic acid and a polyprotic acid. The same general rules apply to these mixtures as to the titration of a single polyprotic acid or base. The K_a of the stronger acid is divided by the K_a of the weaker acid.

If $K_{a(str)}/K_{a(wk)}$ is $\geq 10^4$, the titration curve will exhibit an end point for the stronger acid. If $K_{a(wk)}$ is $> 1 \times 10^{-7}$, an end point for the weaker acid will also be observed.

The same is true for a mixture of two bases.

If a mixture consists of a *strong* acid, such as hydrochloric acid, and a weak acid, the above rule cannot readily be applied. It appears from the titration curve of a mixture of hydrochloric and acetic acids that an end point for a *strong* acid will be observed if the K_a of the weak acid is 2×10^{-5} or less [4].

To calculate the amount of each acid or base present in a two-component mixture, it is necessary to decide which acid or base is titrated first and then use the volume of titrant corresponding to it in the calculation. See the example below.

EXAMPLE: A 1.0000-g sample of trichloroacetic acid ($K_a = 2.2 \times 10^{-1}$) and acetic acid ($K_a = 1.8 \times 10^{-5}$) is titrated with 0.1000M sodium hydroxide. The buret reading at the first end point is 30.00 ml; the reading at the second end point is 55.00 ml. Calculate the percentage of each acid present. The formula weight of trichloroacetic acid is 163.4 and that of acetic acid is 60.0.

Solution: First decide which acid is titrated at the first end point. This is trichloroacetic acid, since its K_a is over 10^4 times as large as that of acetic acid. Now assign volumes at the end point to each acid.

$$ml_{1st} \text{ for neut. of } Cl_3CCO_2H = 30.00 \, ml$$

$$(ml_{2nd} - ml_{1st}) \text{ for neut. of } CH_3CO_2H = 55.00 - 30.00 = 25.00 \, ml$$

Finally, calculate the percentage of each acid.

$$\% \, Cl_3CCO_2H = \frac{(30.00 \, ml)(0.1000M \, NaOH)(163.4 \, mg/mmole)(100)}{1000.0 \, mg} = 49.02\%$$

$$\% \, CH_3CO_2H = \frac{(25.00 \, ml)(0.1000M \, NaOH)(60.0 \, mg/mmole)(100)}{1000.0 \, mg} = 15.0_0\%$$

The calculations for mixtures containing a polyprotic acid or base and a monoprotic acid or base are more complicated. After deciding whether the monoprotic acid is titrated at the first or the second end point, find the volume of titrant it consumes by subtracting the volume consumed by the polyprotic acid or base at the other end point. See the example below.

EXAMPLE: A 1.0000-g sample of sodium carbonate and sodium hydroxide is titrated with 0.1000M hydrochloric acid. The buret reading at the first end point is 25.00 ml; the reading at the second end point is 40.00 ml. Calculate the percentages of sodium carbonate and sodium hydroxide (form wt = 40.00) present.

Solution: First decide whether the sodium hydroxide is titrated at the first or second end point. Because it is a strong base, it will be titrated at the first end point as the CO_3^{-2} ion is neutralized to the HCO_3^- ion. Now assign volumes at the end points to each acid.

$$(ml_{2nd} - ml_{1st}) \text{ for neut. of } Na_2CO_3 = 40.00 - 25.00 = 15.00 \text{ ml}$$

$$ml_{1st} - ml \text{ for neut. of } Na_2CO_3 = ml \text{ for neut. of } NaOH = 25.00 - 15.00 = 10.00$$

Finally, calculate the percentage of each. From the example on p. 217,

$$\% \ Na_2CO_3 = 15.90\%$$

and

$$\% \ NaOH = \frac{(10.00 \text{ ml})(0.1000M \text{ HCl})(40.00 \text{ mg/mmole})(100)}{1000.0 \text{ mg}} = 4.00\%$$

Analysis of Complex Mixtures

Where samples are very complex and contain many components, special methods may be needed before a titration can be employed. The Kjeldahl method for organic nitrogen is a good example of such a method.

The Kjeldahl Method for Nitrogen

The Kjeldahl method is used for the determination of the percentage of nitrogen in many foods and fertilizers. Although much of the nitrogen in these substances is present as the basic amino group, $-\overset{|}{\underset{|}{C}}-NH_2$, some of the nitrogen is present in a form that is not basic. Thus, titration with acid will not give an accurate determination of the percentage of nitrogen. The Kjeldahl method consists of the following three essential steps:

1. *Digestion.* The organic nitrogen is heated with concentrated sulfuric acid, which converts the nitrogen to ammonium bisulfate.

$$H-\overset{|}{\underset{|}{C}}-NH_2 + 3H_2SO_4 \longrightarrow CO_2(g) + NH_4HSO_4 + 2H_2SO_3[\rightarrow SO_2(g)]$$

The organic nitrogen is now in a readily titratable form.

2. *Distillation.* The digestion mixture is cooled and neutralized with sodium hydroxide.

$$NH_4HSO_4 + 2NaOH \longrightarrow NH_3 + 2Na^+ + SO_4^{-2} + 2H_2O$$

The resulting ammonia is distilled into an excess of standard hydrochloric acid.

$$NH_3(g) + HCl(excess) \longrightarrow NH_4^+ + Cl^- \qquad \textbf{(10-22)}$$

3. *Titration.* The unreacted hydrochloric acid is back titrated with standard sodium hydroxide. The amount of ammonia is then calculated.

$$\text{mmole } NH_3 = (ml)(M \text{ HCl}) - (ml)(M \text{ NaOH})$$

In a variation of the method, the ammonia is distilled into a standard solution of boric acid, so instead of **10-22** the following reaction occurs:

$$NH_3(g) + H_3BO_3 \; \rightleftharpoons \; NH_4^+ + H_2BO_3^- \qquad (10\text{-}23)$$

The solution is then titrated with a standard solution of hydrochloric acid. Calculation of the equilibrium constant of the above reaction using $(K_1 K_b)/K_w$ gives a value of only 1.1, so some ammonia does not react with the boric acid. It is titrated along with the $H_2BO_3^-$ ion by the acid, however.

Analysis of Other Mixtures   There are many other methods for analysis of complex mixtures that are found in advanced texts [1, 3, 4]. For example, nitrogen compounds containing the nitrate ion may be reduced by Devarda alloy to ammonia and then distilled. Other types of distillations are also possible.

SELF-TEST | **14. Calculation of Acid-Base Titration Results**

Answers:

Directions: Cover the answers in the left column before beginning the test. Compare your answers with those at the left.

A. A 0.4000-g sample of sodium bicarbonate tablets is analyzed by titration with 40.00 ml of 0.1000M hydrochloric acid. Calculate the percentage of $NaHCO_3$, form wt = 84.01.

84.01%

B. A 5.00-ml sample of milk of magnesia is analyzed by adding 50.00 ml of 0.2000M hydrochloric acid to it. The unreacted hydrochloric acid is back titrated with 10.00 ml of 0.10000M sodium hydroxide. Calculate the weight percent of $Mg(OH)_2$, form wt = 58.3.

$5.24_7\%$ $(0.0524_7$ g/ml)

C. A 1.0000-g sample of phosphoric acid mixed with an unknown monoprotic acid is titrated with 0.1000M sodium hydroxide. The buret reading at the first end point is 25.00 ml; the reading at the second end point is 51.12 ml. Decide whether the unknown acid is titrated at the first or second end point and calculate the percentage of H_3PO_4, form wt = 98.00.

Unknown acid titrated at second end point; 24.50%

D. A 1.0000-g sample of sodium carbonate mixed with an unknown monofunctional weak base is titrated with 0.1000M hydrochloric acid. The buret reading at the first end point is 15.00 ml; the reading at the second end point is 31.22 ml. Decide whether the unknown base is titrated at the first or second end point and calculate the percentage of Na_2CO_3, form wt = 106.0.

Unknown base titrated at second end point; 15.90%

E. A 1.0000-g sample of food containing nitrogen is treated by the Kjeldahl method, and the resulting ammonia is distilled into 50.00 ml of $0.1000M$ hydrochloric acid. The unreacted acid is back titrated with 15.00 ml of $0.1000M$ sodium hydroxide. Calculate the percentage of N, form wt = 14.00.

4.900% N

10-5 | NONAQUEOUS TITRATION METHODS: PHARMACEUTICALS

In the previous section the applications of acid-base titrations in aqueous solution were discussed. In this section we shall discuss acid-base titrations in nonaqueous solvents. Such solvents are usually pure organic solvents, such as glacial acetic acid, dimethyl-formamide, acetone, pyridine, and acetonitrile.

One important advantage of nonaqueous titrations is that they are not limited to acids or bases with a K_a or K_b greater than 1×10^{-7} (Sec. 10-1). Usually acids or bases with a K_a or K_b greater than 10^{-11} can be titrated. The reason for this is that the non-aqueous solvent does not react with the anion or cation produced by the neutralization to the same extent that water does. Thus the end point is much sharper; that is, it has a steeper portion in the titration curve.

Another important advantage is that nonaqueous solvents can dissolve many organic compounds that are insoluble in water. Many important drugs or pharmaceuticals can thus be titrated.

Nonaqueous Solvents

Because the solvent is so important, we will discuss the requirements for a nonaqueous solvent and the types of solvents before we discuss the applications.

Requirements

Fritz [5] has listed a number of requirements for nonaqueous solvents, some of which follow.

Good Dissolving Ability. A solvent should dissolve not only the reactants but also the products, if possible. Many salts produced in the titrations are insoluble in many solvents and are a potential source of error through coprecipitation of the substance being titrated.

Reasonably High Dielectric Constant. A reasonably high dielectric constant favors the attainment of stable potentiometer readings and good potentiometric titration curves. When a nonaqueous solvent is used instead of water, the use of a pH meter or potentiometer to locate the end point is often subject to problems, such as unstable readings, not encountered in water. Many of the solvents listed in Table 10-3 have dielectric constants of 12 or higher, so these solvents have the potential to facilitate a potentiometric titration. A high dielectric constant is also another indication of good ability to dissolve the reactants and products of a titration.

Proper Acidity or Basicity. To facilitate the titration of a weak acid, a solvent should be as weak an acid as possible; that is, it should be basic or *neutral*. (Neutral solvents include those listed for acids and bases in Table 10-3.) For the accurate titration of a

TABLE 10-3. Some Nonaqueous Solvents [5]

Solvent and Formula	Dielectric Constant, 25°C
Used for acids and bases	
Acetone, CH_3COCH_3	20.7
Acetonitrile, CH_3CN	36
Methanol, CH_3OH	33
Methyl isobutyl ketone, $CH_3COC_4H_9$	13
2-Propanol, $CH_3CHOHCH_3$	18.3
Sulfolane, $C_2H_4SOC_2H_4$	44
Used for acids	
tert-Butyl alcohol, C_4H_9OH	10.9
Dimethylformamide, $HCON(CH_3)_2$	27
Dimethylsulfoxide, CH_3SOCH_3	46.7
Ethylenediamine, $H_2NC_2H_4NH_2$	12.5
Pyridine, C_5H_5N	12.3
Used for bases	
Acetic acid, CH_3CO_2H	6.1
Acetic anhydride, $(CH_3CO)_2O$	20.7
1,4-Dioxane, $C_2H_4O_2$	2.2
Nitromethane, CH_3NO_2	36

weak base, a solvent should be as weak a base as possible; that is, it should be acidic or *neutral.*

High Purity. For the titration of bases, a solvent should be as free as possible of basic impurities; for the titration of acids, it should be as free as possible of acidic impurities. A solvent blank should be run by titrating the same amount of solvent used for a titration without any sample added. If necessary, this blank can then be subtracted from the volume of titrant used for the sample.

Stability. The solvent should not decompose during the titration or on standing in the pure state. Obviously, it should not react with the titrant or with any of the various samples to be titrated.

Hydrogen-Bonding Ability. In certain potentiometric titrations, the neutralized product hydrogen bonds to the reactant, causing the shape of the potentiometric titration curve to be abnormal. This makes it difficult to locate the end point, particularly if two end points are to be located for a mixture of two or more acids or bases. Using a solvent having an available hydrogen (such as methanol) or an oxygen to hydrogen bond to a hydrogen of the reactant (such as acetone) prevents the hydrogen bonding of the neutralized reactant to the product.

Reasonable Volatility and Viscosity. If a solvent is too volatile, its odor may be offensive or its fumes may be dangerous. Pyridine is a good example. It has a boiling point over

115°C, yet its odor is offensive and its fumes are not safe. It is best to use this solvent in a well-ventilated room or in a hood. If a solvent is too viscous, it will of course be difficult to stir efficiently during the titration.

Three Different Solvent Types

The three different types of solvents for nonaqueous titrations are the solvents for acids, the solvents for bases, and the so-called neutral solvents which can be used for the titration of acids or bases (Table 10-3).

The neutral solvents are neutral only in the sense that both acids and bases can be titrated in them with good results. These solvents include mainly oxygenated solvents, such as acetone and 2-propanol, as well as acetonitrile. Weak organic acids, such as phenols, and weak organic bases, such as amines, have been titrated in each solvent [5]. For the titration of mixtures by potentiometric detection of the end point, Fritz [5] recommends sulfolane, which has the largest potential range reported so far. This range permits differentiation of a large number of acids or bases. Methyl isobutyl ketone also has a large range.

Neutral solvents are generally referred to as *differentiating* solvents, because they permit mixtures of acids or bases to be differentiated according to their acid or base strength. Potentiometric titration curves of acids or bases may exhibit two or more end points, if there is a factor of about 10^3 difference in their ionization constants.

The solvents for acids (Table 10-3) are generally basic to some degree, with the exception of *tert*-butyl alcohol. The advantage of using a solvent with some basicity is that the basic titrant will react to a greater extent with the acid sample than with the solvent. Thus there is less competition between the solvent and the sample for the basic titrant.

With the exception of *tert*-butyl alcohol, the solvents for acids are *leveling* solvents. They react with acid samples and "level" them to nearly the same acidity. Titration curves of mixtures of such acids exhibit only one end point, not two or more. The advantage of using a leveling solvent is that it enhances the acidity of very weak acidic compounds more than neutral, or differentiating, solvents do, so that these compounds are more readily titrated.

The solvents for bases (Table 10-3) are all acidic to some degree. The advantage of using a solvent having some acidity is that the acid titrant will react to a greater extent with the base sample than with the solvent. With the possible exception of dioxane, the solvents for bases are leveling solvents with the same advantages given for the leveling solvents for acids.

Titrants and End Point Detection

We will first discuss the conditions for titrations of acids and then the conditions for bases.

Titration of Acids

There are two important types of titrants for weak acids—methoxides of sodium, potassium, or lithium, and the tetraalkylammonium hydroxides. The methoxides are not readily available, but sodium methoxide, for example, is made by carefully mixing a weighed amount of sodium metal with methyl alcohol.

$$2\,Na(s) + 2\,CH_3OH \longrightarrow 2\,NaOCH_3 + H_2(g)$$

Such a solution is not a standard solution, since the sodium is not pure; the sodium methoxide is standardized against primary standard benzoic acid. The preparation of tetraalkylammonium hydroxides, such as tetra-*n*-butylammonium hydroxide, is more complicated, but these compounds are available [5]. This type of titrant must also be standardized against primary standard benzoic acid.

End point detection is not much different for nonaqueous solvents than for water. Many of the same types of indicators are used, except that more weakly acidic indicators must be used for the titration of the weaker acids. We can summarize the criteria for the choice of three indicators as follows:

$$\text{Thymol blue: Acids whose } K_a \geqslant 10^{-9}$$
$$\text{Azo violet: Acids whose } K_a \geqslant 10^{-11}$$
$$\text{2-Nitroaniline: Acids whose } K_a < 10^{-11}$$

These are general suggestions; differences may be noted in specific cases. Thymol blue (Table 10-1) is used for the standardization of sodium methoxide against benzoic acid and for the titration of most acids with a K_a larger than 10^{-9} in dimethylformamide solvent. The 2-nitroaniline indicator is used for very weak acids which must be titrated in ethylenediamine solvent.

End points may also be located potentiometrically, using a pH meter with a potential readout. Special types of reference electrodes are used to overcome instability problems associated with the saturated aqueous calomel electrode. The best approach *for acids* is to use a salt bridge containing the nonaqueous solvent (*tert*-butyl alcohol, etc.) containing a soluble organic salt, such as tetrabutylammonium bromide, and to achieve electrical contact between the solution and the saturated calomel electrode with the salt bridge [6]. The potentiometric titration of bases can be performed using a standard saturated calomel reference electrode without a salt bridge [5]. The glass electrode is usually used as the indicating electrode for titration of acids and bases. Fritz [5] recommends storing the glass electrode in water between periods of use to avoid possible dehydration of the outer layer.

Titration of Bases

The most important titrant here is perchloric acid dissolved in glacial acetic acid. Since concentrated perchloric acid contains 28 % water, acetic anhydride is added to the solution to react with the water and produce more acetic acid. The perchloric acid is standardized against primary standard potassium acid phthalate (KHP), which acts as a base in this case.

$$\text{HClO}_4 + \text{KHC}_8\text{H}_4\text{O}_4 \longrightarrow \text{H}_2\text{C}_8\text{H}_4\text{O}_4 + \text{KClO}_4$$
$$\text{(KHP)} \qquad \text{(phthalic acid)} \quad \text{(potassium perchlorate)}$$

The most common indicators used are Methyl violet and crystal violet. Both change colors through several shades—from violet to blue to green to yellow, so the correct color change has to be established for each base. Again a pH meter with special electrodes can be used to locate the end point.

Application to Acids and Pharmaceuticals

Many types of organic compounds may be titrated as acids in nonaqueous solvents. These include carboxylic acids, phenols, imides (—CO—NH—CO—), aliphatic nitro compounds, and sulfa drugs [5]. The reaction of imides is as follows:

$$\text{—CO—NH—CO— + NaOCH}_3$$
$$\longrightarrow (\text{—CO—N—CO—})^{-1} + \text{CH}_3\text{OH} + \text{Na}^+$$

This type of reaction also occurs with many sulfa drugs, such as sulfanilamide and sulfathiazole, whose structures are as follows:

Sulfanilamide   Sulfathiazole

Most sulfa drugs, such as sulfathiazole, are fairly acidic and can be titrated in dimethyl-formamide solvent using thymol blue indicator. Sulfanilamide, however, is very weakly acidic and does not react under these conditions. It can be titrated quantitatively using azo violet indicator and butylamine solvent. Mixtures of acidic sulfa drugs, such as sulfathiazole with sulfanilamide, can be analyzed by two titrations.

1. $NaOCH_3$ + azo violet: Total of sulfanilamide + acidic sulfa drug (sulfathiazole, etc.)
2. $NaOCH_3$ + thymol blue: Acidic sulfa drug only

The amount of sulfanilamide present is found by subtracting the mmoles of acidic sulfa drug from the second titration from the mmoles of the total found in the first titration.

Application to Bases   In general, there are two types of weak organic bases that can be determined—organic amines (aliphatic and many aromatic) and salts of weak acids. The reactions are as follows:

$$RNH_2 + HClO_4 \longrightarrow RNH_3{}^+ClO_4{}^-$$

$$CH_3CO_2{}^-Na^+ + HClO_4 \longrightarrow CH_3CO_2H + NaClO_4$$
(sodium acetate)

Mixtures of bases whose K_b values differ by a factor of 10^3 can also be titrated to obtain the amount of each [5].

| QUESTIONS AND PROBLEMS

(Answers to most even-numbered problems are in Appendix 5.)

Concepts and Definitions

1. List at least two advantages of a direct titration over a back titration, assuming that the end point can be clearly seen by each method.

2. What is the advantage of a back titration over a direct titration for a suspension such as milk of magnesia?

3. Define a primary standard for acid-base titrations. What are some requirements for an acid-base primary standard?

4. Explain why sodium hydroxide and hydrochloric acid cannot be considered primary standard reagents under normal conditions. Is hydrochloric acid ever used as a primary standard?

5. Define what is meant by the pH transition range of an acid-base indicator.

6. Of what use is the middle half of the pH transition range of an acid-base indicator?

7. Explain the three main steps of the Kjeldahl method for determining nitrogen.

8. When boric acid is substituted for hydrochloric acid in the Kjeldahl method, what are the two reactions that are used for the analysis?

9. What are the advantages of a nonaqueous titration over an aqueous titration?

Feasibility of Aqueous Titrations

10. Calculate the actual value of K_b at the end point for the anions of each of the following acids. State whether sodium hydroxide titration of each will be quantitative at the $0.1M$ level.
    a. HF
    b. Boric acid, H_3BO_3 (first H^+ only)
    c. p-nitrophenol
    d. Potassium acid phthalate (KHP)

11. Calculate the actual value of K_b at the end point for the anions of each of the following acids, as well as the maximum theoretical value of K_b at the concentration listed. State whether the sodium hydroxide titration of each will be quantitative.
    a. $0.001M$ HF
    b. $10M$ H_3BO_3 (first H^+ only)
    c. $1.0M$ p-nitrophenol
    d. $0.01M$ Potassium acid phthalate (KHP)

12. Calculate the actual value of K_a at the end point for the cations of each of the following bases, as well as the maximum theoretical value of K_a at the concentration listed. State whether the hydrochloric acid titration of each will be quantitative.
    a. $0.1M$ ammonia (NH_4OH or NH_3)
    b. $0.0001M$ ammonia (NH_4OH or NH_3)
    c. $1M$ hydroxylamine (NH_2OH)

13. It can be shown that the equilibrium constant for the reaction of any weak acid with any weak base is equal to $K_a K_b / K_w$. Calculate this value for the reaction of the first hydrogen of boric acid with ammonia and then interpret its relevance to the Kjeldahl method.

Calculation of Titration Curves

14. Sketch the titration curve for the titration of 100 ml of $0.040M$ benzoic acid and calculate the following points for $0.050M$ sodium hydroxide titrant:
    a. 0 ml
    b. 20 ml
    c. 40 ml
    d. 80 ml
    e. 100 ml

15. Sketch the titration curve for the titration of 50 ml of $0.10M$ potassium acid phthalate and calculate the following points for $0.10M$ sodium hydroxide:
    a. 0 ml
    b. 10 ml
    c. 25 ml
    d. 50 ml
    e. 60 ml

16. Sketch the titration curve for the titration of 100 ml of a solution containing 616.8 mg of potassium acid maleate (form wt = 154.2) and calculate the following points for $0.050M$ sodium hydroxide titrant:
    a. 0 ml
    b. 20 ml
    c. 40 ml
    d. 80 ml
    e. 100 ml

Calculation of Titration Results

17. A 0.5000-g sample of NaH_2PO_4 is neutralized completely to Na_2HPO_4 with 16.10 ml of $0.0984M$ sodium hydroxide. Calculate the percentage of NaH_2PO_4 (form wt = 119.98) in the sample.

18. A 0.1000-g sample of impure sodium carbonate (form wt = 106.0) is neutralized to carbonic acid with 21.00 ml of $0.0450M$

hydrochloric acid. Calculate the percent purity of the sodium carbonate.

19. A 1.0000-g impure sample containing sodium carbonate and sodium bicarbonate is dissolved in water and titrated with 0.1000M hydrochloric acid. The buret reading at the phenolphthalein end point is 15.5 ml and at the methyl red end point, 40.1 ml. Calculate the percentages of sodium carbonate and sodium bicarbonate in the sample.

20. A mixture of bases is titrated with hydrochloric acid first to a phenolphthalein end point, then the titration is continued to a methyl red end point. The buret reading at the first end point is 28.0, and the buret reading at the second end point is 40.0 ml. From this information:
    a. Decide whether the mixture is sodium hydroxide–sodium carbonate or sodium carbonate–sodium bicarbonate.
    b. Compute the number of milliliters of hydrochloric acid needed to titrate the sodium carbonate present.

21. A 1.000-g sample of food is analyzed for nitrogen by the Kjeldahl method. After digestion of the sample, the ammonia is distilled and collected in a receiver containing excatly 50.00 ml of 0.1000M hydrochloric acid. The *unreacted* hydrochloric acid requires 24.60 ml of 0.1200M sodium hydroxide for back titration. Calculate the percentage of nitrogen (N) in the sample.

22. Calculate the pH at the end point in the titration of 500 mg of potassium acid phthalate (form wt $= 204.2$) dissolved in 50 ml of water and titrated with 0.1000M sodium hydroxide. Choose an indicator suitable for detecting the end point. The ionization constants for phthalic acid are $K_1 = 1.2 \times 10^{-3}$ and $K_2 = 3.9 \times 10^{-6}$ ($\mu = 0$).

23. Calculate the pH at the end point in the titration of 0.2M potassium acid maleate (KHM) with 0.2M sodium hydroxide. Choose a suitable indicator for detecting the end point. The ionization constants for maleic acid (H$_2$M) are $K_1 = 1.2 \times 10^{-2}$ and $K_2 = 9 \times 10^{-7}$.

Challenging Problems

24. Titration of 664 mg of a pure unknown organic carboxylic acid gives potentiometric breaks at 40.00 ml and at 80.00 ml of 0.1000M sodium hydroxide titrant. The pH when 20.00 ml of titrant has been added is 2.78; the pH when 60.00 ml of titrant has been added is 5.10. The ionic strength is essentially constant at 0.1 throughout.
    a. Is the acid monoprotic, diprotic, triprotic, or what?
    b. Calculate the value(s) of the ionization constant(s) of the acid.
    c. Calculate the formula weight of this acid.
    d. If the acid is C$_x$H$_y$(CO$_2$H)$_n$, write a possible molecular formula. (*Hint:* assume in turn that it is *saturated*, monounsaturated, and so on, using general formulas such as C$_x$H$_{2x+1}$ for a *saturated* carbon-hydrogen grouping.)
    e. Write a logical structure for the acid that is consistent with all facts.

25. Titration of 172.0 mg of a pure unknown organic acid, R—(CO$_2$H)$_n$, requires 20.00 ml of 0.1000M sodium hydroxide titrant. If the molecular weight of the acid is 172 ± 0.1:
    a. Calculate the value of n. (Is the acid monoprotic or diprotic?)
    b. Calculate a numerical value for R.
    c. Give at least two reasonable but different structures for R.
    d. Given that the acid does not react by addition with bromine or with ozone, suggest a reasonable structure for R.
    e. Given that the titration curve shows only one break and that the pH at 5.00 ml of titrant is 5.60, suggest a reasonable structure for the acid.

26. After consulting appropriate references, state whether each compound in each mixture below can be determined by a nonaqueous differentiating titration, using either a potentiometric or a visual indicator end point. Also state which compound in each mixture will be titrated first and give numerical reasons.
    a. Sulfathiazole and sulfanilamide
    b. Sulfathiazole and sulfapyridine
    c. Phenol and acetic acid

d. Pyridine and *n*-butylamine

e. Phenol and 2,4-dinitrophenol

f. *m*-nitrophenol, 2,4-dinitrophenol, and trinitrophenol (picric acid)

g. 2,4,6-trinitroaniline and 2,4-dinitroaniline

27. A pure unknown organic nitrogen base is titrated with standard perchloric acid in glacial acetic acid. A 372-mg sample is titrated with 32.00 ml of 0.1250M perchloric acid. Assuming that the nitrogen base has just one basic group (one —NH_2 group):

a. Calculate the formula weight of the nitrogen base.

b. Calculate a reasonable formula for the nitrogen base assuming it contains one —NH_2 group and carbon and hydrogen are the only other elements present.

c. Suggest a reasonable structure for the nitrogen base.

| NOTES

[1] H. A. Diehl and G. F. Smith, *Quantitative Analysis*, Wiley, New York, 1952.

[2] G. Marinenko and C. E. Champion, *Anal. Chem.* **41**, 1208 (1969).

[3] G. L. Jenkins, A. M. Knevel, and F. E. DiGangi, *Quantitative Pharmaceutical Chemistry*, McGraw-Hill, New York, 1967, pp. 103–4.

[4] J. S. Fritz and G. H. Schenk, *Quantitative Analytical Chemistry*, Allyn & Bacon, Boston, 1974, pp. 178–79.

[5] J. S. Fritz, *Acid-Base Titrations in Nonaqueous Solvents*, Allyn & Bacon, Boston, 1973.

[6] L. W. Marple and J. S. Fritz, *Anal. Chem.* **34**, 796 (1962).

11 | Oxidation-Reduction Methods

"I flatter myself that I can distinguish at a glance the ash [oxides] of any known brand either of cigar or tobacco," said Holmes.

ARTHUR CONAN DOYLE
A Study in Scarlet, Chapter 4

In this chapter we will consider the key concepts in the vast field of oxidation-reduction reactions and equilibria, and will apply these concepts to real analytical problems. When you have completed this chapter, you should have a useful, *though limited*, understanding of the scope of application of oxidation-reduction in analytical chemistry. Be aware that it is often difficult to "distinguish at a glance" every subtle detail of this complex area.

11-1 | BALANCING OXIDATION-REDUCTION REACTIONS

Reduction involves the gain of electrons by a chemical species, such as the Cu^{+2} ion in the equation below:

$$2Cu^{+2} + 2OH^- + 2e^- \longrightarrow Cu_2O(s) + H_2O \qquad (11\text{-}1)$$

In this reaction, the Cu^{+2} ion is also called an oxidizing agent (oxidant) because it gains electrons from another species being oxidized. As we have just inferred, oxidation involves the loss of electrons from a chemical species, such as the organic molecule (acetaldehyde) in the equation below:

$$CH_3CHO + 2OH^- \longrightarrow CH_3COOH + H_2O + 2e^- \qquad (11\text{-}2)$$

To obtain a complete oxidation-reduction reaction, we add together two *half reactions* that contain, or are made to contain by multiplication, the same number of electrons, as shown below.

$$2Cu^{+2} + CH_3CHO + 4OH^- \longrightarrow Cu_2O(s) + CH_3COOH + 2H_2O \quad \textbf{(11-3)}$$

Note that in this reaction the two electrons gained by the Cu^{+2} ions are given up by the CH_3CHO, which we call a reducing agent (reductant) because it helps to reduce the Cu^{+2} ions to Cu_2O. Reaction **11-3** is the classic Fehling's or Benedict's test for organic aldehydes wherein red, insoluble copper(I) oxide is formed.

It is fundamental to an understanding of oxidation-reduction to be able to obtain the electron change in each of the half reactions **11-1** and **11-2**. In this section we will first briefly review the oxidation number method, and then devote the rest of the discussion to the more useful ion-electron ("half equation") method.

The Oxidation Number Method: A Brief Review

Oxidation Number

Recall that the oxidation number is the charge that an atom in a molecule would have if the bonding electrons were assigned to the more electronegative atom. Often this charge is simply the charge of the ion itself, as for monatomic ions such as Cu^{+2} and Fe^{+2}, which have an oxidation number of $+2$ and are referred to as copper(II) ion and iron(II) ion. For polyatomic ions, we calculate the oxidation number from the overall charge on the ion and from the constant oxidation numbers assigned to oxygen (-2) and to hydrogen ($+1$). Thus for the $Cr_2O_7^{-2}$ ion, we find that *each* chromium has an oxidation number of $+6$. We can go through this procedure for molecules for which the sum of all oxidation numbers of the atoms in a molecule must be zero. This method is a bit more difficult to apply to a molecule such as CH_3CHO, however, so we will defer explanation of how to balance half reaction **11-2** until the discussion of the ion-electron ("half equation") method.

Balancing Half Reactions

The oxidation number method is often used directly to balance a complete oxidation-reduction reaction, but it may also be used to balance a half reaction. Consider half reaction **11-1**. We first calculate the oxidation number of each of the copper species: on the left we have copper(II) ions and on the right copper(I) ions in Cu_2O. We then have to adjust the coefficient of Cu^{+2} to achieve an atom balance.

$$2Cu^{+2} \longrightarrow Cu_2O(s)$$

We then calculate the electron change from the total decrease in oxidation number going from $2Cu^{+2}$ to $Cu_2O(s)$, which gives a two-electron change.

$$2Cu^{+2} + 2e^- \longrightarrow Cu_2O(s)$$

The remaining species in **11-1** are obtained by the usual method.

Balancing Complete Reactions

To apply the oxidation number method to a complete reaction, consider the following oxidation of Fe^{+2} by the dichromate ion in acid solution:

$$Fe^{+2} + Cr_2O_7^{-2} \xrightarrow{\ H^+\ } Fe^{+3} + 2Cr^{+3}$$

Again we begin with the calculation of the oxidation number of each atom—for iron it is $+2$ on the left and $+3$ on the right, and for chromium it is $+6$ on the left and $+3$ on the right. We then calculate the electron change from the total decrease in oxidation number from $Cr_2O_7^{-2}$ to $2Cr^{+3}$, a decrease of 6. For the iron, the total increase in oxidation number is obviously one. We then write down a partial reaction using a coefficient of 6 in front of the two iron ions to balance the electron change of 6 for the chromiums in $Cr_2O_7^{-2}$.

$$6Fe^{+2} + Cr_2O_7^{-2} \xrightarrow{\ H^+\ } 6Fe^{+3} + 2Cr^{+3}$$

Finally, we balance the number of oxygen atoms by adding water molecules to the right side and hydrogen ions to the left.

$$6Fe^{+2} + Cr_2O_7^{-2} + 14H^+ \xrightarrow{\ H^+\ } 6Fe^{+3} + 2Cr^{+3} + 7H_2O$$

This method works very well for many inorganic oxidation-reduction reactions, but it is not the most general method for balancing oxidation-reduction reactions. For example, to find the electron change for *organic* half reactions, such as the two-electron change in **11-2**, it is better to use the so-called ion-electron method. This method is usually applied only to inorganic half reactions in general chemistry texts, so a thorough discussion of its application to organic and biochemical half reactions follows.

The Ion-Electron Method Simply stated, the ion-electron method of balancing half reactions consists of counting the charges on both sides of a *balanced* half reaction and then adding the correct number of electrons needed to balance the charges. This method is a completely general method of finding the electron change in any inorganic, organic, or biochemical half reaction. Although its name might seem to imply that it can be used only for ionic reactions, this is not true. It is equally useful for molecular reactions.

To apply the ion-electron method to a half reaction requires that the half reaction first be balanced with respect to all atoms. Then the charges can be counted and electrons added. The entire method involves all or most of the following steps:

1. Start by writing the half reaction with only the ions and/or molecules actually gaining or losing electrons.
2. Balance all the atoms, including O and H. Underlining identical large groups helps.
   a. If the solution is acid or neutral, on the side deficient in O atoms write H_2O with the proper coefficient needed to balance the O atoms. Only *after* you have balanced the O atoms should you try to balance the H atoms. On the side that is now deficient in H atoms, write H^+ with the proper coefficient needed to balance the H atoms.
   b. If the solution is basic, strictly speaking you should write OH^- on the side deficient in O atoms and H_2O on the side deficient in H atoms. However, this is somewhat more complicated than the method used in acid solution, so some instructors may prefer to have you use step 2a for all situations. (Consult your instructor on this point.)
3. Sum the charges algebraically on each side of the half reaction.
4. Write a *charge-balance equation* to find the electron change. Write down the sum of the charges on each side, and add the proper number of electrons to the more positive side so that the charges balance. For example, if the sum of the charges on the left is $+4$ and that on the right is $+2$, the charge-balance equation is $+4 + 2e^- = +2$.

To see the use of the ion-electron method, consider the oxidation of an aldehyde to an acid **(11-2)**. We will show that in general this is a two-electron change, no matter what aldehyde is oxidized. To show the power of the ion-electron method, we will use the oxidation of vitamin A aldehyde, or retinal, to vitamin A acid as an example.

$$C_{16}H_{23}-\overset{\overset{\displaystyle O}{\|}}{\underset{\underset{\displaystyle CH_3 \; H}{|\;\;\;|}}{C=C}}-CH \longrightarrow C_{16}H_{23}-\overset{\overset{\displaystyle O}{\|}}{\underset{\underset{\displaystyle CH_3 \; H}{|\;\;\;|}}{C=C}}-COH \qquad \textbf{(11-4)}$$

The molecules in **11-4** can be written more simply because they are virtually identical.

$$\underline{C_{19}H_{27}}-CHO \longrightarrow \underline{C_{19}H_{27}}-CO_2H \qquad \textbf{(11-5)}$$

Note that the $C_{19}H_{27}$ group in **11-5** is underlined on both sides, indicating that these groups are identical and need not be counted to balance the atoms on both sides.

To balance this half reaction, steps 1–4 above are followed. Step 1 has already been done in **11-5**. For step 2, we assume the solution is acid or neutral and add water to the left side to balance the O atoms.

$$\underline{C_{19}H_{27}}-CHO + H_2O \longrightarrow \underline{C_{19}H_{27}}-CO_2H$$

Then two H^+ are added to the right side to balance the H atoms.

$$\underline{C_{19}H_{27}}-CHO + H_2O \longrightarrow \underline{C_{19}H_{27}}-CO_2H + 2H^+ \qquad \textbf{(11-6)}$$

All atoms in **11-6** are now balanced, the underlined groups being identical. Since there are no charges on the left side and a total of $+2$ on the right (step 3), the *charge-balance equation* (step 4) is

$$0 = +2 + 2e^-$$

The complete half reaction is

$$\underline{C_{19}H_{27}}-CHO + H_2O \longrightarrow \underline{C_{19}H_{27}}-CO_2H + 2H^+ + 2e^- \qquad \textbf{(11-7)}$$

It is shown by **11-7** that no matter what the underlined group is, any aldehyde with just one —CHO group will lose two electrons in being oxidized to an acid with just one —CO$_2$H group. This includes acetaldehyde **(11-2)**. The reverse process where an acid, such as vitamin A acid, is reduced to an aldehyde, such as retinal or vitamin A aldehyde, is also a two-electron process.

$$C_{19}H_{27}-CO_2H + 2H^+ + 2e^- \longrightarrow C_{19}H_{27}-CHO + H_2O \qquad \textbf{(11-8)}$$

You may be interested to know that the body cannot reduce vitamin A acid to retinal **(11-8)**. Experiments conducted with rats deficient in retinal needed for vision [1] proved this. The rats were fed only vitamin A acid and no other source of retinal. Not only did the rats remain deficient in retinal, but they eventually became night-blind.

To become proficient with the ion-electron method to balance organic and biochemical half reactions, you should work the following self-test. If you have not used the ion-electron method previously, you should balance other equations, such as those given in the problems at the end of the chapter, after you finish with the self-test.

<table>
<tr><td>

SELF-TEST

Answers:

</td><td>

15. The Ion-Electron Method

Directions: Cover the answers in the left column before beginning the test. Balance one half reaction at a time, giving the *charge-balance equation* for each in addition to following the special directions. Check each answer at the left.

A. Balance the oxidation of ethyl alcohol, CH_3CH_2OH, to acetic acid, CH_3CO_2H, using the following steps:

a. Write the molecules actually involved in the oxidation (identical groups underlined) as follows:

</td></tr>
</table>

Correct as written

$$\underline{CH_3}CH_2OH \longrightarrow \underline{CH_3}CO_2H$$

b. Balance all atoms, assuming an acid solution. Start with O atoms, not H atoms.

$CH_3CH_2OH + H_2O$
$\longrightarrow CH_3CO_2H + 4H^+$

$$\underline{CH_3}CH_2OH + \underline{\hspace{1.5cm}}$$
$$\longrightarrow \underline{CH_3}CO_2H + \underline{\hspace{1.5cm}}$$

c. Sum charges on each side and write the charge-balance equation.

$0 = +4 + 4e^-$ (4e^- change)

$$\underline{\hspace{1.5cm}} = \underline{\hspace{1.5cm}} + \underline{\hspace{1cm}}e^-$$

B. Balance the oxidation of $CH_3CH_2CHOHCH_2OH$ to $CH_3CH_2CO_2H + HCO_2H$ in acid.

$CH_3CH_2CHOHCH_2OH$
$\longrightarrow CH_3CH_2CO_2H$
$CH_3CH_2CHOHCH_2OH + 2H_2O$
$\longrightarrow CH_3CH_2CO_2H$
$+ HCO_2H + 6H^+ + 6e^-$

a. Underline the identical groups in the first and second molecules (omit third).

b. Balance all atoms, sum charges, and write charge-balance equation.

C. Balance the reduction of gluconic acid, $CH_2OH(CHOH)_4CO_2H$, to glucose (blood sugar), $CH_2OH(CHOH)_4CHO$, after underlining identical groups.

Charge-balance equation:
$+2 + 2e^- = 0$

$$CH_2OH(CHOH)_4CO_2H + \underline{\hspace{1.2cm}} + \underline{\hspace{1.2cm}}e^-$$
$$\longrightarrow CH_2OH(CHOH)_4CHO + \underline{\hspace{1.2cm}}$$

D. Balance the oxidation in acid of glucose to gluco-saccharic acid, $HO_2C(CHOH)_4CO_2H$.

Charge-balance equation:
$0 = +6 + 6e^-$

$$CH_2OH(CHOH)_4CHO + \underline{\hspace{1.2cm}}$$
$$\longrightarrow HO_2C(CHOH)_4CO_2H + \underline{\hspace{1.2cm}} + \underline{\hspace{1cm}}e^-$$

E. Balance the reduction in acid solution of oxygen, O_2, in the body to water.

$O_2 + 4H^+ + 4e^- \longrightarrow 2H_2O$

$$O_2 + \underline{\hspace{1.2cm}} + \underline{\hspace{1cm}}e^- \longrightarrow \underline{\hspace{0.8cm}}H_2O$$

11-2 | STOICHIOMETRIC REDOX CALCULATIONS

In Section 6-2 we discussed the calculations of volumetric analysis, including redox calculations. Both the mole-molarity system and the equivalent-normality system were discussed there. In this section we will review the equivalent-normality system only. We feel that the equivalent-normality system is easier to use, once understood, than the other system. If you intend to use the mole-molarity system, you should review Section 6-2, particularly the examples dealing with molarity.

The reason for presenting the ion-electron method in the previous section is that now we will apply it to calculate the equivalent weight and the normality N. A change in either will change the number of equivalents (eq) or milliequivalents (meq) of any oxidizing or reducing agent—thus the correct electron change is important.

The Equivalent Weight

The equivalent weight is expressed in units of g/eq or mg/meq and is defined for calculation purposes as follows:

$$\text{Equivalent weight} = \frac{\text{Formula weight}}{e^- \text{ change per formula weight}} \qquad \textbf{(11-9)}$$

Note that the electron change must be the *total change* per formula weight of the substance weighed out or to be calculated, not just the electron change in the simplest half reaction. The equivalent weight depends on the reaction involved. For example, ethyl alcohol can have at least two different equivalent weights—$46.07/4e^-$ for oxidation to acetic acid, and $46.07/2e^-$ as in the next example.

EXAMPLE: A simple half reaction is the oxidation of ethyl alcohol to acetaldehyde.

$$\text{CH}_3\text{CH}_2\text{OH} \longrightarrow \text{CH}_3\text{CHO}$$

Using the ion-electron method, the electron change is

$$\text{CH}_3\text{CH}_2\text{OH} \longrightarrow \text{CH}_3\text{CHO} + 2\text{H}^+ + 2e^- \qquad \textbf{(11-10)}$$

Given that the formula weight of ethyl alcohol is 46.07, the equivalent weight (eq wt) is

$$\text{eq wt} = \frac{46.07}{2e^- \text{ change}} = 23.04 \text{ mg/meq}$$

This oxidation **(11-10)** is the first step in the metabolism of alcohol in the body and is catalyzed by a zinc-containing enzyme. Failure of the body to oxidize acetaldehyde rapidly to carbon dioxide results in a state called acetaldehyde syndrome. Symptoms of this state include nausea, vomiting, sweating, and lowering of blood pressure [2].

A more complicated example is analysis of the purity of Fe_2O_3, iron(III) oxide. In this analysis, the oxide is dissolved and reduced to iron(II) before being titrated with an oxidizing agent to iron(III). The balanced titration half reaction is

$$\text{Fe}^{+2} \longrightarrow \text{Fe}^{+3} + e^-$$

However, the electron change per Fe_2O_3 formula weight is two, because each of the iron(II) ions produced from Fe_2O_3 loses an electron, for a total of two.

$$\text{Fe}_2\text{O}_3 + 6\text{H}^+ + 2e^- = 3\text{H}_2\text{O} + 2\text{Fe}^{+2} \xrightarrow{\text{(oxidation)}} 2\text{Fe}^{+3} + 2e^-$$

The equivalent weight is calculated.

$$\text{eq wt} = \frac{159.7}{2e^- \text{ total change}} = 79.85 \text{ mg/meq}$$

Milli-equivalents and Normality

The number of milliequivalents (meq) or equivalents (eq) is found by dividing the appropriate weight by the equivalent weight expressed as mg/meq or g/eq.

$$\text{meq} = \frac{\text{mg}}{\text{eq wt in mg/meq}} \qquad \text{eq} = \frac{\text{g}}{\text{eq wt in g/eq}} \qquad \textbf{(11-11)}$$

The normality of a titrant in any titration method is always

$$N = \frac{\text{meq}}{\text{ml}} = \frac{\text{eq}}{\text{liter}} \qquad \textbf{(11-12)}$$

Iodine is a common titrant in both inorganic and organic analyses. As an oxidizing agent, it has a half reaction

$$I_2 \longrightarrow 2I^- + 2e^-$$

Iodine may be weighed out and dissolved to make a titrant, as in the following example.

EXAMPLE: The formula weight of iodine, I_2, is 253.8. Calculate its equivalent weight, using the fact that it undergoes a total electron change of two.

$$\text{eq wt} = \frac{253.8}{2e^- \text{ total change}} = 126.9 \text{ mg/meq}$$

Now calculate the normality of 100 ml of solution to which is added exactly 1269 mg of pure iodine, I_2. The number of meq in 1269 mg is found using **11-11**.

$$\text{meq} = \frac{1269 \text{ mg}}{126.9 \text{ mg/meq}} = 10.00 \text{ meq}$$

The normality is then found by using **11-12**.

$$N = \frac{10.00 \text{ meq}}{100 \text{ ml}} = 0.100N$$

Percent Purity and Equivalent Weight Calculations

In general, the percent purity of an impure sample S is found by titrating it with a standard solution of a titrant T. The weight, mg_S, of pure S is found from the ml of the titrant used to reach the titration end point; the normality, N_T, of the titrant; and the equivalent weight, eq wt_S, of the sample as follows:

$$\text{mg}_S = (\text{ml}_T)(N_T)(\text{eq wt}_S) \qquad \textbf{(11-13)}$$

The percent purity, %S, of the impure sample is then calculated by dividing the weight of pure S by the weight, $\text{mg}_{\text{impure}}$, of the impure sample.

$$\%S = \frac{\text{mg}_S}{\text{mg}_{\text{impure}}} (100) \qquad \textbf{(11-14)}$$

EXAMPLE: Calculate the percent purity of impure ascorbic acid, 1000 mg of which is titrated with 40.00 ml of 0.1000N iodine according to the following reaction:

$$I_2 + C_4H_6O_4(OH)C{=}C(OH) \longrightarrow 2I^- + C_4H_6O_4C({=}O){-}C{=}O + 2H^+$$
$$(11\text{-}15)$$

The equivalent weight is 176.12/2, or 88.06, mg/meq of ascorbic acid. The weight of pure acid is found using **11-13**.

$$mg_{asc.\ a.} = (40.00\ ml\ I_2)(0.1000N\ I_2)(88.06\ mg/meq\ asc.\ a.) = 352.2\ mg$$

The percent purity of ascorbic acid is found using **11-14**.

$$\%\ ascorbic\ acid = \frac{352.2\ mg_{asc.\ a.}}{1000\ mg_{impure\ sample}}\ (100) = 35.22\%\ purity$$

It is interesting that ascorbic acid, or vitamin C, is oxidized to the same product, dehydroascorbic acid, in the body as in **11-15**. This half reaction will be considered in more detail in the self-test below.

The equivalent weight of a pure solid or liquid can also be found from a titration, given the ml of titrant, ml_T, used to reach the titration end point; the normality, N_T, of the titrant; and the weight, mg_S, of the *pure* sample of solid or liquid. (It is very important that the sample be as pure as possible or the equivalent weight will be in error.) Then **11-13** can be rearranged to give the appropriate equation.

$$Eq\ wt = \frac{mg_S}{(ml_T)(N_T)} \qquad (11\text{-}16)$$

Since the units of equivalent weight are mg/meq, you can see that the right-hand side of the preceding equation gives the correct units (recall that $(ml)(N) = meq$).

Since all of the above calculations are unique, you should work the following self-test to be sure you become somewhat familiar with all of them. Additional problems given at the end of the chapter should be solved for areas where you need more practice.

SELF-TEST | **16. Equivalent Weights and Other Calculations**

Answers:

Directions: Cover the answers at the left before beginning the test. In addition to following any special directions, find the electron change for each half reaction by the ion-electron method and then calculate the equivalent weight. Check each answer to the left (only the charge-balance equation, not the entire half reaction, is given in the answer).

A. Ascorbic acid, vitamin C, is oxidized to dehydroascorbic acid by iodine.

    a. Use the ion-electron method to find the electron change for its half reaction.

$$C_4H_6O_4(OH)C{=}COH$$
$$\longrightarrow\ C_4H_6O_4C({=}O){-}C{=}O + \underline{\hspace{1cm}} + \underline{\hspace{1cm}}e^-$$

$0 = +2 + 2e^-$ $(2e^-$ change)

    b. The formula weight of ascorbic acid is 176.12. What is its equivalent weight?

88.06

11-3 STANDARD POTENTIALS AND THEIR USES **237**

B. Glycerine, $CH_2OHCHOHCH_2OH$, has several different equivalent weights, depending on what oxidizing agent reacts with it. Using its formula weight of 92.1, calculate the equivalent weight in each half reaction below.

a. Periodic acid, HIO_4, oxidizes it according to this half reaction (identical groups underlined):

$$CH_2OHCHOHCH_2OH + \underline{\hspace{1cm}}$$
$$\longrightarrow \quad 2H_2\underline{CO} + HCO_2H + \underline{\hspace{1cm}} + \underline{\hspace{1cm}}e^-$$

$0 = +4 + 4e^-$; eq wt = 23.0

b. Cerium(IV) oxidizies it according to this half reaction.

$$CH_2OHCHOHCH_2OH + \underline{\hspace{1cm}}$$
$$\longrightarrow \quad 3HCO_2H + \underline{\hspace{1cm}} + \underline{\hspace{1cm}}e^-$$

$0 = +8 + 8e^-$; eq wt = 11.5

C. Arsenious oxide, As_2O_3, dissolves to form two H_3AsO_3 molecules from each molecule of As_2O_3.

a. Find the electron change per H_3AsO_3 molecule if it is oxidized to H_3AsO_4.

$2e^-$

b. Using a formula weight of 197.8, calculate the equivalent weight of As_2O_3.

49.45

c. Calculate the normality of 100 ml of solution to which is added 395.6 mg of As_2O_3, which dissolves to form H_3AsO_3.

$0.0800N$

D. Calculate the percent purity of 1000 mg of impure Fe_3O_4 (form wt = 231.55) which is dissolved to form three Fe^{+2} ions per Fe_3O_4. The resulting Fe^{+2} ions are all titrated with 20.00 ml of $0.1000N$ $K_2Cr_2O_7$, which oxidizes them to Fe^{+3}.

$\% = (154.36 \text{ mg}/1000 \text{ mg})100 =$
15.44%

E. Calculate the equivalent weight of pure As_2O_x, 229.8 mg of which requires 40.00 ml of $0.1000N$ titrant.

eq wt = 229.8 mg/4.000 meq =
57.45 ($x = 5$)

11-3 | STANDARD POTENTIALS AND THEIR USES

Each of the half reactions we have discussed has a definite tendency or potential to gain or lose electrons. Consider the reduction of iodine to iodide ion.

$$I_2 + 2e^- \rightleftharpoons 2I^- \tag{11-17}$$

As the reaction is written, iodine has a certain tendency to gain electrons to form iodide ions; iodide ions also have a certain tendency to lose electrons to form iodine. At equilibrium these tendencies balance at some point.

If all of the known half reactions are written the same way, then their tendencies to gain or lose electrons can be compared by measuring them in the form of electrical potential or voltage. It has been recommended by the International Union of Pure and Applied Chemistry (IUPAC) that all half reactions such as **11-17** be written as reductions for this purpose. However, to complete an electrical cell to measure a potential, the potential must be measured relative to that of a second half reaction that will take up the electrons released by the half reaction being measured. If the same (standard) half reaction is used for all such measurements under a set of standard conditions, then a set of *standard potentials*, or $E°$ values, can be accumulated.

Measurement and Compilation of $E°$ Values

The standard half reaction chosen for the comparative measurement of all $E°$ values is the reduction of hydrogen ions to hydrogen gas.

$$2H^+ + 2e^- = H_2(g) \qquad E° = 0.0 \text{ (defined)}$$

The $E°$ for this reaction is defined to be 0.0 volts at standard conditions of one atmosphere pressure and unit activity of hydrogen ion. (As usual, we will use molarity to approximate the activity of an ion such as the hydrogen ion.)

A special electrical device called a *potentiometer* (Fig. 11-1) is used for measurement of $E°$ values in relation to $E°$ of the reduction of hydrogen ion. The complete cell consists of two electrodes. One is a standard hydrogen *reference* electrode in equilibrium with a $1M$ solution of hydrogen ion. The other electrode is typically a bright platinum *indicating* electrode in equilibrium with the species involved in the half reaction to be measured. The potentiometer measures the voltage difference between the two half reactions in such a way that no appreciable reaction occurs. Because of this, the voltage

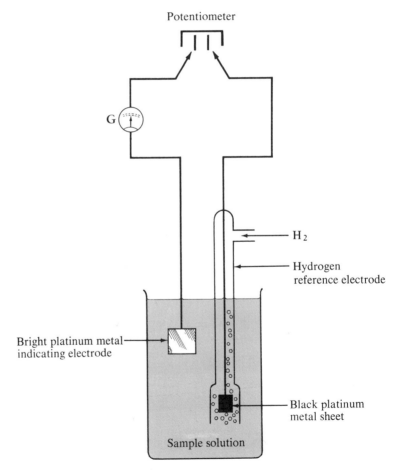

FIGURE 11-1. Potentiometric measurement of $E°$ versus standard hydrogen reference electrode (black platinum sheet covered with hydrogen gas in a $1M$ hydrogen ion solution).

measured is referred to as a potential, or form of potential energy (Sec. 16-1). In general, the potential measured on the potentiometer, E_{pot}, is given by the following equation:

$$E_{pot} = E_{ind \, el} - E_{ref} \tag{11-18}$$

Since E_{ref} is defined to be 0.0 volts for a hydrogen reference electrode, the voltage reading (potential) of the potentiometer is equal to the $E°$ of the half reaction in equilibrium with the indicating electrode.

$$E_{pot} = E_{ind \, el} = E° \quad (\text{at } 1M \text{ concn}) \tag{11-19}$$

Some typical values measured using **11-19** are listed in Table 11-1. These $E°$ values refer to a number of common oxidizing agents; the oxidizing agent is the reactant in each half reaction listed there. The $E°$ values apply to these half reactions only under standard conditions—one molar concentration of all species including H^+ and anions, one atmosphere pressure, and 25°C. Special exceptions are footnoted.

Interpretation of $E°$ Values

Note that all of the $E°$ values for the oxidizing agents in Table 11-1 are *positive* relative to that for the reduction of hydrogen ions. This implies that all these agents have a greater tendency than H^+ to gain electrons. In general, the oxidizing agent with the more positive $E°$ has the greater tendency to react and gain electrons. Thus the most powerful

TABLE 11-1. Common Oxidizing Agents and Their Standard Potentials

Oxidizing agent $+ ne^- \rightarrow$ product	Standard potential, $E°$, in volts (1M H^+)
$Ce(IV) + e^- = Ce^{+3}(\text{in } 1M \text{ HClO}_4)$	1.70[a]
$KMnO_4 + 4H^+ + 3e^- = MnO_2 + 2H_2O + K^+$	1.7 (neutral solution)
$H_5IO_6 + H^+ + 2e^- = IO_3^- + 3H_2O$	1.60
(periodic acid)	
$KMnO_4 + 8H^+ + 5e^- = Mn^{+2} + 4H_2O + K^+$	1.51
$Ce(IV) + e^- = Ce^{+3}(\text{in } 1N \text{ H}_2SO_4)$	1.44[a]
$Cl_2 + 2e^- = 2Cl^-$	1.36
$Cr_2O_7^{-2} + 14H^+ + 6e^- = 2Cr^{+3} + 7H_2O$	1.33
$O_2 + 4H^+ + 4e^- = 2H_2O$	1.229 ($= +0.816$ at pH 7)
$IO_3^- + 6H^+ + 5e^- = \frac{1}{2}I_2 + 3H_2O$	1.195
$VO_2^+ + 2H^+ + e^- = VO^{+2} + H_2O$	1.00
$Cu^{+2} + I^- + e^- = CuI(s)$	0.86
$Ag^+ + e^- = Ag(s)$	0.800
$Fe^{+3} + e^- = Fe^{+2}$	0.771
$O_2 + 2H^+ + 2e^- = H_2O_2$	0.682
$H_3AsO_4 + 2H^+ + 2e^- = H_3AsO_3 + H_2O$	0.559 ($=0.004$ at pH 7)
$I_2 + 2e^- = 2I^-$	0.535
Dehydroascorbic $+ 2H^+ + 2e^- =$ Ascorbic acid	0.39 ($= -0.023$ at pH 7)
$Fe(CN)_6^{-3} + e^- = Fe(CN)_6^{-4}$	0.36
Cytochrome-$a(Fe^{III}) + e^- =$ Cytochrome-$a(Fe^{II})$	0.290
$Hg_2Cl_2 + 2e^- = 2Hg + 2Cl^-$	0.268
$2H^+ + 2e^- = H_2(g)$	0.000 (exact no.)

[a] These are so-called formal potentials for which the measurement is made in the acid in parentheses; a change in acid changes the potential.

oxidizing agent in Table 11-1 is cerium(IV) ion in $HClO_4$; its $E°$ of $+1.70$ volts is the most positive of those in the table.

As the $E°$ values become less positive and approach zero volts, the associated oxidizing agents have a lesser tendency to gain electrons and be reduced. Thus iodine has a relatively small positive $E°$ value and a relatively small tendency to gain electrons (11-17). In fact, compared to other oxidizing agents, it is a weak to moderate oxidizing agent, not a strong oxidizing agent. A useful rule for any two oxidizing agents being considered is in the box below.

> The oxidizing agent with the more positive $E°$ will react spontaneously with the reduced form of the oxidizing agent having the less positive $E°$.

In terms of Table 11-1, the oxidizing agent (species on the left) will react with the reduced form (species on the right) of a half reaction *below* it in the table. This applies to standard conditions only; later we will give a generalization for practical analytical conditions which is based on a difference in $E°$ values.

As an illustration of the above generalization, consider the reaction of iodide with arsenic acid with $[H^+] = 1M$:

$$H_3AsO_4 + 2I^- + 2H^+ \rightleftharpoons H_3AsO_3 + I_2 + H_2O \qquad (11\text{-}20)$$

To decide whether or not these two species will react, we first identify the half reaction that is written in the same direction in the table and in the complete reaction. This is the H_3AsO_4 half reaction, and it has the larger $E°$. Therefore, the reaction should be spontaneous.

To find the numerical difference in the $E°$ values, we combine the half reactions and $E°$ values as follows. First we write the half reaction that is in the same direction as in the table, along with its $E°$ value. Below it we write the other half reaction, after reversing its direction in the table, with its $E°$ value in parentheses and with a negative sign before the parentheses. Then we combine the half reactions and $E°$ values algebraically:

$$
\begin{array}{lll}
H_3AsO_4 + 2e^- \rightleftharpoons H_3AsO_3 + H_2O & & E°_{ox\ agt} = +0.559\ V \\
2I^- \rightleftharpoons I_2 + 2e^- & & -(E°_{red\ agt}) = -(+0.535\ V) \\
\hline
H_3AsO_4 + 2I^- + 2H^+ \rightleftharpoons H_3AsO_3 + I_2 + H_2O & & E°_{rxn} = +0.024\ V
\end{array}
$$

In general, the potential for a complete reaction is always $E°_{ox\ agt} - E°_{red\ agt}$, as is shown for this example. It should be noted that if the number of electrons in each half reaction is not equal, each has to be multiplied by the number of electrons in the other · half reaction before the addition is performed. (This multiplication does not affect $E°_{rxn}$.)

We will show below that the reaction of iodide with arsenic acid (11-20), though spontaneous, is not quantitative. It can be made quantitative by increasing the hydrogen ion concentration above $1M$. Now consider the opposite reaction, the oxidation of arsenious acid by iodine. This is obviously not spontaneous at $1M$ hydrogen ion, but from Table 11-1 we see that the potential for the arsenic couple at pH 7 is 0.004 V, as compared to $+0.559$ V in $1M$ hydrogen ion. To predict whether the oxidation of H_3AsO_3 by iodine is spontaneous at pH 7, we use the general procedure as above:

$$
\begin{array}{ll}
I_2 + 2e^- \rightleftharpoons 2I^- & E^\circ_{ox\ agt} = +0.535\ V \\
\underline{H_3AsO_3 + H_2O \rightleftharpoons H_3AsO_4 + 2H^+} & \underline{-(E_{red\ agt}) = -(0.004\ V)\ at\ pH\ 7} \\
I_2 + H_3AsO_3 + H_2O \rightleftharpoons 2I^- + H_3AsO_4 + 2H^+ & E_{rxn} = +0.531\ V\ at\ pH\ 7
\end{array}
$$

Calculation of the Equilibrium Constant

One approach to use in deciding whether or not a reaction will be quantitative is to calculate the equilibrium constant. After this approach has been explained, we will examine the similar use of a minimum theoretical value of E°_{rxn} in the next section.

Each oxidation-reduction reaction should in theory have a unique equilibrium constant, K. The general form of an oxidation-reduction reaction governed by K is

$$ xA_{ox} + yB_{red} \rightleftharpoons xA_{red} + yB_{ox} $$

in which x molecules of an oxidizing agent, A_{ox}, react with y molecules of a reducing agent, B_{red}, to give the corresponding products. We will assume that each A_{ox} gains y electrons and that each B_{red} loses x electrons; we will also let the total number of electrons involved be n.

For each half reaction, the relationship between E, the potential of the half reaction, and the concentrations of the various species is given by the Nernst equation. The *general* form of this equation at 25°C is

$$ E = E^\circ + \frac{0.059\ V}{n} \log \frac{[Ox]}{[Red]} \tag{11-21} $$

Note that the constant of 0.059 in the Nernst equation has units of volts (V); this results from the units of the constants used to calculate it, $2.3\ RT/F$. See Problem 33 at the end of this chapter for the calculation.

For the A_{ox}/A_{red} half reaction, **11-21** has this form:

$$ E_A = E^\circ_A + \frac{0.059\ V}{y} \log \frac{[A_{ox}]^x}{[A_{red}]^x} \tag{11-22} $$

For the B_{red}/B_{ox} half reaction, **11-21** has this form:

$$ E_B = E^\circ_B + \frac{0.059\ V}{x} \log \frac{[B_{ox}]^y}{[B_{red}]^y} \tag{11-23} $$

After A_{ox} and B_{red} have reacted and reached equilibrium, the potential of A to react further will equal the potential of B to react further; that is, $E_A = E_B$. We can then equate the right-hand side of **11-22** to that of **11-23**.

$$ E^\circ_A + \frac{0.059\ V}{y} \log \frac{[A_{ox}]^x}{[A_{red}]^x} = E^\circ_B + \frac{0.059\ V}{x} \log \frac{[B_{ox}]^y}{[B_{red}]^y} \tag{11-24} $$

Recognizing that the total number of electrons involved is the lowest common denominator of x and y, we combine terms in **11-24** to obtain the following:

$$ \log \frac{[A_{red}]^x [B_{ox}]^y}{[A_{ox}]^x [B_{red}]^y} = \log K = \frac{n(E^\circ_A - E^\circ_B)}{0.059\ V} = \frac{nE^\circ_{rxn}}{0.059\ V} \tag{11-25} $$

The log term on the left-hand side of **11-25** can be recognized as the product of the concentration of the products divided by that of the reactants for our general oxidation-reduction reaction. In the two right-hand terms are given different but equivalent terms for calculating the log of the equilibrium constant for any redox reaction.

It is important to use the correct value of n in **11-25** for calculating log K values. If x and y are not equal, then $n = (x)(y)$. For example, if $x = 1$ and $y = 2$, then $n = 2$. If x and y involve 2 and 3 electrons, respectively, then $n = 6$. If $x = y = 2, 3, \ldots$, then in deriving **11-25** the terms simplify such that $n = x = y = 2, 3, \ldots$. Thus if both half reactions involve two electrons, $n = 2$, not 4. An example to illustrate this is given next.

> **EXAMPLE:** Calculate the equilibrium constant for the reaction of iodide and arsenic acid to produce iodine and arsenious acid **(11-20)**.
> *Solution:* After identifying arsenic acid as the oxidizing agent and iodide as the reducing agent, we calculate E°_{rxn}.
>
> $$E^\circ_{rxn} = (E^\circ_{ox\ agt} - E^\circ_{red\ agt}) = +0.559 - (+0.535) = +0.024\ \text{V}$$
>
> Using **11-25** and recognizing that $n = 2$, log K is calculated as follows:
>
> $$\log \frac{[I_2][H_3AsO_3]}{[I^-]^2[H_3AsO_4][H^+]^2} = \log K = \frac{2(+0.024\ \text{V})}{0.059\ \text{V}} = 0.81_{36}$$
>
> Converting log K to a numerical value for K gives the equilibrium constant.
>
> $$K = \text{antilog of } 0.81_{36} = 6.5_1 \ (\text{or } 6.5_1 \times 10^0)$$
>
> A review of significant figures in Section 3-1 will indicate why only two significant figures are justified for the value of K.

It can be shown that the minimum theoretical value of K for a quantitative reaction (99.9%) should be at least 1×10^6. Thus the reaction is not quantitative at a hydrogen ion concentration of $1M$.

For further practice with calculation of the equilibrium constant, you should work the problems in that group at the end of this chapter.

11-4 | REDOX METHODS OF ANALYSIS

This section is an introduction to oxidation-reduction methods; it is not a complete survey of such methods, since such a survey would be beyond the scope of this book. The basic requirements for redox methods with some titrimetric and spectrophotometric examples are discussed to introduce you to this type of method.

General Principles

Several general principles must be followed if a successful redox method is to be selected or devised. One of these is to choose the proper titrant or reagent for spectrophotometric analysis. If the species to be determined is present in a lower oxidation state and can be oxidized quantitatively, then an oxidizing agent should be chosen as titrant or reagent. Table 11-1 lists many good oxidizing agents and the half reactions they undergo. All the oxidizing agents are found on the *left* in each half reaction. Not all the oxidizing agents given are written as compounds; what compound to choose is also important and will be discussed later under the topic of primary standards.

TABLE 11-2. Common Reducing Agents and Their Standard Potentials

Product + ne^- ← *reducing agent*	*Standard potential* $E°$, *in volts*
$Fe^{+3} + e^- = Fe^{+2}$	$+0.771$
$H_3AsO_4 + 2H^+ + 2e^- = H_3AsO_3$	$+0.559 \ (= 0.004$ at pH 7)
$I_2 + 2e^- = 2I^-$	$+0.535$
Dehydroascorbic + $2H^+ + 2e^-$ = Ascorbic acid	$+0.39 \ (= -0.023$ at pH 7)
$Fe(CN)_6^{-3} + e^- = Fe(CN)_6^{-4}$	$+0.36$
$AgCl(s) + e^- = Ag(s) + Cl^-$	$+0.222$
$Sn^{+4} + 2e^- = Sn^{+2}$	$+0.15$
$S_4O_6^{-2} + 2e^- = 2S_2O_3^{-2}$ (thiosulfate)	$+0.15$
$TiO^{+2} + 2H^+ + e^- = Ti^{+3} + H_2O$	$+0.1$
$2H^+ + 2e^- = H_2(g)$	0.000 (exact no.)
$Pb^{+2} + 2e^- = Pb(s)$	-0.126
$Cr^{+3} + e^- = Cr^{+2}$	-0.41
$2CO_2(g) + 2H^+ + 2e^- = H_2C_2O_4$	-0.49
$Zn^{+2} + 2e^- = Zn(s)$	-0.763
$Mg^{+2} + 2e^- = Mg(s)$	-2.37

If the species to be determined is present in its highest, or a higher, oxidation state and it is known that it can be reduced quantitatively, then a reducing agent should be chosen as titrant or reagent. Table 11-2 lists many good reducing agents and the half reactions they undergo. Because the IUPAC recommends that all half reactions be written as reductions, all the reducing agents are found on the *right* in each half reaction. When each reacts, it will give the product listed on the left. Not all the reducing agents given are written as compounds; what compound to use is also important and will be discussed under the topic of primary standards.

Regardless of the substance chosen as a titrant, it must be known or established that the reaction in question will be quantitative and rapid. If the $E°$ of the substance to be determined is known, then the $E°_{rxn}$ can be calculated and used to predict whether or not the reaction will be quantitative. By substituting a value of K of 1.0×10^6 for a 99.9% reaction (for $x = y$) into **11-25**, we obtain a general equation for calculating a minimum $E°_{rxn}$ to which to compare the actual $E°_{rxn}$.

$$\text{minimum } E°_{rxn} = \frac{0.0590 \text{ V}}{n} \log K = \frac{0.0590 \text{ V}}{n} \log(1.0 \times 10^6) = \frac{0.354 \text{ V}}{n} \quad \textbf{(11-26)}$$

The minimum value of $E°_{rxn}$ for the two most common cases can easily be calculated. These are the case where the reactants each undergo a one-electron change ($n = 1$) and the case where the reactants each undergo a two-electron change ($x = y = n = 2$). The minimum value of $E°_{rxn}$ for each of these cases is given in the box.

For $n = 1$, the minimum $E°_{rxn}$ for 99.9% rxn = $+0.354$ V.

For $n = 2$, the minimum $E°_{rxn}$ for 99.9% rxn = $+0.177$ V.

This means that the oxidizing agents with the larger positive $E°$ values will give the larger $E°_{rxn}$ values and that reactions involving these agents will tend to be 99.9%, or more, complete. However, it must also be known or established that the reaction in question is rapid as well. If a reducing agent is needed for a reaction, then one with as small an $E°$ as possible should be selected. This does not necessarily imply that the $E°$ should be negative; for example, iron(II) is a good reducing agent for the determination of many oxidizing agents, such as cerium(IV). The $E°$ of $+0.77$ V for the iron half reaction is sufficiently small with respect to an $E°$ of $+1.70$, or $+1.44$, for cerium(IV) that the reaction between Ce(IV) and Fe(II) will be quantitative (99.9%).

It is difficult to predict whether or not a given reaction will be rapid. Several observations can be made on the basis of observed rates of reactions.

1. Reactions between cations and anions tend to be fast.
2. Reactions involving the exchange of unequal numbers of electrons tend to be slow (for example, $2Fe^{+3} + Sn^{+2}$ reacting to give $2Fe^{+2}$ and Sn^{+4} is slow).
3. Reactions involving the breaking of bonds in both reactants tend to be slow (for example, MnO_4^- oxidizing $C_2O_4^{-2}$ to CO_2 and forming Mn^{+2} as the other product is slow).

It should be stressed that the most valid check on whether a reaction will be quantitative and rapid is *experimental*. No amount of theory or generalizations about reaction rates will replace actually performing the experiment in the laboratory. If you are following directions for a method that has already been published, you can usually be sure that it is quantitative and rapid. If you are modifying the method or devising a new method, you should check it by using 100% pure compounds, or primary standards.

Primary Standards Although many of the general requirements for primary standards listed in Section 6-1 apply to redox methods, redox primary standards must be chosen more carefully than those for other types of reactions. In many cases, a redox primary standard will not react quantitatively or will not react rapidly with the solution to be standardized. For example, suppose that it is desired to standardize a solution of iodine using one of the primary standard reducing agents listed in Table 11-3. Sodium oxalate (no. 7) can not be used because it does not react rapidly enough with iodine. Ferrous ammonium sulfate (no. 10) can not be used because the reaction would not be quantitative (the $E°_{rxn}$ would be negative).

To avoid the difficulties of choosing a proper primary standard, you are advised to consult Table 11-3, which lists many of the uses of each primary standard. Iodine is a special case. Although it is a primary standard material, iodine is difficult to weigh and dissolve accurately, so that many methods call for standardizing it against arsenic(III) oxide after weighing it out approximately.

End Point Detection As in most other methods, the end point may be detected by using an indicator, finding the end point by potentiometric means, or using a photometric titration. Because some titrants are intensely colored and form essentially colorless products, or vice versa, they can be used as *self-indicators*. For example, potassium permanganate has an intense purple color and forms the nearly colorless manganese(II) ion as a product. The first slight excess of potassium permanganate titrant therefore gives the solution a light purple, or pink, color. Iodine is similar in that the first excess of iodine causes aqueous solutions to appear light yellow and nonaqueous solutions, a light purple. The

TABLE 11-3.   Redox Primary Standards and Their Uses

Primary standard	Uses in standardization
Oxidizing agents	
1. Iodine, I_2	1a. Used as a titrant
	1b. Standardization of reducing agents such as sodium thiosulfate
2. KIO_3	2a. Used as a titrant; titrated into acid and KI to produce I_2
	2b. Standardization of reducing agents $(Na_2S_2O_3)$
3. $(NH_4)_2Ce(NO_3)_6$	3. Used as a titrant
4. $K_2Cr_2O_7$	4. Used as a titrant
5. $KBrO_3$	5. Used as a titrant; titrated into acid and KBr to produce Br_2
Reducing agents	
6. As_2O_3	6. Standardization of I_2, Ce(IV), $KMnO_4$, and H_5IO_6 (periodic acid)
7. $Na_2C_2O_4$	7. Standardization of $KMnO_4$ and Ce(IV)
8. $K_4Fe(CN)_6$	8. Standardization of $KMnO_4$ and Ce(IV)
9. Fe wire	9. Standardization of $KMnO_4$, Ce(IV), and $K_2Cr_2O_7$
10. $Fe(NH_4)_2(SO_4)_2 \cdot 6 H_2O$	10. Standardization of $KMnO_4$, Ce(IV), and $K_2Cr_2O_7$
11. Cu wire	11. Standardization of $Na_2S_2O_3$ after dissolution to Cu^{+2} and reduction to CuI(s) + I_2

iodide ion product is colorless and therefore does not interfere with visual perception of the end point.   In contrast to these is potassium dichromate titrant.   It is an intense orange color but forms the light purple chromium(III) ion, or in HCl the light green $CrCl^{+2}$ ion, as a product.   The concentration of either of these ions at the end point is sufficient in most cases to prevent perception of the first excess of potassium dichromate.

Table 11-4 lists a number of useful *redox indicators*.   There is no difficulty in choosing an indicator for most standard titrants, since redox indicators have been studied in detail. It can be shown that the $E°$ value listed for each indicator in the table should fall within $\pm 0.059/n$ volts of the potential at the end point.   Since most indicators have already been experimentally checked for a particular method, it is usually only necessary to employ this rule when using a new or revised method or a new indicator.

The starch indicator action for iodine titrations is different from the actions of the other indicators listed in Table 11-4 and needs explanation.   Starch forms a *complex* with iodine and iodide ion as follows:

$$I_2 + starch + I^- \rightleftharpoons (starch)I_3^-$$

The color of the starch-triiodide complex is blue to blue-purple.   The first excess of iodine titrant added past the end point reacts with the starch and iodide ion to give the blue color.   This complex is ionic and does not form readily in organic solvents, so starch is used only in aqueous titrations.

TABLE 11-4. Redox Indicators

$E°$, V	Indicator (reduced form)	Color	Color of oxidized form
—	Starch[a]	Colorless	Purple
+1.25	Tris(nitro-1,10-phenanthroline) iron(II) sulfate, or nitroferroin	Red	Weak blue
+1.14[b]	Tris(1,10-phenanthroline) iron(II) sulfate, or ferroin	Red	Weak blue
+1.06	p-nitrodiphenylamine	Colorless	Violet
+0.97	Tris(2,2'-bipyridine) iron(II) sulfate	Red	Weak blue
+0.85[c]	Diphenylaminesulfonic acid	Colorless	Violet or purple
+0.76[c]	Diphenylamine	Colorless	Violet or purple
+0.54	1-naphthol-2-sulfonic acid-indophenol	Colorless	Red
+0.53[b]	Methylene blue	Colorless	Blue

[a] Used only for iodine titrations.
[b] In $1M$ sulfuric or hydrochloric acids, the standard potential is 1.06 V.
[c] These potentials are for $1M$ sulfuric acid.

Potentiometric detection of the end point is possible with a potentiometer, an indicating electrode, and a reference electrode in a setup similar to the one shown in Figure 11-1. Changes in concentration of the ion being titrated are indicated by the indicating electrode; the corresponding changes in potential are followed on the potentiometer. If a reducing agent is being titrated with an oxidizing agent, the potential will increase, as shown in Fig. 11-2. At the end point, a sharp increase in potential occurs,

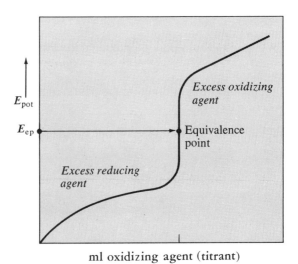

ml oxidizing agent (titrant)

FIGURE 11-2. Potentiometric titration curve using an oxidizing agent (more positive $E°$) as the titrant and a reducing agent as the species titrated.

followed by a leveling off of the increase in potential. (If an oxidizing agent is titrated with a reducing agent, the opposite type of curve is obtained, with a sharp decrease in potential.) It can be shown that the potential at the end point, E_{ep}, of a redox titration can be calculated as follows:

$$E_{ep} = \frac{yE°_{ox\ agt} + xE°_{red\ agt}}{y + x}$$ (11-27)

in which y is the number of electrons gained in the half reaction of the oxidizing agent and x is the number of electrons gained in the half reaction of the reducing agent. Equation 11-27 is valid for all conditions except where the reaction involves hydrogen or hydroxide ions and the $[H^+]$ or $[OII^-]$ is not $1M$, or where one ion is changed into two ions in the reaction $(Cr_2O_7^{-2} \rightarrow 2Cr^{+3})$.

Titrations with Oxidizing Agents

Because so many elements are easily oxidized from a lower oxidation state to a higher one, and because so many organic compounds are readily oxidized to more oxidized forms, many methods have been based on direct or indirect titrations with oxidizing agents. Oxidizing agents containing some oxidation state of iodine are generally the most useful because they are mild, easy to handle, and have few side reactions. Some of these agents are potassium iodate, iodine itself, and periodic acid (HIO_4) or its hydrate (H_5IO_6). Stronger oxidizing agents are cerium(IV) in sulfuric acid, potassium permanganate, and potassium dichromate. These reagents oxidize more elements and compounds than the iodine agents but also undergo more side reactions and are more difficult to control because of this.

Iodine Titrations

Direct titration with iodine is a useful method for the determination of a fairly large number of inorganic and organic substances. The titration is simple, although in many cases the pH of the solution must be adjusted or controlled by buffering. If water is the solvent, starch is added at the beginning of the titration. Near the end point, the purple color of the starch-triiodide complex appears momentarily. As the end point is approached, the mixing time required for its disappearance increases until a stable purple color appears.

Iodine oxidizes a number of inorganic substances via reactions in which the oxidation number of one element is increased by two, that is, via two-electron oxidations (Table 11-5). A good example is the reaction of iodine with arsenious acid, which is arsenic in the $+3$ oxidation state.

$$H_3AsO_3 + I_2 + H_2O \xrightarrow{\text{NaHCO}_3 \text{ buffer}} HAsO_4^{-2} + 2I^- + 4H^+$$

Because the reaction produces hydrogen ions, sodium bicarbonate buffer must be used to keep the pH slightly basic. If the pH falls below about 4, the reaction is not quantitative. This reaction is also important because it can be used for the standardization of iodine. Primary standard arsenic(III) oxide (Table 11-3) is dissolved in basic solution, and the solution is acidified to produce arsenious acid.

Iodine also oxidizes a number of inorganic and organic substances to dimers. The best example of this type of reaction is that with sodium thiosulfate, producing the tetrathionate ion.

$$2S_2O_3^{-2} + I^2 \longrightarrow S_4O_6^{-2} + 2I^-$$
$$(Na_2S_2O_3)$$

TABLE 11-5. **Inorganic and Pharmaceutical Substances Determined by Iodine Titration**

Inorganic or pharmaceutical[a]	*Oxidation state change or reaction*
Two-electron increase in oxidation number	
$\quad$ Sn(II)	Sn(II)–Sn(IV)
$\quad$ H_2SO_3 or SO_3^{-2}	SO_3^{-2}–SO_4^{-2}
$\quad$ As(III)	As(III)–As(V)
$\quad$ Oxophenarsine, $C_6H_6AsNO_2$	As(III)–As(V)
$\quad$ Carbarsone, $C_7H_9AsN_2O_4$	As(III)–As(V)
$\quad$ Sb(III)	Sb(III)–Sb(V)
$\quad$ Antimonic potassium	
$\quad\quad$ tartrate, $KSbOC_4H_4O_6 \cdot \frac{1}{2}H_2O$	Sb(III)–Sb(V)
Oxidation to dimers	
$\quad$ Thiosulfate, $S_2O_3^{-2}$	$2S_2O_3^{-2}$–$S_4O_6^{-2}$
$\quad$ Mercaptans, RSH	2 RSH–(RSSR + $2H^+$)
$\quad$ Dimercaprol, $C_3H_8S_3$	2 RSH–(RSSR + $2H^+$)
$\quad$ Cysteine, $HSC_2H_3(NH_2)CO_2H$	Cysteine–Cystine
Other oxidations	
$\quad$ Ascorbic acid	Ascorbic acid–dehydroascorbic acid
$\quad$ $Fe(CN)_6^{-4}$	$Fe(CN)_6^{-4}$–$Fe(CN)_6^{-3}$

[a] Taken in part from G. L. Jenkins, A. Knevel, and F. E. DiGangi, *Quantitative Pharmaceutical Chemistry*, 6th ed., McGraw-Hill, New York, 1967, pp. 179–88.

The tetrathionate ion consists of two thiosulfate species bonded together as

$$[O_3S_2—S_2O_3]^{-2}$$

and is essentially a dimer. This reaction can be used for the standardization or analysis of either reaction; since sodium thiosulfate is not a primary standard, it must be first standardized before it is used to standardize iodine. Other substances listed in Table 11-5 are oxidized in similar fashion.

Oxidation and Determination of Ascorbic Acid

The determination of ascorbic acid (vitamin C) involves a rather unusual oxidation reaction:

$$C_4H_6O_4(OH)C{=}C(OH) + ox\ agt$$
$$\longrightarrow\quad C_4H_6O_4—C({=}O)—C{=}O + 2H^+ + red\ agt \quad \textbf{(11-28)}$$

Both iodine and 2,6-dichloroindophenol have been used as oxidizing agents for this purpose. Both being mild oxidants, they cause no further reactions. Both reagents oxidize the enediol functional group of ascorbic acid to an alpha-diketone group, shown in the following:

enediol $\longrightarrow$ alpha-diketone

Stronger oxidizing agents are not suitable for this determination because of incomplete side reactions that occur involving other functional sites in ascorbic acid, leading to high results.

Determination of Ascorbic Acid with Iodine. Iodine is an excellent reagent for the determination of ascorbic acid for reasons other than the fact that it oxidizes ascorbic acid quantitatively without any side reactions. For one, iodine is very stable. Also, it is easily standardized against primary standard arsenic(III) oxide (while 2,5-dichloro-indophenol is not). In addition, iodine can be utilized in two different ways; it can be used as the titrant as is, for the direct titration [3], or it can be generated in the solution from a measured excess of potassium iodate and iodide in acid [4]. The iodate and iodide react in acid solution to produce iodine.

$$IO_3{}^- + 5I^- + 6H^+ \longrightarrow 3I_2 + 3H_2O$$

The iodine reacts with the ascorbic acid, and the unreacted iodine is determined by back titration with standard sodium thiosulfate. Both this approach and the direct iodine titration can be used for the determination of vitamin C in tablets (Exp. 10); in fruit drinks, such as Hi-C; and in dehydrated juice solids, such as Tang.

The oxidation of ascorbic acid by iodine (11-28) is not quite quantitative in $1M$ acid solution, as can be predicted by calculating the equilibrium constant or by subtracting the standard reduction potentials as follows:

$$E^\circ_{rxn} = +0.535 \text{ V} - (+0.39 \text{ V}) = +0.14_5 \text{ V} (< +0.177 \text{ V for } 99.9\% \text{ rxn})$$

The oxidation is quantitative at lower acid concentrations; for example, at pH 7, the potential of ascorbic acid is -0.023 V (Table 11-2), and it is obvious that the E°_{rxn} is then greater than the theoretical minimum of $+0.177$ V.

Determination of Ascorbic Acid with Dichloroindophenol. Another standard reagent for the determination of ascorbic acid is the organic compound 2,5-dichloroindophenol [5]. It is a powerful, though somewhat unstable oxidizing agent. The reaction is as follows:

(red Na$^+$ salt) (colorless)

Because the dichloroindophenol is not stable over a long period of time, it must be standardized against a known standard of ascorbic acid at the time it is used for analysis. The end point is taken as the first appearance of a light red (pink) color from the first excess of the sodium salt of the dichloroindophenol.

Oxidation of Ascorbic Acid: Air Oxidation. In both the iodine and dichloroindophenol methods for determination of ascorbic acid, the air oxidation of ascorbic acid is a potential source of error.

$$2C_4H_6O_4(OH)C=C(OH) + O_2 \longrightarrow C_4H_6O_4C(=O)-C=O + 2H_2O$$

The E°_{rxn} for this reaction is calculated as follows:

$$E^\circ_{rxn} = +1.229 \text{ V} - (+0.39 \text{ V}) = +0.83_9 \text{ V}$$

This result indicates that the reaction should be quantitative. Fortunately however, it is very slow, allowing ascorbic acid to be oxidized quantitatively by reagents such as iodine or dichloroindophenol, as long as proper precautions are taken.

Periodic Acid Oxidations

Oxidation with periodic acid is a very selective method for organic and biochemical molecules having a 1,2-diol grouping. The periodic acid cleaves the carbon-carbon bond and takes each remaining fragment to the next higher oxidation state. For example, it oxidizes ethylene glycol to two molecules of formaldehyde as follows:

$$\begin{array}{l} H_2C-OH \\ | \qquad\quad + HIO_4 \\ H_2C-OH \end{array} \longrightarrow \begin{array}{l} H_2C=O \\ \qquad\quad + HIO_3 + H_2O \\ H_2C=O \end{array}$$

Glycerine, an important ingredient in many pharmaceutical preparations, is oxidized by periodic acid to two molecules of formaldehyde plus formic acid as follows:

$$\begin{array}{l} H_2C-OH \\ | \\ HC-OH + 2HIO_4 \\ | \\ H_2C-OH \end{array} \longrightarrow \begin{array}{l} H_2C=O \\ \\ + HCOOH + 2HIO_3 + H_2O \\ \\ + H_2C=O \end{array}$$

Sugars, such as glucose and fructose, are also oxidized by periodic acid.

A measured excess of periodic acid is usually added to the compound to be determined, and the unreacted periodic acid is reduced to iodine, which is then titrated with standard sodium thiosulfate.

$$HIO_4 + 7I^- + 7H^+ \longrightarrow 4H_2O + 4I_2 \text{ (titrated with Na}_2S_2O_3\text{)}$$

Ferricyanide Oxidations and Colorimetric Analysis

The ferricyanide ion, $Fe(CN)_6^{-3}$, is a mild oxidizing agent ($E^\circ = +0.36$ V). It can therefore be used only for the quantitative oxidation of very easily oxidized substances— for example, aldehydes (RCHO) oxidized to acids (RCO$_2$H). Because it has an intense yellow color, it can be used for the *colorimetric determination* (Ch. 12) of such aldehydes. A measured excess of the ferricyanide is added, and the unreacted ferricyanide is determined colorimetrically. The difference between the amount of ferricyanide added before reaction and the amount unreacted is a measure of the amount of the aldehyde present.

A good example of this method is the colorimetric determination of glucose in body fluids. A measured excess of ferricyanide oxidizes the glucose to gluconic acid.

$$2Fe(CN)_6^{-3} + C_5H_6(OH)_5CHO + H_2O \xrightarrow{90°C}$$
$$2Fe(CN)_6^{-4} + C_5H_6(OH)_5CO_2H + 2H^+$$

The reaction is slow at room temperature, so heat is needed to increase the rate of the reaction. This reaction is used for the automated colorimetric determination of glucose in many clinical and hospital laboratories. The details are discussed in Section 13-4.

Titrations with Strong Oxidizing Agents

Strong oxidizing agents are those that have an $E°$ greater than $+1$ V and that frequently oxidize a substance with two or more oxidation products to the highest oxidation state or product. The most common strong oxidizing agents are potassium permanganate, cerium(IV), and potassium dichromate. While all three can sometimes be used for the same determination, each has special advantages and disadvantages.

Strong oxidizing agents are more suitable for the determination of inorganic substances, which often cannot be oxidized by a mild oxidizing agent, than for the determination of organic substances. Two of the main reasons for this are given in the box.

---

1. The oxidation of an organic compound *with one functional group* by a strong oxidizing agent often does not reach 99.9 % completion because of side reactions and/or large (>2) electron changes.
2. The oxidation of an organic compound *with two or more functional groups* by by a strong oxidizing agent usually does not reach 99.9 % completion because the chances of the second functional group's being 99.9 % oxidized are usually less than that of the first.

In spite of the above difficulties, a number of organic compounds having just one functional group can be determined using strong oxidizing agents. The oxidation of alcohols is one such example; the three types of alcohols and their possible reactions are as follows:

primary: $R—CH_2OH$ + ox agt $\longrightarrow$ $R—CO_2H$
secondary: $R_2—CHOH$ + ox agt $\longrightarrow$ $R_2C{=}O$
tertiary: $R_3—COH$ + ox agt $\longrightarrow$ no oxidation to higher state (dehydration to olefin may give rise to reaction)

Potassium dichromate ($K_2Cr_2O_7$), a primary standard, is a less reactive strong oxidizing agent ($E° = 1.33$ V) than either potassium permanganate or cerium(IV), and therefore often has fewer interferences and causes fewer side reactions. A case in point is its use in the oxidation of alcohols. The oxidation of a secondary alcohol, such as isopropyl alcohol, to a ketone is virtually quantitative [6]. The oxidation of a primary alcohol, such as ethyl alcohol, is more complicated because the oxidation proceeds from the alcohol to an aldehyde to the final carboxylic acid product.

$$CH_3CH_2OH \xrightarrow[Cr(VI)]{-2e} CH_3CHO \xrightarrow[Cr(VI)]{-2e} CH_3CO_2H$$

The second step of the above reaction is often incomplete. Nevertheless, the reaction has been conducted in a reproducible manner as a basis for the Breathalyzer method for analyzing alcohol in the breath. The decrease in color of the potassium dichromate is used as an indication of the amount of ethyl alcohol present.

Potassium dichromate is also used frequently in inorganic analysis, especially for the determination of iron(II) following dissolution of metallic iron or iron ore in hydrochloric acid.

$$Fe^\circ(s) + 2\,HCl \longrightarrow H_2(g) + Fe^{+2}$$

$$Fe_2O_3(s) + 6\,HCl \longrightarrow 3\,H_2O + 2\,Fe(III)$$

Stronger oxidizing titrants oxidize the chloride ion to chlorine, but potassium dichromate is too weak an oxidizing agent ($E^\circ = +1.33$ V) to oxidize chloride to chlorine ($E^\circ = +1.36$ V); that is, chlorine is a stronger oxidizing agent. If an iron ore is to be analyzed for its percentage of iron, the dissolution of the ore yields iron(III), which must then be reduced using tin(II) chloride.

$$2\,Fe(III) + SnCl_2 \xrightarrow{\text{HCl solvent}} 2\,Fe^{+2} + Sn(IV) + 2\,Cl^-$$

The titration of iron(II) is then the same whether metallic iron or iron ores are being analyzed.

$$6\,Fe^{+2} + K_2Cr_2O_7 + 14\,H^+ \longrightarrow 6\,Fe^{+3} + 2\,Cr(III) + 7\,H_2O + 2\,K^+$$

The solution is somewhat green from the chloro complexes of chromium(III), but the purple to violet color of the diphenylamine indicator is readily seen.

Potassium permanganate, $KMnO_4$, is a convenient titrant in that it serves as a self-indicator. The end point is indicated by the pink to light purple color imparted to the solution by the first excess of the permanganate ion. Since potassium permanganate is not available in primary standard purity, it must be standardized against one of the primary standards listed in Table 11-3.

Potassium permanganate is used for the analysis of hydrogen peroxide solutions, such as the 3% solution sold for medicinal purposes. The hydrogen peroxide is oxidized to oxygen in the titration.

$$5\,H_2O_2 + 2\,KMnO_4 + 6\,H^+ \longrightarrow 5\,O_2(g) + 2\,Mn^{+2} + 2\,K^+ + 8\,H_2O$$

Note that all of the products in the reaction are essentially colorless, so the first excess of permanganate indicates the end point.

Potassium permanganate is also used for the titration of iron(II), either in the form of ferrous sulfate tablets or after it has been produced following dissolution of iron ores or iron metal. The reaction is as follows:

$$5\,Fe^{+2} + KMnO_4 + 8\,H^+ \longrightarrow 5\,Fe^{+3} + Mn^{+2} + K^+ + 4\,H_2O$$

The iron(III) ion is slightly colored in the sulfuric acid solvent used for the reaction, but the end point is readily seen.

Other uses for potassium permanganate are listed in Table 11-6.

Cerium(IV), or the *ceric ion*, is a more useful titrant than potassium permanganate for the following reasons. Cerium(IV) is available in primary standard form as ammonium hexanitratocertate(IV) salt (Table 11-3). It can be used in perchloric acid ($E^\circ = +1.70$ V) or in sulfuric acid ($E^\circ = +1.44$ V). Although cerium(IV) is yellow in sulfuric acid, its color is not suitable for indication of the end point. Instead, ferroin or one of the other indicators in Table 11-4 is used.

TABLE 11-6.   Inorganic and Pharmaceutical Substances Determined by Potassium Permanganate and Cerium (IV) Titration

Inorganic or pharmaceutical[a]	Oxidation state change or reaction
KMnO$_4$ titrant	
Fe(II)	Fe(II)–Fe(III)
Fe(CN)$_6^{-4}$	Fe(CN)$_6^{-4}$–Fe(CN)$_6^{-3}$
As(III)	As(III)–As(V)
Na$_2$C$_2$O$_4$ or H$_2$C$_2$O$_4$	C$_2$O$_4^{-2}$–2CO$_2$(g)
H$_2$O$_2$ or Na$_2$O$_2$	O$_2^{-2}$–O$_2$(g)
Cerium(IV) titrant	
Fe(II)	Fe(II)–Fe(III)
Fe(CN)$_6^{-4}$	Fe(CN)$_6^{-4}$–Fe(CN)$_6^{-3}$
Ferrous fumarate tablets	Fe(II)–Fe(III)
Ferrous sulfate tablets	Fe(II)–Fe(III)
Ferrous gluconate tablets	Fe(II)–Fe(III)
As(III)	As(III)–As(V)
Na$_2$C$_2$O$_4$ or H$_2$C$_2$O$_4$	C$_2$O$_4^{-2}$– 2CO$_2$(g)
Hydroquinone	Quinone
Tocopherol (vitamin E)	Quinone form of vitamin E

[a] Taken in part from G. L. Jenkins, A. Knevel, and F. E. DiGangi, *Quantitative Pharmaceutical Chemistry*, 6th ed., McGraw-Hill, New York, 1967, pp. 173–77.

Cerium(IV) can be used for the determination of a large number of substances, a few of which are listed in Table 11-6. The reactions for the determination of iron(II) in various forms are the following:

$$Fe^{+2} + Ce(IV) \xrightarrow{H_2SO_4 \text{ solvent}} Fe^{+3} + Ce^{+3}$$

$$\text{Ferroin} + Ce(IV) \longrightarrow \text{Ferriin} + Ce^{+3}$$
$$\text{(red)} \qquad\qquad\qquad \text{(weak blue)}$$

Although the oxidized form of the ferroin indicator, ferriin, is a weak blue color, the end point is easily perceived from the disappearance of the intense red color of ferroin itself.

Cerium(IV) can also oxidize many organic compounds; for example, it oxidizes hydroquinone to quinone.

Because vitamin E, or tocopherol, has a hydroquinone-type structure, cerium(IV) can oxidize vitamin E to a quinone form.

Titrations Involving Reducing Agents Not as many methods are available for titrations involving reducing agents as are for oxidizing agents because the oxygen in the air oxidizes many reducing agents, making it difficult to prepare standard solutions. The most useful procedure is reduction with the iodide ion; it is used for the determination of many oxidizing agents.

Iodide Reductions The iodide ion itself is not used for direct titration of many substances; instead, an excess of the iodide ion is added to the substance, which must be capable of quantitatively oxidizing the iodide to iodine. The iodine is then titrated with a standard solution of sodium thiosulfate. The amount of sodium thiosulfate titrant is equivalent to the amount of the oxidizing substance originally present. The method for the determination of an oxidizing substance, A_{ox}, that is reduced to the form A_{red} is summarized by the following equations:

$$A_{ox} + 2I^- \text{(excess)} \longrightarrow A_{red} + I_2$$

$$I_2 + 2Na_2S_2O_3 \longrightarrow 2I^- + 2Na^+ + Na_2S_4O_6$$
$$\text{(titrant)}$$

Starch is used as the indicator for the titration. It is usually added quite close to the end point because the starch-triiodide complex breaks apart somewhat slowly.

A number of substances that can be determined by the above method are listed in Table 11-7. An important example for both inorganic and pharmaceutical analysis is the determination of copper or copper sulfate (National Formulary method). If copper is present in metallic form, it is first dissolved in nitric acid as follows:

$$3Cu°(s) + 2NO_3^- + 8H^+ = 3Cu^{+2} + 2NO + 4H_2O$$

TABLE 11-7. Inorganic and Pharmaceutical Substances Determined by Reduction with Iodide Ion

Inorganic or pharmaceutical[a]	Oxidation state change or reaction
Halogen-containing substances	
Cl_2	Cl_2–$2Cl^-$
OCl^-, hypochlorite (bleach)	OCl^-–Cl^- + H_2O
Br_2	Br_2–$2Br^-$
IO_3^-, iodate ion	$IO_3^- + 5I^- = 3I_2 + 3H_2O$
HIO_4 or IO_4^-	$IO_4^- + 7I^- = 4I_2 + 4H_2O$
Iodine pentoxide	$I_2O_5 + 10I^- = 6I_2 + 5H_2O$
Other reductions	
Cu^{+2}, $CuSO_4$ (N.F.)	Cu^{+2}–$CuI(s)$
As(V)	As(V)–As(III)
Arsanilic acid, $C_6H_8AsNO_3$	As(V)–As(III)
Drocarbil, $C_8H_{10}AsNO_5$	As(V)–As(III)
Halazone tablets, $C_7H_5Cl_2NO_4S$	R—$Cl + 2I^- = RH + I_2 + Cl^-$

[a] Taken in part from G. L. Jenkins, A. Knevel, and F. E. DiGangi, *Quantitative Pharmaceutical Chemistry*, 6th ed., McGraw-Hill, New York, 1967, pp. 191–98.

Then the copper(II) ion, or copper sulfate, is treated with excess iodide ion.

$$2Cu^{+2} + 4I^- \longrightarrow 2CuI(s) + I_2$$

The reaction produces insoluble copper(I) iodide and iodine, which is titrated with standard sodium thiosulfate.

Other Reductions
Methods based on other reducing agents are sometimes used, but they are fairly specialized or are not used as frequently as iodide reduction because the reducing agents are oxidized by oxygen in the air. An example of a specialized type of reducing agent is sodium nitrite, $NaNO_2$. It is used for the analysis of sulfa drugs and other organic compounds that contain an amino group ($-NH_2$) bonded to an aromatic ring. The reaction is conducted in acid, which converts the sodium nitrite to nitrous acid, HNO_2. The product is a diazonium salt, $RN_2^+Cl^-$. For example, with sulfanilamide, the reaction is as follows:

$$H_2NSO_2C_6H_4-NH_2 + HNO_2 + HCl = H_2NSO_2C_6H_4-N_2^+Cl^- + 2H_2O$$
(sulfanilamide)

The end point is detected by dipping starch-iodide paper into the solution; a blue color is produced when excess nitrous acid is present to oxidize iodide to iodine.

11-5 | SIMPLE POTENTIOMETRIC ANALYSIS

Introduction
Titrations are not the only techniques that utilize oxidation-reductions for analysis. It is also possible to use a potentiometer to measure ion concentrations with a single measurement. Potentiometric measurements were described in general in Section 11-3. A potentiometric measurement requires the general equipment picture in Figure 11-1—a reference electrode and an indicating electrode in addition to the potentiometer. Equation **11-18** gives the relationship between the voltage measured on the potentiometer, E_{pot}, and the potential, or voltage, at each of the two electrodes. If the standard hydrogen electrode is used as the reference electrode, then its potential is zero and **11-18** simplifies to:

$$E_{pot} = E_{ind\ el} \tag{11-30}$$

Since the potential at the indicating electrode is equal to the potential of the half reaction, which is in equilibrium with this electrode, an equation can be developed for analysis. The potential at the indicating electrode must be equal to E, the potential of the half reaction. E is defined by the Nernst equation **(11-21)**, where n is the number of electrons in the half reaction.

$$E_{ind\ el} = E = E^\circ + \frac{0.059\ V}{n} \log \frac{[Ox]}{[Red]} \tag{11-31}$$

Substituting from **11-31** into **11-30** gives the following:

$$E_{pot} = E^\circ + \frac{0.059\ V}{n} \log \frac{[Ox]}{[Red]} \tag{11-32}$$

Equation **11-32** can be rearranged to:

$$\log \frac{[Ox]}{[Red]} = \frac{n(E_{pot} - E^\circ)}{0.059\ V} \tag{11-33}$$

Equation **11-33** shows that the ratio of the concentration of the oxidized form to the concentration of the reduced form in a half reaction can be measured simply by measuring the voltage, on a potentiometer, of a solution containing these forms. If one or the other of these forms is insoluble, then the concentration of the other form can be measured directly.

The classical indicating electrode half reaction that can be used to measure the chloride ion concentration is the following:

Measure-
ment of
Chloride Ion

$$AgCl(s) + e^- = Ag°(s) + Cl^-$$

The Nernst equation expression for this half reaction is

$$E = E° + 0.059 \text{ V log } 1/[Cl^-] \tag{11-34}$$

Note that the only concentration that affects the potential of the half reaction is the chloride ion concentration. If a silver/silver chloride indicating electrode is used in a measurement setup, such as the one shown in Figure 11-1, with a hydrogen reference electrode, then **11-32** becomes for this system:

$$E_{pot} = E° + 0.059 \text{ V log } 1/[Cl^-] \tag{11-35}$$

Equation **11-35** can be rearranged to give the following:

$$-\log[Cl^-] = \frac{(E_{pot} - E°)}{0.059 \text{ V}} \tag{11-36}$$

Thus the concentration of chloride ion can be found by a single measurement of the potential of the solution. Unfortunately, the measurements give only two significant figures under routine conditions (three under carefully controlled conditions), so the molarity of the chloride ion is known to only one significant figure (see Sec. 3-1).

More modern electrodes have been designed to replace the classical silver/silver chloride electrode; these are the so-called *specific-ion* electrodes. They function somewhat differently but give the same type of result (Ch. 16).

A more complete discussion of potentiometric measurements, including pH measurement, is given in Chapter 16.

| QUESTIONS AND PROBLEMS

(Answers to most even-numbered problems are in Appendix 5.)

Balancing with the Ion-Electron Method

1. Using oxidation numbers, complete and balance the following half reactions involving the oxidation of the reducing agents to the species at the right. Assume an acid solution unless otherwise indicated.
   a. $H_3AsO_3 \rightarrow H_3AsO_4$
   b. $Mn^{+2} \rightarrow MnO_4^-$
   c. $I^- \rightarrow I_2$
   d. $NH_2OH \rightarrow N_2(g) + H_2O$
   e. $H_2O_2 \rightarrow O_2(g)$

2. Using the ion-electron method, complete and balance the following inorganic half reactions involving the reduction of the oxidizing agents to the species at the right. Assume an acid solution unless otherwise indicated.
   a. $Cr_2O_7^{-2} \rightarrow 2Cr^{+2}$
   b. $IO_4^- \rightarrow IO_3^-$
   c. $IO_3^- \rightarrow I_2$
   d. $Fe(CN)_6^{-3} \rightarrow Fe(CN)_6^{-4}$
   e. $MnO_4^- \rightarrow Mn^{+2}$

3. Using the ion-electron method, complete and balance the following organic half reactions involving the oxidation of the reducing agents to the species at the right. Assume an acid solution unless otherwise indicated.

a. $C_2H_5OH \rightarrow CH_3CO_2H$

b. $CH_2OHCHOHCH_2OH \rightarrow 2H_2CO + HCO_2H$

c. $C_4H_6O_4(OH)C=COH \rightarrow C_4H_6O_4C(=O)-C=O$

d. $[C_{19}H_{27}]CHO \rightarrow [C_{19}H_{27}]CO_2H$

e. $H_2C_2O_4 \rightarrow 2CO_2(g)$

4. Combine the balanced half reactions of the inorganic reducing agents in Problem 1 with those of the oxidizing agents in Problem 2 to obtain five balanced equations. Combine part a of Problem 1 with part a of Problem 2, etc.

5. Combine the balanced half reactions of the organic reducing agents in Problem 3 with those of the oxidizing agents in Problem 2 to obtain five balanced equations. Combine part a of Problem 3 with part a of Problem 2, etc.

Quantitative Reaction

6. Calculate the E_{rxn} of each of the following reactions and decide whether or not the reaction is quantitative (the oxidizing agent is the first species).

a. $Ce^{IV} + Fe^{+2} \rightarrow Ce^{III} + Fe^{+3}$ (H_2SO_4)

b. $Ag^+ + Fe^{+2} \rightarrow Ag°(s) + Fe^{+3}$

c. $I_2 + H_3AsO_3 + H_2O \xrightarrow{1MH^+}$ $2I^- + H_3AsO_4 + 2H^+$

d. $I_2 + H_3AsO_3 + H_2O \xrightarrow{pH7}$ $2I^- + H_3AsO_4 + 2H^+$

e. $I_2 + $ ascorbic acid $\xrightarrow{pH7}$ $2I^- + $ dehydroascorbic acid $+ 2H^+$

f. $2Cu^{+2} + 4I^- \rightarrow 2CuI(s) + I_2$

g. $I_2 + 2S_2O_3^{-2} \rightarrow 2I^- + S_4O_6^{-2}$

h. $2Fe^{+3} + Sn^{+2} \rightarrow 2Fe^{+2} + Sn^{+4}$

7. Periodic acid (H_5IO_6) or potassium periodate is frequently used to oxidize Mn^{+2} to MnO_4^- to determine manganese in steel spectrophotometrically. Answer the following questions involving this method:

a. Assuming that the minimum difference in standard potentials $(E°_{ox} - E°_{red})$ for this

reaction is $+0.124$ V, decide whether an *equivalent* amount of either periodate compound will oxidize Mn^{+2} quantitatively $(> 99.9\%)$ to MnO_4^-.

b. In the laboratory, either compound can be used for quantitative oxidation of Mn^{+2} to MnO_4^-. How can you rationalize this fact with your answer to part a?

8. Calculate the equilibrium constant for the following reaction carried out in $1M$ acid:

$$VO_2^+ + Fe^{+2} + 2H^+ \rightleftharpoons VO^{+2} + Fe^{+3} + H_2O$$

Is the oxidation of Fe^{+2} to Fe^{+3} quantitative $(> 99.9\%)$? How might conditions be adjusted to make the oxidation of iron more complete?

9. Calculate the equilibrium constant for the following oxidation-reduction reaction:

$$2Cu^{+2} + 4I^- \longrightarrow 2CuI(s) + I_2$$

Calculation of Iodine Titration Results

10. A 0.5000-g sample of impure ascorbic acid (form wt $= 176.12$) in tablet form is oxidized to dehydroascorbic acid by 45.10 ml of $0.1000N$ $(0.0500M)$ I_2. Calculate the percent purity of the tablets.

11. A 1.2500-g sample of pure As_2O_3 is weighed into a 250-ml volumetric flask, dissolved, and diluted to volume. A 25.00-ml aliquot of this solution requires exactly 26.00 ml of iodine solution for titration to the starch end point. Calculate the percentage of As in a 1.0000-g ore sample that requires 15.00 ml of this same iodine solution for titration.

12. Calculate the percentage of copper in 1000 mg of a copper ore that requires 12.10 ml of $0.1000M$ sodium thiosulfate to titrate the iodine produced by the reaction of copper(II) with KI.

13. A 1.0000-g sample containing As_2O_3, As_2O_5, and an inert salt is dissolved and titrated in neutral solution to the starch end point with 20.00 ml of $0.2000N$ iodine. The resulting solution is made strongly acid and excess KI is added, releasing iodine. The iodine is titrated with 40.00 ml of $0.1500N$ sodium thiosulfate. Calculate the percentage of

As_2O_3 and the percentage of As_2O_5 (form wt = 229.84) in the sample.

14. Calculate the normality (or molarity) of an iodine titrant, 24.10 ml of which is required to react with 25.00 ml of $0.0300N$ ($0.0150M$) arsenious acid (H_3AsO_3).

15. Calculate the normality of an iodine titrant, if 40.00 ml of it are needed to oxidize 0.505 g of H_3AsO_3 (form wt = 125.94).

16. Calculate the normality (or molarity) of an iodine titrant, if 21.32 ml are needed to react with 139.3 mg of arsenic(III) oxide, As_2O_3 (form wt = 197.8).

17. Arsenic in an ore sample is sometimes present in elemental form, but when it is dissolved, it is oxidized to arsenic(III), which exists as H_3AsO_3 in solution. Calculate the percentage of elemental arsenic (form wt = 74.92) in 0.9000 g of an ore sample, if after dissolution the arsenic(III) requires 31.00 ml of $0.1100N$ iodine for titration.

Calculation of Vitamin Tablet Content

18. A vitamin C tablet, which contains 500 mg *according to the label* on the bottle, is titrated with 28.54 ml of $0.2000N$ iodine titrant. Calculate the actual mg of ascorbic acid (form wt = 176.12) in the tablet. Also, comment on the label's claim from the point of view of significant figures and from the point of view of the consumer.

19. A vitamin C tablet, which contains 500 mg *according to the label* on the bottle, is titrated with 43.54 ml of $0.130N$ iodine. Calculate the actual mg of ascorbic acid (form wt = 176.12) in the tablet. Also, comment on the label's claim from the point of view of significant figures and from the point of view of the con-consumer.

20. Two 250-mg vitamin C tablets are dissolved in a 100.0-ml volumetric flask. A 25.00-ml aliquot is removed and titrated with 28.44 ml of $0.500N$ iodine. Calculate the mg of ascorbic acid (form wt = 176.12) per tablet.

21. Four vitamin C tablets are dissolved in a 100.0-ml volumetric flask. A 10.00-ml aliquot is removed and titrated with 40.00 ml of $0.0300N$ iodine. Calculate the mg of ascorbic acid (form wt = 176.12) per tablet.

Calculations with Strong Oxidizing Agents

22. Potassium dichromate (form wt = 294.2) is a primary standard material. Calculate the normality and molarity of a solution made by dissolving 2.460 g of $K_2Cr_2O_7$ in a volumetric flask holding 500.0 ml solution. Assume reduction to the Cr(III) ion.

23. Calculate the percent purity of a sample of isopropyl alcohol (form wt = 60.1) if for 0.2000 g of the sample 31.00 ml of $0.2000N$ potassium dichromate is required to oxidize the isopropyl alcohol to acetone quantitatively.

24. A 1.0000-g sample of impure hydrogen peroxide is analyzed by titration with 28.10 ml of $0.1000M$ cerium(IV) titrant (Table 11-6). Calculate the percent purity of the hydrogen peroxide.

Devising New Methods

25. Hairwaving preparations contain organic mercaptans (thiols) as the active ingredient. Suggest an oxidation-reduction method for the determination of the mercaptan (RSH).

26. Outline an oxidation-reduction method for each of the following determinations:
    a. Cu in a CuS ore containing some Fe_2O_3 and CaO
    b. Ti in ferrotitanium (Fe-Ti)
    c. Cr in a stainless steel (Fe-Cr-Ni)
    d. Sodium hypochlorite (NaClO) in a laundry bleach
    e. U in a uranium-aluminum alloy. (The alloy dissolves in HNO_3, forming UO_2^{+2} and Al^{+3}.)

27. Many modern laundry bleaches are powdered oxidants containing positive halogen atoms. Suggest a way to determine the relative amounts of oxidizing power of competing brands of laundry products.

28. It is desired to titrate Fe^{+2} using one of the titrants listed below. Assume in each case that the second compound listed is present with the Fe^{+2} and that it must be decided whether it will cause an error by reacting with the titrant. (Recall that the reaction need only be *spontaneous* to cause an error.)
    a. $KMnO^4$ titrant and HCl
    b. $K_2Cr_2O_7$ titrant and HCl

c. Ce(IV) titrant and $Ce_2(SO_4)_3$
d. $K_2Cr_2O_7$ titrant and $SnCl_2$
e. Ce(IV) titrant and H_3AsO_3

Challenging Problems
29. An iron ore is mistakenly calculated to contain 10.00% Fe_2O_3 (form wt = 159.7) when the result should have been calculated as % Fe_3O_4 (form wt = 231.5). Calculate the percentage of Fe_3O_4 without knowing the sample weight or any other data from the analysis.
30. A silver indicating electrode and a standard hydrogen reference electrode are dipped into a saturated aqueous solution of silver bromide. The potential measured on the potentiometer (E_{meas}) is +0.434 V. From the standard potential of the $Ag°(s)/Ag^+$ electrode given below, calculate:
    a. The $[Ag^+]$ using the correct number of significant figures.
    b. The solubility product constant for silver bromide.

$$Ag^+ + e^- = Ag°(s) \qquad E° = +0.800 \text{ V}$$

31. Suppose that a new buret has been invented that has a relative accuracy of 0.1 ppt (1 part in 10,000), so that a *quantitative* reaction is

now defined as one in which no more than one part in ten thousand of A_{ox} and B_{red} remains when the titration reaction below is complete.

$$A_{ox} + B_{red} \rightleftharpoons A_{red} + B_{ox}$$

Assuming that both reactants undergo a one-electron change, calculate the minimum difference in standard potentials ($E°_{ox} - E°_{red}$) for a *quantitative* reaction.
32. The minimum value of $E°_{rxn}$, or $E°_{ox\,agt} - E°_{red\,agt}$, for a quantitative reaction in which both reactants undergo a one-electron change is +0.354 V at 25°C. What would this minimum value be at 83°C?
33. At any temperature, the general form of the Nernst equation is $E = E° + 2.303(RT/nF)\log[Ox]/[Red]$. At 25°C, the $2.303(RT/F)$ term has the value of 0.0591 V. Show that this term does indeed have the units of volts (V), using $R = 1.9872$ cal/°K mole or $R = 8.314$ J/°K mole.
34. A 1.0000-g sample of limonite iron ore ($2Fe_2O_3 \cdot 3H_2O$) is dissolved, reduced to Fe^{+2}, and titrated with 20.00 ml of 0.2000N cerium(IV). Calculate the percentage of $2Fe_2O_3 \cdot 3H_2O$ (form wt = 373.38) in the sample.

| **NOTES**

[1] D. C. Neckers, *J. Chem. Educ.* **50**, 166 (1973).
[2] D. A. Labianca, *Chemistry* **47**, 21 (Oct., 1974).
[3] J. W. Stevens, *Ind. Eng. Chem., Anal. Ed.* **10**, 269 (1938).
[4] D. N. Bailey, *J. Chem. Educ.* **51**, 488 (1974).
[5] C. G. King, *Ind. Eng. Chem.* **13**, 225 (1941).
[6] K. B. Wiberg, *Oxidation in Organic Chemistry*, Academic Press, New York, 1965, pp. 142–53.

PART TWO	# Some Instrumental Methods of Analysis

This part of the book is an elementary treatment of some selected instrumental methods of analysis. We do not attempt to cover every method because the volume of material necessary to do so would mean that little useful information could be included.

In Chapter 12 we introduce spectrophotometry via the Beer–Lambert law and certain spectrophotometric instruments. In Chapter 13 we discuss applications of qualitative and quantitative methods for ultraviolet, visible, and infrared absorption.

Chapter 14 covers mainly the use of fluorescence, although refractive index measurements and light scattering measurements are also included. Chapter 15 is an overview of both flame emission spectrometry and atomic absorption spectrometry.

In Chapter 16 we have consolidated all material on potentiometric measurements, pH measurements, and the use of ion-selective electrodes. Chapter 17 covers electrodeposition, coulometric methods, and polarography, as well as amperometric titrations.

12 Introduction to Spectrophotometry and Spectrophotometric Instruments

Holmes shut the slide across the front of his lantern and left us in *pitch darkness*. . . . The smell of hot metal remained to assure us that the *light* was still there ready to flash out at a moment's notice.

ARTHUR CONAN DOYLE
The Red-Headed League

Nearly any physical property can be the basis for an instrumental method of analysis. This chapter and the three succeeding chapters concern absorption, emission, refraction or scattering of some form of radiation or light. This chapter and Chapter 13 both deal with the absorption of radiation as measured on an instrument called the *spectrophotometer*. This chapter is an introduction and covers the law governing the absorption of radiation as measured on the spectrophotometer. Chapter 13 covers primarily the applications of ultraviolet, visible, and infrared spectrophotometry.

Absorption spectrophotometry, or just spectrophotometry, is the science of the measurement of the amount of electromagnetic radiation absorbed at a particular wavelength or set of wavelengths. If only visible radiation (light) is measured, the science is called colorimetry. Instruments capable of measuring the absorption of visible and ultraviolet or infrared radiation at any desired wavelength are called *spectrophotometers*. Instruments that have no capability for selecting a given wavelength and are capable of measuring only the absorption of light are called *colorimeters*. It is interesting that many of these instruments are adjusted to a zero reading as in the quotation above—the sample and detector are kept in "pitch darkness" with a sort of "slide" in front of the source, which remains "ready to flash out at a moment's notice."

12-1 | LIGHT AND OTHER ELECTROMAGNETIC RADIATION

The Nature of Electromagnetic Radiation

Light is but one form of electromagnetic radiation. Other types are ultraviolet (UV) radiation, infrared (IR) radiation, and X rays. As long as a beam of radiation is characterized by a wavelike motion and consists only of photons, it is considered electromagnetic radiation. We next consider the wave and photon characteristics of such radiation.

Waves and Photons

Light *appears* to have a dual nature. On one hand, it behaves as if it consisted of particles, which we call photons. But then it travels through space as if it consisted of waves. However, it has been said [1] that light *does not really* possess a dual nature because under no circumstance does it behave in both ways at the same time. In other words, when it is absorbed or emitted by matter, it behaves only as a photon; when it travels through space or through a prism, it behaves only as a wave and does not travel in a straight line as a particle does.

The relationship between these two natures is given by the following equation:

$$E = hv = \frac{hc}{\lambda} \tag{12-1}$$

where E = energy in joules (J)
v = frequency in number of waves per second, or s^{-1}
h = Planck's constant, 6.62×10^{-34} J-s
c = speed of light, 3.0×10^{17} nanometers/second, or $nm s^{-1}$
λ = wavelength of radiation in nanometers (nm)

Equation 12-1 specifies that radiation which passes a given point at a rate of a certain number of waves per second must have a particular energy. Thus each photon in that radiation must also have that particular energy.

Note that the *nanometer* (nm) is the unit of wavelength most commonly used by chemists. It is equal to 10^{-9} meter. (It is essentially the same as the millimicron (mμ) unit used before 1968.) We will characterize all radiation in terms of nanometers; some texts use wavenumbers ($1/\lambda$), which are symbolized as $\bar{v}$. Both the cm^{-1} and the μm^{-1} (reciprocal micrometer) are common wavenumber units.

The Electromagnetic Spectrum

We can now discuss the various regions of the electromagnetic spectrum in terms of wavelength and energy. (Recall from 12-1 that wavelength is inversely proportional to energy.) The various regions are summarized in Figure 12-1. At the top are microwaves and radio waves which have very large wavelengths and very small energies. In the middle is the very useful infrared-visible-ultraviolet radiation region.

The infrared region consists of two subdivisions—the middle infrared (or simply, infrared) region from 2500 to 15,000 nm and the near infrared from 750 to 2500 nm. The near infrared borders the visible region which extends from 380 to 750 nm. It is in this region that we perceive color. Ultraviolet radiation borders the higher-energy side of the visible region and extends from 185 to 380 nm. At the bottom of Figure 12-1 are the vacuum ultraviolet region (10–185 nm) and other high-energy radiation, such as X rays and gamma rays.

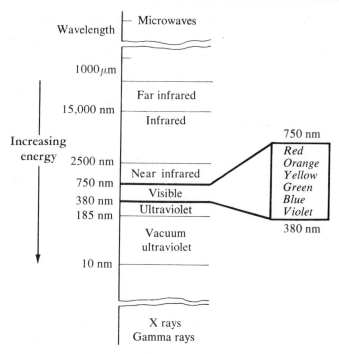

FIGURE 12-1. Types of radiation in the electromagnetic spectrum. Note that as the wavelength decreases, the energy of the radiation increases. (The micrometer (μm) wavelength unit is equal to 1000 nm.) At the right is shown the division of the visible region into its respective colors from 380 to 750 nm.

Ultraviolet, Visible, and Infrared Radiation

Ultraviolet radiation extends from 185 nm to 380 nm and is the highest-energy radiation of the three types to be discussed (Table 12-1). Its energy is such that wavelengths below 280 nm can seriously harm the eye as well as cause electronic energy changes. Visible radiation extends from 380 to 750 nm; it causes electronic energy changes responsible for the perception of color. (Visible radiation will be discussed in detail below.) Useful infrared radiation is divided into two regions—the near infrared and the infrared

TABLE 12-1. Wavelength and Energy Ranges for Ultraviolet, Visible, and Infrared Radiation

Type of radiation	λ range, nm	Energy range, Joules[a]
Ultraviolet (UV)	185[b]–380	1.07×10^{-18}–5.23×10^{-19}
Visible (light)	380–750	5.23×10^{-19}–2.65×10^{-19}
Near infrared	750–2500	2.65×10^{-19}–7.94×10^{-20}
Infrared (IR)	2500–15,000	7.94×10^{-19}–1.32×10^{-20}

[a] Energy calculated from **12-1** (1 Joule = 0.24 calorie).
[b] This is a limit imposed by optical materials and atmospheric absorption.

(Table 12-1). This radiation is of the lowest energy of the three; its energy is so low that it cannot cause electronic changes and so is invisible to the eye.

The range of the visible region is somewhat arbitrary, since it depends on the *average* eye response. At the ultraviolet-violet borderline, the response of the eye is limited by the absorption of radiation by the lens [2]. At the infrared-red borderline, the eye's response is limited by the absorption of radiation by water in the aqueous humor and other parts of the eye.

The response of the eye within the visible region also varies with wavelength, as shown in Table 12-2. This table lists the response of the eye relative to a value of 1.0000, or 100%, for yellow-green radiation at 555 nm. It can be seen from the table that the eye has a large response to green and yellow light. The color that a solution appears to the eye depends first of all on whether or not yellow and/or green (as well as yellow-green) are transmitted. If light of 495–590 nm is transmitted, the solution will appear yellow-green. If light of 495–570 nm is absorbed, the solution will appear yellow; if light of 550–590 nm is absorbed, the solution will appear primarily green; and so on. This response is that of the cones of the eye, not that of the rods [2].

The responses listed in Table 12-2 are for the *average* eye, assuming a *normal* flux, or flow, of photons striking the eye. It should not be assumed that one cannot see radiation below 380 nm just because the response listed is 0.0000. The eye does respond, but the magnitude of the response will have a value in the fifth or sixth decimal, so the eye will only see radiation with a higher than normal flux. Color has been discerned as low as 313 nm, and even X rays (~ 1 nm) have been reported to appear bluish to monkeys in certain experiments. The eye can also see high fluxes of red photons at above 750 nm; for example, the eye can observe the "red" flame emission of potassium at 766–69 nm [3].

Although Table 12-2 implies that the response of the eye varies smoothly with wavelength, this is not true. The eye behaves as though it has three independent color receptors. There are two major receptors—one in the yellow-green at 540 nm and the other in the yellow at about 580 nm. There is also a minor receptor at about 450 nm in the blue region.

Color blindness is apparently due to the absence of one of these receptors. In one class of color-blind subjects, the 540-nm receptor is missing as a result of a deficiency of a certain pigment. In another class of color-blind subjects, the 580-nm receptor is missing as a result of a deficiency of a different pigment in the eye [2].

TABLE 12-2. Response of the Eye to Various Colors

Color	Wavelength	Response of eye in middle of range[a]
Ultraviolet	185–380 nm	0.0000 (at 380 nm)
Violet	380–450 nm	0.0022
Blue	450–495 nm	0.10
Green	495–550 nm	0.83
Yellow-green	550–570 nm	0.995
Yellow	570–590 nm	0.87
Orange	590–620 nm	0.57
Red	620–750 nm	0.10
Near infrared	750–2500 nm	0.0001 (at 750 nm)

[a] Relative to a maximum response of 1.0000 at 555 nm.

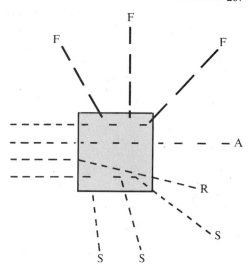

FIGURE 12-2. Hypothetical representation of four different interactions of photons with sample solution as follows:

F = **Photons are absorbed and emitted as longer wavelength fluorescence photons (note longer lines representing photons) in all directions.**

A = **Photons are partially absorbed; the remaining photons travel straight through the solution.**

R = **Photons are bent or refracted in a different direction, but none are absorbed and their wavelength remains the same.**

S = **Photons are scattered in all directions; none are absorbed and their wavelength is the same as that before scattering.**

When a beam of radiation is focused on a sample solution, various things can happen (Fig. 12-2). One is absorption (A) of some of the photons. The other three interactions pictured are such that the resulting photons cannot be measured by a detector set at the wavelength of the original photons. These are fluorescence (F), which produces photons of longer wavelengths; refraction (R), in which the photons are bent away from their original pathway; and scattering (S), in which the photons are scattered away in all directions at the same wavelength as the original photons. Not shown here is the special case of emission of photons by gaseous atoms (Ch. 15).

Absorption of Photons

The absorption of electromagnetic radiation can be explained in terms of the interaction of photons with electrons of a chemical species or with the bond vibrations of this species. Although such processes are complicated, they can be briefly described in general terms. First, it is necessary to define the ground state and the excited state or states of a chemical species with reference to the electrons and to the bond vibrations. An electronic ground state is one in which all of the electrons are in their most stable orbital. A vibrational ground state is one in which all of the bonds possess the smallest possible amount of vibrational energy. An electronic excited state is one in which an electron occupies a higher energy orbital than it does in the ground state. A vibrational excited state is one in which one bond (or more) has a larger amount of vibrational energy than it does in the ground state. Although there is only one electronic ground state and one vibrational ground state, there may be a number of excited states.

To understand the absorption of photons by electrons, consider the usual one-photon absorption by the gaseous sodium atom. The ground state of gaseous sodium, Na_0, absorbs appreciable amounts of radiation at both 330 nm and 589 nm. Since 589-nm radiation is the lowest energy radiation absorbed, it promotes the atom to the first, or lowest energy, excited state, Na_1. Absorption at 330 nm is assumed to promote the atom to the second excited state, Na_2. These processes can be represented as follows:

$$Na_0 + photon_{589\,nm}(10^{-15}\,s) \longrightarrow Na_1$$

$$Na_0 + photon_{330\,nm}(10^{-15}\,s) \longrightarrow Na_2$$

The time of 10^{-15} second for either process is the time required for a sodium electron to "jump" from its most stable orbital to a higher energy orbital. In terms of orbitals, the Na_1 and Na_2 excited states differ from one another in that the excited electron occupies a $3p$ orbital in the Na_1 state and a $4p$ orbital in the Na_2 excited state.

The absorption of photons by molecules or ions in solution is more complicated than that by atoms in the gaseous state. There is still just one ground state, but each electronic excited state is split into many (n) sublevels, each of slightly different energy. Each sublevel will be populated by absorption of photons of slightly different energy. For example, the absorption of light by oxyhemoglobin in the ground state, symbolized as $[Hb(O_2)_x]_0$, can be represented as follows:

$$[Hb(O_2)_x]_0 + n \text{ photons}_{560-620\,nm} \longrightarrow n[Hb(O_2)_x]_1 \qquad (12\text{-}2)$$

where the symbol $n[Hb(O_2)_x]_1$ represents the many sublevels of the first excited state of the oxyhemoglobin molecule. Of the molecules in this first excited state, there will be molecules whose excess energies range from a low that corresponds to the energy of 620-nm light to a high that corresponds to the energy of 560-nm light.

To rationalize the well-known red color of oxyhemoglobin with the previous absorption process, we must examine the *absorption spectrum* of oxyhemoglobin. This is a plot of the amount of light absorbed, or absorbance, versus wavelength and is shown in Figure 12-3. It can be seen that oxyhemoglobin absorbs nearly all visible radiation up to 620 nm; above this wavelength, it transmits most, but not all, of the red light (plus a small amount of orange light).

We also see in Figure 12-3 that oxyhemoglobin has three absorption *maxima*, or *bands*. The highest energy band is not completely shown but is close to 450 nm. The most intense absorption band is at 540 nm; the lowest energy band is at 575 nm. Since absorption at 575 nm gives the first excited state of oxyhemoglobin, absorption at 540 nm gives the second excited state, and absorption below 450 nm gives the third excited state of oxyhemoglobin. The energy differences between these states may be described in the following manner. As long as any two molecules are in the same excited state, they must possess the same amount of excess electronic energy and can vary only in the amount of excess vibrational energy possessed by their bonds. If any two molecules are in different excited states, they will possess different amounts of excess electronic energy and probably also different amounts of excess bond vibrational energy.

It should be emphasized that although **12-2** describes the interaction of n photons with hemoglobin, the equation is only describing a one-photon *absorption* process. No more than one photon is absorbed by a single hemoglobin molecule at a time. The reason for this is that most light (photon) sources are not very intense, so only one photon can be absorbed from the source during the 10^{-15} second needed to reach the excited state. If a laser, such as a ruby laser, is used as a source, the very concentrated beam of photons emitted by the laser can achieve a two-photon absorption process.

(A two-photon absorption process has also been postulated to show how photosynthesis can occur via absorption of relatively low energy redlight [4]. Red photons have such low energies that scientists have been unable to explain satisfactorily how a relatively high energy process such as photosynthesis could utilize a single red photon per chlorophyll molecule. Absorption of two red photons would of course double the energy available.)

It should be mentioned in conclusion that a molecule such as hemoglobin does not remain in the excited state very long. It returns to the ground state in less than 10^{-9} second by one of two processes. Once it has reached the ground state, it can again absorb a photon and become excited.

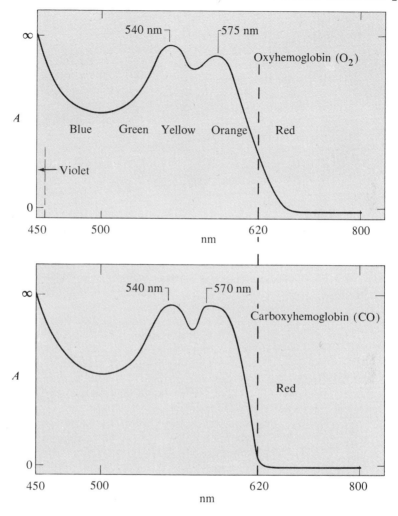

FIGURE 12-3. The absorption spectra of oxyhemoglobin (top) and carboxyhemoglobin (bottom). The left axis is absorbance, *A*, or the amount of light absorbed. Note that when hemoglobin is combined with oxygen, it absorbs almost all colors except most of the red, so it appears red. When hemoglobin is combined with carbon monoxide, it does not absorb any of the red, so it appears redder or pinker than oxyhemoglobin.

12-2 | MEASUREMENT OF CONCENTRATION: THE BEER–LAMBERT LAW

The spectrophotometer and the colorimeter are instruments used to measure the amount of absorption of UV, visible, or IR radiation by solutions of chemical substances. The basic law governing such measurements is the combined Beer–Lambert law. Before we discuss this law, it is helpful to visualize the flow of photons through the spectrophotometer cell containing a chemical solution.

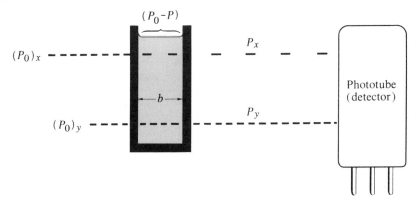

Figure 12-4. **A diagram of the cell of a spectrophotometer containing a solution that absorbs radiation at wavelength x, but not at wavelength y. The internal cell length is designated as b. The radiant power of either beam striking the solution is P_0; the radiant power of either beam transmitted by the solution (and striking the detector) is P; the radiant power of the absorbed radiation is $(P_0 - P)$.**

To begin with, a solution containing a chemical substance is poured into a spectrophotometer cell and inserted into the spectrophotometer. Various wavelengths of radiation are directed at the solution in the cell. Figure 12-4 pictures how beams of photons of two different wavelengths pass through the solution. At a certain wavelength, x, the solution will absorb some of the photons striking it. As shown in the figure, about half of the photons are absorbed and half pass through. The number of photons in a beam passing a given point per unit time (one second, for example) is called the *radiant power* of the beam. The radiant power of the beam striking the solution is P_0, and that transmitted by the solution is P. At wavelength x (Fig. 12-4), P_0 is about twice P. Radiation at other wavelengths, such as wavelength y, is not absorbed by the solution. Note that P_0 is the same as P at this wavelength.

If you understand the difference between P_0 and P, you are ready to read about absorbance and transmittance, the two sets of units used on the scale of the spectrophotometer and the colorimeter.

Absorbance, Transmittance, and Percent Transmittance

The terms covered in this section are absorbance, A; transmittance, T; and percent transmittance, $\%T$. The amount of radiation absorbed by a solution is measured in *absorbance units*, ranging from 0.00 to 2.0. The fraction of radiation transmitted is measured in percent transmittance (or units of transmittance), ranging from $100\%T$ ($T = 1.00$) to $0\%T$ ($T = 0.00$). Both appear on the spectrophotometer scale; the key scale values are as follows:

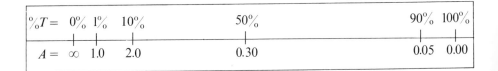

Both the absorbance and percent transmittance readings can be better understood and related to one another by using the concept of the radiant power of the beam of photons striking a solution and being transmitted by a solution.

First, consider a solution of an absorbing species shown in a spectrophotometer cell in Figure 12-4. Let the radiant power of the beam in photons/second striking the cell be P_0. The beam then passes through the solution, and part of its radiant power is absorbed by ions or molecules in the solution. Let the radiant power of the beam leaving the cell be symbolized as P. (P is less than or equal to P_0.) Absorbance is then defined as follows:

$$A = \log P_0 - \log P = \log \frac{P_0}{P} \qquad (12\text{-}3)$$

Note that absorbance is a *logarithmic difference* and is best understood in the mathematical form of the middle terms. For mathematical manipulations, the right-hand term is more convenient, as will now be illustrated.

Mathematically stated, transmittance is defined as a fraction.

$$T = \frac{P}{P_0}$$

This is termed *transmittance*. From **12-3** it is easily shown that

$$A = -\log T \qquad (12\text{-}4)$$

Since percent transmittance is simply $100 \times T$, absorbance may be calculated from $\%T$ in a manner similar to **12-4**.

$$A = -\log(\%T/100) \qquad (12\text{-}5)$$

Significant Figures and Spectro- photometer Readout

The number of justifiable significant figures for a spectrophotometer reading depends on whether the spectrophotometer has a *scale readout* or a *digital readout*. Rules for recording readings from a scale are in the box.

Absorbance, scale readout
0.00–0.99 units: Two significant figures to the right of the decimal. Since this is a log term, *all* zeroes to the right of the decimal are significant. Thus readings of 0.05 and 0.00 both have two significant figures.
1.0–2.0 units: One significant figure to the right of the decimal.

% Transmittance, scale readout
Two significant figures to the left of the decimal about 9%. One decimal to the right is sometimes estimated but is uncertain by ± 0.3 to $0.4\%T$.

To calculate absorbance from percent transmittance, or vice versa, using **12-4** or **12-5**, the rules in Section 3-1 for log terms must be used. Thus absorbance must be expressed with the same number of significant figures to the right of the decimal as the total number of significant figures in the transmittance or percent transmittance readings.

If you use a digital spectrophotometer, record your readings as in the box below.

0.0–3.0 + absorbance units: All digits are significant. All zeroes to the right of the decimal are significant.

0.0–99.9% transmittance units: All digits are significant. Usually only one digit to the right of the decimal is displayed. (The zero in 9.0%T on a digital spectrophotometer is significant even though it is not on a scale readout spectrophotometer.)

EXAMPLE: Calculate the absorbance corresponding to a 9.0%T reading if the reading is from (a) a digital readout and (b) a scale readout.

Solution to a: Using **12-5,**

$$A = -\log(9.0\%/100) = -\log 9.0 - \log 10^{-2}$$

$$A = -0.95_4 + 2 = 1.04_6$$

Solution to b: The zero in 9.0%T is not significant on a scale readout; therefore, $A = 1.0_{46}$ or 1.0 (one significant figure to the right of the decimal).

SELF-TEST

Answers:

$T = 0.20$
$\%T = 20\%$
$A = 0.70$

$A = 0.39_8$
$A = 1.3$
$A = 2.0$ (last zero in 0.010 not significant)
$A = 1.1$ (last zero in 0.080 not significant)
$A = 0.04_1$ (two sig. figs.)

17. Absorbance, Transmittance, and Radiant Power

Directions: Cover the answers in the left column before beginning the test. Check your answers with those given at the left. If necessary, review significant figures for logs in Section 3-1.

A. Yellow (575-nm) light with a radiant power of 2.00×10^{14} photons/second strikes a solution of hemoglobin; the solution transmits yellow light with a radiant power of 4.0×10^{13} photons/s. Calculate:
   a. Transmittance
   b. % Transmittance
   c. Absorbance

B. The absorption of 575-nm light by various solutions of hemoglobin is read as percent transmittance and transmittance from a spectrophotometer with a *scale readout*. Decide whether all the digits are significant and then calculate the absorbance to the correct number of significant figures.
   a. $\%T = 40\%$; $A =$ _____
   b. $\%T = 5\%$; $A =$ _____
   c. $T = 0.010$ $(1.0\%T)$; $A =$ _____

   d. 0.080 $(8.0\%T)$; $A =$ _____

   e. $\%T = 91\%$; $A =$ _____

C. The solutions in the problem above are read in a spectrophotometer with a *digital readout*. Calculate the

absorbance to the correct number of significant figures if
the percent transmittance readings are as follows:

$A = 0.398$ a. $\%T = 40.0\%$; $A = $ _____

$A = 1.30$ b. $\%T = 5.0\%$; $A = $ _____

$A = 2.00$ c. $\%T = 1.0\%$; $A = $ _____

$A = 1.09$ d. $\%T = 8.0\%$; $A = $ _____

$A = 0.041$ e. $\%T = 91.0\%$; $A = $ _____

D. The following absorbance values were read from a
spectrophotometer with a *scale readout*. Calculate the
transmittance, T, to the correct number of significant
figures.

$T = 0.02$ a. $A = 1.7$; $T = $ _____

$T = 0.89$ (0.05 has two sig. figs.)   b. $A = 0.05$; $T = $ _____

$T = 0.25$ c. $A = 0.60$; $T = $ _____

The Beer–Lambert Law

The law governing the relation between concentration of a solution and the amount of
light or radiation absorbed by it is the Beer–Lambert law (sometimes inaccurately called
Beer's law). It relates the absorbance to the concentration of the absorbing species and
to the path length of the solution through which radiation must pass. This law states
that if the concentration of the absorbing species increases, the absorbance must increase.
It also states that if the path length of the solution increases, the absorbance must
increase.

The symbol usually used for the path length of a solution is b. As shown in Figure
12-4, b is the internal cell *length*; it can also be the internal tube *diameter* if a test tube
is used instead of a square cell. The symbol usually used for concentration is c.

It is important to recognize that c is the equilibrium concentration of a single
absorbing chemical species, not the total of all. Thus, in a mixture of $FeSCN^{+2}$
and $Fe(SCN)_2{}^+$, c is the concentration of $FeSCN^{+2}$ and not of both ions.

If the concentration, c, of the compound is stated in terms of molarity, then the Beer–
Lambert law gives the relation between A and c as follows:

$$A = \varepsilon bc$$

where b is given in cm and ε is a proportionality constant known as the *molar absorptivity*.
The dimensions of ε are either $cm^{-1}M^{-1}$ or liter $\cdot$ mole^{-1} cm^{-1}. Hypothetically
speaking, ε is the absorbance of a $1M$ solution in a 1-cm cell.

If the concentration is given in units other than molarity, then the Beer–Lambert law
is stated in the following form:

$$A = abc$$

where the proportionality constant is symbolized as a and is called the *absorptivity*.
The dimensions of a depend on the concentration unit used. If c is in g/liter, then the
dimenstions of a are liter$\cdot$g^{-1} cm^{-1}.

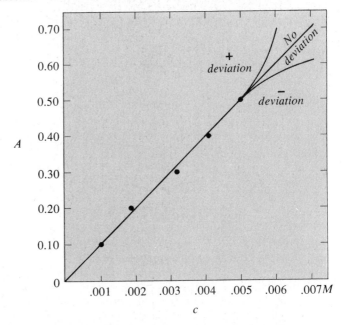

FIGURE 12-5. A typical Beer–Lambert law plot. Below 0.005M, the plot of A vs. c obeys the Beer–Lambert law.

If A is plotted against c, a straight line passing through the origin is the usual result (Fig. 12-5). Such a system is said to obey the Beer–Lambert law. If the line curves upward at some point, it is said to exhibit a *positive* deviation from the law. If it curves downward at some point, it is said to show a *negative* deviation from the law. Both situations are shown in Figure 12-5. The slope of the straight line portion of the curve in Figure 12-5 will be ε if a 1-cm cell is used.

It should be stressed that ε is a constant for a *specific* wavelength only. For example, the $Co(H_2O)_6^{+2}$ ion has an ε of 10 at 530 nm. If the wavelength is changed to 500 nm, the ion has a different ε. The ε at 500 nm is, in fact, lower than that at 530 nm, because the peak of the absorption band of $Co(H_2O)_6^{+2}$ is at 530 nm.

Frequently, values of ε are reported in logarithmic form. To evaluate or compare such values, one should convert them to semiexponential form.

EXAMPLE: The amino acid phenylalanine has a log ε of 2.30 at 258 nm. What is the semiexponential value of ε?

Solution: Convert the characteristic and mantissa separately to antilogs (numbers).

$$\varepsilon_{258nm} = \text{antilog of } 2.30 = (\text{antilog of } 0.30) \times (\text{antilog of } 2) = 2.0 \times 10^2$$

Beer–Lambert Calculations   The simplest situations involve the calculation of one variable, such as A or c, given the other three variables.

EXAMPLE: The amino acid phneylalanine, $C_6H_5CH(NH_2)CO_2H$, has a log ε of 2.30 at 258 nm. Calculate the absorbance of a $2.00 \times 10^{-3}M$ solution in a 1.00-cm cell.
Solution: Use the Beer–Lambert law.

$$A = \varepsilon bc = (2.0 \times 10^2 M^{-1}\, cm^{-1})(1.00\, cm)(2.00 \times 10^{-3}M)$$

$$A = 0.40 \text{ (2 sig. figs.)}$$

Somewhat more complicated calculations arise if the concentration of an unknown is to be found by comparing the absorbance of the unknown (u) with the absorbance of a standard (s) solution. The general equation is derived by writing a ratio of the Beer–Lambert law for each solution.

$$A_u = \varepsilon b_u c_u$$

$$A_s = \varepsilon b_s c_s$$

Then, assuming that

$$b_u = b_s$$

$$c_u = c_s \frac{A_u}{A_s} \qquad\qquad\qquad \textbf{(12-6)}$$

EXAMPLE: The absorbance of a $1.0 \times 10^{-4}M$ solution of the amino acid tyrosine at 275 nm is 0.200. The absorbance of an unknown solution at 275 nm is 0.500. Calculate the concentration of tyrosine in the unknown.
Solution: Use **12-6** to calculate c.

$$c_u = 1.0 \times 10^{-4}M \frac{(0.500)}{(0.200)} = 2.5 \times 10^{-4}M$$

A Beer–Lambert plot, such as in Figure 12-5, can also be used to find the concentration of an unknown. This will be discussed in Chapter 13.

SELF-TEST | **18. Beer–Lambert Law Calculations**

Answers:

| | *Directions:* Cover the answers in the left column before beginning the test. Work one problem at a time and check your answers before proceeding to the next problem. If necessary, review significant figures in Section 3-1. |

A. Given the value for molar absorptivity, calculate the log ε in each case.

3.30
-1.67_7
1.00

a. $\varepsilon = 2.0 \times 10^3$
b. $\varepsilon = 0.021$
c. $\varepsilon = 10$

B. Given the value for log ε, calculate ε in each case.

$2._5$
$1.00_9 \times 10^1$
4×10^{-1}

a. log $\varepsilon = 0.4$
b. log $\varepsilon = 1.004$
c. log $\varepsilon = -0.4$

C. Calculate the designated variable using the Beer–Lambert law. Assume that $b = 1.000$ cm. (Use only Rule 2 on p. 38 for multiplication and division.)

0.030

a. Given $\varepsilon = 1.0 \times 10^4$ and $c = 3.00 \times 10^{-6}M$, calculate A.

0.0_2 (no sig. figs.)

b. Given $\varepsilon = 10^4$ and $c = 2.00 \times 10^{-6}M$, calculate A.

0.060

c. Given $\log \varepsilon = 4.30$ and $c = 3.00 \times 10^{-6}M$, calculate A.

$\varepsilon = 2.0 \times 10^4$, $\log \varepsilon = 4.30$

d. Given $A = 0.400$ and $c = 2.0 \times 10^{-5}M$, calculate ε and $\log \varepsilon$.

D. The aromatic amino acid phenylalanine is to be determined by comparing the absorbance of the unknown solution with that of a single standard solution of phenylalanine. In each case, calculate the concentration of phenylalanine in the unknown, assuming all absorbance measurements are at 259 nm. Assume $b_u = b_s$ unless otherwise stated.

$1.5 \times 10^{-2}M$

a. The standard concentration is $1.0 \times 10^{-2}M$, and its absorbance is 0.200. The absorbance of the unknown is 0.300.

$1.50 \times 10^{-2}M$

b. The standard concentration is $2.00 \times 10^{-2}M$, and its absorbance is 0.400. The absorbance of the unknown is 0.300.

$1.8 \times 10^{-2}M$

c. The standard concentration is $1.0 \times 10^{-2}M$, and its absorbance in a 2.00-cm cell is 0.400. The absorbance of the unknown in a 1.00-cm cell is 0.360.

12-3 | SPECTROPHOTOMETERS AND COLORIMETERS

The spectrophotometer is an instrument that can measure the amount of visible, UV, and/or IR radiation absorbed by a solution at a given wavelength. The colorimeter is an instrument that essentially measures only light, using a filter to select a range of wavelengths. There are other less important differences between the many types of these instruments, some of which will be discussed in this section. First, we will describe the major components of both types of instruments and then some specific examples of these instruments.

The Major Components

The four major components of a spectrophotometer are the source, the monochromator, the cell, and the detector (Fig. 12-6). We will begin with the first component, the source.

The Source

The most common sources of radiation in spectrophotometers are listed in Table 12-3. Beginning with the ultraviolet region, the *deuterium discharge lamp* is currently the preferred source because of its stability, even though it has only a moderate output of photons. It emits a continuum of radiation from about 185 to 400 nm from an electric arc passing through a tube containing deuterium gas. It has replaced the *hydrogen discharge lamp*, which was used in the first commercial ultraviolet spectrophotometers.

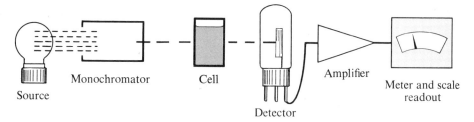

FIGURE 12-6. A diagram of a simple spectrophotometer. The dashes represent numbers of photons emitted by the source.

Where a high output of ultraviolet photons is needed, as in fluorescence instruments (Ch. 14), a xenon arc or mercury arc lamp is used.

From about 320 to above 750 nm, *tungsten lamps* are used as sources. The usual tungsten lamp has a glass envelope, limiting its lower wavelength to about 320 nm. Its photon output depends on temperature—as the temperature increases, the photon output increases and the wavelength at which the maximum output occurs decreases. At about 3000°K the wavelength of maximum output is near 1000 nm (Fig. 12-7). A modern improvement in the tungsten lamp is the *tungsten halogen lamp*, with a quartz envelope. It has a longer lifetime and can readily be used for wavelengths as low as 250 nm, or even 220 nm at a higher temperature setting. Either of these tungsten sources is useful for visible region spectrophotometers or colorimeters.

For infrared measurements, either the *Nernst glower* or the *globar* is used, even though both have lower photon outputs than any of the other sources in Table 12-3. The Nernst glower is used in most standard infrared instruments, since it has a higher output than the globar throughout most of the 2–15 μm infrared region. Its wavelength of maximum photon output is 1.5 μm (1500 nm); that of the globar is 1.9 μm (1900 nm). Both of these sources consist of electrically heated rods, rather than a filament in an envelope as in the tungsten lamp.

TABLE 12-3. Sources of Ultraviolet, Visible, and Infrared Radiation

Source	Wavelength Range	Photon Output
Hydrogen discharge lamp	185–375 nm	Weak; should not be used above 360 nm
Deuterium discharge lamp	185–400 nm	Moderate (3 to 5 times that of hydrogen lamp)
Tungsten halogen lamp	250– <2000 nm	No output below 320 nm unless envelope is quartz
Tungsten lamp	320–2500 nm	Best output from 400 to 1200 nm
Nernst glower	1000–35,000 nm (1–35 μm)	Weak output above 10 μm
Globar	1000–50,000 nm (1–50 μm)	——

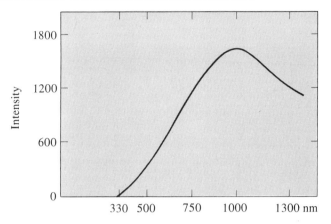

FIGURE 12-7. The intensity of a tungsten filament source at various wavelengths. The intensity units on the left axis are realtive.

The Mono-chromator
Radiation from the source passes through an entrance slit, which defines its path, and then enters the monochromator (Fig. 12-6). The key element of a monochromator is the dispersion device, or wavelength-selection device. We will discuss three types: the optical filter, the prism, and the diffraction grating.

The *optical filter* is either a small piece of glass containing a colored chemical that transmits over a range of wavelengths or a dielectric material, such as magnesium fluoride, sandwiched between two semitransparent metallic films (interference filter). Figure 12-8 is a plot of the transmittance of a typical violet glass filter at various wavelengths. Such a *band-pass filter* is a poor dispersion device for spectrophotometers,

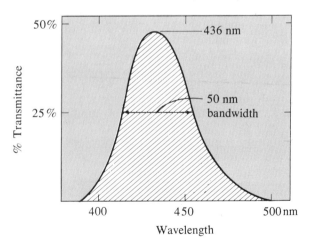

FIGURE 12-8. A plot of percent transmittance against wavelength for a typical violet filter (color specification 47B). The 50-nm bandwidth refers to the width of the band at half of the height of the curve at maximum transmittance.

because its bandwidth of 50 nm is much larger than required for monochromatic radiation ($\leqslant 1$ nm). If the bandwidth of a dispersion device is significantly larger than 1 nm (~ 20 nm), then deviations from the Beer–Lambert law, which is defined for monochromatic radiation, can become significant. For this reason, the interference filter is preferred when bandwidths from 1 to 20 nm are desired in a filter. Both the band-pass filter and the interference filter are often used in fluorescence analysis (Ch. 14), but cannot be used to obtain an absorption spectrum.

The *prism* functions as a dispersion device by refracting radiation. Different wavelengths of radiation are bent to different degrees as they travel through a prism; for example, violet-blue light is bent more than red light. A greater degree of dispersion occurs at shorter wavelengths. The prism is thus a good dispersion device in the ultraviolet and the violet-blue regions, and it was formerly used in spectrophotometers to obtain absorption spectra. When the wavelength dial of a spectrophotometer with a prism is set in the ultraviolet region, for instance at 300 nm, it is possible to obtain monochromatic radiation (bandwidth < 1 nm). However, radiation at long wavelengths, especially in the red (620–750 nm) and the near infrared, exhibits a large bandwidth and deviates seriously from monochromaticity.

The *grating*, or diffraction grating, is a piece of aluminized glass (original grating) or plastic (replica grating) that has a large number of accurately spaced lines ruled on the surface. When radiation strikes the unruled portion of its surface, most wavelengths are destroyed (via a destruction interference process) and do not reach the cell. The selected wavelength and a few wavelengths on either side of the selected wavelength undergo constructive interference, resulting in a related series of wavelengths, known as first-order diffraction (same wavelength as the selected wavelength), second-order diffraction (twice the wavelength of the selected wavelength), and so on.

The grating is a good dispersion device, and thus it is possible to utilize it in a spectrophotometer to obtain an absorption spectrum. The desired wavelength is obtained simply by setting the wavelength dial. An advantage of the grating over the prism is that the bandwidth is constant (for example, 20 nm) over the entire range of wavelengths. Because of this advantage, most spectrophotometers employ gratings instead of prisms.

The Cell   After leaving the monochromator through an exit slit, radiation of only one wavelength or in a band of several wavelengths strikes an absorbing liquid sample in a cell or cuvette (Fig. 12-6). Construction of these cells varies, depending on whether UV, visible, or IR radiation is measured. For UV or visible region measurements, the cell is commonly a rectangular type with a 1.00-cm length and width so the internal cell length, b, is always 1.00 cm (Fig. 12-4). Cells for the IR region are different and will be discussed separately. As will be stressed, a cell used for a particular spectral region must be as transparent as possible to that region.

Cells used in the visible region can be made of optical quality borosilicate glass. Note in Figure 12-9 that the transmittance of this material is minimal at wavelengths below 310 nm, so it cannot be used for most of the UV region. Some colorimeters or inexpensive spectrophotometers use carefully selected cylindrical cells (test tubes) made of borosilicate glass or soft glass which transmits only to about 350 nm (Fig. 12-9). To ensure a reproducible cell length, these cylindrical cells are usually marked so that they can be inserted into the cell holder always in the same position.

Cells used in the UV region must be made of an optical "glass" that is free of constituents in ordinary glass that absorb UV radiation. Fused standard silica may be used to give transparency from 1000 to 220 nm and is satisfactory for most UV measurements.

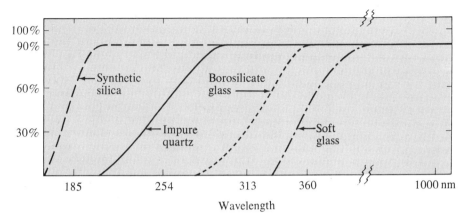

FIGURE 12-9. Plot of transmittance of various optical materials against wavelength.

Ordinary quartz is impure and is not as useful for UV work (Fig. 12-9), but pure quartz or synthetic silica (Fig. 12-9) may be used down to 180 nm.

Cells used in the IR region are a special type of sealed cell unit. The optical material is either polished sodium chloride or polished potassium bromide, both in the form of transparent "windows." Two of these windows are joined to an amalgamated lead or Teflon spacer and sealed together with metal plates having apertures somewhat smaller than the windows. Glass and quartz cannot be used for IR measurements because they do not transmit over the entire IR region. Cells made from sodium chloride or potassium bromide must be handled carefully, since these materials are soft, easily scratched, and easily damaged by water.

The Detector   After the radiation has passed through the sample cell, the unabsorbed part impinges on some type of detector (Fig. 12-6). The type of detector varies depending on the wavelength region (and the cost of the instrument), but usually the detector converts radiant energy into electrical current or voltage which can easily be measured.

In single-beam instruments, the cell is first filled with the pure solvent and the $100\%T$ ($A = 0$) setting is established at the desired wavelength. Then the solvent is replaced with the sample, and the detector measures the amount of radiation absorbed compared to the pure solvent. In double-beam instruments, two cells (for the solvent and the sample) are used, eliminating the need to change cells. Six types of detectors are the following:

1. The vacuum photodiode (phototube)
2. The semiconductor photodiode
3. The photomultiplier tube
4. The lead sulfide photoconductive cell
5. The bolometer
6. The thermocouple

We will discuss each detector briefly, noting the photosensitive element, the optical range, and whether current, voltage, or resistance is measured.

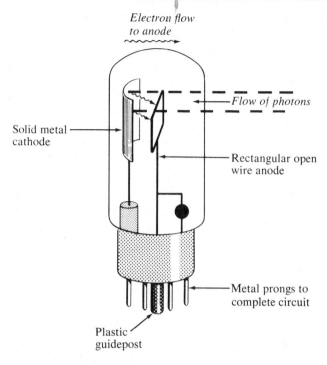

Electron flow to anode

Flow of photons

Solid metal cathode

Rectangular open wire anode

Metal prongs to complete circuit

Plastic guidepost

FIGURE 12-10.   A diagram of a phototube.   Photons of light flow past the open wire anode, strike the cathode, and are converted to electrons which are attracted to the positive anode.   (In some phototubes the anode is a single wire.)

The Vacuum Photodiode (*Phototube*).   The vacuum photodiode has a glass (or quartz) envelope enclosing an anode and a photocathode (cathode) as the photosensitive element (Fig. 12-10).   When a high voltage is impressed on the photocathode, photons striking it can displace electrons.   In the air these electrons would react with oxygen, but in the evacuated photodiode they are attracted to the positive anode, causing a measurable increase in electrical current.   *Blue* phototubes cover the 320–600 nm range (glass envelope) or the 200–600 nm range (quartz envelope); *red* phototubes cover the 600–1000 nm range.

The Semiconductor Photodiode.   The semiconductor photodiode is a solid-state detector consisting of a *p-n* junction, typically a small (1.5 × 1 cm) silicon chip on a circuit board.   It also has a voltage source hooked into the circuit (Fig. 12-11).   Photons strike the semiconductor material (silicon, etc.), creating electron-hole pairs; this causes an increase in electrical current, just as in the phototube.   To form an electron-hole pair, photons must possess a certain minimum energy (band-gap energy), which is characteristic of the semiconductor material used.   This limits the long wavelength response of silicon, for example, to 1170 nm.   The usual silicon photodiode covers the 350–1170 nm range; a "UV-enhanced" silicon photodiode covers the 200–1170 nm range.

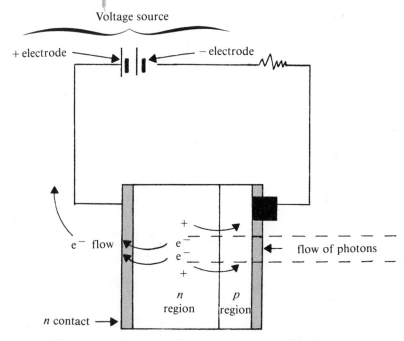

FIGURE 12-11. A diagram of a semiconductor photodiode, showing photons forming electron-hole pairs. The electrons flow through the *n* region to the positive electrode of the voltage source in the circuit.

The Photomultiplier Tube. The photosensitive element of the photomultiplier tube consists of a quartz (or glass) envelope enclosing an anode, about 10 dynodes, and a photocathode (Fig. 12-12). As in the vacuum photodiode, photons strike the photocathode, releasing electrons to the first dynode, maintained at a less negative voltage than the cathode (Fig. 12-12). Each electron from the photocathode striking the first dynode generates up to 4 or 5 "secondary" electrons. This multiplication of electrons continues until 10^6 to 10^8 electrons for every original one ultimately strike the anode. Because of this internal amplification process, no external amplifier is needed. The standard blue-sensitive photomultiplier tube covers the 200–600 nm range; others are available to cover different ranges.

The Lead Sulfide Photoconductive Cell. The lead sulfide photoconductive cell consists of a solid-state lead sulfide *p-n* junction, but without the voltage source as used in the photodiode. Photons strike the lead sulfide, creating electron-hole pairs and causing a measurable voltage to flow. The range that lead sulfide covers is from about 700 to 3000 nm, making it very useful for the near infrared region.

The Bolometer and the Thermocouple. Since IR radiation does not possess enough energy to ionize electrons from a photocathode, thermal detectors such as the bolometer and the thermocouple must be used instead to measure it. The bolometer is a temperature-sensitive resistor consisting of two thin foils. IR radiation strikes one of the foils

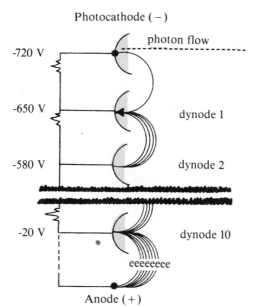

Photocathode (−)

photon flow

-720 V

-650 V — dynode 1

-580 V — dynode 2

-20 V — dynode 10

eeeeeeee

Anode (+)

FIGURE 12-12.   Diagram of a photomultiplier tube with 10 dynodes.

while the other is shielded to act as a reference; this causes an imbalance in the Wheatstone bridge circuit of the bolometer. The bolometer is thus a resistance device. The thermocouple is a blackened noble-metal foil welded in two thermoelectric substances and enclosed in a rectangular housing to minimize heat losses. Absorption creates a voltage. The thermocouple covers a range from 1 μm to beyond 30 μm, whereas the bolometer covers from 1 μm to 15 μm or 30 μm.

Types of Colorimeters and Spectrophotometers

Various types of colorimeters and spectrophotometers are available; most commercially marketed instruments fall into one of the general types below. The colorimeters are the least expensive and are, strictly speaking, limited to the visible region. Recording spectrophotometers of the UV-visible type or IR type are the most expensive.

Filter Colorimeters (Photometers)

A filter colorimeter, or photometer, measures light absorption by using different filters rather than a monochromator. This type of instrument is used for routine measurements by unskilled operators. The filter is frequently inserted into the instrument before the operator uses it. Glass test tubes are used for cells, and a simple phototube is used for detection. The Coleman Nepho-Colorimeter and the Klett-Summerson Colorimeter are examples of this type. Although they are inexpensive, they are rugged and dependable for routine work.

"Visible" Spectrophotometers

Although the spectrophotometer is unique in that it can measure outside the visible range of the colorimeter, it is also unique in that it is equipped with a good monochromator that allows one to select any desired wavelength of radiation. The so-called visible spectrophotometer actually covers a wavelength range much wider than the visible

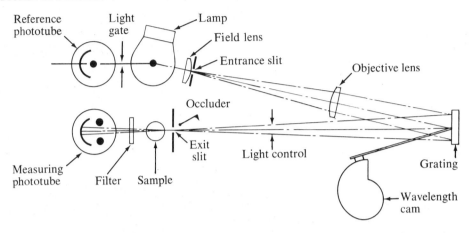

FIGURE 12-13. Optical diagram of the Spectronic 20 Spectrophotometer, top view. (Courtesy of Bausch and Lomb, Incorporated, Rochester, N.Y.)

range and utilizes an inexpensive grating as a monochromator. For example, the Bausch & Lomb Spectronic 20 Spectrophotometer (formerly called a colorimeter) covers the 330–950 nm range, the Coleman Junior II Spectrophotometer covers the 325–825 nm range, and the Turner Model 350 Spectrophotometer spans the 335–1000 nm region. All of these relatively inexpensive spectrophotometers measure part of the UV, all of the visible, and some of the near IR regions.

The optical diagram of the Bausch & Lomb Spectronic 20 in Figure 12-13 is typical of this type of spectrophotometer. Radiation emitted by the tungsten lamp is directed by two lenses and a slit onto the diffraction grating. The radiation emitted by this lamp (Fig. 12-7) starts at about 325–330 nm and ends well above 1000 nm.

The tungsten lamp in Figure 12-14 is a very small version of the usual household tungsten lamp, except that it has a clear, not frosted, glass envelope, and it is a 6-volt, not a 110-volt, lamp. The clear glass permits optimum transmittance of radiation in the 325–1000 nm range covered by the visible spectrophotometer.

The grating selects radiation with a bandwidth of 20 nm centered around the nominal wavelength setting. Thus, if that setting is 550 nm, the bandwidth of radiation actually falling on the sample is 540–560 nm. Any nominal wavelength from 330 to 950 nm may be selected by turning the wavelength control knob, which controls the wavelength cam (Fig. 12-13), which in turn moves the grating. The selected band of radiation is then reflected off the grating through the light control and the exit slit onto the sample solution. Any radiation not absorbed by the sample impinges on the measuring phototube, where the light energy is converted into an electrical signal in terms of absorbance units.

The Spectronic 20 Spectrophotometer is also equipped with a reference phototube (Fig. 12-13). Since a change in the line voltage would change lamp intensity, the voltage fed to the lamp is monitored by the reference phototube so that it can be changed to keep lamp intensity, and scale readings, constant.

Ultraviolet-Visible Spectro-photometers   The ultraviolet-visible spectrophotometer covers the UV region starting around 200 nm, the entire visible region, and part of the near IR through at least 1000 nm. Examples of this type are the Beckman DU-2 Spectrophotometer, the Perkin-Elmer Model 139 Spectrophotometer, and the Spectronic 21 (UV model) Spectrophotometer. In addition

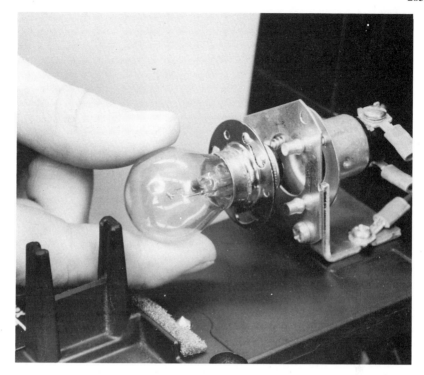

FIGURE 12-14. The 6-volt tungsten lamp of the Spectronic 20 Spectrophotometer. The lamp is fitted to the housing by a clockwise turn, and the housing is connected below by leads to the power system. Directly above the lamp is the holder for the red filter used above 625 nm. (Courtesy of Bausch and Lomb, Incorporated, Rochester, N.Y.)

to covering the ultraviolet region, an important advantage of these spectrophotometers is that many are equipped with recorders that provide a plot of absorbance vs. wavelength. For this plot to be accurate, the radiation striking the sample must be as close to monochromatic as possible. Thus these spectrophotometers use prisms or gratings that have an effective bandwidth of one nanometer or less.

Another feature of this type of spectrophotometer is that it has two interchangeable sources—the hydrogen (or deuterium) discharge tube and the tungsten lamp. Both can be operated simultaneously, so the hydrogen tube is used from 200 to 325 or 330 nm, after which a simple mirror adjustment directs emission from the tungsten source to the sample. The cells are usually fused silica or synthetic silica (pure quartz), so radiation from 200 to 1000 nm can be measured without changing cells. Finally, two interchangeable phototubes (or photomultipliers) are used in some such spectrophotometers, the changeover occurring at 625 nm where the red–near IR phototube is used. In others, a phototube containing two photosensitive surfaces (195–625 and 625–950 nm) is used. The Spectronic 21 utilizes the silicon photodiode detector, which covers the 200–1100 nm range.

Infrared Spectro- photometers   The infrared spectrophotometer covers the IR region from about 2 μm (5000 cm^{-1}) to about 15 μm (660 cm^{-1}). Because of the complicated infrared spectra of most molecules, automatic recording of IR spectra is essential; thus almost all such spectrophotometers

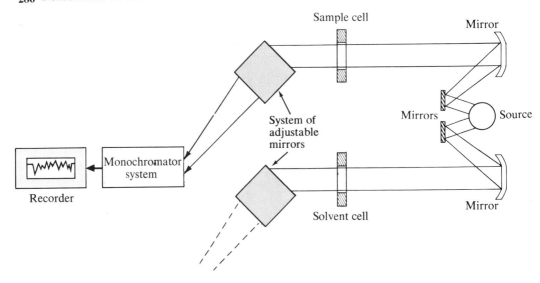

FIGURE 12-15. A diagram of a double-beam infrared spectrophotometer. A single source is split by mirrors so that its beams strike the sample cell and the solvent cell simultaneously. The signals are then fed alternately to the monochromator (the sample beam is shown being fed and the solvent beam diverted), and the difference between the signals is recorded as the infrared spectrum.

are equipped with recorders. Because of the sensitivity of the instruments to many variables and because of the complicated nature of IR measurements, it is desirable to employ a *double-beam* spectrophotometer, or its equivalent, to cancel out variations when comparing the solvent to the sample.

The essential parts of the double-beam spectrophotometer are shown in Figure 12-15. Radiation from the Nernst glower or globar source is split into two beams falling simultaneously on sealed cells containing the pure solvent and the sample in the same solvent. Thus, any variation in the intensity of the source will not affect the comparison of the solvent and the sample. After each beam is passed through the respective cell, it is caused to travel on alternate cycles to the thermocouple or the bolometer detector, automatically establishing $100\%T$.

12-4 | ERRORS IN SPECTROPHOTOMETRY

A theoretical discussion of the errors in spectrophotometry is beyond the scope of this text. However, we will mention some common errors from experimental variables and measurements.

**Experi-
mental
Variables**

The cells are an important potential source of error, whether a single-beam visible spectrophotometer or a double-beam infrared spectrophotometer is used. All cells should be carefully cleaned, especially where samples such as proteins might adsorb on cell walls. Leaving fingerprints on the outside of cells should be avoided; the organic compounds present may absorb ultraviolet radiation, and any aqueous residue may absorb infrared radiation.

The positioning of the cells in the cell compartment must be reproducible; this is the reason for the line etched on test tubes used as cells in inexpensive visible spectrophotometers. (This line is matched with a line in the cell compartment.) Finally, use of cells whose diameters are carefully matched is important in double-beam instruments for qualitative analysis and in all instruments for good quantitative work.

Other variables that must be controlled include the solvent composition, which should be the same in the standards and the unknown sample in quantitative analysis. The pH in aqueous solutions should be controlled by a buffer if an acid-sensitive substance is measured. The reagent concentration used to produce an absorbing species in quantitative work should likewise be controlled. In all these cases, deviations from the Beer–Lambert law (Fig. 12-5) may result.

**Measure-
ments**

In general, the wavelength setting and the absorbance reading are the two instrumental settings most subject to errors and deviations. The wavelength setting is subject to an *error* in the actual wavelength selected by the monochromator, and subject to *deviations* when the wavelength must be reset between two or more separate measurements. The possibility of the error can be eliminated by calibrating the instrument to obtain an *accurate* wavelength; the deviations can be minimized by careful wavelength selection and checking to obtain good *precision*.

Like the eye, the phototube is subject to large errors in discerning differences between two faintly colored solutions and between two intensely colored solutions. It can be shown theoretically that the minimum concentration error will result when the absorbance reads between 0.12 and 1.0 ($\%T$ of 75 to 10%). If an error of $\pm0.4\%T$ is made in a transmittance reading, it can also be shown that the relative error in concentration will be of the order of ±2 pph. This assumes that the percent transmittance scale is divided by 100 lines, each representing 1%, and that a third significant figure can be estimated between any two lines with an uncertainty of $\pm0.4\%T$.

12-5 | LOOKING BACK AND AHEAD

After reading this chapter, you should have a certain understanding of the absorption of ultraviolet, visible, and infrared radiation and how this is measured. Until you actually *use* the instruments for this measurement, your understanding will not be complete. To review the concepts involved and to improve your comprehension of the next chapter, you may want to reread this chapter before you read the next on applications of spectrophotometric measurements to chemistry, biochemistry, and medicine.

To review, you should also work as many of the following problems as possible. Concentrate on the problems in those groups where you have the most difficulty with the type. If you are unsure when calculating absorbance, transmittance, or Beer–Lambert law variables, don't forget to go over the self-tests in this chapter.

| QUESTIONS AND PROBLEMS

(Answers to most even-numbered and some odd-numbered problems are in Appendix 5.)

Definitions and the Absorption Process

1. Define each of the following terms first by using only words—no symbols or equations. Then define each by using an equation, if possible.
   a. Absorbance
   b. Radiant power
   c. Transmittance
   d. Beer–Lambert law
2. Explain the difference between each of the following pairs of terms:
   a. Absorbance and absorption
   b. Positive deviation and negative deviation (from Beer–Lambert law)
   c. Molar absorptivity and absorbance
   d. A colorless solution and a black solution
3. Explain how a molecule absorbs a photon of radiation. Include a definition of both the ground state and the excited state in your explanation.
4. There are three possible excited states of oxyhemoglobin, which can be symbolized as $[Hb(O_2)_x]_{1, 2, \text{ and } 3}$. Write three different equations showing the absorption of radiation by the ground state to give these states.
5. Review the response of the eye in answering the following (recall also that purple = blue and red).
   a. What color would oxyhemoglobin appear if it transmitted radiation in the range of 450–495 nm in addition to that shown in Figure 12-3?
   b. What color would deoxyhemoglobin appear (transmits in the ranges of 450–495 nm and 620–750 nm)?
   c. Oxyhemoglobin can be oxidized to Fe^{+3}-containing methemoglobin, which transmits in the range of 590–620 nm and absorbs weakly in the range of 620–700 nm. What color would it appear?
   d. What color would butter appear if it absorbed all light in the range of 380–570 nm and reflected all other light? Draw an absorption spectrum.
   e. What color would lettuce appear if it absorbed all light in the range of

550–750 nm and reflected all other light? Draw an absorption spectrum.

6. The maximum response of the eye at night (rod vision) is shifted to 507 nm; in addition, the response of the eye to radiation above 620 nm at night is extremely weak. What transmitted colors will the eye respond the most to at night? (Assume the actual color would appear gray.)
   a. Oxyhemoglobin solution
   b. A purple (day) solution of permanganate (MnO_4^-)
   c. A red-violet (day) solution
   d. A solution transmitting at wavelengths from 380 to 495 nm

Scale Readout Absorbance and Transmittance Calculations

(Answers to the first two parts of *every* problem in this group are given in the appendix.)

7. Calculate the absorbance for each $\%T$ value from a scale readout, observing the significant figure rules.
   a. $20\%T$
   b. $10\%T$
   c. $9\%T$
   d. $7\%T$
8. After deciding whether all the digits (scale readout) are significant, calculate the absorbance for each $\%T$ value, using the correct number of significant figures.
   a. $20.0\%T$
   b. $10.0\%T$
   c. $9.0\%T$
   d. $1.8\%T$
9. Calculate the absorbance for each $\%T$ value from a scale readout, observing the significant figure rules. Each value should have at least one zero to the right of the decimal point (explain why such a zero is significant).
   a. $82\%T$
   b. $97\%T$
   c. $95\%T$
   d. $99\%T$ (explain your answer)

10. Calculate the absorbance of each solution below after it is transferred to a 2.0-cm cell.
    a. A solution with $A = 0.20$ in a 1.0-cm cell.
    b. A solution with a $\%T$ of 28% in a 5.0-cm cell.
    c. A solution with a T of 0.40 in a 4.0-cm cell.

11. Calculate the absorbance of each solution below after the specified change has occurred.
    a. A solution with $A = 0.40$ (1.0-cm cell) is diluted to one-half (0.50) the original concentration and placed in a 3.0-cm cell.
    b. A solution with $A = 0.76$ (3.0-cm cell) is doubled in concentration and placed in a 2.0-cm cell.
    c. A solution with $A = 0.92$ (1.0-cm cell) is diluted from 10 to 50 ml and placed in a 6.0-cm cell.

12. Calculate the correct absorbance of each solution below from the sample $\%T$ as measured with the $\%T$ settings specified for infinite absorbance and zero absorbance.
    a. Sample $\%T$: $40\%T$; infinite A: $0\%T$; zero A: $90\%T$
    b. Sample $\%T$: $50\%T$; infinite A: $10\%T$; zero A: $100\%T$
    c. Sample $\%T$: $40\%T$; infinite A: $5\%T$; zero A: $95\%T$

Beer–Lambert Law Calculations with Single Solution
(Answers to the first two parts of *every* problem in this group are in the appendix.)

13. Calculate the log of each molar absorptivity given to the correct number of significant figures.
    a. $\varepsilon = 2.0 \times 10^4$
    b. $\varepsilon = 0.02$
    c. $\varepsilon = 11$
    d. $\varepsilon = 1.02$

14. Calculate the molar absorptivity from each log given to the correct number of significant figures.
    a. $\log \varepsilon = 4.0$
    b. $\log \varepsilon = 4$
    c. $\log \varepsilon = 0.06$
    d. $\log \varepsilon = -0.02$

15. Calculate the molar absorptivity of each species below, assuming a 1.00-cm cell.

    a. At 540 nm, a $2.0 \times 10^{-4}M$ solution of $KMnO_4$ has an A of 0.40.
    b. At 400 nm, a $2.00M$ solution of $Mn(H_2O)_6^{+2}$ has an A of 0.07.
    c. A $2.0 \times 10^{-4}M$ solution of acetylsalicylic acid has an A of 0.28 at 280 nm and an A of 0.22 at 235 nm.

16. Calculate the designated variable using the Beer–Lambert law. Assume $b = 1.000$ cm in each case. Report c in molarity.
    a. Calculate A, given $\varepsilon = 1.5 \times 10^4$ and $c = 2.00 \times 10^{-6}M$.
    b. Calculate A, given $\log \varepsilon = 4.30$ and $c = 3.00 \times 10^{-6}M$.
    c. Calculate c, given $\log \varepsilon = -0.02$ and $A = 0.31$.

17. Calculate the concentration in each case as both molarity and mg per deciliter (mg/dL), assuming $b = 1.00$ cm.
    a. Calculate c, given $A = 0.10$, $\varepsilon = 2.0 \times 10^2$, and form wt $= 200$.
    b. Calculate c, given $A = 0.10$, $\log \varepsilon = 3.30$, and form wt $= 200$.
    c. Calculate c, given $\%T = 40\%$, $\log \varepsilon = -0.40$, and form wt $= 50$.

18. After calculating the molarity of the copper(II) ion in each of the following solutions, calculate the molar absorptivity of the $Cu(H_2O)_6^{+2}$ to the correct number of significant figures. Form wts are $CuSO_4 \cdot 5H_2O$, 250; $CuSO_4$, 160; H_2O, 18.
    a. A solution of 0.400 g of $CuSO_4 \cdot 5H_2O$ in 100 ml of water has an absorbance of 0.576 at 790 nm in a 3.00-cm cell.
    b. A solution containing 600 mg/dL of $CuSO_4 \cdot 5H_2O$ has a $\%T$ of 33% at 820 nm in a 2.00-cm cell.

19. After calculating the molarity of the amino acid phenylalanine in each solution below, calculate the molar absorptivity of this molecule at 258 nm.
    a. A solution of 5.19 mmoles of phenylalanine in 1.000 L has a $\%T$ of 9.5% at 258 nm in a 1.00-cm cell.
    b. A solution of 2.475 mg/dL of phenylalanine (form wt $= 165$) has a $\%T$ of 81% in a 3.000-cm cell.

Beer–Lambert Law Calculations with Standard Solution and Unknown Solution

20. The concentration of an unknown solution of permanganate ion, MnO_4^-, is to be found in each case below by comparing its absorbance with that of a standard solution of permanganate. Calculate the concentration of MnO_4^- in each unknown, assuming that the same cell is used for standard and unknown; the cell diameter is not known. The wavelength is 525 nm.
   a. A $1.00 \times 10^{-4}M$ permanganate standard has an A of 0.20; the unknown $A = 0.70$.
   b. A $2.0 \times 10^{-4}M$ permanganate standard has a $\%T$ of 40%; that of the unknown is 5%.

21. At 760 nm on a digital readout spectrophotometer, a $5.3 \times 10^{-2}M$ standard solution of copper(II) ion has an A of 0.577; an unknown copper(II) solution measured in the same cell (1.00 cm) has an A of 0.233.
   a. Calculate the molarity of copper(II) in the unknown to the correct number of significant figures.
   b. Calculate the mg/dL of copper(II) in the unknown (form wt of $Cu = 63.54$).

22. A 10-ml aliquot of an Fe^{+3} unknown is diluted to 50 ml. A 5.0-ml portion is then reduced and treated with 1,10-phenanthroline; the resulting complex ion has an absorbance of 0.233 at 512 nm. A $2.2 \times 10^{-5}M$ standard solution of Fe^{+3} treated with 1,10-phenanthroline has an absorbance of 0.577 at the same wavelength in the same size cell.
   a. Calculate the molarity of Fe^{+3} in the 10-ml aliquot of the unknown.
   b. Calculate the ppm of the Fe^{+3} in the 10-ml aliquot of the unknown.

23. The following absorbances of potassium permanganate solutions were measured at 540 nm in a 1.0-cm cell. Prepare a plot of absorbance vs. concentration, and then use it to find the concentration of an unknown that has a percent transmittance of 50% at 540 nm in a 1.0-cm cell.

$0.000050M, A = 0.10$
$0.00020M, A = 0.40$
$0.00030M, A = 0.61$
$0.00040M, A = 0.81$

24. A $0.00010M$ solution of potassium permanganate has a $\%T$ of 69% at 540 nm in a 1.0-cm cell. Enter this point on the concentration axis of the plot from Problem 23 and characterize the type of deviation from the Beer–Lambert law.

Absorption Spectra

25. Plot the absorption spectrum of a $1.0 \times 10^{-4}M$ acetylsalicylic acid solution from the following data on wavelength and absorbance. $215: A = 0.10$; $225: A = 0.15$; $235: A = 0.22$; $245: A = 0.16$; $255: A = 0.19$; $275: A = 0.24$; $280: A = 0.28$; $285: A = 0.25$; $295: A = 0.20$; $305: A = 0.15$; $315: A = 0.05$.
   a. Calculate the molar absorptivity for each absorption band ($b = 1.00$ cm).
   b. Decide whether or not the solution is colored.

26. In 1966, inorganic nitrate was reportedly found in tobacco. This ion absorbs at 203 nm ($\varepsilon = 1 \times 10^4$) and at 300 nm ($\varepsilon = 7.5$).
   a. What kind of spectrophotometer must be used to determine nitrate in tobacco?
   b. Calculate the minimum detectable concentration of nitrate at each wavelength, if the minimum detectable absorbance is 0.01.

27. The permanganate ion absorbs at 225 nm ($\varepsilon = 3 \times 10^3$), 310 nm ($\varepsilon = 1.5 \times 10^3$), and 525 nm ($\varepsilon = 2 \times 10^3$).
   a. What type of spectrophotometer would have to be used to measure it at each of the wavelengths above?
   b. Calculate the minimum detectable concentration of permanganate at each wavelength in a 1.0-cm cell, assuming a minimum detectable absorbance of 0.01.
   c. In nitric acid, iron(III) appears yellow from an absorption band that covers the entire ultraviolet region as well; in phosphoric acid, iron(III) is colorless and the absorption band is only in the ultraviolet.

1. Can manganese in steel be determined by dissolving the steel in nitric acid and oxidizing the manganese to permanganate? Explain.
2. Can the manganese be determined if phosphoric acid is added to the above mixture? Explain and suggest wavelength(s).

Spectrophotometric Components

28. List four sources of ultraviolet radiation and the wavelength range emitted by each. List one disadvantage of using each in the 330–350 nm region.
29. List three sources of 800-nm radiation. What type of spectrophotometer should be used to measure this radiation?
30. Arrange the following sources in increasing order according to the number of photons emitted at 950 nm: tungsten lamp, globar, hydrogen lamp, and Nernst glower.
31. List three detectors of ultraviolet-visible radiation, and the wavelength of the response of each. Which detector does not need to operate in a vacuum?
32. Suggest a specific detector that will respond to the entire wavelength range given in each case:
    a. 300–600 nm
    b. 700–1000 nm
    c. 500–700 nm
33. Explain the effect of temperature on the photon output of the tungsten lamp. Also explain why the tungsten halogen lamp is often preferred over the tungsten lamp.

Challenging Problems

34. The flame test for potassium yields two lines: at 404 nm (relative no. of photons = 1), and at 766 nm (relative no. of photons = 100). The relative response of the eye to 404-nm radiation is 0.0006 and to 766-nm radiation is 0.00004. Which color will the eye perceive predominantly? (*Hint:* calculate a ratio.)

35. Assume that a liter of $0.010M$ copper(II) has been excited for 10 hours without any excited copper(II) being able to return to the ground state. Assuming that 1×10^3 copper(II) ions are excited every 1×10^{-15} second, calculate the "molarity" of excited copper(II) after 10 hours and decide whether or not the solution is colored.
36. It can be shown that the difference between any two absorbance readings, $A_1 - A_2$, is equal to the log of the inverse ratio of the corresponding transmittances ($= \log(T_2/T_1)$). Verify mathematically that this is true for the following values of A; you must calculate the corresponding T's.
    a. $A_1 = 1.00$ and $A_2 = 0.10$
    b. $A_1 = 0.60$ and $A_2 = 0.20$
37. Using two Beer–Lambert law equations for A_1 and A_2, derive the following relation:

$$A_1 - A_2 = \log\left(\frac{T_2}{T_1}\right)$$

38. A certain spectrophotometer scale is to be imprinted with a $\%T$ scale from 1% to 100%, and an absorbance scale with values opposite each $\%T$ value. Calculate the *distance* between absorbance units ($A_1 - A_2$) for each of the following sets of adjoining $\%T$ values without calculating either of the absorbance values.
    a. 3% and 4%T
    b. 9% and 10%T
    c. 49% and 50%T
    d. 90% and 91%T
    e. 94% and 95%T
39. A solution with absorbance A_1 is diluted to give a solution with A_2 such that $A_1 - A_2 = 0.50$. The second solution is diluted to give a solution with A_3 such that $A_2 - A_3 = 0.25$. Calculate T_3/T_1 using just one equation.
40. Terbium(III) has one of the lowest molar absorptivities of all the rare earth *f-f* absorption bands. Calculate the molar absorptivity at 284 nm of the Tb^{+3} ion from the following data: 0.20 g Tb_4O_7 (form wt = 748) in 10 ml solution; absorbance of solution in a 5.00-cm cell is 0.20 at 284 nm.

| NOTES

[1] G. H. Begbie, *Seeing and the Eye*, Anchor Press/Doubleday, Garden City, N.J., 1973.

[2] R. K. Clayton, *Light and the Living Matter, Vol. 2, The Biological Part*, McGraw-Hill, New York, 1971, pp. 93–127.

[3] F. C. Strong III, *J. Chem. Educ.* **45**, 178 (1969).

[4] *Chemical & Engineering News* **52**, 23 (Apr. 8, 1974).

13 Applications of Spectrophotometry

"Now I add this small quantity of blood to a litre of water," Holmes said. "The proportion of blood cannot be more than one in a million. I have no doubt, however, that we shall be able to obtain the *characteristic* (*color*) *reaction*."

ARTHUR CONAN DOYLE
A Study in Scarlet, Chapter 1

In this chapter we will discuss the absorption of radiation by *chromophores* in a molecule, qualitative analysis by spectrophotometric methods, and quantitative analysis of solutions containing as little as "one in a million" molecules or ions per "litre of water."

The chromophore concept is basic to an understanding of which molecules or ions absorb radiation that can be measured to identify them. A chromophore is an atom or a group of atoms that absorbs radiation at roughly the same wavelength regardless of the structure of the molecule containing it (often the wavelength agrees to two significant figures).

We will show that qualitative analysis can involve either spectral identification of an unknown compound, or simply spectral screening for the presence or absence of a known compound.

In the treatment of quantitative analysis by spectrophotometry, we extend the Beer–Lambert law to situations involving a comparison of one unknown, such as manganese in steel, to one standard, such as standard steel with a known percentage of manganese. We will also consider multicomponent analysis—the use of two equations in two unknowns to analyze a two-component mixture, etc.

13-1 | ABSORPTION BY SOME COMMON CHROMOPHORES

The two general classes of chromophores are (1) *electronic*, in which certain electrons absorb ultraviolet or visible radiation, and (2) *vibrational*, in which certain bonds absorb infrared radiation. We will discuss the former first.

Electronic Chromophores The electronic ground state and excited state were defined in Section 12-1. When an electron of a ground state species absorbs a photon, the species undergoes a transition to an electronic excited state, as shown in Figure 13-1. In that state the electron occupies a different type of orbital. It helps to think of chromophores in terms of what type of *electron* will absorb radiation and also what type of *orbital* the electron will occupy in the excited state. Remember this when reading the following detailed discussion; it will help you see the forest in spite of the trees.

Unsaturated molecules or ions with double- or triple-bond chromophores form a broad class of compounds having π electrons that absorb UV or visible radiation. One of the bonds in a double bond is a stable sigma (σ) bond; the other bond, the π bond, is weaker,

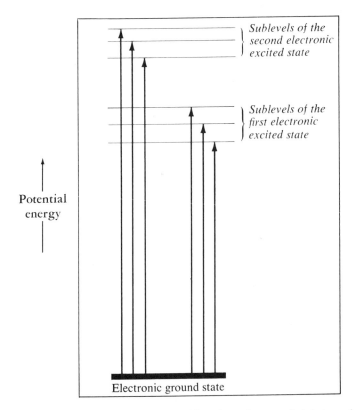

FIGURE 13-1. Two typical electronic transitions from the ground state to various sublevels of the first excited state, and from the ground state to various sublevels of the second excited state. The energy absorbed in going to the second excited state is larger either because a different electron is jumping to the same orbital, or because the same electron is jumping to a higher energy orbital.

and either of its two electrons is easily excited by a UV or visible photon to a higher energy empty orbital.

Whether the double- or triple-bond chromophore is part of an organic molecule or in an inorganic ion or molecule does not matter. The radiant energy absorbed may differ somewhat, but in most cases a π electron absorbs in the UV-visible region. For example, one resonance structure of the purple $MnO_4{}^-$ ion can be written as follows:

$$
\begin{array}{c}
O^- \\
| \\
O{=}Mn{=}O \\
\| \\
O
\end{array}
$$

By its structure the permanganate ion should absorb in the UV-visible region, and it does, both at 525 nm and at 227 nm, as a result of absorption by π electrons [1].

Ozone (Table 13-1) is another example. It can be represented by a resonance structure having a double bond.

$$
\overset{..}{:}O{=}O\overset{\overset{..}{\cdot}}{{-}\overset{..}{O}\!:}
$$

TABLE 13-1. **Ultraviolet-Visible Absorption for π Electron Chromophores**

Name of chromophore	Formula of chromophore	Example	λ_{max}, nm	ε
Oxygen	O=O	O_2	~100 to ~180	—(gas)
Ozone	O=O$^+$—O$^-$	O_3	~200 to ~270	—(gas)
Nitroso	—N=O	$\begin{cases} NO_3{}^- \\ NO_2 \end{cases}$	203 ~290 to ~360	1×10^4 —(gas)
—	Mn=O	$MnO_4{}^-$	$\begin{cases} 227 \\ 525 \end{cases}$	1.6×10^3 2.2×10^3
Carbonyl	C=O	$(CH_3)_2C{=}O$	166	1.6×10^4
Ethylene	C=C	$\begin{cases} \text{Ethylene} \\ \text{2-Hexene} \end{cases}$	160 193	2×10^4 10^4
Acetylene	C≡C	Acetylene	173	6×10^3
Polyene	H(HC=CH)$_n$H	$\begin{cases} \text{Butadiene} \\ \text{Hexatriene} \end{cases}$	217 268	2×10^4 3.5×10^4
Benzene		$\begin{cases} \text{Benzene} \\ \text{Salicylic acid} \\ \text{Acetylsalicylic acid} \end{cases}$	254 308 280	2×10^2 4×10^3 1.5×10^3
Phenanthrene		Phenanthrene	251 293	5×10^4 1.5×10^4
Anthracene		Anthracene	253 340, 357, 375	1.5×10^5 all 9×10^3

O_3 in the atmosphere absorbs UV radiation from the sun in the range from below 240 nm to about 310 nm and thereby protects the eyes and skin from harmful damage by such high energy UV radiation. (It is interesting that ozone itself is formed by oxygen molecules absorbing still higher energy UV radiation below 240 nm.)

As can be seen from Table 13-1, there are many molecules and ions that have the double-bond chromophore and a few that have the triple-bond chromophore: Most absorb UV radiation rather than the UV-visible radiation absorbed by the permanganate ion.

One other point is important—molecules having two or more *conjugated* double bonds have one absorption band at a longer wavelength than those which have only one double bond. For example, hexatriene and benzene, each of which has three conjugated double bonds, absorb at 268 nm and 254 nm, respectively; these absorption bands are at much longer wavelengths than the 160-nm band of ethylene (Table 13-1). The substitution of certain functional groups on conjugated double-bond chromophores commonly shifts absorption bands to longer wavelengths also. For example, acetylsalicylic acid (Table 13-1), the active ingredient in aspirin tablets, absorbs at longer wavelengths than benzene. The two functional groups responsible will be discussed in the next section.

Unsaturated molecules or ions with a pair of nonbonding (*n*) electrons on one of the two atoms in a double-bond chromophore form a second class of compounds that absorb in the UV-visible region, as seen in Table 13-2. The *n* electrons that absorb in this manner are found on atoms such as oxygen, sulfur, and nitrogen, but not on carbon. Two examples of such chromophores are acetone and the nitrate ion:

In each compound, the pair of *n* electrons on oxygen is obviously not involved in bonding and can be easily excited by UV radiation to a higher energy empty orbital. It does not

TABLE 13-2. Ultraviolet-Visible Absorption for *n* Electron Chromophores

Name of chromophore	Formula of chromophore	Example	λ_{max}, nm	ε
Carbonyl	C=O	Acetone	270	18
		$(C_6H_5)_2C=O$	330	180
Carboxyl	R—C=O(OH)	Acetic acid	204	60
Nitro	$^-O—^+N=O$	NO_3^-	300	7.5
Nitroso	—N=O	Nitrosobutane	300	100
			665	30
Thiocarbonyl	C=S	CS_2	318	108
		$(C_6H_5)_2C=S$	620	70
Azomethine	C=N	$(C_6H_5)_2C=NH$	340	125
Azo	N=N	$C_6H_5—N=N—C_6H_5$	448	425

matter whether the oxygen is part of an ion or an organic molecule; the process is the same.

As can be seen from Table 13-2, there are a number of different chromophores in which n electrons absorb UV or visible radiation. The nonbonding electrons are not shown, but are present on the atom at the right of the double bond. The molar absorptivity of each chromophore in Table 13-2 is quite low, of the order of 10^2, compared to the molar absorptivities of the chromophores listed in Table 13-1. This is characteristic for absorption of radiation by n electrons as compared to π electrons.

It is obvious that all of the chromophores in Table 13-2 have n electrons on an atom with a double bond, and you should be wondering why this is so. The reason has to do with the empty orbital to which the n electron is excited. In each case, the n electron jumps to a so-called π^* orbital associated with the double bond. This orbital has four lobes, two of which are centered on each of the atoms of the double bond. For acetone, the n-π^* electronic transition looks like this:

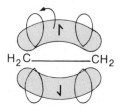

(In the excited state the electron can occupy any of the four lobes of the π^* orbital.) This explains the absorption of radiation by all of the chromophores listed in Table 13-2 at the wavelengths listed.

The π electron in the chromophores in Table 13-1 jumps to the same type of π^* orbital as the n electron. The π-π^* transition in ethylene can be represented by the following:

$$H_2C \text{———} CH_2$$

The shaded area represents the π electron density, and the line represents the sigma bond. Of course, the excited electron can occupy any of the four lobes of the π orbital, not just the one shown. The same kind of transition occurs in all of the chromophores in Table 13-1. Also note that some of these chromophores can undergo both π-π^* and n-π^* transitions. For example, in a solution of nitrate ions, some ions absorb at 203 nm (π-π^* transition) while others absorb at 300 nm (n-π^* transition). (Review Fig. 13-1 to picture this.) This is typical of small molecules or ions; the n-π^* transition occurs at longer wavelengths.

Inorganic ions or chelates form the last broad class of chromophores that absorb in the UV-visible region. Many of the common inorganic ions are colored, because a d electron in a lower energy d orbital absorbs light and jumps to a higher energy unoccupied or half-filled d orbital. Recall that there are five d orbitals; in ions of the type $M(H_2O)_6^{+x}$ or ML_6^{+x}, the lower energy d orbitals are the d_{xy}, d_{xz}, and d_{yz} orbitals, and the higher energy d orbitals are the d_{z^2} and $d_{x^2-y^2}$ orbitals.

A *typical ion* that is colored because of *d-d* transitions is the green $Fe(H_2O)_6^{+2}$ ion. This ion has six *d* electrons, so four of the *d* orbitals hold one electron each, and the fifth, let us assume the d_{xy} orbital, is filled. A *d-d* transition would look like this:

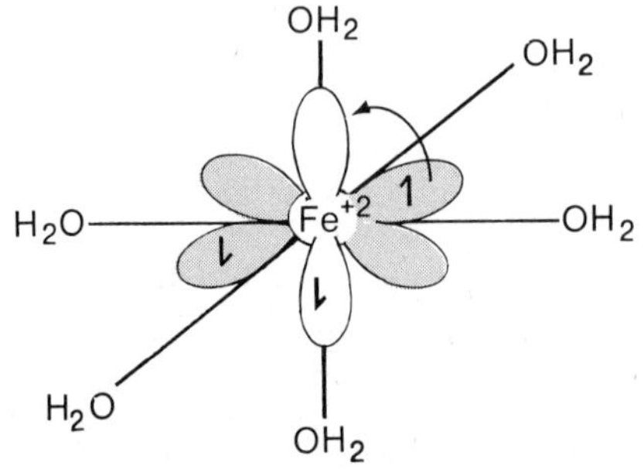

wherein an electron in the filled d_{xy} orbital (shaded) absorbs a photon of red light (690 nm) and jumps to the half-filled d_{z^2} orbital.

The d_{z^2} orbital has higher energy than the d_{xy} orbital because of greater repulsion from electron pairs on oxygen which focus directly at the former orbital. An electron in the d_{xy} orbital is between the regions of greater repulsion by oxygen electron pairs; it uses the light energy to overcome the greater repulsion it encounters briefly in the excited state.

Similar *d-d* electronic transitions are responsible for the colors of the common metal ions listed in Table 13-3. As long as a metal ion possesses at least one *d* electron, it will absorb in the UV-visible region, although it may not always be colored. For example, the $Mn(H_2O)_6^{+2}$ ion does absorb light at 532 and 435 nm, but the molar absorptivities of these bands are so low that a $0.1M$ solution is colorless. (A $2M$ solution is a distinct pink color.) A metal ion having a filled subshell of ten *d* electrons of course does not exhibit any *d-d* transitions and frequently is colorless.

Inorganic chelates are often colored because of slightly different types of electronic transitions. A good example is the chelate of iron(II) and 1,10-phenanthroline (phen), which has the formula $Fe(phen)_3^{+2}$. Each 1,10-phenanthroline molecule bonds to the iron(II) via two nitrogen donor atoms as follows:

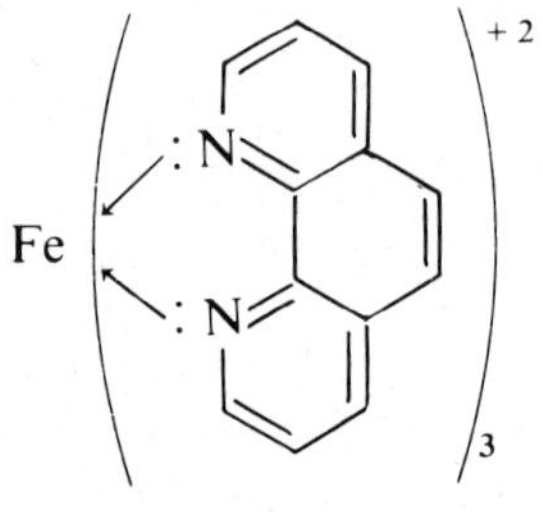

Note that a total of six electron pairs from six nitrogen donor atoms are coordinated to iron(II), so this chelate is equivalent to a ML_6^{+x} complex ion. A close look at the structure reveals that a five-membered *chelate ring* is formed by the iron(II), the two nitrogen donor atoms, and two carbon atoms connecting the nitrogen atoms. When light of 512-nm wavelength is absorbed by this chelate, a d electron such as a d_{xy} electron of iron(II) is excited to an empty π^* orbital on the unsaturated ring as follows:

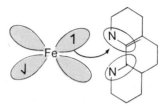

Since the two π^* orbitals shown are equivalent, the d electron can jump to either orbital. Although the d electron may be excited to an unfilled d_{z^2} or $d_{x^2-y^2}$ orbital, as for $Fe(H_2O)_6^{+2}$, this requires much more energy than that of a photon of 512 nm (green light), because the 1,10-phenanthroline exerts more repulsion on these orbitals than does a water molecule.

The light absorption process for at least one of the three absorption bands of oxyhemoglobin (Fig. 12-3) is probably very similar to that for $Fe(phen)_3^{+2}$. Oxyhemoglobin is a chelate of iron(II), being bonded to the unsaturated porphyrin molecule in the heme group as follows:

As shown above, a d electron such as a d_{xy} electron can absorb a photon and jump to one of the empty π^* orbitals of the heme nitrogens bonded to the iron(II). Other transitions are also possible.

Not all important ions of iron involve iron(II). Iron(III), the ferric ion, forms many important colored ions. Strangely enough, $Fe(H_2O)_6^{+3}$ is essentially colorless in dilute solution. It absorbs radiation so weakly in the red–near infrared and yellow-green regions (Table 13-3) that only concentrated solutions appear faintly colored. If ions such as the hydroxide or chloride ion (see $FeCl^{+2}$ in Table 13-3) are mixed with $Fe(H_2O)_6^{+3}$, it forms yellow complex ions. The absorption process probably involves an electron of the hydroxide or chloride being excited to an unfilled d orbital of the iron(III) ion.

Another important colored complex ion of iron(III) is that formed with the thiocyanate (SCN^-) ion. The appearance of a red color when thiocyanate is added to a solution is an important qualitative analytical test for the presence of iron(III). The

TABLE 13-3. Ultraviolet-Visible Absorption for *d* Orbital Chromophores

Formula of ion	Number of d electrons	λ_{max}, nm (color)	ε
Ions with d-d transitions			
$Ti(H_2O)_6^{+3}$	1	500 (purple)	4
$Mn(H_2O)_6^{+2}$	5	435, 532 (colorless)	0.02
$Fe(H_2O)_6^{+3}$	5	794, 540 (colorless)	0.1
$Fe(H_2O)_6^{+2}$	6	450, 690, 962 (lt. green)	2
$Fe(CN)_6^{-4}$	6	420 (lt. yellow)	1
$Ni(H_2O)_6^{+2}$	8	400, 740 (green)	5
$Cu(H_2O)_6^{+2}$	9	790 (green-blue)	12
Ions and chelates with transitions involving d orbitals			
$Fe(phen)_3^{+2}$	6	512 (orange)	1.1×10^4
Oxyhemoglobin [Iron(II)]	6	575, 540, 415 (red)	—
$Co(NCS)_4^{-2}$	7	312, 620 (blue)	10^3
$Fe(SCN)^{+2}$ [Fe(SCN)(H_2O)_5^{+2}]	5	450 (red)	5×10^3
$FeCl^{+2}$ [FeCl(H_2O)_5^{+2}]	5	340 (yellow)	2×10^3
$Fe(CN)_6^{-3}$	5	415 (yellow)	1×10^3

test is so selective that few ions interfere with it. Iron(II) does not form a colored complex with thiocyanate; hence the test is specific for iron in the +3 oxidation state.

The material discussed above is rich, but the fundamentals are not difficult to understand. Once you complete the following self-test, you should have a good grasp of the fundamentals and you will then recognize that the remaining material is intended as enrichment. For additional applications of the fundamentals, you should proceed to the problems at the end of the chapter.

SELF-TEST

19. Electronic Absorption

Answers:

Directions: Cover the answers in the left column before beginning the test. Work one problem at a time and check the answer at the left.

A. Recognition of chromophores. Indicate whether each of the following molecules or ions should absorb in the UV-visible region and explain why.

   a. $[:O{=}C(-O)_2]^{-2}$, carbonate ion; $:O{=}C{=}O:$, carbon dioxide; $[(:O{=})_2\,Cr(-O)_2]^{-2}$, chromate ion;

     $CH_3C{=}O:$, acetic acid
        |
       OH

All absorb because each has a double bond and because each has a pair of nonbonding electrons on an atom that is part of a double bond. In each case the *n* electrons are on oxygen. Both π-π^* and n-π^* transitions can occur.

None should absorb because none of the atoms with n electrons are part of a double bond.

b. $:NH_3$, $CH_3{-}\ddot{N}H_2$, $H{-}\ddot{\underset{..}{C}l}:$, $[:\ddot{\underset{..}{F}}:]^-$

B. Metal ions with d orbitals. Indicate whether each of the following ions should absorb in the UV-visible region or whether it should be colorless and explain why.

All should absorb because they have d electrons for a d-d transition.

a. $Cr(H_2O)_6^{+3}$, a d^3 ion; $Co(H_2O)_6^{+2}$, a d^7 ion; $Mn(H_2O)_6^{+3}$, a d^4 ion

None should absorb because they have no d electrons; all are colorless.

b. $Ca(H_2O)_6^{+2}$, a d^0 ion; $Ti(H_2O)_6^{+4}$, a d^0 ion; $K(H_2O)_6^{+}$, a d^0 ion

None should absorb because their d orbitals are filled; all are colorless.

c. $Zn(H_2O)_6^{+2}$, a d^{10} ion; $Cu(H_2O)_6^{+}$, a d^{10} ion; $Ag(H_2O)_6^{+}$, a d^{10} ion

All should absorb in that an electron on the ligand may be excited to an unfilled d orbital on iron(III) or iron(II).

C. Ions and chelates with d orbitals. Indicate whether each of the ions or chelates should absorb in the UV region and explain why.

$FeBr^{+2}$ ion; $Fe(C_2O_4)_3^{-3}$ oxalate chelate; $Fe(CN)_6^{-4}$ ion

Chromo-phores Undergoing Vibrational Absorption

Absorption of infrared radiation by a molecule or by a certain bond in a molecule leads only to vibrational, not electronic, energy changes. Hence during infrared absorption, all the electrons remain in their most stable orbitals, and only the vibrational energies of the bonds in the molecule increase. The electronic ground state is divided into many sublevels, starting with the lowest energy sublevel, called the GV_0 sublevel, and continuing with sublevels of increasing vibrational energy—GV_1, GV_2, $\ldots$, GV_n. These are shown in Figure 13-2.

Nearly all molecules exist in the GV_0 sublevel, the ground vibrational sublevel. The most probable infrared absorption process, as shown in Figure 13-2, is

$$GV_0 + \text{IR photon} \longrightarrow GV_1$$

in which the GV_1 sublevel is the first excited vibrational sublevel. The resulting absorption band is called the *fundamental*. The molecule does not remain in the GV_1 sublevel very long but returns to the ground vibrational sublevel.

$$GV_1 \longrightarrow GV_0 + \text{Heat}(\rightarrow \text{solvent})$$

The return to the GV_0 sublevel is accompanied by the loss of the absorbed energy as heat, given up by the molecule during collision with a solvent molecule. Weaker absorption bands, called overtones, also occur in the near infrared region (as shown in Fig. 13-2), but they are beyond the scope of our discussion.

Although infrared absorption is quite complicated, a helpful generalization can be made. The generalization is this: molecules with similar bonds, or *functional groups*, do exhibit similar fundamental infrared absorption bands. Thus all molecules with a $C{-}O{-}H$ group have similar $C{-}O$ and $O{-}H$ absorption bands, etc. The locations of the absorption bands listed in Table 13-4 illustrate this fact in more detail. To understand the *relative* locations of these bands, we will discuss a few types of bonds that absorb fairly close to one another.

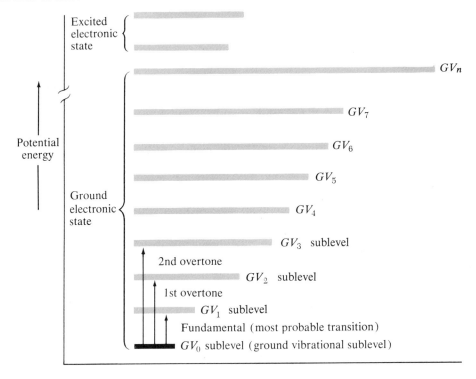

FIGURE 13-2. Subdivision of the ground electronic state into vibrational sublevels—GV_0, GV_1, GV_2, GV_3, ..., GV_n. Typical transitions occurring upon absorption of infrared radiation are shown. The first and second overtones occur with a mucn lower probability than that of the fundamental; they are generally found in the near infrared region at higher energies than the fundamental absorption band. (The length of each horizontal line symbolizes the internuclear distance for the bond absorbing the infrared radiation.)

Bonds consisting of a hydrogen atom and another light atom have the largest vibrational energies in the GV_0 sublevel and therefore absorb the highest energy (smallest wavelength) infrared radiation to jump to the GV_1 sublevel. Examples of such bonds are the O—H bond, the N—H bond, the $\overset{\diagdown}{\underset{\diagup}{C}}$—H bond, and the =C—H bond. The essential information concerning absorption by these bonds is given in Table 13-4; only a few points will be elaborated.

Absorption by the O—H bond is complicated both by the nature of the OH molecule and by the concentration of the solution measured. For alcohols, phenols, and acids, the O—H bond absorbs at different wavelengths depending on the concentration. In dilute solution, O—H-containing molecules tend not to interact with one another and exist as *monomers*. The O—H absorption band is sharp. In moderate to concentrated

TABLE 13-4. IR Absorption Bands of Bonds Having a Hydrogen Atom

Bond or functional group	Wavelength, μm	Comment and confirming absorption
CO—H (alcohol or phenol)	2.74–2.82	Sharp O—H band only in dilute solution. Confirm by band at 7–8 and C—O band at 8.3–9.5.
	2.82–3.1	Broad O—H band from hydrogen bonding if solution is not dilute.
$\overset{O}{\overset{\|}{C}}$O—H (acid)	2.8–3.0	O—H band is sharp only in very dilute solution. Confirm by C=O band at 5.8–6.
	3.3–4.0	Broad O—H band from hydrogen bonding in most solutions.
CN—H (amide or amine)	2.82–3.0	Weak band from N—H. Confirm amide by C=O band at 5.9–6.06. Confirm amine by medium band at 6.1–6.7 or C—N band at 8.1–9.7.
C=C—H (aromatic or alkene)	3.29–3.32	Sharp C—H band of varying intensity. Confirm aromatic by strong C=C band at 6.7–6.8. Confirm alkene by variable C=C band at 6.0–6.2.
C—H (aliphatic carbon)	3.4–3.6	Usually strong C—H band because of large number of C—H bonds in organic molecules. Almost always present.
$\overset{O}{\overset{\|}{C}}$—H (aldehyde)	Two bands at 3.5 and 3.65	The 3.5 band is usually overlapped by the C—H band, but the 3.65 band is usually sharp and clearly separated from the C—H band. Confirm by C=O band at 5.75–6.0.

solutions (0.03–1M), O—H-containing molecules hydrogen bond to one another as follows:

$$R-\overset{H}{\overset{/}{O}}: \longrightarrow H-\overset{R}{\overset{/}{O}}$$

This lowers slightly the vibrational energy of the O—H bond and the energy of the infrared radiation absorbed by this bond. You can see in Table 13-4 that the hydrogen bonding absorption is at longer wavelengths and is a broad, not a sharp, band. In fact, in solutions of moderate concentrations of alcohols or phenols, both absorption bands will occur in the infrared spectrum. Hydrogen bonding in acids is much stronger, so only the hydrogen bonded O—H absorption band is observed except for very dilute solutions.

Because bonds having hydrogen atoms are *all* of relatively high energy, they absorb at wavelengths quite close together in the infrared region. Consider the absorption spectrum of acetaldehyde, CH_3CHO, shown in Figure 13-3. The 3.5-μm absorption

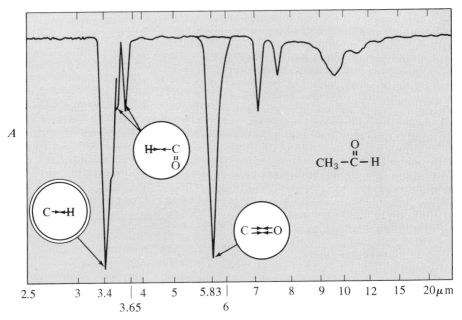

FIGURE 13-3. The infrared absorption spectrum of acetaldehyde. Note that the 3.5-μm aldehyde C—H band is overlapped by the 3.4-μm CH₃ band, but that the 3.65-μm aldehyde C—H band is clearly separated.

band arising from hydrogen bonded to the C=O group is barely discernible because it is overlapped by the C—H absorption band of the CH₃ group. Fortunately, the 3.65-μm band is clearly separated from the intense C—H absorption band of the methyl group.

Bonds consisting of atoms other than hydrogen have smaller vibrational energies than bonds having a hydrogen atom in the GV_0 sublevel. For a jump to the GV_1 sublevel, such bonds therefore absorb lower energy (longer wavelength) infrared radiation. The essential infrared wavelengths absorbed by such bonds are given in Table 13-5, and only a few further comments will be added.

Triple bonds are the highest energy bonds of those listed in Table 13-5, so such bonds absorb at the shortest wavelengths listed. Note that this includes the nitrile C≡N, the acetylenic C≡C, and even the C≡O triple bond of carbon monoxide.

Double bond energies are less than those of triple bonds but about twice those of single bonds, so their absorption bands are located between those of the triple bonds and single bonds. Note the various carbonyl (C=O) absorption bands; slight differences in location indicate an aldehyde (see Fig. 13-3), a ketone, an acid, or an ester. Gaseous air pollutants such as carbon dioxide and sulfur dioxide absorb in the double-bond region.

The absorption bands of C—C single bonds are not very useful because they are so common. However, other single bonds such as C—Cl, C—Br, C—N, and C—O are more significant.

No self-test is included for infrared absorption because the discussion has been brief. However, you can check your understanding by working the pertinent problems at the end of this chapter.

TABLE 13-5. IR Absorption Bands of Bonds Without a Hydrogen Atom

Bond or functional group	Wavelength, μm	Comment
Moderate energy bonds		
C≡N, nitrile	4.40–4.5	—
C≡C, acetylenic	4.42–4.76	—
C≡O, carbon monoxide	4.65–4.7	—
O=C=O, carbon dioxide	4.30	Confirm at 2.6 & 2.7 μm (weak bands)
—C=O(OR), ester	5.71–5.76	—
—CH, aldehyde ‖ O	5.75–5.95	Confirm at 3.65 μm
—COH, acid ‖ ＼O	5.80–5.95	Confirm by broad band at 3.3–4.0 μm
—C=O, ketone	5.80–6.01	—
—CO⁻, acid anion ‖ O	6.21–6.45	—
C=C, alkene	5.95–6.17	Very weak
O=S—O, sulfur dioxide	7.25–7.45	Confirm at 8.60–8.85 μm
Low energy bonds		
C—N	7.46–9.8	—
C—O	8.7–9.5	—
C—Cl	12.5–16.6	—
C—Br	16.6–20.0	—

13-2 | QUALITATIVE ANALYSIS

Qualitative analysis can involve the *identification* of a compound whose identity is unknown or only partially known or *screening* on a routine basis for the presence or absence of significant amounts of a suspected compound. Since both of these facets of qualitative analysis depend on obtaining the proper absorption spectrum or spectra, we will discuss this point first.

Obtaining the Proper Absorption Spectrum

The nature of the sample will help you decide whether to obtain ultraviolet, visible, and/or infrared spectra. For example, if the sample is an aqueous solution of primarily inorganic substances, it is difficult to obtain an infrared spectrum because the typical cell (sodium chloride, etc.) would be partially dissolved by the water. A *colorless* aqueous solution will obviously not yield any information in the visible region but might give some information in the UV region. The 200–400 nm region should therefore be scanned on the spectrophotometer for one or more absorption bands. A *colored* solution of inorganic and/or organic compounds may absorb in the UV as well as in the visible region, so it should be scanned from 200 to 750 nm. Any organic sample, whether liquid, solid, or gas, should yield an infrared absorption spectrum giving some valuable structural information. The infrared spectrum of an inorganic gaseous sample is also frequently useful.

The best and fastest way to obtain an absorption spectrum is to use a spectrophotometer that automatically scans and records the spectrum. If a visible-UV spectrophotometer is not so equipped, then the absorbance should be measured throughout the desired regions in steps of 10 or 20 nm. The absorbance is then plotted on graph paper against wavelength, and a curve is drawn to connect all the points. In the region of the absorption band maximum, enough measurements should be taken to locate the maximum within ± 1 to 2 nm.

The exact wavelength location of the *peak* of each absorption *band*, as well as the relative absorbance of each band, is much more important than the appearance of the entire absorption spectrum. The absorption spectrum of a mixture of several compounds may not resemble the absorption spectrum of any of the individual compounds present, so the peak wavelength(s) are the most important evidence for the presence of one or more compounds.

Identification

Identification of an unknown compound is usually based on the measurement of *several* physical and/or chemical properties. *One* absorption band or one other physical property is seldom considered to be *rigorous proof* of an identity, although it may be enough for *tentative identification*.

If a compound can be isolated in sufficient quantity, a molecular weight determination, an elemental analysis (for C, H, etc.), and at least one absorption spectrum are usually sufficient for rigorous identification. If the compound cannot be isolated in sufficient quantity to carry out the two former measurements, then at least three other measurements should be sufficient. Not all three of these measurements should be taken from the same spectrum; other confirming measurements should be obtained. (For example, the fluorescence spectra described in the next chapter may be used to confirm UV absorption spectral evidence.) Of course, the spectrum of the unknown sample should be carefully compared to the spectrum of the pure compound run on the same instrument whenever possible.

Ultraviolet absorption spectra are especially useful in identifying aromatic hydrocarbons in complex mixtures. A good example is the study by Giger and Blumer [2] of the aromatic hydrocarbons present in near-shore marine sediments. The analysis of these sediments is an important indication of the effects of industrial pollution and oil spills on the coastal environment. Because the samples were so complex, Giger and Blumer had to separate the large number of compounds into fractions which then had to be identified. They used a *mass spectrometer* to obtain the equivalent of a molecular weight measurement and a characteristic mass spectrum having several peaks for each aromatic hydrocarbon. Two of the aromatic hydrocarbons that they identified with the help of ultraviolet absorption spectra and mass spectra were phenanthrene and anthracene. The structure and peak wavelengths of these compounds are given in Table 13-1.

Their samples were especially rich in phenanthrene; the top spectrum in Figure 13-4 is a hypothetical representation of an UV absorption spectrum from their Sampling Station *A*. The identity of phenanthrene is confirmed by the sharp 293-nm peak. This peak is unique among the aromatic hydrocarbons that could possibly be present with the phenanthrene after separation. The 251-nm peak, although part of the same absorption band as the 293-nm peak, is not unique. Anthracene, for example, also absorbs strongly in this region. However, its presence does support the identification in that phenanthrene is expected to absorb at this wavelength. If the absorption were weaker at 251 nm than at 293 nm, this would lead us to suspect that phenanthrene was

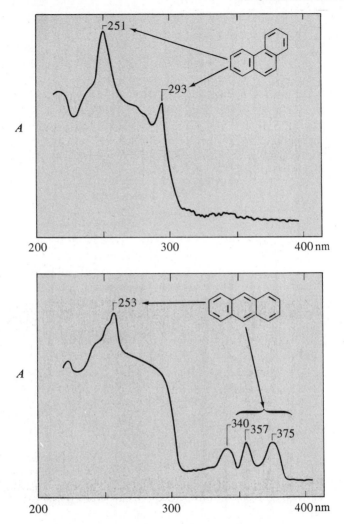

FIGURE 13-4. Hypothetical ultraviolet absorption spectra of some aromatic hydrocarbon separated fractions from near-shore marine sediments analyzed by Giger and Blumer [2]. The spectrum at the top is that of a fraction rich in phenanthrene, whose identity is confirmed by the normally sharp peak at 293 nm and supported by absorption at 251 nm. The spectrum at the bottom is that of a fraction rich in anthracene, whose identity is confirmed by the intense peak at 253 nm as well as the three weak peaks at longer wavelengths.

not present, since its molar absorptivity at 251 is larger than that at 293 (Table 13-1). The phenanthrene is identified, then, *on the basis of several peaks in the mass spectrum, the two ultraviolet absorption peaks, and its molecular weight.*

The bottom spectrum in Figure 13-4 is a hypothetical representation of an UV absorption spectrum of a possible phenanthrene fraction from Sampling Station *B* of Giger and Blumer. The spectrum is "smoothed out" in the 293-nm region, so phenanthrene cannot be identified. However, anthracene is confirmed by the strong peak at 253 nm and three weaker peaks from a different band at 340, 357, and 375 nm. Although the latter peaks are quite weak, they are unique to anthracene. Anthracene is identified *on the basis of the mass spectrum, the four ultraviolet absorption peaks, and its molecular weight.*

Infrared absorption spectra are very useful for identifying solids, liquids, or gases. A good example of the use of infrared spectrophotometry in identifying a compound is in the analysis of impure samples suspected of containing heroin. Heroin itself is a tertiary amine.

It also can exist as a hydrochloride salt in which the pair of electrons on the nitrogen is protonated, giving it a characteristic $R_3 N^+$—H bond.

As stated above, the best way to identify a compound is to compare its infrared spectrum with that of a pure sample, as is done in Figure 13-5. You observe that most of the absorption bands of the impure and pure heroin correspond, even though their absorbance values are different. A key band is that at 3.9 μm, indicating a protonated tertiary amine N^+—H bond. Most N—H bonds absorb in the 2.82–3.3 μm range (Table 13-4). Another key band is the 5.75-μm band, indicating an ester carbonyl group (Table 13-5). Heroin has two of these groups, both CH_3C=O groups. A third key band is the 7.4-μm band, indicating a CH_3 group bonded to the tertiary nitrogen; the C—H bonds actually absorb at this wavelength.

Routine Screening

Screening on a routine basis is done for compounds whose identity is already known. All that is desired is that a definite positive test or a definite negative test be obtained. The following factors are important.

1. The test should be rapid to handle large numbers of samples.
2. The test should be simple and easy for technicians to interpret.
3. The test should involve a simple instrument or a single wavelength measurement.

Screening is especially important in drug analysis and in drug abuse. For example, in the drug abuse area, the identities of most drugs, such as barbiturates, are already known.

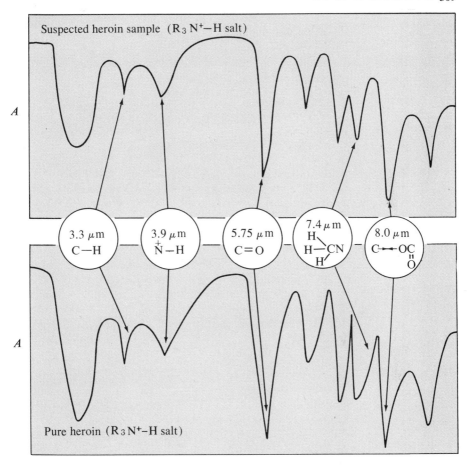

FIGURE 13-5. The infrared absorption spectra of a sample suspected to contain heroin (top) and of a sample of pure heroin (below), both in the salt form so the tertiary amine (R_3N) of heroin exists in a protonated from (R_3N^+—H), which has a characteristic absorption band for the N^+—H bond at 3.9 μm.

It is only a matter of establishing whether a certain drug is present in a body fluid, such as blood or urine. Both barbiturates (I = acid form) and phenylethylamines such as amphetamine (II, R=CH$_3$)

I

II

can be identified by routine screening using their ultraviolet and/or infrared absorption spectra. In alkaline solution, most barbiturates form a -2 anion having a conjugated π electron system with a strong $\pi\text{-}\pi^*$ absorption band at 250–275 nm ($\varepsilon = 10^4$). The phenylethylamines are, of course, substituted benzenes and have a moderate $\pi\text{-}\pi^*$ absorption band at 247–290 nm ($\varepsilon > 300$).

An interesting use of screening in the area of drug analysis is in the test for the presence of salicylic acid in aspirin tablets. Acetylsalicylic acid (ASA) is slowly hydrolyzed by water in the air or by absorbed water to salicylic acid (SA) as follows:

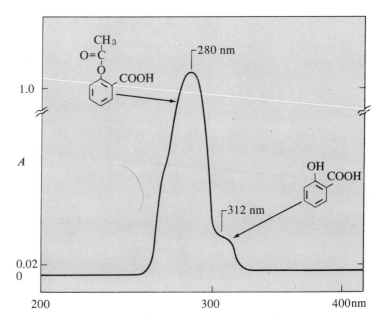

Since acetic acid is somewhat volatile, the main impurity that remains in the tablets is salicylic acid. Screening for this impurity can be done using ultraviolet spectrophotometry. Although ASA and salicylic acid both have a benzene chromophore, salicylic acid has a $\pi\text{-}\pi^*$ absorption band at 312 nm, a much longer wavelength than that of ASA (Table 13-1). A measurable absorbance at 312 nm should be a *definite positive test* for salicylic acid (Fig. 13-6).

FIGURE 13-6. Hypothetical ultraviolet absorption spectrum of a solution of aspirin tablets that contains more than 0.10% of salicylic acid. The salicylic acid impurity absorbs more than the allowable 0.02 units at 312 nm.

The Food and Drug Administration (FDA) has established a *tolerance* of 0.10% salicylic acid in unbuffered aspirin. To test whether a particular brand meets this tolerance, a specified quantity of that brand of aspirin is dissolved and its absorbance is measured at 312 nm. If the absorbance exceeds 0.02 (Fig. 13-6), then that brand exceeds the tolerance. It is not necessary to run the entire spectrum as shown in Figure 13-6, although it may be done to have a record.

The reason the tolerance for salicylic acid in aspirin is so low is that salicylic acid is very irritating to the stomach lining, much more so than ASA. Once ASA passes through the stomach, it is hydrolyzed to salicylic acid, so the function of the CH_3—C=O group on ASA is simply to protect the stomach lining. The active, pain-killing form of aspirin is thus salicylic acid, not ASA.

13-3 | QUANTITATIVE ANALYSIS

Quantitative analysis involves the determination of the amount of a substance present by spectrophotometric measurement. The substance may previously be known to be present, or qualitative analysis using spectrophotometry, etc., may have established its presence. Before discussing the one-time analysis of a sample and the continuous monitoring of a sample whose composition changes, we will first discuss some of the important steps in a quantitative analysis.

Steps in Quantitative Analysis

A quantitative method may be based on measuring the radiant energy absorbed by a species as it exists or on measuring the radiation absorbed by a species after it has undergone a chemical reaction. In the latter case, careful control of the solvent, the reagents added, the reaction time, and the pH is necessary. If these variables are not controlled, deviations from the Beer–Lambert law may result. Each variable should be tested by obtaining a complete spectrum at several values of the quantity being varied.

The choice of wavelength is governed by the necessity to measure at the wavelength at which one can achieve the maximum absorbance of the desired constituent with the minimum error from other constituents that may be present and that absorb in the same general region. If nothing else absorbs at the peak wavelength of the absorption band, then this wavelength would be chosen for analysis.

Figure 13-7 illustrates a situation where choosing the proper wavelength is more complex. Assume first that you are only determining oxyhemoglobin. If you choose either 415 nm or 540 nm as the proper wavelength, your measurement will be in error. The error at 415 nm will be large, because bilirubin absorbs strongly there; the error at 540 nm will be small, because it absorbs weakly at this wavelength. If there were no other choice, measurement at 540 nm could be used, provided accuracy is not needed. However, the best choice for the analysis would obviously be 575 nm, where bilirubin does not absorb at all. If the situation were reversed and bilirubin had to be determined, a correction would have to be made for the absorbance of oxyhemoglobin at 461 nm. This will be discussed later.

A plot of absorbance against concentration should be prepared before the proposed method and wavelength are finally adopted. If such a plot is linear, then the Beer–Lambert law is said to be obeyed (Fig. 12-5). Even if the plot deviates, as shown in Figure 12-5, it may still be used if the deviation is reproducible. As long as an absorbance reading gives an accurate value corresponding to only one concentration, the plot can be used for quantitative work. Occasionally not all of the points will fall exactly on a straight line. In that case, the line should be drawn so that an equal number of the points are above and below the line. Any point(s) far away from the line should be ignored or else rechecked.

During preparation of the above plot, conditions should be adjusted to be the same for each solution of known concentration of the constituent to be measured. For example, if iron(III) is to be determined by adding anion X^- to form the colored complex ion FeX_2^+, enough of X^- should be added so that all of the iron(III) is present as FeX_2^+ and not as FeX^{+2}, etc.

One-Time Analysis

A one-time determination of a substance is a "one-shot" analysis in that information is required about a substance whose composition does not change with time or whose composition at a particular time is all that is needed. An example of the former is the analysis of aspirin tablets that have just been bottled at the plant; an example of the latter is a blood analysis performed in the morning before breakfast foods disturb the levels of chemicals in the blood. A one-time determination is different from the continuous analysis or monitoring of a substance whose composition may change with time. The latter will be discussed separately.

Many metal ions are determined on a one-time basis because of their stability or because their concentrations do not change rapidly with time. An example of a simple one-time analysis is the colorimetric determination of iron(III) in water by forming a chelate with 1,10-phenanthroline (phen). The characteristics of this chelate were discussed in the first section and listed in Table 13-3. Because this chelate forms best with iron(II), hydroxylamine is first added to reduce the iron(III) as follows:

$$2Fe^{+3} + 2NH_2OH \longrightarrow 2Fe^{+2} + N_2(g) + 2H_2O + 2H^+$$
$$(-1\,N\text{ ox. st.})$$

The resulting iron(II) is then chelated by the 1,10-phenanthroline.

$$Fe^{+2} + 3\text{ phen} \longrightarrow \text{orange }Fe(phen)_3^{+2}$$

A few other ions, such as copper(I), also complex with 1,10-phenanthroline, but none of the complexes is as intensely colored or absorbs as strongly at 512 nm, so the method is essentially free of interferences for water analysis.

Inorganic Oxidation in Spectro-photometric Analysis

A number of metal ions in low oxidation states can be oxidized to stable, intensely colored higher oxidation states. For example, manganese(II) can be oxidized to manganese(VII) as MnO_4^-; chromium(III) can be oxidized to chromium(VI) as the yellow CrO_4^{-2} ion; and vanadium(II) can be oxidized to the blue VO^{+2} ion.

Manganese in steel is readily determined by dissolving the steel and converting the resulting manganese(II) ion to the purple permanganate ion. The nitric acid solvent for the steel converts metallic manganese to the nearly colorless $Mn(H_2O)_6^{+2}$ ion. This ion does not form any colored simple complex ions with chloride, for example (although it does form a few colored chelates with organic ligands). Hence, its oxidation to the intensely purple permanganate ion by oxidation with periodate in hot acid is often used.

$$2Mn^{+2} + 5IO_4^- + 3H_2O \longrightarrow 2MnO_4^- + 5IO_3^- + 6H^+$$

This method is almost specific for manganese (dichromate ion interferes) and can be used to determine traces of permanganate as low as $5 \times 10^{-6}M$ (at $A = 0.01$).

The actual procedure for manganese in steels involves first dissolving the entire steel sample in hot nitric acid, resulting in manganese(II) and yellow iron(III). Then the manganese(II) is oxidized to manganese(VII) by periodate at the boiling point of the solution. Because the yellow color of the iron(III) would interfere with spectrophotometric measurement, about $2 M$ phosphoric acid is added to form a colorless complex with iron(III) as follows:

$$Fe(III) + xH_3PO_4 \longrightarrow [Fe(HPO_4)_x]^{+3-2x} + 2xH^+$$

For a rapid analysis when high accuracy is not required, the absorbance of the permanganate in the unknown steel is compared to that of a *single* standard steel (known %Mn). To simplify calculations, both the unknown and the standard steel are dissolved and diluted to the same volume and measured in the same diameter spectrophotometer cell. In this case, the ratio of two Beer–Lambert law equations simplifies to

$$\text{wt Mn}_u = \left(\frac{A_u}{A_{std}}\right)\text{wt Mn}_{std} \tag{13-1}$$

Then we substitute the product of the weight of the steel and the percentage of manganese in the steel for the weight of the manganese in both sides of **13-1**.

$$(\%\text{Mn}_u)(\text{wt steel}_u) = \left(\frac{A_u}{A_{std}}\right)(\%\text{Mn}_{std})(\text{wt steel}_{std}) \tag{13-2}$$

If the sample weights of the unknown steel and the standard steel are adjusted before analysis to be within ± 0.01 g of each other, then **13-2** can be further simplified by omitting the weights. The error from this simplification should be less than 1%. The student should verify the derivation of **13-1** and **13-2**.

EXAMPLE: Calculate the %Mn in an unknown steel, 1.202 g of which gives an absorbance of 0.460. The unknown is compared to 1.212 g of a standard steel containing 0.800%Mn. The standard steel is dissolved to the same final volume of solution as the unknown; its absorbance is 0.390. Calculate the %Mn without using the weights of the steels, and then calculate the percent error incurred by ignoring the weights.
Solution: When the weights are omitted from **13-2**, we calculate %Mn as follows:

$$\%\text{Mn}_u = \frac{0.460}{0.390}(0.800\%\text{Mn}) = 0.943_6\%$$

If the weights are retained in **13-2**, we obtain the correct %Mn.

$$\%\text{Mn}_u = \frac{0.460}{0.390}(0.800\% \text{ Mn})\frac{(1.212 \text{ g})}{(1.202 \text{ g})} = 0.951_4\%$$

$$\%\text{error} = \frac{(0.951_4 - 0.943_6)}{0.951_4}[100] = 0.82\% (<1\% \text{ error})$$

Condensation Reactions in Organic-Clinical Analysis

A number of colorless aldehydes and ketones can be condensed with aromatic amines to form colored condensation products, called Schiff bases. The general reaction is as follows:

$$R_2C{=}O \quad + H_2N{-}R' \longrightarrow R_2C{=}N{-}R' + H_2O$$

(aldehyde or ketone)   (amine)

For example, 2,4-dinitrophenylhydrazine, an amine, is used to form orange-red Schiff bases with many aldehydes and ketones. Other amines, such as *o*-toluidine, are used for certain special compounds.

Glucose in blood serum is readily determined by condensation with *o*-toluidine, C_7H_7—NH_2, to form a green Schiff base, $C_5H_{11}O_5$—HC=N—C_7H_7 (Ch. 22). The Schiff base has an absorption band at 630–640 nm. After *o*-toluidine reacts with the glucose, the absorbance at 630 nm is directly proportional to the glucose concentration. For accurate analysis, a Beer–Lambert law plot is used; for a quick, less accurate analysis, the sample (unknown) absorbance, A_u, is compared with the absorbance of a standard, A_{std}. By writing a ratio of two Beer–Lambert law equations, where the cell path lengths are equal, we obtain a general equation for this situation.

$$c_u = \frac{A_u}{A_{std}} (c_{std}) \qquad (13\text{-}3)$$

Since glucose concentrations are usually expressed in mg/dL ($=$ mg/100 ml), the results are expressed in this unit rather than molarity. The normal range for glucose in blood serum is from 65 to 90 mg/dL.

Multi-component Analysis

Many times spectrophotometric analyses must be made of a mixture of components, two or more of which absorb at the analytical wavelength. If not more than two or three components absorb at the analytical wavelength, it is often possible to calculate the concentration of any or all of them from two or more absorbance values at different wavelengths. Two calculation situations are the most common: (1) only one component must be determined, so that the absorbance at the analytical wavelength is simply corrected for the absorbance of the other components(s); and (2) two or three components must be determined, so that two equations in two unknowns or three equations in three unknowns must be developed.

Determination of One Component in a Mixture

In the former situation, only one component, component *y*, is to be determined in a mixture where both *y* and a second component, *z*, absorb at the analytical wavelength, wavelength 1. The simplest way to correct for the absorbance of *z* at this wavelength is to find a second wavelength, wavelength 2, at which *y* does not absorb but at which *z* has the same absorbance ("isoabsorptive"), so that

$$A_{2z} = A_{1z}$$

Then the absorbance of *y* alone at wavelength 1, A_{1y}, is found by measuring the total absorbance at wavelength 1, A_1, and measuring A_{2z} and subtracting as follows:

$$A_{1y} = A_1 - A_{1z} = A_1 - A_{2z}$$

The determination of bilirubin in the blood of infants is an example of this type of analysis, provided the infants have not started to eat solid food containing carotene (the carotene absorbs in the same region as bilirubin). Bilirubin is a yellow degradation product of hemoglobin in the body and is responsible for the color of the bile and the urine. An abnormally high level of bilirubin in the blood causes the yellow color associated with jaundice in newborn infants; this is the reason the level of bilirubin must be measured frequently. Spectrophotometric measurement of bilirubin at 461 nm is complicated by the absorption of oxyhemoglobin at this wavelength (Fig. 13-7).

Examination of the spectrum of oxyhemoglobin alone reveals that there is a wavelength at 551 nm which is "isoabsorptive" with the wavelength at 461 nm; that is, they have the same molar absorptivities and thus the same absorbances for a given solution.

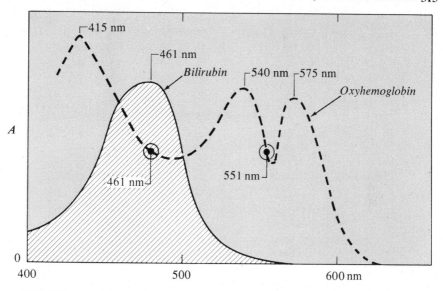

FIGURE 13-7. **Absorption spectra of separate solutions of bilirubin (shaded area), 30 mg/100 ml, and oxyhemoglobin, 800 mg/100 ml. The circled points at 461 nm and 551 nm have equal absorbances for any concentration of oxyhemoglobin.**

Note that bilirubin does not absorb significantly at 551 nm. (If it did, another isoabsorptive wavelength of oxyhemoglobin at about 590 nm could be chosen.) The absorbance of bilirubin at 461 nm, $A_{461(b)}$, can be found by subtracting the absorbance of oxyhemoglobin at 551 nm, $A_{551(o)}$, from the total absorbance of both at 461 nm.

$$A_{461(b)} = A_{461} - A_{551(o)} \tag{13-4}$$

The value of $A_{461(b)}$ can then be entered onto a Beer–Lambert law plot of the absorbance of bilirubin at 461 nm against concentration. For quick analysis, the absorbance of a bilirubin standard, $A_{461(std)}$, can be used with $A_{461(b)}$ in an equation analogous to **13-3** to calculate the unknown concentration, $c_{u(b)}$.

$$c_{u(b)} = \frac{A_{461(b)}}{A_{461(std)}} (c_{std})$$

A special instrument made by the American Optical Corporation enables medical laboratories to perform the above analysis automatically. It splits the light transmitted by the bilirubin-oxyhemoglobin mixture into two beams, measures the absorbances at 461 and 551 nm, and automatically corrects the final readout for the oxyhemoglobin.

Determination of Two or More Components in a Mixture

If two or more components must be determined in a mixture, then two or more Beer–Lambert law equations must be used to calculate their concentrations. Of course, the Beer–Lambert law must be obeyed by all components involved in the calculations; that is, the absorbances must be additive. This requirement can be checked experimentally by mixing standard solutions of known absorbance and measuring the absorbances of the mixture.

Once additivity is verified, two components, y and z, can be determined by measuring the absorbances at two wavelengths, 1 and 2. Since the total absorbance at wavelength 1,

A_1, and the total absorbance at wavelength 2, A_2, are the sums of the absorbances of y and z, two Beer–Lambert law equations can be written.

$$A_1 = \varepsilon_{1y} b c_y + \varepsilon_{1z} b c_z$$

$$A_2 = \varepsilon_{2y} b c_y + \varepsilon_{2z} b c_z$$

The first character in the subscripts of ε refers to the wavelength and the second refers to the component. Assuming that $b = 1.0$ cm, we simplify the above equations to the following:

$$A_1 = \varepsilon_{1y} c_y + \varepsilon_{1z} c_z \tag{13-5}$$

$$A_2 = \varepsilon_{2y} c_y + \varepsilon_{2z} c_z \tag{13-6}$$

Once the values of the molar absorptivities are inserted, **13-5** and **13-6** reduce to two equations in two unknowns. These are easily solved by substituting one equation into the other to obtain one equation in one unknown.

The choice of wavelengths in this kind of situation is important. Each wavelength chosen should be one at which there is a significant difference between the absorbances of each of the components. As shown in Figure 13-8, component z absorbs more intensely at wavelength 1 and component y absorbs more intensely at wavelength 2. (The actual absorbance spectrum that will result from a mixture of these two is shown as the sum of y and z.)

For a three-component mixture, three equations involving absorbance measurements and known molar absorptivities at three wavelengths are necessary to solve for the three concentration unknowns. This calculation is not cumbersome once it has been reduced to two equations in two unknowns (see Problem 34 at the end of this chapter), but handling four or more components is unwieldy. Fortunately, some modern spectrophotometers are equipped to handle such situations.

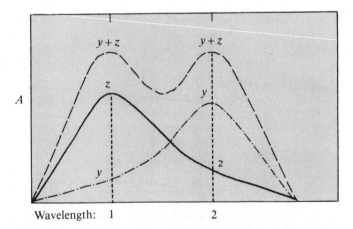

FIGURE 13-8. Choice of wavelength for analysis of a two-component mixture. Wavelength 1 is chosen so that component z absorbs more intensely than component y; wavelength 2 is chosen so that y absorbs more intensely than z.

One such spectrophotometer is the Hewlett-Packard 8450A UV-visible spectrophotometer with a built-in microcomputer. Because it is equipped with 400 diode detectors,
it can measure the absorbance of a mixture at 400 wavelengths simultaneously. This
capability, along with the microcomputer, means that absorption curves of standard
solutions of up to seven components can be "fitted" to the absorption spectrum of the
mixture. Once a "fit" has been achieved, the concentration of each component in the
mixture is then printed out automatically. A practical analysis of a headache preparation for salicylic acid, salicylamide, acetaminophen, and caffeine is a routine procedure
with such a spectrophotometer.

Continuous Monitoring of a Changing Sample

The composition of the air we breathe and the water we drink is constantly changing,
however slightly. It is now recognized that the amounts of certain constituents in each
must be determined continuously, or monitored continuously, to protect our health.

Monitoring Public Water Supplies

The continuous monitoring of public water supplies is much simpler than monitoring
the air, because all of the water supply from a given plant must pass through its pumping
system. A sample of the water flowing past a certain point can be withdrawn continuously and analyzed for a desired constituent by continuous spectrophotometric measurement. Examples of constituents in the water that can be measured continuously
are chlorine and the fluoride ion (added as sodium fluoride). The concentration of the
chlorine must be monitored so that it does not fall below the level below which the
bacteria content of the water would make it unsafe to drink. The concentration of the
fluoride must be monitored so that it does not fall below the 1-ppm level needed for
protection of the teeth, and so that it does not rise appreciably above this level and poison
the water. The monitoring for both chlorine and fluoride is usually done by some type
of *repetitive* colorimetric analysis. The analyses are usually done at definite intervals
throughout the day; the absorbance readings from the colorimeter may be traced on a
circular chart representing a given 24-hour interval.

Monitoring Air

The continuous monitoring of the air in any city is very difficult, because changing wind
direction and speed make the sampling of the air difficult. Metropolitan areas such as
Los Angeles and New York have as many as 10 to 12 stations at which levels of sulfur
dioxide, carbon monoxide, particles in the air, etc., are measured spectrophotometrically
and reported to a central headquarters. Air pollution *alerts* are announced on the basis
of these rapid spectrophotometric measurements.

In addition to the air supply, potential sources of pollution are also monitored directly.
Infrared absorption spectra are very useful for identifying and measuring harmful
constituents in flue gases from plant exhaust stacks. A rapid-scan infrared spectrophotometer gives a complete spectrum in 12.5 seconds, providing almost continuous
checks on the composition of the flue gas (Fig. 13-9).

Harmful gases, such as sulfur dioxide and carbon monoxide, are both unsaturated
molecules and absorb in the *medium energy* infrared region (Table 13-5). Even carbon
dioxide can be identified from the spectrum in Figure 13-9. The fact that the spectrum is
run many times means that any increase in the absorbance gives a quantitative estimation
of an increase in concentration of any of the gases in the flue.

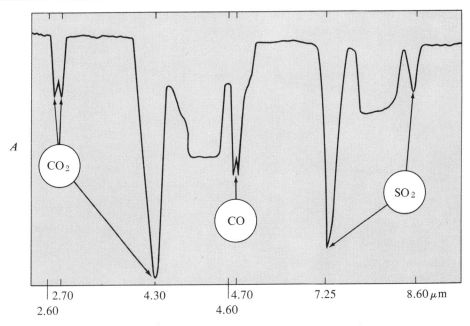

A

FIGURE 13-9. **The hypothetical infrared spectrum of a flue gas similar to that reported by Comberiati [3]. The identity of carbon dioxide is confirmed by the intense band at 4.30 μm and supported by weaker bands at 2.60 and 2.70 μm. The identity of carbon monoxide is established by the two peaks at 4.60 and 4.70 μm. The identity of sulfur dioxide is confirmed by the intense band at 7.25 μm and supported by the weaker band at 8.60 μm.**

13-4 | AUTOMATED ANALYSIS

In laboratories where large numbers of samples must be analyzed for the same constituent(s), quantitative analytical methods have been automated to obtain the results rapidly. This is especially true in the clinical or hospital laboratory. Many types of instruments are available; we will discuss the common Technicon® AutoAnalyzer® system.

In any automated spectrophotometric measurement, the problem is how to keep the samples separated until the absorbance is measured, and then to measure one sample at a time automatically. In the above-mentioned instrument, air bubbles, with or without a wash solution, are used to keep the samples separate. If necessary, the bubble-separated samples are first treated to form a light-absorbing species; they are then pumped to a one-piece debubbler and flowcell (Fig. 13-10). The sample "falls" into an elbow, as the bubble of air is pumped out through a debubbler exhaust. It then passes into a narrow flowcell with a volume equal to that of only one sample. Pumping action is interrupted long enough for light to pass through the flowcell and for the absorbance to be measured by the detector. The sample is then pumped out through a waste exit, and the next sample enters and fills the flowcell to be measured.

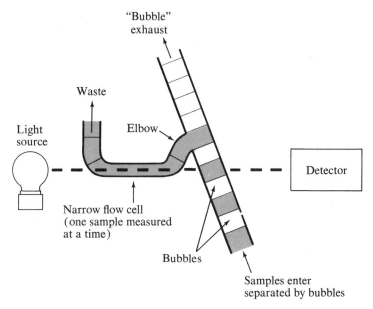

FIGURE 13-10. A typical Technicon® AutoAnalyzer® one-piece debubbler and flowcell. Bubbles leave the light path (dotted line) before the absorbance is measured by the detector. The flowcell is bent so that the light only passes through it and not through the sample in the elbow behind it.

Automated Determination of Glucose

One of the most common constituents determined in body fluids is glucose,

$$C_5H_6(OH)_5CHO$$

which contains an easily oxidized aldehyde. Many of the reagents employed to determine glucose either oxidize it to gluconic acid or condense with the glucose aldehyde group (Sec. 22-2).

glucose + o-toluidine reagent $\xrightarrow{\text{8 min boiling}}$ colored product (630 nm)

glucose + copper neocuproine $\xrightarrow{\text{heat at } 90°}$ colored product (460 nm)

In the ferricyanide method, glucose reacts with yellow ferricyanide and destroys the color of the ferricyanide by forming the virtually colorless ferrocyanide ion as follows:

glucose + $2\,Fe(CN)_6^{-3}$ $\xrightarrow{\text{heat at } 95°}$ $2\,Fe(CN)_6^{-4}$ + gluconic acid
(colorless)

The decrease in the absorbance of the yellow ferricyanide ion is proportional to the concentration of the glucose originally present.

A typical AutoAnalyzer® used to determine glucose by any of the above methods is shown in Figure 13-11.

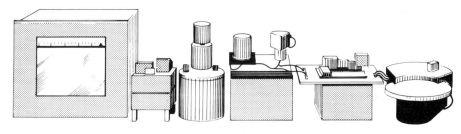

TYPICAL EXAMPLE OF A SINGLE CHANNEL
(here a Blood Urea Nitrogen Analysis)

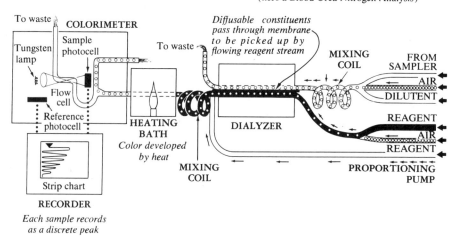

FIGURE 13-11. **A single-channel Technicon® AutoAnalyzer® used for analysis of blood glucose or blood urea nitrogen. Each sample is kept separate from the others by means of air bubbles. The absorbance measured by the colorimeter is recorded for each sample by the recorder at bottom left.**

Automated Determination of Protein-Bound Iodine (PBI)

Protein-bound iodine is present at such low levels in blood samples that it is measured by first decomposing the blood sample to inorganic iodide ion.

$$R-CH_2-I \xrightarrow{\text{decomp.}} I^- + CO_2 + H_2O$$

The aqueous part of the sample (which contains the iodide ion) is added to a mixture of yellow cerium(IV), arsenic(III), and sulfuric acid. In this acid, these two reagents react extremely slowly.

$$2\,Ce(IV) + As(III) \xrightarrow[\text{catalyst}]{\text{slow without}} 2\,Ce(III) + As(V)$$

Iodide is an effective catalyst for this oxidation-reduction reaction, so as soon as the sample is added, the reaction rate increases. The decrease in the color of cerium(IV) is measured at 420 nm; this decrease is proportional to the amount of iodide present. None of the other substances present absorb at 420 nm, so there are no interferences.

The AutoAnalyzer® can also be used to follow this reaction; a modification of the instrument shown in Figure 13-11 is used. The automated analytical results using this method are much more reliable than those possible before automation.

| QUESTIONS AND PROBLEMS

(Answers to even-numbered problems are in Appendix 5.)

Definitions and Concepts

1. List and explain the two types of:
   a. Qualitative spectrophotometric analysis.
   b. Quantitative spectrophotometric analysis.
2. Explain how each of the following chromophores absorbs radiation:
   a. Electronic chromophore
   b. Vibrational chromophore
3. Explain the difference between an n-π^* and a π-π^* transition.
4. Indicate which bond in each of the following pairs has the higher vibrational energy and would therefore absorb at shorter wavelengths in the infrared.
   a. C—H and C—C
   b. O—H and C—H
   c. C=O and C=C
   d. C≡O and C=O
   e. Ester C=O and ketone C=O
5. Explain how protein-bound iodine is measured spectrophotometrically.

Electronic Chromophores

6. As discussed in this chapter, when purple MnO_4^- absorbs light, a π electron is excited to a higher energy orbital.
   a. It is known that the π electron is not excited to a π^* orbital. To what other orbital could the electron be excited? (*Hint:* The manganese in MnO_4^- has how many d electrons?)
   b. Why doesn't the manganese in MnO_4^- act as a d orbital chromophore similar to those ions in Table 13-3?
7. The yellow chromate ion has the following structure:

$$:O=\overset{\overset{\displaystyle O^-}{|}}{\underset{\underset{\displaystyle O^-}{|}}{Cr}}=O:$$

   a. What type of electron can absorb the visible radiation that results in the yellow color?
   b. What two different kinds of orbitals could the excited electron occupy in the excited state?

If chromate is like permanganate in the previous problem, which orbital is more likely?
   c. Why doesn't the chromium(VI) in CrO_4^{-2} act as a d orbital chromophore similar to those ions in Table 13-3? (*Hint:* Find the number of d electrons in chromium(VI).)
8. The gas nitrogen dioxide can be represented by the structure O=N⁄⁄⁄O.
   a. Would you predict that it would absorb in the UV-visible region? Why or why not?
   b. Absorption of UV radiation by this molecule is thought to be the first step in formation of smog. Nitrogen dioxide decomposes to form a molecule and a reactive atomic species. Suggest what the molecule and atomic species are.
   c. Some components of smog are irritating aldehydes. What two species could react to produce aldehydes?

Vibrational Chromophores

9. As a solution of ethanol is diluted from $1.2M$ to $0.025M$, the 3.0-μm band of the $1.2M$ solution decreases in size and a new band appears at 2.8 μm. At $0.025M$ concentration, the 3.0-μm band disappears. Explain the appearance of the new band and the disappearance of the first band.
10. Indicate whether or not each of the following molecules will absorb in the 2.7–3.0 μm region. Also explain which bond is primarily responsible.
    a. CH_3CO_2H
    b. CH_3CHO
    c. CH_3CH_2OH
    d. $CH_3-\overset{\overset{\displaystyle O}{\|}}{C}-CH_3$
11. Indicate whether or not each of the following molecules will absorb in the 4.4–5.90 μm region. Also explain which bond is primarily responsible.
    a. CH_3CN
    b. $CH_3CH_2NH_2$

c. $CH_3C \equiv CH$

d. $CH_3CH = CH_2$

12. The infrared spectrum of

$$CH_3 - \underset{\underset{O}{\parallel}}{C} - CH_2 - \underset{\underset{O}{\parallel}}{C} - CH_3 \text{ has a broad}$$

absorption band at 3.3 μm. Explain why this band is observed when both oxygens appear to be double-bonded to carbons; also explain why the band is broad.

Qualitative Analysis

13. 0.400 g of regular aspirin has an absorbance of 0.012 at 312 nm and meets the FDA tolerance for salicylic acid. A special 0.600-g time-release aspirin tablet has an absorbance of 0.040 at 312 nm. Does the time-release tablet meet the salicylic acid tolerance for the regular weight of aspirin tablet?

14. A bottle of aspirin tablets is analyzed spectrophotometrically for salicylic acid. First 20 tablets are weighed out to obtain a representative sample and an average tablet weight. The 20 tablets weigh a total of 8.080 g. The tablets are then analyzed for their aspirin content and are found to contain 80.2% acetylsalicylic acid. Analysis of a 500-mg portion of the 20 tablets (taken after grinding them to a fine powder) gives 0.15 mg of salicylic acid.
   a. Calculate the weight of the average tablet.
   b. Calculate the average weight of aspirin in the average tablet.
   c. Calculate the percentage of salicylic acid impurity in an average tablet by using the weight of the aspirin alone in the denominator. Does the aspirin meet the FDA tolerance?

15. A screening test is to be devised for the species that absorbs at the longer wavelength from each pair below. State which species this is, whether it absorbs in the visible or ultraviolet, and what type of electron (d, π, etc.) absorbs the radiation.
   a. $Na(H_2O)_6{}^+$, $Fe(H_2O)_6{}^{+2}$
   b. $Ca(H_2O)_6{}^{+2}$, $Cr(H_2O)_6{}^{+3}$
   c. $Ag(H_2O)_6{}^+$, $Ag(H_2O)_6{}^{+2}$
   d. $MnO_4{}^-$, $NO_3{}^-$
   e. $NO_3{}^-$, F^-

16. Identify the two ultraviolet chromophores and the two *unique* infrared chromophores that can be used in the identification of the compound below in the presence of $CH_3 - CH_2 - CH_2 - NH_2$.

$$\underset{N(CH_3)_2}{\underset{|}{\overset{H}{\overset{|}{\underset{}{C}}}}} - CH_2 - \overset{H}{\underset{}{C}} - CO_2H$$

17. A sample suspected of containing heroin exhibits UV absorption in the 250–280 nm region and IR absorption at 3.9 μm. Do these results:
   a. Rigorously indicate heroin?
   b. Tentatively indicate heroin?
   c. Fail to indicate heroin?

18. A sample is suspected of being a barbiturate or a phenylethylamine. It exhibits UV absorption at 260 nm and IR absorption at 3.0 μm and at 5.8–6.0 μm. Suggest what might be present.

Quantitative Analysis Involving One Absorbing Species

(Answers to the first two parts of every problem in this group are given in the appendix.)

19. In each case below, calculate the %Mn in the unknown steel after allowing for the difference in weights between the steels (Sec. 13-3). Unless stated otherwise, the solution volumes and cell path lengths are the same.
   a. The %Mn in the standard steel is 0.78%, and the weight of the sample of standard steel is 1.292 g. The sample weight of the unknown steel is 1.301 g. At 520 nm, the absorbance of the standard is 0.60; the %T of the unknown steel is 40%.
   b. The %Mn in the standard steel is 0.715%, and the weight of the sample of standard steel is 1.294 g. The sample weight of the unknown steel is 1.494 g. At 520 nm, the absorbance of the standard is 0.55; the %T of the unknown steel is 14%.
   c. The %Mn in the standard steel is 0.78%, and the weight of the sample of standard steel is 1.292 g. The sample weight of the unknown steel is 1.322 g. At 520 nm, the absorbance of the standard in a 1.00-cm

cell is 0.60; the $\% T$ of the unknown steel in a 2.00-cm cell is 16%.

20. In each case below, a 0.200-g sample of a manganese-containing alloy is dissolved in acid and diluted to exactly 200 ml in a volumetric flask. The specified aliquot is withdrawn and oxidized to permanganate. After correcting for the dilutions specified, calculate the percentage of manganese in each alloy.

   a. A 25.0-ml aliquot is taken from the 200 ml of alloy A; spectrophotometric measurement of permanganate gives a concentration of 1.85 ppm of manganese in the 25.0 ml.

   b. A 25.0-ml aliquot from the 200 ml of alloy B is diluted to 100.0 ml before spectrophotometric measurement; the concentration of manganese is found to be 1.80 ppm of Mn in the 100 ml.

   c. A 10.0-ml aliquot from the 200 ml of alloy C is diluted to exactly 250 ml before spectrophotometric measurement; the concentration of manganese is 1.60 ppm Mn in the 250 ml.

21. In each case below, calculate the mg/dL of glucose in the blood sample, using the absorbances and the concentration of the given standard solution of glucose. Unless stated otherwise, the wavelength is 630 nm and the cell path lengths are the same. Also state whether the blood sample in each case is within the normal range.

   a. The absorbance of the blood sample is 0.33. The absorbance of a 100 mg/dL standard of glucose is 0.41.

   b. The absorbance of the blood sample is 0.21. The absorbance of a 900-ppm (mg/liter) glucose standard is 0.12.

   c. The percent transmittance of the blood sample is $40\% T$ (1.00-cm cell). The absorbance of a 550-ppm standard of glucose in a 2.00-cm cell is 0.51.

22. Iron(III) in vitamin tablets can be determined by first wet-ashing the organic matter to convert all of the iron to a soluble form, and then adding a color-forming ligand. The absorbance of the resulting solution is compared with that of a solution of standard iron(III) whose concentration is given in ppm (mg/liter). Calculate the mg of Fe^{+3} per average vitamin tablet in each of the following samples:

   a. Five tablets (total wt = 5.75 g) are wet-ashed and diluted to a final volume of 2.00 liters. Then a 5.00-ml aliquot is mixed with 25.00 ml of color-forming ligand, and the mixture is diluted to 50.0 ml, which gives an A of 0.50 (1.00-cm cell). A 5.00-ml aliquot of 17.5-ppm iron(III) standard when treated similarly (diluted to 50.0 ml) gives an A of 0.61 (1.00-cm cell).

   b. Ten tablets (total wt = 11.50 g) are wet-ashed and diluted to a final volume of 2.00 liters. A 5.00-ml aliquot is mixed with 25.00 ml of color-forming ligand, and the mixture is diluted to 50.0 ml, which gives an A of 0.30 (1.00-cm cell). A 5.00-ml aliquot of 17.5-ppm iron(III) standard when treated similarly (diluted to 50.0 ml) gives an A of 0.61 (1.00-cm cell).

   c. Five tablets are ground together, giving 5.75 g of powder; a 1.725-g sample of the powder is wet-ashed and diluted to 2.00 liters. A 5.00-ml aliquot is mixed with 25.00 ml of color-forming ligand and then diluted to 50.0 ml, which gives an A of 0.30 (1.00-cm cell). A 2.00-ml aliquot of a 17.5-ppm iron(III) standard when treated similarly (diluted to 50.0 ml) gives an A of 0.244 (1.00-cm cell).

Quantitative Multicomponent Analysis (for One or Two Components)

23. Calculate the concentration of bilirubin in each bilirubin-oxyhemoglobin mixture below, using the data for the given standard bilirubin solution. Unless otherwise stated, assume that the cell path lengths are the same in each case.

   a. A bilirubin-oxyhemoglobin mixture has an A of 0.412 at 461 nm and an A of 0.103 at 551 nm. A $1.0 \times 10^{-4} M$ standard bilirubin solution (no oxyhemoglobin) has an A of 0.400 at 461 nm.

   b. A bilirubin-oxyhemoglobin mixture has an A of 0.70 at 461 nm and an A of 0.30 at 551 nm. A $3.185 \times 10^{-4} M$ standard bilirubin solution (no oxyhemoglobin) has a %T of 31.6% at 461 nm.

c. A bilirubin-oxyhemoglobin mixture has a %T of 19.9% at 461 nm and an A of 0.30 at 551 nm in a 2.00-cm cell. A $3.185 \times 10^{-4} M$ standard bilirubin solution (no oxyhemoglobin) has an A of 0.50 at 461 nm in a 1.00-cm cell.

24. Calculate the concentration of bilirubin in each of three mixtures of bilirubin (b) with the respective components I, II, and III, using the data for a given standard bilirubin solution (second column) for all three mixtures. The concentrations of I, II, and III alone are different from their concentrations in the respective mixtures. To begin, choose the two isoabsorptive wavelengths for I, II, and III before calculating the concentration of bilirubin. Graph the data if necessary, assuming 1.0-cm cells. See table below.

25. Calculate the concentration of both o-nitroaniline and p-nitroaniline in each of the two-component mixtures below. Use 1.00 cm for the path length and the following molar absorptivities: at 347 nm, $\varepsilon_o = 1.28 \times 10^3$ and $\varepsilon_p = 9.20 \times 10^3$; at 285 nm, $\varepsilon_o = 5.26 \times 10^3$ and $\varepsilon_p = 1.40 \times 10^3$.

a. The mixture has an A of 0.916 at 347 nm and an A of 1.040 at 285 nm.

b. The mixture has an A of 0.711 at 347 nm and an A of 0.200 at 285 nm.

c. The mixture has an A of 0.690 at 347 nm and an A of 0.114 at 285 nm.

26. Calculate the concentration of both $Co(H_2O)_6^{+2}$ and $Cr(H_2O)_6^{+3}$ in each of the two-component mixtures below. Use 1.00 cm for the path length and the following molar absorptivities: at 530 nm, $\varepsilon_{Co} = 10.0$ and $\varepsilon_{Cr} = 1.00$; at 410 nm, $\varepsilon_{Co} = 2.00$ and $\varepsilon_{Cr} = 20.0$.

a. The mixture has an A of 0.53 at 530 nm and an A of 0.70 at 410 nm.

b. The mixture has an A of 0.24 at 530 nm and an A of 0.84 at 410 nm.

c. The mixture has an A of 0.35 at 530 nm and an A of 1.06 at 410 nm.

Ultraviolet Spectrophotometric Analysis

27. German researchers have found that inorganic nitrate in tobacco can be determined by UV spectrophotometric measurement after removal of all but one of the ultraviolet-absorbing impurities. The remaining impurity absorbs both at 203–210 nm and at 232 nm. Suggest two possible approaches for determining the nitrate ion without error from the impurity.

28. Most proteins have an absorption band at 280 nm because of the three aromatic amino acids; however, nucleic acids also absorb there and must be corrected for by taking the absorbance at 260 nm. An equation for calculating protein in the presence of nucleic acids is

$$\text{Protein(mg/ml)} = 1.55A_{280} - 0.76A_{260}$$

λ, nm	A, bilirubin $(3.185 \times 10^{-4} M)$	A_I	A_{I+b}	A_{II}	A_{II+b}	A_{III}	A_{III+b}
560	0.00	0.65	0.20	0.10	0.08	0.20	0.09
550	0.00	0.55	0.16	0.40	0.30	0.30	0.14
540	0.00	0.80	0.25	0.45	0.36	0.20	0.09
530	0.00	0.90	0.28	0.55	0.44	0.66	0.32
490	0.20	0.85	0.40	0.65	0.67	0.75	0.51
480	0.50	0.70	0.60	0.70	0.96	0.90	0.83
460	0.75	0.65	0.80	0.65	1.12	0.75	0.96
450	0.70	0.50	0.64	0.50	0.96	0.66	0.93
440	0.50	0.40	0.50	0.40	0.70	0.55	0.65
430	0.40	0.30	0.39	0.15	0.44	0.45	0.53

Calculate the protein concentration in both mg/ml and mg/dL in a sample with an A of 0.35 at 280 nm and an A of 0.20 at 260 nm.

29. The inorganic species below are all colorless, but some in each group absorb in the ultraviolet region. Predict which species will do so and draw the formula of each chromophore.
  a. NO_2, CS_2, O_3, HF, He, Ne
  b. NO_3^-, NH_4^+, NO_2^-, F^-, SCN^-, $H_2BO_3^-$

30. State which molecule of each pair below absorbs at the longer wavelength (and can thus be determined without interference from the other molecule). If no wavelengths are available in the text, estimate each λ_{max} to two significant figures.
  a. Salicylic acid and acetylsalicylic acid
  b. $H(HC=CH)_2H$ and $H(HC=CH)_3H$
    (butadiene)   (hexatriene)
  c. $H(HC=CH)_3H$ and $H(HC=CH)_4H$
    (hexatriene)   (octatetrene)
  d.

  (benzene)   (naphthalene)
  e.

  (anthracene)

  and

  (tetracene)

31. Two mixtures containing anthracene and phenanthrene (Fig. 13-4) are to be analyzed by ultraviolet spectrophotometry. For each of the mixtures below, select the best analytical wavelength, and calculate the detection limit at that wavelength for $A = 0.01$.
  a. Determine anthracene mixed with 1/100th as much phenanthrene.
  b. Determine phenanthrene mixed with 1/100th as much anthracene.

Challenging Problems

32. The molar absorptivity of the $FeSCN^{+2}$ ion is 5×10^3, but the molar absorptivity of the $Fe(SCN)_2^+$ ion is 9.8×10^3. Suggest why the molar absorptivity of the second ion is almost twice that of the first. What are the

implications of this for the determination of iron(III)?

33. A $0.01M$ solution of $Fe(H_2O)_6^{+3}$ has two absorption bands in the visible at 406 and 540 nm ($\varepsilon_{540} = 0.1$; $\varepsilon_{406} = 0.5$) and a trough from 620 to 750 nm.
  a. Evaluate the spectrophotometric response to such a solution in a 1.0-cm cell.
  b. Describe the appearance to the eye of such a solution in a 1.0-cm cell.

34. Calculate the concentration of $Co(H_2O)_6^{+2}$, $Cr(H_2O)_6^{+3}$, and $Ni(H_2O)_6^{+2}$ in the three-component mixture described below. Use 1.00 cm for the cell path length. Use the molar absorptivities and the absorbance values in the table to set up three equations in three unknowns.

	410 nm	530 nm	575 nm
A of mixture	1.20	0.570	0.910
ε_{Co}	2.00	10.0	5.00
ε_{Cr}	20.0	1.00	15.0
ε_{Ni}	5.00	0.50	1.00

35. The infrared spectrum of 2,4-pentanedione has a broad absorption band at 3.3 μm and bands at 5.8 and 6.1 μm.
  a. Explain why two bands instead of one are observed in the carbonyl region (5.8–6.1 μm) and suggest what type of carbonyl group is absorbing at each wavelength.
  b. This compound consists of what two tautomers? Which of these would absorb at 3.3 μm, and what chromophore is responsible?

36. The following compounds absorb at the given wavelengths (the molar absorptivity follows each in parentheses): CH_3I—257 nm (378); CH_3Br—202 nm (80); CH_3Cl—177 nm ($\varepsilon < 80$).
  a. Suggest a reasonable electronic transition that might account for all of the bands.
  b. Rationalize the trends in the location of the bands and the molar absorptivities.

37. In the mixture of compounds below, identify the compound absorbing at the longest wavelength and the one absorbing at the next-to-longest wavelength. Indicate the

number of conjugated double bonds and electron pairs to support your answers.

a.

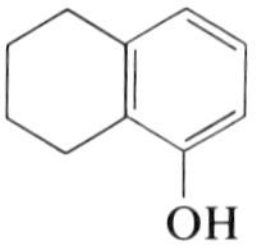

OH

b.

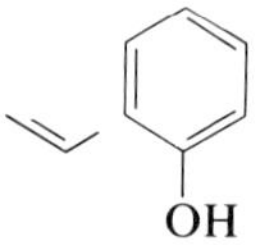

OH

c.

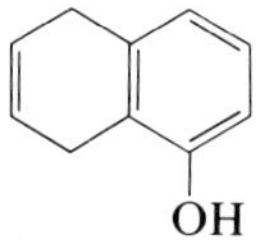

OH

d.

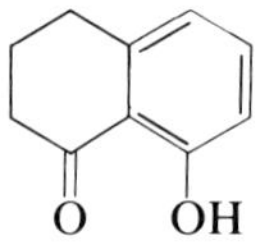

O   OH

e.

OH   OH

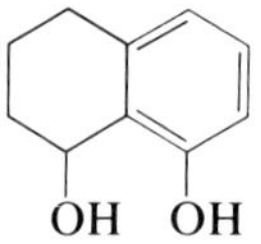

38. Hooke's law indicates that the stretching frequency (measured in cm^{-1}) of a bond that is absorbing infrared radiation is proportional to the square root of the quotient of the force constant, k, of the bond and the reduced mass of the respective masses, M_1 and M_2, of the atoms of the bond.

$$\bar{v} \propto \sqrt{\frac{k(M_1 + M_2)}{M_1 M_2}}$$

Given that the force constant for a 2980 cm^{-1} C—H stretching is the same as the analogous C—D force constant, calculate:
a. The ratio of the C—H to C—D stretching frequencies to four significant figures, using 2.0140 for the mass of deuterium.
b. The value in cm^{-1} of the C—D stretching frequency.

| NOTES

[1] A. Viste and H. B. Gray, *Inorg. Chem.* **3**, 1113 (1964); and G. H. Schenk, *Rec. Chem. Progr.* **28**, 135 (1967).

[2] W. Giger and M. Blumer, *Anal. Chem.* **46**, 1663 (1974).

[3] J. R. Comberiati, *Anal. Chem.* **43**, 1497 (1971).

14

Fluorometry, Refractometry, and Light Scattering

Even now in the stillness of death, the huge jaws
seemed to be dripping with a bluish flame. . . . I
placed my hand upon the *glowing* muzzle and as
I held them up my fingers *gleamed in the darkness.*
  "Phosphorus," I said.
  "A cunning preparation of it," said Holmes.

ARTHUR CONAN DOYLE
The Hound of the Baskervilles, Chapter 14

In the previous chapters we described the measurement of absorption of radiation. In this chapter three other measurements of the interaction of radiation with solutions will be discussed—fluorometry, refractometry, and the measurement of light scattering.

These measurements are related in that each is a measurement of radiation after it has been *affected* in some way by interaction with a species in solution. In contrast, spectrophotometry involves measurement of radiation that has *not* interacted with a species in solution.

Fluorometry is concerned with the measurement of the fluorescence emitted by a species that has absorbed incident radiation. The fluorescence is emitted in *all directions* and is of longer wavelengths and lower energies than the absorbed radiation. As shown in Figure 14-1, not all of the incident radiation is absorbed; some of it passes through the solution unchanged. Only that which is emitted as fluorescence changes in wavelength.

Refractometry involves measuring the *change in direction* of incident radiation as it passes from air into a solution or solid. As shown in Figure 14-1, the incident radiation is not absorbed; it interacts with the solution species in such a way that only its direction, not its wavelength, is changed. If only refraction occurs, we can say that none of the photons of the incident radiation are absorbed; they merely undergo a change in direction.

Measurement of light scattering is the measurement of the scattering of photons of the incident radiation by solid particles suspended in the solution. Some of the incident

327

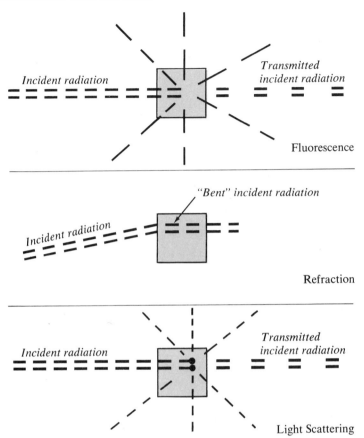

FIGURE 14-1. Schematic representations of fluorescence (top), refraction (middle), and light scattering (bottom). The length of the line (photon) represents the wavelength. Only in the case of fluorescence (top) does the wavelength change, becoming longer. The solid dots in the bottom portion of the figure represent suspended solids which scatter the incident radiation.

radiation passes through the solution (Fig. 14-1), but much of it is scattered in all directions, *without a change in wavelength.*

We will begin our discussion of these topics with a discussion of fluorometry.

14-1 | THE FLUORESCENCE OF MOLECULES AND IONS

The nature of fluorescence and the measurement of fluorescence are both complicated; in this chapter we present only the essentials—a simplified picture of the energy changes occurring during fluorescence, the fluorescence-concentration relation, and a few applications selected from thousands.

You should review both the discussion of absorption of photons in Section 12-1 and the descriptions of the components of a spectrophotometer in Section 12-3 before reading the material below.

Energy Changes Occurring During Fluorescence

The absorption process discussed in the two previous chapters is the first step in fluorescence. In this step a molecule or ion is *excited* by a *photon* of radiation; this is referred to as *photoexcitation*. The emission of photons following photoexcitation is in general called *photoluminescence*, or, less accurately, luminescence. Fluorescence is one type of photoluminescence; phosphorescence, which is beyond the scope of this chapter, is another type. In a photoluminescent process, such as fluorescence, the emitting species always reaches the excited state by absorbing a photon. Chemiluminescence and bioluminescence are processes that differ from a photoluminescent process like fluorescence in that chemical energy is used to promote the emitting species to the excited state. For example, the firefly exhibits bioluminescence as a result of chemical reactions producing a luminescent species.

The exciting radiation in fluorescence is usually ultraviolet (UV) radiation, but occasionally visible radiation is used. The excitation promotes the ground state ion or molecule to the excited state. The entire process is summarized simply as follows:

$$\text{ground st.} + \text{UV} \xrightarrow{\text{excitation}} \text{excited st.} \xrightarrow{\text{emission}} \text{ground st.} + \text{fluor.}$$

In general, the energy of fluorescent radiation is much lower than that of the exciting radiation. Since wavelength is inversely proportional to energy, fluorescent radiation is located at longer wavelengths than the exciting radiation. Usually fluorescent emission is visible to the eye, but it may also occur in the UV above 300 nm. Just as substances in solution absorb over a range of wavelengths (see the absorption *bands* of hemoglobin in Fig. 12-3), fluorescent ions or molecules in solution emit a range of wavelengths. A fluorescence spectrum is obtained using a *spectrofluorometer*, just as an absorption spectrum is obtained using a spectrophotometer.

Typical fluorescence spectra are shown in Figure 14-2 for the organic molecule *quinine*, an alkaloid used to treat malaria. The left spectrum is a plot of the UV radiation

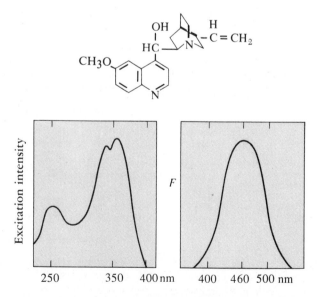

FIGURE 14-2. The fluorescence excitation spectrum (left) and fluorescence emission spectrum (right) of the organic molecule quinine. The structure of quinine is given at the top.

that will excite quinine and is called an *excitation spectrum*. The right spectrum is a plot of the radiation emitted by excited quinine and is called a fluorescence emission spectrum or just a *fluorescence spectrum*.

To explain why species emit fluorescence at longer wavelengths (lower energies) than those of the radiation they absorb, it is necessary to employ a molecular energy diagram. As shown in Figure 14-3, such a diagram displays the electronic ground state and its vibrational sublevels and the electronic excited state and its vibrational sublevels in order of increasing energy. (Recall from Fig. 13-2 and the discussion of IR absorption that an electronic state is divided into many vibrational energy sublevels.)

The Excitation Process

The excitation process is the same as absorption; it begins with all molecules in the ground state, symbolized as S_0 in Figure 14-3. The electronic transition can be represented by one equation as follows:

$$S_0 + \text{UV photons} \longrightarrow S_1$$

where S_1 is the symbol for the first excited state. If vibrational sublevels are included in a description of this process, then *many* transitions must be written to describe excitation accurately. Just three are shown in Figure 14-3—the lowest energy transition to the EV_0 sublevel of S_1, a middle energy transition to any middle energy EV_x sublevel of S_1, and the highest energy transition to the highest energy EV_n sublevel of S_1. The many transitions to the many sublevels of the S_1 state account for the excitation *band*, such as that of quinine in the left spectrum of Figure 14-2.

Relaxation: Loss of Energy

What follows the excitation process largely explains why fluorescence is a lower energy process than excitation (absorption). Molecules or ions in all S_1 vibrational sublevels except EV_0 *relax* to the EV_0 sublevel, losing part of their energy as heat. Since the relaxation occurs through the vibrational motions of the bonds, it is called vibrational relaxation (VR in Fig. 14-3). At this point all the S_1 molecules have less energy than when they were excited, except for the few originally excited to the EV_0 sublevel. The energy lost as heat is transferred to solvent molecules during collisions.

The Fluorescence Emission Process

Any molecule in the EV_0 sublevel of S_1 potentially can fluoresce by emitting a photon; the electronic transition can be represented by one equation as follows:

$$S_1 \longrightarrow S_0 + \text{photons (fluorescence)}$$

However, as can be seen in Figure 14-3, S_1 molecules undergo *many* transitions, emitting photons of varying energies. Just six are shown in Figure 14-3, and all but one emit photons of lower energy than needed to excite an S_0 molecule. This shows why fluorescence is of lower energy and longer wavelengths than absorption (Fig. 14-2) and why a fluorescence *band* is observed.

Molecules that drop to sublevels such as GV_3, GV_2, etc., do not remain in these sublevels but lose their excess vibrational energy by relaxation. This energy is lost as heat to the solvent.

Internal Conversion: A Competing Process

Not all molecules, and very few ions, fluoresce in solution. There are many reasons for this, including processes that compete with fluorescence. Internal conversion, shown as *IC* in Figure 14-3, is one such process. In this process, an S_1 species interacts strongly with the solvent, converting its electronic energy to vibrational energy which is lost to the solvent.

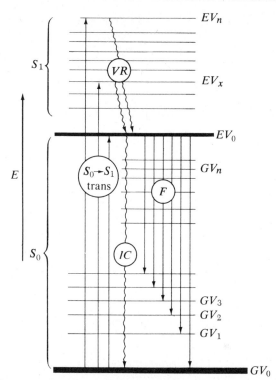

FIGURE 14-3. A molecular energy diagram showing excitation (absorption) of the
ground state, S_0, to the excited state, S_1, and fluorescence (F) from the lowest
vibrational sublevel, EV_0, of the excited S_1 state. After excitation to various
vibrational sublevels of S_1, all molecules relax vibrationally (VR) to the EV_0 sublevel
of S_1. Many of the molecules emit fluorescence and return to various ground
vibrational sublevels (GV_2, GV_1, GV_0, etc.). Many molecules do not fluoresce but
internally convert (IC) their energy to vibrational energy and return to the S_0 state with
loss of heat.

**Fluores-
cence
Instrumen-
tation**

The two types of fluorescence measuring instruments are the filter fluorometer and the
spectrofluorometer. The chief difference between them is that the former utilizes filters
as monochromators and the latter uses gratings as monochromators. Both of these
components were discussed in Chapter 12.

   The basic design of both instruments is similar, and their four essential components
are shown in Figure 14-4. The source is one of the various mercury arc lamps or a xenon
arc lamp; the latter is used in the spectrofluorometer. The radiation from the source
passes through an excitation grating or primary filter, either of which allows only a
certain wavelength range to excite the sample. As the radiation is absorbed by a substance in the sample cell, the substance emits fluorescence in all directions. Fluorescence
emitted at right angles to the exciting radiation passes through a fluorescence grating
or secondary filter, either of which transmits only a certain wavelength range. The
transmitted fluorescence is measured by a detector, and the detector output is recorded
or displayed on a meter.

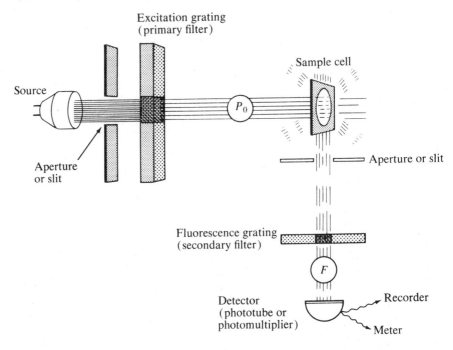

FIGURE 14-4. Schematic (top view) of the components of a fluorometer (filter fluorometer or spectrofluorometer). The source is either a mercury arc or a xenon arc lamp. As shown, the excitation grating or primary filter transmits only a portion of the radiation emitted by the source. Most of the exciting radiation passes through the sample cell without being absorbed. The radiation that is absorbed causes the sample to fluoresce in all directions, but only the emission that passes through the aperture or slit and through the secondary filter or fluorescence grating is measured by the phototube or photomultiplier. The output of the detector is either measured on a meter or plotted on a recorder.

Filter Fluorometric Measurements

Filter fluorometers are primarily used for quantitative analysis; because they are simple to use, they are well suited for routine work. The source is a mercury arc lamp, either an 85-watt medium-pressure lamp (366, 405, 436 nm) or a 4-watt low-pressure lamp. The latter may be of the clear quartz-envelope type for emitting the *254-nm mercury line*, or of the white phosphor-coated type for emitting a *300–400 nm band* of radiation. A filter fluorometer is limited to the above wavelengths for excitation.

Exciting radiation is selected by using a *primary filter*. The bandwidth transmitted by the band pass filters, which are used as primary filters, is similar to that shown in Figure 12-8—about 40–60 nm. The wide bandwidth and higher transmittance of such a filter is one of the advantages of a filter fluorometer over a spectrofluorometer—P_0, the radiant power of the radiation striking the cell, is larger in the filter fluorometer than in the spectrofluorometer. (Recall that P_0 is measured in photons/second.)

The sample cell may be a glass test tube, square glass cell, optical-grade quartz cell, or synthetic silica cell. Glass cells transmit ultraviolet radiation down to about 320 nm; quartz or silica cells must be used below 320 nm.

As shown in Figure 14-4, fluorescence is emitted in all directions, but only that emitted *at right angles* to the exciting radiation is measured. This minimizes scattering of the exciting radiation through the slit and secondary filter to the detector. Since scattering cannot be eliminated, the secondary filter is chosen so it will transmit at longer wavelengths than the exciting radiation. Thus the secondary filter *absorbs* any scattered exciting radiation.

Secondary filters can be of two types—the *band pass* filter, also used as a primary filter, and the *sharp cut* filter. The latter transmits nearly all radiation above a given wavelength. The sharp cut filter thus transmits a larger number of fluorescent photons/second than the band pass filter. As shown in Figure 14-4, the symbol F is used for the number of fluorescent photons/second passing through the secondary filter and striking the detector. F is also used to refer to the intensity of fluorescence read from the meter or recorder. The units for F are usually arbitrarily set from 0 to 100.

Spectro-
fluorometric
Measurements

A spectrofluorometer has basically the same design as a filter fluorometer, so only the differences will be mentioned. The source is a xenon arc lamp; this emits a continuous range of radiation from 200 nm to above 600 nm. It can thus be used to obtain spectra like those in Figure 14-2, because the amount of excitation at each wavelength and the intensity of fluorescence at each wavelength can be measured. This cannot be done using mercury arc excitation, because none of the mercury arcs emits continuously in the 200–600 nm range.

Both the exciting radiation and the emitted fluorescence are selected by using gratings. Usually the grating bandwidth is small, such as 10 nm, and constant. Thus the radiant power, P_0, of the radiation striking the cell in the spectrofluorometer is less than that in a filter fluorometer. The same will be true for F, the number of fluorescent photons transmitted by the emission grating. The advantage of a grating is that you can select any given wavelength for excitation and for emission. Thus if two fluorescent molecules A and B are present in a mixture, A can be measured in the presence of B using one of the following techniques:

1. If A absorbs radiation in a region where B does not, select a unique excitation wavelength so that only A is excited. Only A will fluoresce and be measured. (This can sometimes be done on a filter fluorometer by choosing the correct primary filter.)
2. If both A and B absorb in the same region, but A emits fluorescence at different wavelengths than B, select an excitation wavelength that will be more favorable for A, if possible. Then select a unique fluorescence emission wavelength (where only A emits). When the mixture is excited, only the fluorescence of A will be measured. (This can sometimes be done on a filter fluorometer by using a secondary filter.)

For example, suppose you must determine molecule A in the presence of quinine (Fig. 14-2). Assume that molecule A is excited in the same regions as quinine, but that it emits a fluorescence band from 450–560 nm with a peak at 520 nm. Since quinine does not emit above 520 nm, molecule A can be measured at 520 or 530 nm without error from quinine.

As with the filter fluorometer, the fluorescence intensity readout on the spectrofluorometer is symbolized as F. The units for F also range from 0 to 100 on the recorded fluorescence spectrum or meter readout.

The Relation Between Concentration and F

The symbol F has already been defined as the intensity of fluorescence measured by an instrument and read from the meter or graphical output of the instrument. F is directly proportional to the molarity, c, of a fluorescent species even though not all of the fluorescence from that species is actually measured by any instrument. This is because at low concentrations F is a *constant* fraction, k, of the total fluorescence transmitted by a specific grating or secondary filter.

$$F \cong k \text{ (total fluorescence)} \qquad (14\text{-}1)$$

The total fluorescence is related approximately to c as follows:

$$\text{total fluorescence} \cong \varepsilon b c (2.3 P_0 \phi_f) \qquad (14\text{-}2)$$

The ϕ_f term is the *quantum efficiency*, or *fraction* of excited states that actually emit fluorescence. Substitution of **14-2** into **14-1** gives the usual equation relating concentration and F.

$$F \cong k \varepsilon b c (2.3 P_0 \phi_f) \qquad (14\text{-}3)$$

Equation **14-3** can be derived using the Beer–Lambert law by methods beyond the scope of this text [1].

Equation **14-3** shows mathematically how the various experimental variables affect the values of F read from an instrument. If concentration is varied and all other variables are held constant, then a plot of F vs. c should give a straight line (Fig. 14-5). As is the case for plots of absorbance vs. c (Fig. 12-5), deviations from linearity do occur at higher concentrations in plots of F vs. c. The reason for the downward curvature in Figure

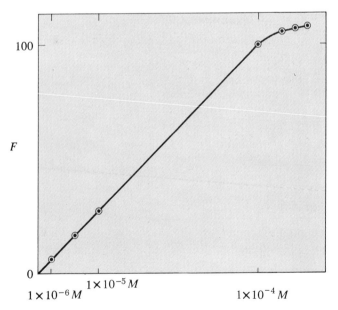

FIGURE 14-5. A typical plot of F, fluorescence intensity, vs. concentration. The plot is linear for most compounds until the solution becomes too concentrated; it bends downward typically above $10^{-4} M$ for weak fluorescers, but not for all compounds. Concentrations much lower than $10^{-6} M$ can be measured for many intensely fluorescent compounds.

14-5 is that the approximation in **14-1** is no longer valid. What happens is that the solution is so concentrated ($\geq 10^{-4}M$) that a disproportionate amount of excitation occurs at the front portion of the cell, where the fluorescence cannot reach the detector. This effect is called the *inner filter effect*; it is discussed in detail in more advanced texts [1, 2].

It can be seen from **14-3** that *increasing* any of the variables on the right side of the equation will increase F. For example, k can be increased by using a wider bandwidth grating or secondary filter, ε can be optimized by exciting at a wavelength where ε is at a maximum, b can be increased by using a somewhat larger cell, and P_0 can be increased by using a wider bandwidth grating or primary filter or by using a source with a higher radiant power. The effect of these variables is best explored by working the following self-test. This self-test is *not* recommended unless the general details of fluorescence have been mastered first.

SELF-TEST

Answers:

20. Fluorescence-Concentration Relation

Directions: Cover the answers in the left column before beginning the test. Work one problem at a time and check the answers given at the left.

A. Predict the effect of b, the cell path length, on F if all other variables in **14-3** are held constant and:

F is doubled.

    a. The cell path length is increased from 1.0 cm to 2.0 cm.

F is reduced to 0.1 of that in a 1-cm cell.

    b. The cell path length is decreased from 1.0 cm to 1.0 mm.

B. Predict the effect of ε on F if all other variables in **14-3** are held constant and:

    a. Excitation is changed to a wavelength at which

F is increased tenfold.

        $\varepsilon = 1 \times 10^4$ from a wavelength at which $\varepsilon = 1 \times 10^3$.

F is reduced to 0.06 of that at 253 nm.

    b. Excitation of anthracene (Table 14-1) is changed to 375 nm ($\varepsilon = 9 \times 10^3$) from 253 nm ($\varepsilon = 1.5 \times 10^5$).

F is increased by a factor of 3.3.

    c. Excitation of phenanthrene (Table 13-1) is changed to 251 nm from 293 nm.

C. Predict the effect of k of F if all other variables in **14-3** are held constant and:

    a. Fluorescence emission is measured using a grating

F is decreased.

        instead of a secondary filter.

    b. Fluorescence emission is measured using a sharp cut

F is increased.

        secondary filter instead of a band pass secondary filter.

D. Predict the effect of P_0 on F if all other variables in **14-3** are held constant and:

    a. Excitation is done using a primary filter instead of a

F is increased.

        grating.

    b. Excitation is done using a spectrofluorometer instead of

F is decreased.

        a filter fluorometer.

F is increased, since P_0 of the mercury arc at 254 nm is very large compared to P_0 of the xenon arc.

    c. A sample is excited at 254 nm with a mercury arc source instead of a xenon arc source, which is very weak in the 200–270 nm region.

Molecules and Ions That Fluoresce

A number of organic molecules and inorganic ions fluoresce. In most cases it is species that are *rigid to some degree* that will fluoresce. This factor will be emphasized as each of the two classes is discussed.

Organic Molecules

Most of the organic molecules that fluoresce are aromatic hydrocarbons or molecules that contain aromatic rings. Molecules with rings having one or two double bonds, such as cyclohexadiene, do not fluoresce. If at least one benzene ring is present in a molecule, it has the potential to fluoresce. Apparently, internal conversion (Fig. 14-3) to the ground state is too rapid for fluorescence to be significant until the carbon-carbon bonds in the conjugated π system are *all rigid*. Then when the π system is excited, it can lose *some* of its energy by fluorescence instead of by increases in bond vibrational energy during the internal conversion process.

Aromatic hydrocarbons, being rigid molecules in general, are usually intensely fluorescent. Benzene is the least rigid because it has only one ring; its quantum efficiency (Table 14-1) is only 0.01 in a solvent exposed to air. This means that only 1% of excited benzene molecules ever emit fluorescence. Because of this, benzene is rarely determined by fluorescence; however, the larger, more rigid aromatic hydrocarbons are commonly determined by fluorescence. These include others listed in Table 14-1—anthracene, phenanthrene, and benzo[a]pyrene.

TABLE 14-1. Some Fluorescent Organic Molecules (Solvent = Ethanol)

Organic molecule	ϕ_f, room temp.	λ_{ex}, nm	λ_{em}, nm
Aromatic hydrocarbons			
Benzene	0.01 (in air)	250	270–310
Anthracene	0.3	254, 375	380–450
Phenanthrene	0.1	251, 265, 293	340–420
Benzo[a]pyrene	>0.4	366 (520[a])	400–500 (540–550[a])
Aromatic amino acids (H_2O solvent)			
Phenylalanine	<0.04	254	270–310
Phenylalanine	Chem. rxn[b]	365	500–530
Tyrosine	0.2	225, 280	280–340
Tryptophan	0.2	220, 280	310–420
Substituted benzenes			
Aniline	0.1	280	310–405
Phenol	0.2	270	285–365
Salicylic acid	≥0.1	310	400–550[c]
Acetylsalicylic acid	0.02[d]	280[d]	310–390[d]
Miscellaneous			
Quinine	0.55	366	400–500
Morphine	Chem. rxn[e]	254	390–420

[a] Sulfuric acid solvent.
[b] Ninhydrin reaction produces a more fluorescent product.
[c] Solvent is 1% acetic acid-chloroform.
[d] Only fluorescent in acetic acid-chloroform type solvents.
[e] Oxidation produces fluorescent pseudomorphine.

The three aromatic amino acids listed in Table 14-1 are all determined by fluorescence. Phenylalanine, whose structure and determination will be discussed in detail later, is a very weakly fluorescent molecule like benzene. Tyrosine and tryptophan have the following structures:

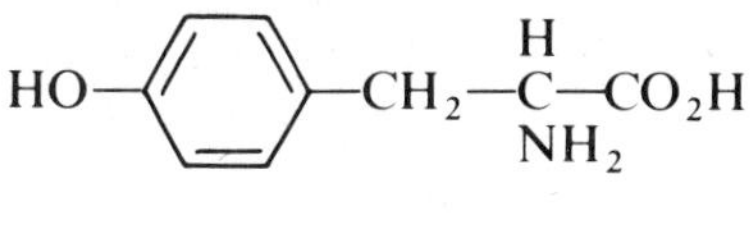

Tyrosine 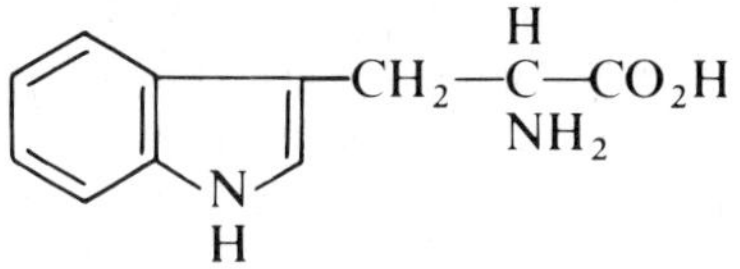   Tryptophan

Tyrosine is very similar structually to phenol, which is strongly fluorescent (Table 14-1). It is not surprising therefore that tyrosine is also strongly fluorescent. Tryptophan consists of a two-membered ring, indole, substituted by an amino acid group. Such a ring has a great deal of rigidity and explains why tryptophan is also strongly fluorescent. Proteins containing the above amino acids are also known to fluoresce [3].

Although benzene itself is very weakly fluorescent at room temperature, substitution of electron-donating groups on the benzene ring increases the fluorescence, probably as a result of increased rigidity, among other factors. Molecules such as aniline, phenol, and salicylic acid fluoresce intensely. In contrast, acetylsalicylic acid (ASA) does not fluoresce under most conditions, so salicylic acid impurities in aspirin tablets may be measured fluorometrically as well as spectrophotometrically (Ch. 13). When a small percentage of acetic acid, such as 1%, is added to a solvent such as chloroform, the ASA also fluoresces [4].

Many molecules that are nonfluorescent or weakly fluorescent can be converted to fluorescent species by adding a reagent to change their structure. Morphine is very weakly fluorescent, but can be oxidized to a highly fluorescent molecule called *pseudomorphine*. This will be discussed in detail later.

Inorganic Ions   Very few unchelated metal ions fluoresce in solution, because they lose any absorbed energy to water molecules that bond strongly to them. The O—H bonds "soak up" the absorbed electronic energy by converting it to vibrational energy. The inorganic ions that do fluoresce in solution at room temperature are UO_2^{+2}, the uranyl ion; Ce^{+3}, the cerium(III) ion; and Tl^+, the thallium(I) ion. Boric acid, H_3BO_3, also reacts with benzoin to form a rigid, green fluorescent addition compound.

The largest class of fluorescent inorganic ions is made up of those that form relatively rigid chelates with organic ligands. The electrons on the organic ligand are excited, not the electrons on the metal ions. However, the metal ion chelated by the organic ligand enables the system to be rigid enough to fluoresce rather than to undergo internal conversion. Metal ions that form such fluorescent chelates are those in Group IIA—Mg(II), Ca(II), and Sr(II)—and those in Group IIIA—Al(III), Ga(III), and In(III). Details of the determination of such metal ions may be found in advanced references [1, 5].

Al(III) can be determined using many different organic ligands. One of these is 8-hydroxyquinoline, which forms the following fluorescent chelate:

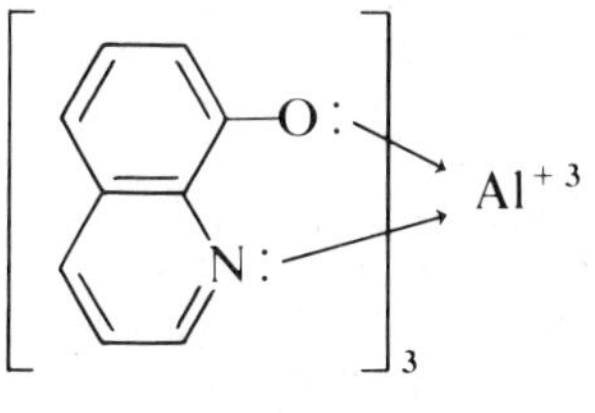

Such chelates have been used for the determination of from 10^{-6} to $10^{-8}M$ Al(III) in drinking water, boiler water, and other water samples.

Qualitative Fluorometric Analysis

As we pointed out in Chapter 13, qualitative spectral analysis consists of both *identification* and *screening*. *Identification* of a specific compound usually involves obtaining the fluorescence excitation and fluorescence emission spectra on a spectrofluorometer and comparing them with those from a pure sample of the compound. For example, if a compound is suspected of being the alkaloid quinine, then the excitation and emission spectra can be obtained and compared with those in Figure 14-2. Excitation bands at 250 nm and 350 nm, as well as a fluorescence emission band at 460 nm, would be good evidence that the compound is indeed quinine.

Identification in some cases is accomplished by the observation of fluorescence without obtaining fluorescence spectra. The detection of oil spills on oceans or lakes is an example. All oils contain many fluorescent aromatic hydrocarbons. Since natural water does not fluoresce, an oil spill can be located by observing fluorescence with the proper excitation. Investigations have been made of the use of lasers operated from airplanes to excite oil spills, and also of the use of buoys containing a low-pressure mercury arc source [6]. In the case of the buoys, the mercury arc emits 254-nm UV radiation which excites fuel oil spilled during unloading at a tanker dock. A detector in the buoy is adjusted to give an alarm only when the fluorescence emission resembles that of the fuel oil, not when fluorescence from outboard motor oil, etc., is detected.

Screening for specific compounds can in many cases be done very simply by using a filter fluorometer rather than a more expensive spectrofluorometer. The screening for morphine in body fluids, such as urine, is a good example. Morphine has the following structure:

(Note that its structure is similar to that of heroin; see the discussion of IR identification in Sec. 13-2.)

Morphine contains a *phenol chromophore*; that is, the left ring is a substituted phenol. It would therefore be expected to fluoresce like phenol (Table 14-1) and to be excited with 254-nm radiation from a mercury arc source in a filter fluorometer. The difficulty is that body fluids contain other compounds that also fluoresce with 254-nm excitation. Because morphine is a metabolite of heroin, its concentration is not always such that its fluorescence intensity is significant compared to that of the other compounds in the urine. This difficulty is overcome by oxidizing the morphine with ferricyanide ion to the intensely fluorescent pseudomorphine. The probable reaction is

$$2\,\text{Fe}\,(\text{CN})_6^{-3} + 2\,\text{Morphine} \longrightarrow \text{Pseudomorphine} + 2\,\text{Fe}(\text{CN})_6^{-4}$$

The pseudomorphine is a dimer of morphine; ferricyanide is known to oxidize phenols (R—OH) to dimers of the type R(OH)—R(OH).

The actual screening test is based on filter fluorometric measurement of the pseudo-morphine after a blank reading has been obtained. The blank reading, taken before ferricyanide is added, gives the fluorescence of morphine itself plus other compounds present in the urine. The ferricyanide reagent is then added, and the fluorescence intensity is again measured after two minutes. Since pseudomorphine is twenty times as fluorescent as morphine, a significant increase in fluorescence intensity definitely indicates morphine. No increase in fluorescence intensity indicates the absence of morphine.

A simple morphine analyzer that is commercially available utilizes disposable glass cells rather than the expensive quartz cells that are normally used with 254-nm excitation. This analyzer is a filter fluorometer with a mercury source positioned above the glass cell so that the 254-nm radiation passes through the open top of the cell. Thus although glass absorbs 254-nm radiation in the usual fluorometer, it cannot in this case. The fluorescence emission, being in the 400-nm range, is transmitted through the glass and is measured by the detector.

Quantitative Fluorometric Analysis

The quantitative analysis of a fluorescent compound in a mixture is accomplished by using either one of the two techniques discussed in connection with fluorescence instrumentation. Either a unique excitation band or a unique fluorescence emission band must be employed for accurate analysis. A good example is the determination of the carcinogenic aromatic hydrocarbon benzo[a]pyrene in samples of particles from polluted air. The structure of benzo[a]pyrene is

It is a very rigid, highly fluorescent aromatic hydrocarbon (Table 14-1). Unfortunately, polluted air contains many other aromatic hydrocarbons that emit fluorescence in the same region as benzo[a]pyrene. To determine it without separation, a solvent of sulfuric acid is used. In this acid, benzo[a]pyrene is excited by light of 520 nm rather than by the ultraviolet (Table 14-1). Although a few other aromatics in polluted air absorb very weakly at this wavelength, none emit at 540–550 nm [7]. Benzo[a]pyrene was determined in the presence of 40 other compounds by this technique of measuring a *unique fluorescence band*.

Quantitative analysis may also be carried out using a unique excitation band. An example of this is the determination of phenylalanine, whose structure is

Fluorometric measurement of phenylalanine is useful in testing for phenylketonuria, a hereditary metabolic disorder that can result in mental retardation. Phenylketonuriac babies cannot convert their phenylalanine efficiently to tyrosine, so a high phenylalanine blood level indicates this disorder. Since phenylalanine is not fluorescent enough (Table 14-1), it is treated with ninhydrin, copper(II) ion, and L-leucyl-L-alanine [8] to give a fluorescent species that is excited at 365 nm. This is a *unique excitation band*, because tyrosine and tryptophan are not excited at this wavelength (Table 14-1). The fluorescence from the species is at 515 nm, in the green region.

Fluorometric
Screening for
Lead
Poisoning

Lead poisoning is always a serious threat to children living where a lead-based paint has been used. Although flame methods (Ch. 15) measure lead *directly*, they also require a relatively large sample, since interferences have to be removed. (It is important to use as small a sample as possible when testing the blood of children.) An unusual indirect method [9] measures lead using only one drop of blood.

The drop of blood is transferred to a glass slide, which is inserted into a specially designed fluorometer using a horizontal, rather than a vertical, sample compartment. One such fluorometer is shown in Figure 14-6. It is called a "Hematofluorometer" by the manufacturer, but basically it is a filter fluorometer. In Figure 14-6, the glass slide holder is shown in position (upper right) within the sample compartment. The holder has compartments for a high- and a low-level lead calibration slide, as well as a compartment for the blood sample. Radiation from a tungsten halogen lamp (Table 12-3) is filtered to isolate exciting wavelengths at 424 nm. The red fluorescence (625 nm) that is emitted directly down into the instrument is measured by a photomultiplier *below* the glass slide, after passing through a secondary filter.

The method is *indirect* for lead(II) ion; it does not measure lead fluorescence but rather the fluorescence of a zinc(II)-protoporphyrin complex produced by the action of lead(II) in the blood. The fluorescence of the zinc(II)-protoporhyrin is so intense that it is routine to detect levels of lead(II) well below the toxic level of 70 micrograms of lead(II) per 100 ml (70 μg/dL). The instrumental readout shown in Figure 14-6 indicates a level of 101 μg lead(II) per 100 ml, a highly toxic level.

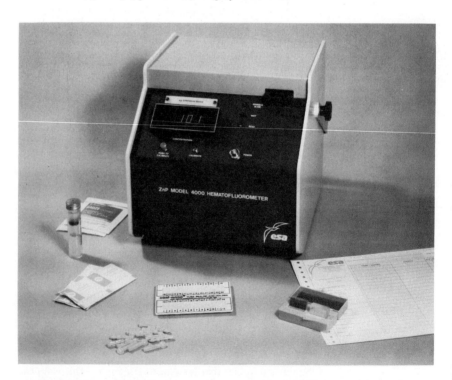

FIGURE 14-6. A filter fluorometer for the determination of lead(II) in blood. (Courtesy of Environmental Science Associates, Bedford, Mass.)

The zinc(II)-protoporphyrin is produced in the blood because the action of an iron-insertion enzyme, Enz-Fe, is blocked by lead(II) ion. In normal blood, this enzyme catalyzes the reaction of iron(II) and protoporphyrin.

$$Fe(II) + protoporphyrin \xrightarrow{Enz\text{-}Fe} Fe(II)\text{-}protoporphyrin$$

When lead(II) ion is present, it reacts with the Enz-Fe so that iron(II) and the protoporphyrin cannot react at a rapid rate. Because zinc(II) ion is present in the blood and can react much faster than iron(II) in the absence of the enzyme catalyst, it reacts first under these conditions.

$$Zn(II) + protoporphyrin \longrightarrow Zn(II)\text{-}protoporphyrin$$

Since zinc(II) is a d^{10} metal ion, the zinc(II)-protoporphyrin chelate will fluoresce. On the other hand, since iron(II) has a partially filled d sublevel, the iron(II) chelate will not.

For other examples of quantitative analysis, work the following self-test.

SELF-TEST | **21. Quantitative Fluorometric Analysis Problems**

Answers:

Directions: Cover the answers in the left column before beginning the test. Suggest a fluorometric *technique*, specifying *both* excitation wavelength and range of fluorescence emission wavelengths for the determination of the stipulated compound. One of the two must be unique (see preceding and Table 13-1).

A. Aspirin tablets contain mainly acetylsalicylic acid (ASA) and salicylic acid (SA). After the tablets have been dissolved in 1% acetic acid-chloroform, suggest a fluorometric technique for the determination of:

Excite both at 280 nm; measure ASA only at 310–390 nm.

a. ASA (see Table 14-1).

Excite only SA at 310 nm; measure SA only at 400–550 nm.

b. SA (see Table 14-1).

B. A mixture of aromatic hydrocarbons contains both benzene and anthracene (see Table 14-1). Suggest a fluorometric technique for the determination of anthracene by means of:

Excite at 375 nm, where benzene does not absorb; measure at 380–450 nm.

a. A unique excitation wavelength.

Excite both at 250–254 nm, where both absorb; measure anthracene only at 380–450 nm.

b. A unique fluorescence wavelength range only.

C. A mixture of aromatic hydrocarbons contains both benzene and phenanthrene (see Table 14-1). Suggest a fluorometric technique for the determination of phenanthrene by means of

Excite at 293 nm, where benzene does not absorb; measure at 340–420 nm.

a. A unique excitation wavelength.

Excite both at 250–251 nm, where both absorb; measure phenanthrene only at 340–420 nm.

Determine benzo[a]pyrene by dissolving the mixture in sulfuric acid, exciting at 520 nm, and measuring it only at 540–550 nm. Determine anthracene by exciting it plus benzo[a]pyrene at 375 nm and measuring it only at 380–395 nm

b. A unique fluorescence wavelength range only.

D. A mixture of aromatic hydrocarbons contains benzene, anthracene, phenanthrene, and benzo[a]pyrene. Suggest fluorometric techniques for the determination of as many of these compounds as possible without separation.

14-2 | THE MEASUREMENT OF REFRACTION AND REFRACTIVE INDEX

Refraction of radiation occurs when a beam of photons changes in velocity as it passes from air into a solution. As a result of the change in velocity, the direction of the beam changes. Refractometry is the measurement of this change in direction; it is mainly concerned with the measurement of the refractive index, n_i, of a solution or pure solvent. This is defined as follows:

$$n_i = \frac{c}{v_i} \qquad (14\text{-}4)$$

where c is the velocity of radiation in air (strictly speaking, in a vacuum) and is constant, and v_i is the velocity of the beam of photons in the solution or pure solvent.

This subject is a large one, like fluorescence, and we will present only the essentials—a simplified picture of what occurs during refraction, a brief introduction to the refractometer, and a few applications chosen from many.

Changes Occurring During Refraction

The refraction of a beam of photons does not involve a change in energy of the photons or of the solution through which the photons pass. As shown in Figure 14-1, the photons in the beam are not absorbed. The interaction of the photons with the molecules in solution is such that only the direction and the velocity of the photons are changed, not the wavelength. Since the interaction always slows down the photons traveling through a solution, the index of refraction is larger than one (14-4).

Since the refractive index depends on the wavelength of radiation passing through a solution and on the temperature of that solution, the wavelength is specified with a subscript and the temperature with a superscript. The yellow D line (589 nm) from a sodium vapor lamp is most often used as the source of photons, and measurements are commonly taken at 20° or 25°C, so n_D^{20} is the common expression for refractive index.

The refraction of radiation by molecules in a liquid depends on the number of atoms per molecule, their atomic weights, and the structure of the molecule. The larger the atomic weight of an atom or the larger the number of atoms in a molecule, the larger the refraction and the refractive index. Evidently a beam of photons interacts more with a larger electronic system, thus reducing its velocity to a smaller value, which in turn results in a larger index of refraction (14-4). Thus in a series of *pure* liquids, such as

TABLE 14-2. Refractive Indices of Some Pure Liquids

Compounds	Formula	$n_D{}^{20}$
Chloroalkanes		
Carbon tetrachloride	CCl_4	1.4573
Chloroform	$CHCl_3$	1.4426
Methylene chloride	CH_2Cl_2	1.4237
Alcohols		
Ethyl alcohol	C_2H_5OH	1.3590
Ethylene glycol	$C_2H_4(OH)_2$	1.4318
Glycerine (glycerol)	$C_3H_5(OH)_3$	1.4740
Others		
Acetic acid	CH_3CO_2H	1.3715
Propronic acid	$C_2H_5CO_2H$	1.3874
Butyric acid	$C_3H_7CO_2H$	1.3979
Water	H_2O	1.3328

CCl_4, $CHCl_3$, and CH_2Cl_2, the larger chlorine atoms, rather than the smaller hydrogen atoms, have the greatest effect on the index of refraction. As you can see in Table 14-2, the order of decreasing refractive index is $n_D{}^{20}$ CCl_4 > $n_D{}^{20}$ $CHCl_3$ > $n_D{}^{20}$ CH_2Cl_2. The same is true for the series of alcohols listed in the table; again, the relatively large oxygen atoms cause the index of refraction of glycerine, $C_3H_5(OH)_3$, to be larger than that of the other two alcohols.

It follows from the preceding discussion that a mixture of two liquids will have a different refractive index than either of the pure liquids. The actual refractive index will depend on the relative concentration of the two liquids. In a more general sense, a solution has a different refractive index than the pure solvent because the solute molecules refract the beam of photons to a different degree than the solvent.

Since water is the most common solvent, let us consider the refraction of radiation in aqueous solutions. Note that water, being a small molecule, has a rather small refractive index compared to the organic molecules in Table 14-2. Aqueous solutions of any of the organic molecules in Table 14-2 should have larger refractive indices than that of pure water. The same should be true for most inorganic solutes, since they are composed of larger atoms than water. As the concentration of most solutes in water is increased, we expect the refractive indices of such solutions to increase also. This is true, as can be seen for the plot of refractive index vs. concentration of glycerine in Figure 14-7. In fact, the refractive index in this case increases in direct proportion to the concentration of glycerine, as indicated by the straight line.

It is not always true that the refractive index vs. concentration plot is a straight line, as in Figure 14-7. Because of interactions between water and solute molecules, such plots are frequently curved. However, the refractive index generally increases as the concentration of the solute increases, even in these cases.

The Refractometer

The refractometer is used to measure the refractive index of liquids. Most refractometers are based on the use of so-called *Amici prisms* by which the light beam is bent to give a dark and a light field separated by a sharp boundary as viewed through a telescope-like eyepiece. The most common refractometer is the Abbe refractometer shown in

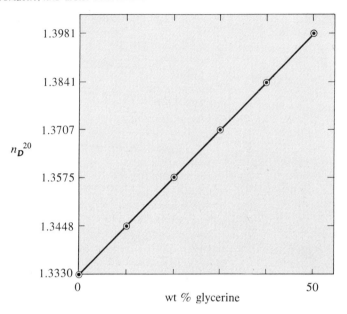

FIGURE 14-7. A plot of $n_D{}^{20}$ vs. weight percent glycerine for aqueous solutions of glycerine. Under the conditions shown, water is the solvent. Above 50 wt %, glycerine would be in excess and would become the solvent.

Figure 14-8. It is equipped with hollow prism casings through which water at a constant temperature, such as 20°C, is pumped to maintain a constant temperature. (Recall that the refractive index depends on the temperature.) Instead of having a yellow sodium source of 589-nm radiation, the Abbe refractometer has a white light source and a special prism arrangement which gives a beam with the same wavelength as that of 589-nm radiation.

To measure the refractive index of a solution or pure liquid, one puts a drop of it on the lower prism and adjusts the prisms to the closed position shown in Figure 14-8. A beam of photons is reflected off the mirror through the prisms and liquid sample and thence to the eye. After an adjustment of the viewer, the eye sees a dark and a light field with a sharp boundary. The prism angle is then changed until the sharp boundary coincides with the intersection of the cross hairs on the eyepiece. The refractive index is then read from a scale (not shown) to four decimal places.

Other types of refractometers are also utilized. A *dipping refractometer* is used for large amounts of samples in a beaker or other large container. For the continuous measurement of fractions from a liquid chromatographic column (Ch. 21), a *differential* type of refractometer is employed. It compares the refractive index of the solvent with solute leaving the column with that of the pure solvent. The difference in the two refractive indices is plotted by a recorder to indicate the relative concentrations of various solutes leaving the column.

Qualitative Analysis The refractive index of a *pure* unknown liquid is very useful for the identification of the compound when used in conjunction with other data such as boiling point, elemental

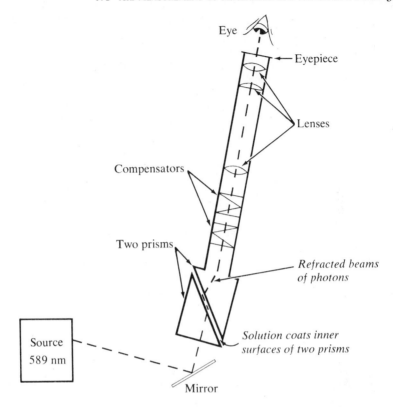

FIGURE 14-8. A schematic of an Abbe refractometer. The liquid whose refractive index is to be measured is added dropwise to the lower prism in the open position; the two prisms are then adjusted to the closed position seen above.

analysis, and absorption spectra. Since the presence of another compound can change the refractive index appreciably, any compound should be carefully purified before the measurement is taken. If the presence of a functional group can be verified, this can make the analysis much simpler. For example, suppose that it has been established that the unknown contains one carboxylic acid group, $-CO_2H$, per molecule and that the refractive index is 1.3698. It is very possible that the compound is acetic acid (Table 14-2), even though the measured refractive index is lower than that of pure acetic acid and it can be seen from the table that higher molecular weight acids have higher, not lower, refractive indices. Since water has a lower refractive index than acetic acid does, it is possible that a trace of water is present, causing the refractive index of the unknown to be lower than that of pure acetic acid.

Quantitative Analysis   For two-component mixtures, the measurement of the refractive index is often the simplest and most rapid method of analysis. If more than two components are present, analysis is probably not accurate by this means, however. All that is needed is a plot of refractive index vs. concentration of the solute component, such as that shown in

Figure 14-7. After the refractive index of the unknown sample has been carefully measured, its concentration is read from the plot.

Since larger molecules in solution refract light to a greater degree than small molecules, it is not surprising that many high molecular weight compounds have been determined in solution by refractometry. The classic example is the analysis of aqueous sugar solutions for dissolved sucrose, $C_{12}H_{22}O_{11}$. The sucrose content of many commercial preparations, such as syrups, is routinely found by measuring their refractive indices. The presence of other disaccharides, such as lactose or maltose, does not cause a significant error, because they have the same formulas and about the same refractive index.

Refractometry is used in clinical chemistry to estimate the specific gravity of urine. The refractive index of the urine is approximately linearly related to the total solids per unit volume, and is therefore related to the specific gravity [10].

14-3 | MEASUREMENT OF LIGHT SCATTERING

Scattering of radiation occurs when a beam of photons changes its direction after colliding with solid particles in solution (Fig. 14-1). If there are a large number of particles, more of the photons are scattered and fewer of the photons pass through the solution in the original direction. If there are relatively few particles, fewer of the photons are scattered and more of them pass through the solution. In any case there is *no change* in the wavelength of the scattered radiation, such as occurs in the fluorescence process (Fig. 14-1).

Two methods of analysis are based on scattering of light. *Turbidimetric* analysis is based on the measurement of the decrease in radiant power $(P_0 - P)$ of the incident radiation. *Nephelometry* is based on the measurement of the radiant power of the radiation scattered at right angles to the incident radiation. Both methods will be discussed briefly.

Measure-
ment of
Changes
Resulting
from
Scattering

Light scattering occurs in a *reproducible manner* when a beam of photons of constant radiant power, P_0, passes into a *stable* suspension of solid particles in solution. As shown in Figure 14-9, a certain fraction of the photons, of radiant power P, pass through the solution without being scattered. A detector placed in the path of these photons can measure both P_0 and P; such an instrument will function as a *turbidimeter*. The loss in radiant power $(P_0 - P)$ is related to the concentration, c, of the suspended particles as follows:

$$\log(P_0 - P) = \log \frac{P_0}{P} = kcb \qquad (14\text{-}5)$$

where b is the path length in cm of the cell, and k is a proportionality term which is ideally constant over the concentration range of interest. A plot of the log term in **14-5** against c will be linear if k is constant, but will be curved if k varies with concentration.

Equation **14-5** is utilized in turbidimetric analysis in the same way as the Beer-Lambert law is utilized in spectrophotometric analysis. Values of $\log(P_0/P)$ are measured for several standard solutions of varying concentrations of suspended particles, and a plot is made of $\log(P_0/P)$ against c. The unknown sample's $\log(P_0/P)$ is then measured, and the concentration of the suspended particles in the unknown sample is read from the plot.

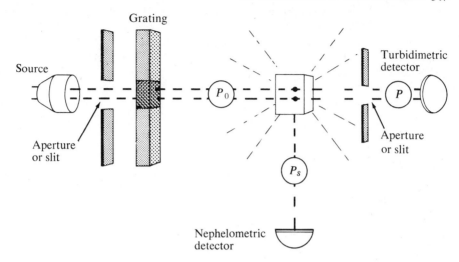

FIGURE 14-9. A schematic of a turbidimeter-nephelometer. When the detector measures the transmitted radiation, P, the instrument functions as a turbidimeter. When the detector measures the scattered radiation, P_s, the instrument functions as a nephelometer.

For measurements in a *nephelometer*, the relationship between concentration and the radiant power, P_s, of the scattered radiation is different from that in **14-5**. In this type of measurement (Fig. 14-9), the radiant power reaching the detector is directly proportional to the concentration.

$$P_s = k'bc \qquad (14\text{-}6)$$

where k' is a proportionality term constant over a limited range of concentration. This expression is similar to **14-3** for fluorescence in that the measured radiant power in each case is directly proportional to the concentration of the species involved. A plot of P_s against c will be linear if k' remains constant over the range of concentrations involved.

Analytical Applications

The relative magnitudes of P, P_0, and P_s determine whether a suspension is analyzed by nephelometric or turbidimetric means. If relatively few particles are present so that the difference between P and P_0 is very small, turbidimetric analysis will be inaccurate and a nephelometric measurement should be used. If a relatively large number of particles are present, then the difference between P and P_0 will be large enough to be measured accurately using a turbidimetric method.

Both of the above methods are useful for the determination of traces of sulfate ion, for which there are few convenient spectrophotometric methods. Barium(II) is added to produce a suspension of barium sulfate. Nephelometry is used to measure sulfate in concentrations as low as a few parts per million, and turbidimetry is utilized to measure higher concentrations. The lead(II) ion interferes by forming insoluble lead sulfate.

Turbidimetric methods are important in pollution analysis. Water, whether polluted or used for drinking, contains many suspended particles which may be measured by turbidimetry.

| QUESTIONS AND PROBLEMS

(Answers to even-numbered problems are in Appendix 5.)

Definitions and Concepts

1. Explain the difference between:
   a. Absorption and fluorescence.
   b. Fluorescence and light scattering.
   c. Light scattering and refraction.
   d. Absorption and light scattering.
2. Explain the difference between a spectrofluorometer and a filter fluorometer.
3. What is the inner filter effect in fluorescence? Does it have anything to do with an optical filter?
4. Explain how morphine can be determined in a glass fluorescence cell, even though it is excited by 254-nm radiation, which is absorbed by glass.
5. Which compound in each of the following groups should have the smallest index of refraction, and why?
   a. CH_3CH_2Cl, $CHCl_2CH_3$, CCl_3CH_3
   b. $CHBr_3$, CBr_4, CH_2Br_2
6. Explain the difference between a turbidimetric and a nephelometric method of analysis.

Fluorometry

7. Which instrument will give the most efficient excitation of anthracene (absorbs as in Table 14-1)—a filter fluorometer with a clear-envelope low-pressure mercury arc lamp or a spectrofluorometer?
8. A standard solution of phenylalanine containing 3.0 mg/100 ml has an F reading of 25 units. On the same fluorometer, an unknown solution containing phenylalanine gives an F reading of 21 units. Calculate the concentration of phenylalanine in the unknown solution.
9. When a solution of anthracene is excited at 375 nm using a xenon arc in a spectrofluorometer, it gives a low F reading. Name several ways to increase that reading.
10. Calculate the ratio of F of anthracene to F of the same concentration of phenanthrene at 252 nm, where their respective molar absorptivities are 1.5×10^5 and 5×10^4.

Challenging Problem

11. Atomic cerium has an outer electronic structure of $4f^1 5d^1 6s^2$.
    a. Which of these outer electrons is excited to cause cerium(III) to fluoresce? What orbital should the excited electron occupy?
    b. Cerium(III) is excited to fluoresce at 250 nm. Its fluorescence is colored. What is the most likely color and why?

| NOTES

[1] G. H. Schenk, *Absorption of Light and Ultraviolet Radiation: Fluorescence and Phosphorescence Emission*, Allyn & Bacon, Boston, 1973, pp. 159–78, 219–42, 267–68.

[2] C. A. Parker, *Photoluminescence of Solutions*, Elsevier, New York, 1968, pp. 220–34; A. L. Conrad, *Treatise on Analytical Chemistry*, Part I, Vol. 5, Wiley Interscience, New York, 1964, pp. 3057–78.

[3] F. W. J. Teal, *Biochem. J.* **76**, 381 (1960).

[4] C. I. Miles and G. H. Schenk, *Anal. Chem.* **42**, 656 (1970).

[5] C. E. White and R. J. Argauer, *Fluorescence Analysis—A Practical Approach*, Dekker, New York, 1970.

[6] *Chem. & Eng. News* **52**, 30 (Mar. 25, 1974).

[7] E. Sawicki et al., *Int. J. Air. Poll.* **2**, 273 (1960).

[8] P. K. Wong, *Clin. Chem.* **10**, 1098 (1964).

[9] *Chem. & Engr. News* **53**, 18 (Feb. 3, 1975); W. E. Blumberg, J. Eisinger, A. A. Lamola, and D. M. Zuckerman, *J. Lab. Clin. Med.* **89**, 712 (1977).

[10] R. J. Henry et al., *Clinical Chemistry*, Harper and Row, New York. 1974. p. 1545.

15

Emission and Absorption of Radiation by Gaseous Atoms

"Really, Watson, you excel yourself," said
Holmes. . . . "It may be that you are not yourself
luminous, but you are a conductor of *light*."

ARTHUR CONAN DOYLE
The Hound of the Baskervilles, Chapter 1

In this chapter we will show how the interaction of *light* or ultraviolet radiation with gaseous atoms is used for chemical analysis. We will present two of the most common of a number of analytical methods based on the excitation of atoms in the gaseous state. Since radiation is measured in both methods, they are types of *spectrometry*. (We will use this term whenever we refer to measurement of radiation involving atoms in the *gaseous state*, reserving the term *spectrophotometry* for the measurement of absorption of radiation by species *in solution*.)

The two spectrometric methods are flame emission spectrometry (an atomic emission method) and atomic absorption spectrometry. Flame emission spectrometry was formerly called flame photometry and is still referred to by that name in much of the literature. It is based on the measurement of the amount of radiation emitted by atoms in a flame. It is widely used in clinical laboratories for the determination of sodium and potassium ions in body fluids.

Atomic absorption spectrometry is a more recent and a more general method for the determination of metal ions than flame emission spectrometry. It is not as applicable to the determination of alkali metal ions, such as sodium and potassium. It is based on the measurement of the amount of radiation absorbed by gaseous atoms, usually, but not always, in a flame.

15-1 | EXCITATION AND EMISSION PROCESSES IN ATOMS

Before we consider flame emission spectrometry and atomic absorption spectrometry, we will discuss briefly the nature of excitation and emission processes in atoms. Because the determination of sodium ions is so important and so common an analysis, we will describe the excitation and emission of sodium atoms. In our discussion it is assumed that you are familiar with the terms *electronic ground state* and *electronic excited state* as introduced in Section 13-1.

Events Occurring in a Flame Since both flame emission and atomic absorption spectrometry usually involve spraying or atomizing an aqueous sample into a flame prior to analysis, we will describe the entire sequence of events in a flame for both methods. Briefly, once a solution of sodium chloride is placed into an atomizer below the flame, the following events take place:

1. The sodium chloride solution is sprayed into the flame in the form of small droplets.
2. The water is vaporized, leaving a fine suspension of solid sodium chloride.
3. The solid sodium chloride is partially vaporized to gaseous sodium atoms and gaseous chlorine atoms.
4. Some of the sodium atoms ionize and form oxides.
5. a. Some sodium atoms are thermally excited to the excited state.
   b. In atomic absorption spectrometry, some sodium atoms absorb photons to reach the excited state.
6. The excited sodium atoms emit photons or heat and return to the ground state.

A more detailed description of these events follows. At stage 1 (Fig 15-1), sodium chloride solution passes through the capillary of a *nebulizer* located inside the burner. The burner gas helps sweep the sample into the flame, and the solution is dispersed in the form of fine droplets, so a fog or aerosol is produced. In each droplet, each sodium chloride is of course surrounded by water molecules.

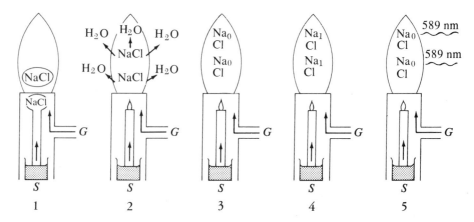

FIGURE 15-1. Schematic of the sequence of events in a flame for sodium chloride sample, S. (1) Sodium chloride sample, S, transported into flame by burner gas, G. (2) Water evaporates from around sodium chloride. (3) Atomic ground state sodium and chlorine produced. (4) Atomic excited state sodium produced. (5) Atomic sodium returns to ground state, emitting radiation.

At stage 2 (Fig. 15-1), the droplets enter the flame and the water is completely vaporized, momentarily leaving gaseous sodium chloride in ionic form. This is quickly dissociated (stage 3) to gaseous atomic chlorine and gaseous atomic sodium in its ground state (Na_0).

$$NaCl(g) \longrightarrow Cl(g) + Na_0$$

It is important to recognize that it is atomic sodium, with its outer $3s$ electron intact, that is produced rather than ionic sodium, which has no outer $3s$ electron.

Some of the atomic sodium may ionize and may possibly form oxides, but these reactions are incidental at this point. At stage 4 (Fig. 15-1), atomic sodium in its electronic ground state is excited to its first excited state (Na_1). In flame emission spectrometry, this is accomplished thermally; that is, heat from the flame excites the sodium.

$$Na_0 + heat \longrightarrow Na_1 \qquad \textbf{(15-1)}$$

This is not very efficient, and most of the sodium atoms remain in the ground state. In atomic absorption spectrometry, the excitation is accomplished more efficiently by exciting the sodium with a beam of photons, some of which are absorbed.

$$Na_0 + 589\text{-nm photons} \longrightarrow Na_1 \qquad \textbf{(15-2)}$$

In atomic absorption, the measurement occurs at this point; the amount of photons absorbed is measured and related to concentration. Light of 589-nm wavelength is used, because this wavelength has the exact amount of energy needed to promote ground state sodium to its first excited state. (Some thermal excitation occurs as in **15-1**, but this is not measured.)

At stage 5 (Fig. 15-1), atomic sodium returns to its electronic ground state. The excess electronic energy of the excited state is dissipated in two ways. Some of the sodium atoms emit photons, primarily at 589 nm, and some lose energy thermally, that is by transferring heat to the surroundings.

$$Na_1 \longrightarrow Na_0 + 589\text{-nm photons} \qquad \textbf{(15-3)}$$

$$Na_1 \longrightarrow Na_0 + heat \qquad \textbf{(15-4)}$$

In flame emission spectrometry, the measurement is performed at this point **(15-3)**. The 589-nm emission is classified as a *line emission*, because a single wavelength is emitted, rather than a *band*. This line is sometimes called the *principal* line of sodium, since it is the most intense line of those emitted by sodium. (It is actually split into two very close lines, at 589.0 and 589.6 nm, but for convenience we refer to the two as one line.)

In atomic absorption spectrometry, the measurement has already been made at this point, and it is immaterial what happens to the excited sodium atoms. Of course, they also return to the ground state via the reaction in **15-3** and **15-4**.

In flame emission spectrometry, the temperature of the flame is more critical than in atomic absorption spectrometry. The flame must be hot enough not only to form atoms but also to excite the atoms thermally to an excited state. In a given type of flame, the relative amounts of ground state and excited state atoms vary widely from element to element. Whether flame emission or atomic absorption is the better method to use depends in part on the element in question.

Other
Sodium
Lines
Atoms can exist in a flame in more than one electronic excited state. This means that atoms, such as sodium, often exhibit other lines in addition to the principal (most intense) line. Some of the other sodium lines involve both the ground state and one excited state. These are termed *resonance* lines, since the atom *resonates* back and forth directly between the ground state and the excited state by absorbing or emitting at this line. Other sodium lines involve two different excited states and are rather weak lines. They can only be used in flame emission spectrometry, because atomic absorption spectrometry involves absorption by the ground state atoms only.

A hypothetical sodium emission in a hot flame is shown in Figure 15-2. The principal, or first, resonance line at 589 nm is of course the most intense, giving the flame a yellow, or yellow-orange, color. Other resonance lines are at 330 and 285 nm. Weaker, non-resonance lines are at 819 and 569 nm. Many of the weaker lines are not observed in a cooler flame, such as the Bunsen flame, because of the smaller thermal energy available for excitation in such a flame. (Most of the lines are split, like the 589-nm line, into two very close lines.) Atomic absorption measurements of sodium are generally made only at 589 and 330 nm, these being the two most intense resonance lines.

Electronic
Orbital
Picture
A simple electronic orbital picture of the 589-nm line of sodium is shown in Figure 15-3.

The electron that is excited either thermally by the flame or by absorbing 589-nm radiation is the $3s$ electron of atomic sodium. (Note that the sodium ion does not have a $3s$ electron and thus cannot undergo such an electronic excitation.) Of the various higher energy orbitals available to it, the excited electron can most easily occupy a $3p$ orbital. This represents the electronic configuration of the first excited state—$1s^2 2s^2 2p^6 3s^0 3p^1$. (The electron can occupy higher energy orbitals, such as the $4p$, if more

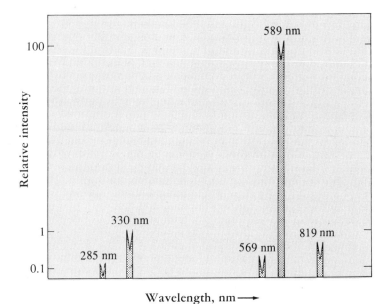

FIGURE 15-2. **Hypothetical flame emission spectrum of the sodium atom in a hot flame.**

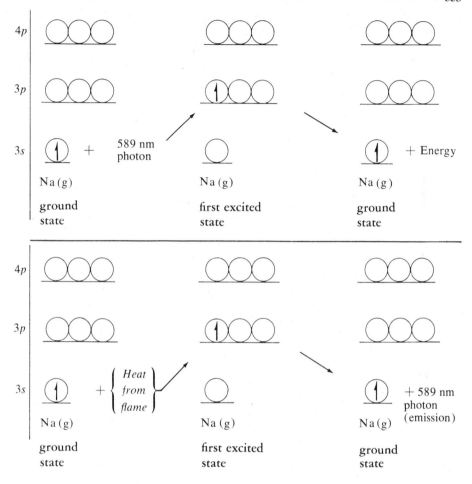

FIGURE 15-3. Electronic orbital diagram of the excitation of sodium. Above: excitation by 589-nm light (as in atomic absorption spectrometry). Below: thermal excitation of sodium atoms (as in flame emission spectrometry) with emission of 589-nm light.

heat can be absorbed or if higher energy radiation can be absorbed.) Because the excited state is unstable, the electron loses its excess electronic energy quickly and again occupies the 3s orbital in the ground state.

15-2 | FLAME EMISSION SPECTROMETRY (FLAME PHOTOMETRY)

Flame emission spectrometry (formerly known as flame photometry) is the instrumental measurement of the amount of radiation emitted by excited atoms in a flame. It is one of the most general methods of analysis, because flames excite atoms of essentially all the

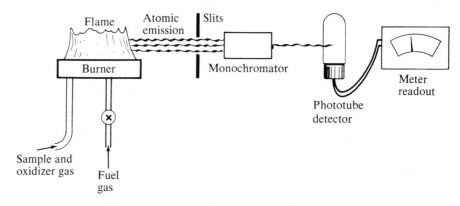

FIGURE 15-4. Schematic of a flame emission spectrometer.

metals, including the rare earths. Also, this method can be used with very simple instrumentation for thousands of rapid, routine determinations of sodium and potassium ions in body fluids every day in clinical laboratories.

The instrument used to measure the amount of emission is generally known as a *flame atomic emission spectrometer*. (Some older models are still called flame photometers.) The spectrometer (Fig. 15-4) consists of a nebulizer, a burner, a monochromator, a detector system, and a readout device. The sample is sprayed by the nebulizer into the burner, carried into the flame, atomized, and excited. The emission from the excited atoms passes into the monochromator, where the selected wavelength is passed through for measurement. The intensity of that emitted wavelength is measured by the detection system, and the units of intensity are indicated on the readout meter. We will discuss each component in more detail.

Instrumentation

We begin our descriptions of the components of a flame emission spectrometer with a discussion of the nebulizer and burner system.

The Nebulizer-Burner System

The nebulizer and burner are operated as one unit. The purpose of the nebulizer is to spray a sample into the gas stream and nebulize (fragment) it into a fine mist or aerosol. The burner serves to mix the *fuel gas* and the *oxidizer gas* in proper proportions to produce a stable flame hot enough to atomize and excite the sample.

There are two types of burners, each of which utilizes a nebulizer in a slightly different manner. These are the *premixed burner* and the *turbulent burner* (or sprayer-burner) [1]. A schematic of the premixed burner and its nebulizer is shown in Figure 15-5. The sample is sucked into the nebulizer and sprayed into a mixing chamber, where it is mixed with the oxidizer gas. The larger drops are trapped and removed in this chamber; only the fine drops are swept into the burner by the oxidizer gas. Before entering the burner, the sample–oxidizer gas mixture is premixed with the fuel gas. The mixture then ignites as it passes into the flame. The flame produced is quite stable, but this apparatus has the disadvantage that not all of the sample enters the flame.

In the turbulent burner, the fuel gas and the oxidizer gas are mixed at the point where they enter the flame. The sample passes through its own channel via a capillary tip into the flame. Because all of the sample enters the flame, the burner is a *total consumption* burner. However, nebulization is less efficient than in the premixed burner and some large drops of sample are carried into the flame.

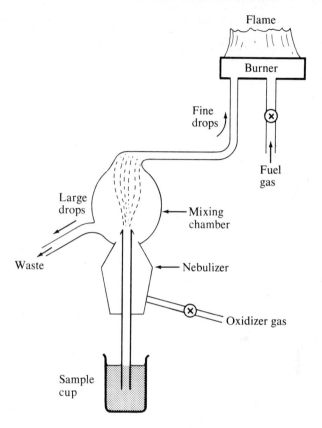

FIGURE 15-5. **Schematic of premixed burner with nebulizer.**

The type of fuel gas–oxidizer gas mixture used in either burner is important for achieving proper flame temperature. The fuel gas must be a compound that can be burned in a controllable manner to give a reproducible, as well as a hot, flame. Table 15-1 lists a few typical mixtures.

The natural gas used in Bunsen burners has such a low flame temperature that it is only capable of exciting the alkali metals (Group IA) and some of the alkaline earth metals (Group IIA). For that reason, hydrogen and acetylene have replaced natural gas as the fuel gas. The acetylene–nitrous oxide flame is used for the determination of

TABLE 15-1. Typical Gases for Flames [1]

Fuel gas	Oxidizer gas	Maximum flame temp., °K
Natural gas	air	1700°
Acetylene (C_2H_2)	air	2500°
H_2	O_2	2940°
Acetylene	N_2O	2970°
Acetylene	O_2	3400°

many metals, because it prevents oxide formation and other interfering reactions. This gas mixture is not used for alkali metals however, because it produces a low fraction of atomic metal in the flame (only 0.3 of all sodium species exist as atomic sodium, for example). Instead an acetylene-air flame is used for the alkali metals (over 0.6 of all sodium species exist as atomic sodium).

The Mono-chromator   There are two types of monochromators used—the optical filter and the grating (Sec. 12-3). The optical filter is used in simpler instruments, such as those made for the clinical determination of sodium and potassium in body fluids. The grating is used in more sophisticated instruments for its better resolution of emission lines in more complicated mixtures. The latter instruments usually also have an adjustable slit width to aid in resolving lines.

The Detector System   The simpler instruments are equipped with a photocell or a phototube (Sec. 12-3). The more sophisticated instruments are equipped with an electron multiplier phototube, or photomultiplier.

The Meter Readout   The meter readout is usually in arbitrary units, such as 0 to 1000 or 0 to 100 intensity units. On flame spectrometers consisting of a nebulizer-burner attached to a spectrophotometer, the percent transmittance scale is used for the readout. A reading of $0\%T$ corresponds to no flame emission, and that of $100\%T$ corresponds to a maximum flame emission intensity.

Flame Emission Analysis   We will first discuss the various methods of determining the concentration of metal ions in solution, and then cover the various groups of metal ions that can be determined.

Direct Determination   In the direct determination method, the emission intensity of the unknown, I_u, is directly compared with the emission intensity of one or more standards, I_s. One approach is to measure the emission intensity of just one standard solution that is fairly close in concentration to that of the unknown. Then, the concentration of the unknown, c_u, is calculated from the known concentration of the standard, c_s, as follows:

$$c_u = c_s \frac{I_u}{I_s}$$

This approach is not often used, because the emission intensity does not always vary in a linear manner as the concentration increases. Unless the concentration of the unknown is quite close to that of the standard, a serious error can result.

A better approach is to plot the emission intensity of a series of standard solutions against the concentration of these solutions (Fig. 15-6). As you can see, the emission intensity of sodium at 589 nm increases in a linear manner up to a concentration of about 4 ppm sodium ion; then the emission intensity shows a negative deviation from linearity. Above this concentration, some of the sodium atoms in the outer cone of the flame are absorbing some of the 589-nm emission from excited sodium atoms in the inner cone of the flame. Error is avoided if the concentration of the unknown is read from the calibration plot.

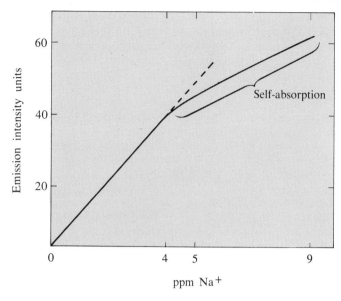

FIGURE 15-6. Calibration plot for sodium ion, showing self-absorption deviation from linearity above 4 ppm sodium ion.

Internal
Standard
Method

If there are interferences present which cannot be eliminated, or whose effect on the emission cannot be corrected for, the internal standard method can be used. This involves adding to each unknown a fixed amount of another metal ion, called the internal standard. The emission wavelength of the internal standard should be close to that of the metal ion being measured, and the emission process should be similar. The *ratio* of the emission intensity of the ion determined to that of the ion used as internal standard is plotted against the concentration of the metal ion being measured, using log-log graph paper. This procedure usually cancels out the effects of any variations in flame and in compositions of the standards and unknown.

Standard
Addition
Method

In the standard addition method, intensity readings are obtained for two solutions—a solution containing an aliquot of the unknown only and a solution containing the same aliquot of the unknown with measured volume of a standard solution of the same metal ion. The concentration of the metal ion in each solution is then read from a calibration plot, such as that in Figure 15-6. If there is no interference, then subtracting the concentration found for the unknown from the concentration should give the concentration of the standard added. If there is an interference, then the foregoing will not be true. A correction factor of the ratio of the concentration of the added standard, S_{added}, to the concentration of the standard found, S_{found}, is multiplied times the concentration of the unknown found, U_{found}, to give the correct concentration of the unknown, $U_{correct}$:

$$U_{correct} = \frac{S_{added}}{S_{found}} (U_{found})$$

Metal Ions Commonly Determined Flame emission spectrometry can be used to determine essentially any of the metallic elements, but not nonmetals such as chloride, bromide, sulfide, etc. There are a number of metal ions for which flame emission is more sensitive [2] than atomic absorption spectrometry (Sec. 15-3). These are as follows:

> Group IA: Li, Na, K, Rb
> Group IIA: Ca, Sr, Ba
> Group IIIA: Al, Ga, In, Tl
> Transition metals: Ru, W, Re
> Rare earths: Eu, Ho, La, Lu, Nd, Pr, Sm, Tb, Tm, Yb

A few of these metal ions are deserving of further comment.

Determination of Sodium and Potassium The flame emission determination of sodium and potassium ions in the clinical laboratory is by far the most important flame measurement. These ions are determined in body fluids, such as blood serum and urine. Their concentrations in urine vary widely, but the concentration ranges of these ions in the blood serum of normal adults are within certain limits.

$$Na^+: 1.38 \times 10^{-1}\text{--}1.46 \times 10^{-1}M \text{ (138--146 meq/liter)}$$

$$K^+: 0.038 \times 10^{-1}\text{--}0.050 \times 10^{-1}M \text{ (3.8--5.0 meq/liter)}$$

The ratio of serum (extracellular) sodium to potassium is thus about 30:1. The sodium line from a serum sample is therefore much more intense than the potassium line (Fig. 15-7).

Since sodium and potassium principal emission lines are so far apart, only a simple optical filter is necessary to enable one line to be measured in the presence of the other. In some situations the sodium and potassium are determined by the direct determination method. A separate calibration plot (Fig. 15-6) is made up for each.

Other samples are analyzed by the internal standard method, with the lithium ion used as the internal standard ion. The ratio of the intensity of sodium to the intensity of lithium is measured and plotted against the concentration of sodium; the same is done for potassium. This minimizes variations in the flame and the interfering effects of other substances in the blood samples. Lithium is suitable for an internal standard because it behaves in nearly the same way as sodium and potassium, but is not found in the blood.

Large concentrations of sodium also enhance the emission intensity of potassium in certain flames. This causes a serious error in acetylene-air flames, but a smaller error in natural gas–air flames. The errors in both types of flames decrease as the concentrations of both ions decrease.

Determination of Other Metal Ions Whereas the alkali metals are generally measured in an air-acetylene flame, or an air-natural gas flame, most other metals are measured in a nitrous oxide–acetylene flame. This flame minimizes oxide formation and other interferences and makes it possible to determine some metals that cannot be measured in an air-acetylene flame. For example, from Table 15-2, aluminum(III), barium(II), and magnesium(II) are all readily measured in the nitrous oxide–acetylene flame, but not as readily in the air-acetylene flame.

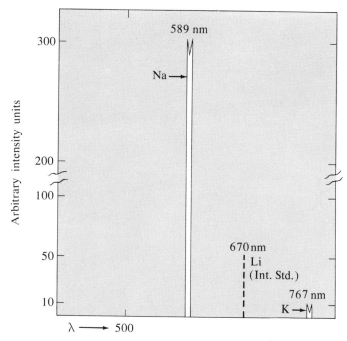

FIGURE 15-7. **Hypothetical flame emission spectrum of a mixture of sodium and potassium ions in blood serum. The internal standard line of lithium ion is shown with a dotted line.**

TABLE 15-2. **Detection Limits in ppm of Metal Ions in a Flame [2]**

		Detection limit, ppm	
Metal	Wavelength, nm	N_2O—C_2H_2	Air—C_2H_2
Al	396.2	0.005[a]	—
Ba	553.6	0.001[a]	—
Ca	422.7	0.0001[a]	0.005[a]
Mg	285.2	0.005	—
K	766.5	—	0.0005
Na	589.0	—	0.0005

[a] Determined in the presence of a high concentration of potassium chloride.

15-3 | ATOMIC ABSORPTION SPECTROMETRY

Atomic absorption spectrometry is the instrumental measurement of the amount of radiation absorbed by unexcited atoms in the gaseous state. It, like flame emission spectrometry, is one of the most general methods of analysis, since atoms of all the metals,

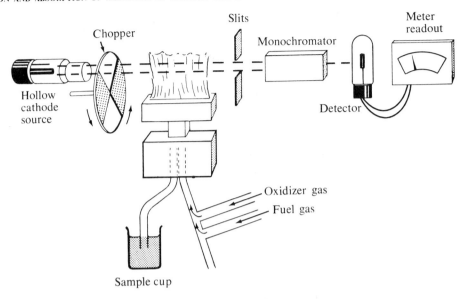

FIGURE 15-8. **Schematic of an atomic absorption spectrometer.**

plus some nonmetals, absorb radiation in the gaseous state. This method can be adapted with special sample handling for the determination of traces of mercury in fish and in other foods.

The instrument used to measure the amount of absorption is known as the atomic absorption spectrometer. There are two types of spectrometers—flame and nonflame. We will first discuss the flame spectrometer and later describe the flameless spectrometer, where atoms are vaporized without a flame.

The flame atomic absorption spectrometer (Fig. 15-8) consists of a premixed burner system, a source of radiation, a monochromator, a detector system, and a readout device. The sample is sprayed in fine drops by the nebulizer into the burner and carried into the flame, where it absorbs some of the radiation from the source. The remaining radiation from the source passes into the monochromator, where the appropriate wavelength is selected for measurement. The radiant power of the transmitted radiation is measured by the detector system, and the amount of radiation absorbed indicated in absorbance units on the readout meter. (The Beer–Lambert law, Sec. 12-2, is applied to atomic absorption measurements.) We will discuss each component in more detail next.

**Instrumen-
tation**

We begin our description of the components with a discussion of the nebulizer and burner system.

**The
Nebulizer-
Burner System**

The Beer–Lambert law states that absorption depends on b, the path length traveled by the radiation through the absorbing medium. For this reason, a long flame path length is needed to increase the absorbance. The *slot burner* shown in Figure 15-8 is used for this purpose. It is similar to the premixed burner (Sec. 15-2) used in flame emission spectrometry, except for the longer flame path length. Although the slot burner is used on most commercial instruments, the turbulent burner is usually available as optional equipment. The premixed slot burner gives a quieter flame but is generally limited to fairly low-burning velocity flames. The nebulizer systems for these burners are similar to those described in Section 15-2.

Insofar as flames are concerned, the air-acetylene flame (Table 15-1) is most commonly used in atomic absorption spectrometry.

The Source The most common source is the hollow cathode tube (Fig. 15-9). Most often, the cathode is made of the metal ion to be determined. The emission from the lamp is an emission line, in contrast to the continuum emission of sources used in spectrophotometry (Sec. 12-3). This means that a separate lamp is needed for each metal ion to be determined.

The name of this source derives from the lamp's cylindrical hollow cathode, which emits the line of the metal from which it is made. (In the case of sodium and other such metals, an alloy with a less malleable metal is used.) A high voltage current (240–400 V) flows between the hollow cathode and the anode. This strikes atoms of an inert filler gas, such as neon, forming positive ions. The ions bombard the negative cathode, causing "sputtering" of cathode metal atoms in vapor form into the inner hollow of the cathode; the outer surface of the cathode is covered with glass to confine emission to the inner hollow. The gaseous metal atoms become excited and emit radiation at the characteristic line(s) of that metal. The radiation passes through the window at the end of the tube. For general purposes, the window is made of quartz, but for some elements, such as sodium, a pyrex glass window can be used since pyrex transmits visible lines, such as 589 nm.

Although a number of lines may be emitted by a given source, the *principal* (or first resonance) line is generally used for analytic purposes (p. 351). This line is the most intense line, because it is the result of a transition between the first excited state and the ground state. The reason that a source emitting at a line is preferred over other types is that the width of the absorption line is quite narrow, similar to the emission lines in Figure 15-7. Thus most of the photons emitted by the source will be absorbed by the atoms in the flame, rather than just a few photons emitted by a continuum source.

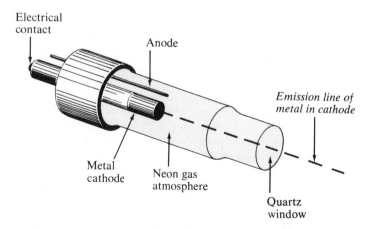

FIGURE 15-9. Schematic of a hollow cathode tube used as a source for an atomic absorption spectrometer.

Choppers and Modulation

After atoms have been excited via atomic absorption, some of them return to the ground state via emission of photons of the same wavelength used for excitation **(15-3)**. This emission causes an error in the measurement of absorption by producing an apparent increase in the number of photons reaching the detector. This causes an apparent decrease in the absorbance read from the meter.

The above error can be eliminated by using a chopper or by employing modulation. The chopper is shown in Figure 15-8. During its rotation, it alternately transmits and blocks the hollow cathode beam. The detector is synchronized and tuned to the intermittent source radiation and does not respond to the constant emission of radiation from excited atoms in the flame. If, instead of a chopper, modulation is used, the hollow cathode is run from a modulated power supply. The transmitted photons from the source are then measured by a detector system employing an AC amplifier at the same frequency as the modulated power supply. Emission from excited atoms in the flame is, of course, not modulated and therefore cannot affect the output of the AC amplifier.

Monochromator and Detector

The monochromator and detector are generally the same as for a sophisticated fluorescence emission spectrometer. A grating is used as the monochromator, and a photomultiplier tube is used as the detector. An AC amplifier is employed as previously mentioned.

Atomic Absorption Analysis

We will first discuss the direction determination method of analysis, and then some examples of the determination of various metal ions.

The Direct Determination Method

The basic law relating the absorbance, A, of radiation to the concentration of the metal ion in solution, c, is the Beer–Lambert law (Sec. 12-2):

$$A = abc \qquad \textbf{(15-5)}$$

Here b is the path length over which absorption occurs in the flame, and a is a proportionality constant. The path length is obviously difficult to measure, so a cannot be readily calculated and reported in the literature as the molar absorptivity is (Sec. 12-2). It is necessary to employ a proportionality constant in the above equation to reflect the proportionality between the number of absorbing atoms in the flame and the concentration of the ions in the solution sprayed into the flame. Thus, a is defined as follows:

$$a = k \left[\frac{\text{no. of atoms in flame}}{c} \right]$$

where k is a constant depending on the type of flame and other conditions. This definition of a keeps **15-5** "honest" in that the left side represents an absorbance in a flame, and the right side also represents an identity that is valid for the flame.

As **15-5** predicts, a plot of absorbance vs. concentration is linear, as long as there are no interactions that cause deviations from the Beer–Lambert law. Two such lines are shown for atomic absorption by strontium in Figure 15-10. Since it is known that the potassium ion represses the ionization of strontium atoms in a nitrous oxide–acetylene flame, it is obvious that the concentration of the strontium atoms *in the flame* is higher when potassium ions are present. Interpreting this in terms of **15-5**, we say that a for the plot where potassium is present is greater than a where potassium is absent.

Although they will not be discussed here, the internal standard method and the standard addition method described in flame emission spectrometry (Sec. 15-2) are

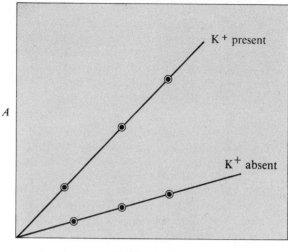

FIGURE 15-10.   Beer–Lambert law plot of atomic absorption by strontium in a nitrous oxide–acetylene flame.

applicable to some degree in atomic absorption, particularly if the direct determination cannot be made to give a useful Beer–Lambert law plot.

Metal Ions Commonly Determined

Atomic absorption spectrometry can be used to determine essentially any of the metallic elements, given the proper flame.   One important limitation of *flame* atomic absorption is that it cannot be used for the determination of mercury; however, mercury can be determined by flameless atomic absorption, which we will discuss later.   Atomic absorption can also be used for certain nonmetals that cannot be determined at all, or only with difficulty, by flame emission spectrometry.   These nonmetals include selenium, tellurium, arsenic, and silicon.

In addition, there are a number of metal ions for which atomic absorption spectrometry is more sensitive [2] than flame emission spectrometry.   These are the following:

Group IB: Ag, Au
Group IIA: Be, Mg
Group IIB: Zn, Cd, Hg
Group IVA: Si, Pb
Group VA: As, Sb, Bi
Transition metals: Co, Fe, Pt

Determination of Lead

One toxic metal that has been prominent in the news is lead.   Not only does it cause lead poisoning in children who eat paint from older buildings, but it also endangers anyone who drinks liquids or eats food stored in pottery and ceramic ware coated with lead-based paints.   It had been previously assumed that the protective glazes used on these vessels prevented any dissolution of lead into liquids or food stored in them.   However, this was found not to be true.

The U.S. Food and Drug Administration (FDA) developed a procedure to simulate the leaching of lead(II) ion into materials stored in these vessels. The FDA procedure involves soaking the glazed ware with a dilute 4% solution of acetic acid (similar to vinegar) for 24 hours at room temperature. The solution is then tested for lead(II) ion to see if it exceeds the allowable maximum of 0.007 mg of lead(II) ion per ml of test solution. Atomic absorption measurement of lead at 217 nm is used for the analysis, since it can detect as little as 1×10^{-5} mg of lead per ml (0.01 ppm lead).

Determination of Toxic Metals in Biological Samples

Flame atomic absorption spectrometry is sensitive enough in most cases to detect the very low concentrations of toxic metals found in biological samples. Table 15-3 lists some of these metals with the flame atomic absorption detection limit in ppm (mg/liter), as reported by Moffitt [3], and the biological threshold limit (BTL). The given BTL indicates excessive toxic *exposure* to the toxic metal but not necessarily toxicity.

The detection limit for mercury listed in Table 15-3 is fairly high compared to its BTL. However, the detection limit for flameless atomic absorption is much lower (see following), so flameless atomic absorption is preferred for the determination of mercury.

Flameless Atomic Absorption

Although the flame method is a good method of atomizing samples, it is not the only one. For metal-containing samples, a *carbon-rod* atomizer has been used with excellent results. This is an electrically heated graphite tube, open at both ends and placed axially in the optical pathway of the hollow cathode lamp. Liquid or solid samples are heated gently in a cavity in the rod to volatilize water and organic compounds; this is followed by strong heating to vaporize the remaining metals. The beam from the lamp passes through the vapor of the metal atoms just above the sample cavity, and absorption is measured as for flame atomic absorption. Most metals can be determined with detection limits as good as or better than those of flame atomic absorption. Mercury, however, is determined by a different type of flameless atomic absorption instrument.

Determination of Mercury

Mercury is a unique metal in that it is so toxic that very sensitive methods are needed to detect the low levels where toxicity starts. For example, the FDA tolerance for

TABLE 15-3. Atomic Absorption Detection Limits and Biological Threshold Limits for Toxic Metals

Toxic metal	AA detection limit, ppm	BTL[a], ppm	Biological sample
Arsenic	0.1	0.3	Blood
		0.4 mg/g	Hair
Cadmium	0.002	0.05	Blood
Lead	0.01	0.8	Blood
		0.024 mg/g	Hair
Mercury	0.5	0.06	Blood
Selenium	0.1	0.07	Urine
Tellurium	0.1	0.02	Urine

[a] BTL is the biological threshold limit in mg/liter (=ppm).

mercury in fish is 0.5 ppm (0.5 mg/kg fish). Flame atomic absorption is usually not sensitive enough to cover all samples needing testing.

In 1968, Hatch and Ott [4] devised a highly sensitive flameless atomic absorption method for mercury that is based on a simple volatilization of free mercury by sweeping air through the liquid sample. The actual method varies according to the type of sample, but in general the sample is treated with an oxidizing acid to convert mercury to its highest oxidation state, the mercury(II) ion. Next, an acidic solution of tin(II) ion is added to reduce the mercury(II) ion as follows:

$$Hg^{+2} + Sn^{+2} \longrightarrow Hg°(soln.) + Sn^{+4}$$

At this point, the free mercury is dispersed in the solution. A flow of air is then passed through the solution, sweeping the free mercury in vapor form out of the reaction flask (Fig. 15-11). The air stream containing the mercury is dried by passing it through a desiccator, and then it passes through a 15-cm long optical cell. The intense 254-nm mercury line from the source passes through the optical cell, and the absorbance at 254 nm is measured, as in flame atomic absorption.

Note that, in contrast to flame atomic absorption, the flameless determination of mercury employs an optical cell to confine the mercury vapors. Because the concentration of gaseous mercury atoms is quite low in the air, the cell is 15 cm long rather than the 1 cm used in solution spectrophotometry. This increases the absorbance **(15-5)** and permits analysis of samples containing as low as 0.005–0.05 ppm mercury.

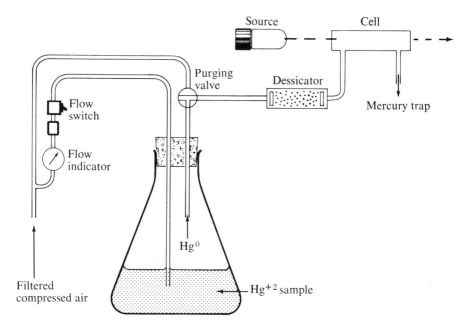

FIGURE 15-11. Schematic of apparatus for the flameless atomic absorption determination of mercury.

| QUESTIONS AND PROBLEMS

(Answers to even-numbered problems are in Appendix 5.)

Concepts and Definitions

1. Explain the role of the flame in both flame emission and atomic absorption spectrometry.
2. Contrast and compare flame emission and atomic absorption spectrometry with respect to the following steps:
   a. Flame excitation
   b. Loss of excitation energy
   c. Measurement
3. Explain the difference between flame and flameless atomic absorption spectrometry.
4. Explain the differences in the following flame emission spectrometric methods of analysis:
   a. The direct determination method
   b. The standard addition method
   c. The internal standard method
5. State whether flame emission or atomic absorption spectrometry is the preferred method for the determination of each element in each of the following groups:
   a. Li, Na, Ag, Au
   b. Be, Mg, Ca, Sr
   c. Al, As, Sb, Tl
   d. Co, Ru, Fe, Re
   e. Zn, Ga, Cd, Tl
6. Explain why flame emission is preferred over atomic absorption spectrometry for the determination of sodium(I) and potassium(I) ions in most clinical laboratories.
7. Draw orbital diagrams to explain the excitation and emission processes for the following elements in a flame (see Fig. 15-3):
   a. Li
   b. K

Quantitative Analysis Problems

8. A standard solution containing $1.50 \times 10^{-1} M$ sodium(I) ion gives an emission intensity of 50 units at 589 nm in a flame emission spectrometer. Calculate the sodium(I) ion concentration in each blood serum sample below, and state whether the concentration falls in the normal adult range.
   a. Sample with emission intensity of 70 units
   b. Sample with emission intensity of 46 units
   c. Sample with emission intensity of 42 units
9. A standard solution containing $1.50 \times 10^{-1} M$ sodium(I) ion and $5.0 \times 10^{-3} M$ potassium(I) ion gives an emission intensity of 100 units at 589 nm and of 10 units at 766 nm in a flame emission spectrometer. Calculate the concentrations of these ions in the following blood serum samples, and indicate if possible whether the concentrations fall in the normal adult range.
   a. Sample with 97 units at 589 nm and 9.0 units at 766 nm
   b. Sample with 80 units at 589 nm and 9.5 units at 766 nm
   c. Sample with 96 units at 589 nm and 6.0 units at 766 nm
   d. Sample with 82 units at 589 nm and 6.5 units at 766 nm
10. A sample is to be analyzed for $\%Na_2O$. The following readings are obtained on the flame emission spectrometer at 589 nm:

mg/liter Na_2O	Emission intensity units
0	3
10	22
25	46
50	70
Sample (10 mg/ml)	28

Calculate the $\%Na_2O$ in the sample after constructing a calibration plot for intensity vs. $\%Na_2O$.

11. The following absorbance readings were obtained for cadmium(II) solutions on an atomic absorption spectrometer. Plot absorbance vs. concentration and determine the concentration of an unknown having an absorbance of 0.205.

ppm Cd^{+2}	A
0	0.000
3.80	0.104
5.80	0.160
8.00	0.220
11.20	0.310

| NOTES

[1] R. D. Dresser, R. A. Mooney, E. M. Heithmar, and F. W. Plankey, *J. Chem. Educ.* **52**, A403 (Sept., 1975).

[2] E. E. Pickett and S. R. Koirtyohann, *Anal. Chem.* **41**, 28A (Dec., 1969).

[3] A. E. Moffitt, Jr. et al., *Amer. Lab.* **3**, 8 (1971).

[4] W. R. Hatch and W. L. Ott, *Anal. Chem.* **40**, 2025 (1968).

16 | Direct Potentiometric and pH Measurements

> ". . . the triggers *were wired together* so that, if you pulled on the hinder one, both barrels were discharged." [he said to Holmes.]
>
> ARTHUR CONAN DOYLE
> *The Valley of Fear*, Chapter 4

In Chapter 11 a potential was defined as the electrically measured tendency of a chemical species to lose or to gain electrons. By measuring this potential instrumentally, the concentration, or more accurately the activity, of a chemical species is determined. This measurement is made using a *potentiometer* connected to a pair of electrodes immersed in the sample solution; the measured tendency of the species to undergo an energy-producing reaction is then related to the concentration or activity.

Direct potentiometric and pH measurement is one of the truly fascinating methods of chemical analysis. Just through inserting two appropriate electrodes *wired* to a potentiometer into a solution, a positive result is obtained (in contrast to the negative result achieved above through *wiring* two guns together). Even more interesting is the fact that the analytical result is obtained without any appreciable reaction; not even the *temporary* formation of an excited state takes place as it does in light absorption.

How is it possible to make a measurement without using a chemical reaction or forming an excited state? The answer lies in an understanding of the potentiometer and the electrodes used for measurement of a particular chemical species, as discussed below.

16-1 | TYPE OF POTENTIOMETERS AND ELECTRODES

We will first discuss the *classical potentiometer* to help you gain a basic understanding of the potentiometer. Next we will describe the general types of electrodes. We will conclude with a discussion of the *modern potentiometer*.

368

The Classical Potenti- ometer

The concentrations of the chemical species in an electric, or galvanic, cell determine the voltage of the cell. Unfortunately, if we try to measure this voltage, the voltmeter draws enough current from the electrical cell to change the concentrations of the ions. Thus we do not obtain an accurate measure of the initial concentrations. What is needed is a *potentiometric measurement*, not a measurement with a voltmeter.

The difference between these two types of measurements is illustrated in Figure 16-1. The reaction between the chemical species in each cell is

$$H_2(g) + Ox\ agt = 2H^+ + reductant$$

In this reaction, hydrogen gas gives up two electrons to some oxidizing agent, becoming two hydrogen ions in the process. The upper half of Figure 16-1 shows the setup of the cell to be used with a voltmeter (not shown) for measurement of cell voltage. Note that electrons released from the hydrogen gas flow through the electrode, as indicated by the current-measuring galvanometer. The electrons flow through the wire and combine with the oxidizing agent at the other electrode. (The oxidizing agent and hydrogen gas parts of the cell are divided from each other.) Thus as soon as the cell is hooked up, the concentrations of the reacting species change, and the voltage measured is already an inaccurate measure of the initial concentrations.

The lower half of Figure 16-1 shows the cell setup for potentiometric measurement. An *external* cell whose voltage can be varied to known values is connected to the external circuit. When the circuit is completed, a few electrons released from the hydrogen gas begin to flow through the electrode. However, the voltage of the external cell is quickly adjusted to equal the voltage of the hydrogen–oxidizing agent cell; the circuit is arranged so that the external voltage *opposes* the voltage of the latter cell. Thus no current flows, as indicated by the current-measuring galvanometer. However, the voltage of the external cell is known and equals the value of the voltage of the hydrogen gas–oxidizing agent cell at the *initial* concentrations of all species. This value is labeled a cell voltage or cell emf, even though it can be viewed as a measure of *potential energy*. It is the value of the voltage that would be measured if a voltage-producing reaction were allowed to occur.

Because a few electrons flow before the external voltage can be adjusted to be equal to that of the cell, no potentiometric device can ensure that the current will be exactly zero. For the accuracy needed for most measurements, however, there will be no appreciable error.

Indicating and Reference Electrodes

Both classical and modern potentiometric measurements are made with an electro- chemical cell consisting of two electrodes called the indicating and the reference elec- trodes (see also Sec. 11-3). The indicating electrode responds to changes in the concen- tration of the chemical species to be measured in the electrochemical cell, according to the Nernst equation **(11-20)**. The reference electrode is a self-contained system whose potential remains constant; it is isolated from reaction with the solution in the cell, although it is constructed to allow electrical contact.

The Reference Electrode

The reference electrode is a self-contained system whose potential is constant and independent of concentrations of ions in the sample solution. It is equipped with a fiber tip (Fig. 16-2), which provides contact with the sample solution; diffusion of the sample

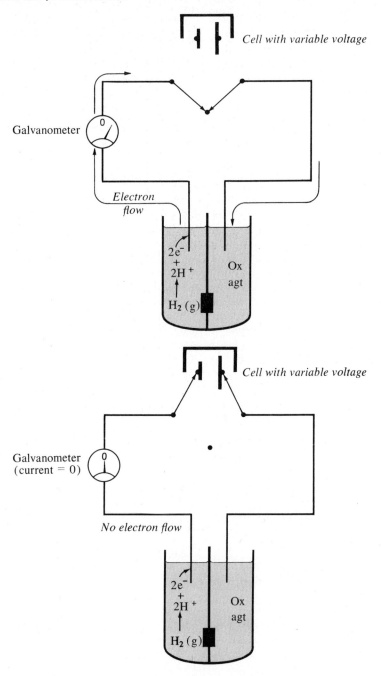

FIGURE 16-1. Above: measurement of cell voltage with a galvanometer. Note that current is flowing. Below: potentiometric measurement of cell voltage as potential. Note that no current is flowing through the galvanometer.

Potentiometer (variable voltage cell)

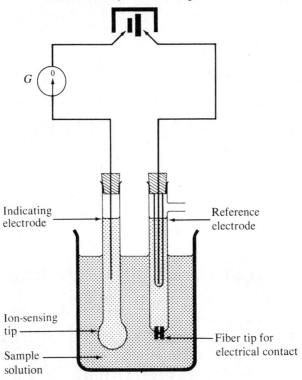

FIGURE 16-2. Apparatus for direct potentiometric or pH measurement. The ion-sensing tip of the indicating electrode responds to changes in the activity (corrected concentration) of the ionic sample. The galvanometer, G, indicates zero current flow.

solution into the electrode is negligible, so the sample does not affect the composition of the fluid inside the electrode.

The most common reference electrode is the saturated calomel electrode (SCE); its construction is shown in Figure 16-3. It is composed of metallic mercury and insoluble mercury(I) chloride (calomel) in equilibrium with a saturated solution of potassium chloride. It has a standard reduction potential of $+0.2458$ V at 25°C. The half reaction that governs its potential is as follows:

$$Hg_2Cl_2(s) + 2e^- \rightleftharpoons 2Hg(l) + 2Cl^-$$

The Nernst equation (Sec. 11-5) for this reaction is as follows:

$$E = E° + \frac{0.059\,V}{2} \log \frac{1}{[Cl^-]^2}$$

Since the potential of the half reaction depends only on the concentration of the chloride ion, and since this is constant in the SCE, the potential of this electrode theoretically is constant.

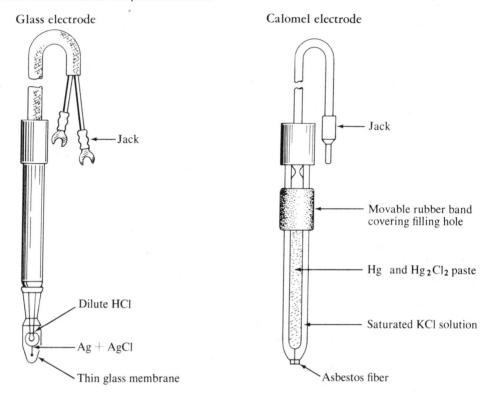

FIGURE 16-3.   The glass electrode (left) and the calomel electrode (right).

The Indicating Electrode

The indicating, or indicator, electrode must respond rapidly to the activity (corrected concentration) of the chemical species to be measured. There are various types of indicating electrodes—metallic, glass membrane, liquid membrane, and solid-state. Each type responds in a somewhat different manner at its ion-sensing tip (Fig. 16-2). All, in general, measure E, the potential of the half reaction involving the chemical species to be measured. The potential of the indicating electrode, $E_{\text{ind el}}$, is related to E and the activity of the chemical species through the Nernst equation (25°C) as follows:

$$E_{\text{ind el}} = E = E^\circ + \frac{0.0590 \text{ V}}{n} \log \frac{a_{\text{ox}}}{a_{\text{red}}} \tag{16-1}$$

where a_{ox} and a_{red} are the activities of the oxidized form and the reduced form, respectively, of the chemical species to be measured. If the total ionic content of the solution is below $10^{-4}M$, the activities are essentially the same as the concentrations of the two forms; otherwise, the activities are somewhat smaller. Of course, only one of the two forms can usually be measured; the activity of the other form is assumed to remain constant for all samples measured.

The *metallic indicating electrode* consists of an active metal, such as silver, mercury, or copper, or an inert metal, such as platinum. Most active metal electrodes consist of a wire made of the active metal dipping into solution containing the metal ion. For the active silver metal electrode, the half reaction is the following:

$$Ag^+ + e^- = Ag^\circ(s) \qquad E^\circ = +0.800 \text{ V}$$

Some metallic electrodes consist of a wire made of the active metal coated with a layer of an insoluble salt containing the metal ion and surrounded by a solution of constant concentration of the anion of the salt. A silver–silver chloride electrode can be made this way; its half reaction is the following:

$$AgCl(s) + e^- = Ag^\circ(s) + Cl^- \qquad E^\circ = +0.222 \text{ V}$$

The silver electrode can be used to measure pAg directly and pCl, pBr, etc., indirectly using the solubility equilibria of the corresponding insoluble silver(I) salt. The silver–silver chloride electrode can be used to measure pCl directly.

Inert metal electrodes consist of a wire, such as platinum, which dips directly into the sample solution without being enclosed in a tube. The function of the inert electrode is simply to transfer electrons; it does not take part in the reaction itself. It can thus be used to measure activity ratios of soluble ions such as Fe^{+3} and Fe^{+2}, which cannot form a solid metal electrode.

The *glass membrane electrode* has a thin, fragile, glass ion-sensing tip. A typical glass electrode is pictured in Figure 16-3. Such an electrode usually has an internal silver–silver chloride reference electrode to serve as a medium for the current flow. A solution of the ion to be measured is placed inside the thin ion-sensing tip. A potential difference develops across the glass membrane, and it is a function of the ratio of the activities of of the ions inside the tip and outside the tip in the sample solution.

This potential difference arises because of an ion-exchange reaction that occurs at the surface of the glass tip. For example, with the type of glass electrode used for pH measurement, the ion-exchange site is either $-SiO^-Na^+$ (older electrodes) or $-SiO^-Li^+$ (newer electrodes). In the sample solution whose pH is to be measured, an equilibrium ion-exchange reaction is established; for the older electrodes, it is the following:

$$-SiO^-Na^+ + \quad H^+ \quad \rightleftharpoons \quad -SiO^-H^+ + \quad Na^+ \qquad \textbf{(16-2)}$$
$$\text{(glass)} \qquad \text{(solution)} \qquad\qquad \text{(glass)} \qquad \text{(solution)}$$

The establishment of an equilibrium between the ion-exchange sites on the glass is thought to be responsible for the potential difference that results. Different glasses with different types of sites are used for the measurement of monovalent cations, such as Na^+ and K^+.

The *liquid membrane electrode* has a thin, porous polymer membrane tip instead of the glass tip of the glass membrane electrode. It is similar to the latter in that it has an internal silver–silver chloride reference electrode to serve as an element for current flow. A solution of the ion to be measured is placed inside the ion-sensing tip (Fig. 16-2). A potential difference develops across the porous polymer membrane; it is a function of the difference of the activities of the ions inside the tip and outside the tip in the sample solution.

The porous polymer membrane is saturated with a solution of a liquid ion-exchange resin dissolved in a water-insoluble organic solvent. One such ion-exchange resin has phosphate groups [$=P(O)O^-H^+$], which can react via an ion-exchange equilibrium with ions such as calcium(II) as follows:

$$2=P(O)O^-H^+ + Ca^{+2} \rightleftharpoons Ca^{+2}[=P(O)O^-]_2 + 2H^+$$
$$\text{(polymer)} \qquad \text{(solution)} \qquad \qquad \text{(polymer)} \qquad \text{(solution)}$$

The change in the ratio of the two different sites on the ion-exchange resin is thought to be responsible for the potential difference observed. Different ion-exchange resins with different types of sites are used for the measurement of divalent cations, such as Ca^{+2}, Mg^{+2}, Cu^{+2}, etc.

The *solid-state electrode* has a single crystal or pellet membrane tip instead of a glass tip or porous polymer tip. It may have either an internal silver–silver chloride reference electrode or merely a simple electrical contact to the inner surface of the membrane. A solution of a compound containing the anion (or cation) to be measured is placed inside the ion-sensing tip (Fig. 16-2). The membrane itself is a conducting solid and contains as a lattice ion the anion (or cation) to be measured. Electrical conduction occurs by a lattice defect mechanism by which only the lattice ion (ion to be measured) moves into vacant positions in the lattice. This type of electrode is used mainly for the measurement of anions, such as F^-, Cl^-, Br^-, SCN^-, and S^{-2}.

The Modern Potentiometer

The classical potentiometer (Fig. 16-1, bottom) utilizes an external cell having a variable voltage which can be read to ±0.001 volt (±1 millivolt). This type of potentiometer is rarely used in modern laboratories. A more popular type is the *electronic voltmeter* with a special amplifier designed to have a high input resistance. This instrument has the same effect as the classical potentiometer in that it minimizes the current flow to an insignificant value. (This arrangement cannot be shown in a simple schematic fashion and thus has been omitted from Fig. 16-2). Thus the concentration of the ionic species interacting with the indicating electrode does not change (**16-1**).

The potentiometer reading, E_{pot}, is now an accurate indication of the potential of the indicating electrode, $E_{ind\ el}$. Once the constant potential of the reference electrode, E_{ref}, is subtracted from $E_{ind\ el}$, the difference is equal to E_{pot}.

$$E_{ind\ el} - E_{ref} = E_{pot} \qquad \qquad (16\text{-}3)$$

Since E_{ref} and E_{pot} are the known parameters in any potentiometric measurement, **16-3** is better rearranged for calculation purposes to the following:

$$E_{ind\ el} = E_{pot} + E_{ref} \qquad \qquad (16\text{-}4)$$

Once $E_{ind\ el}$ is known, **16-1** can be used to calculate the activity of the species to be measured. Frequently, however, the scale of the potentiometer is precalibrated to read directly in units such as pH, pAg, pCl, etc.

Since the voltage scale of the potentiometer is uncertain to ±0.001 V, only three significant figures are justified for reporting $E_{ind\ el}$ (or pH, pAg, pCl, etc.). When $E_{ind\ el}$ is used with **16-1** to calculate an activity, this usually limits the accuracy of the activity to two significant figures because of the log values involved. See the following example.

EXAMPLE: A silver metal indicating electrode and a saturated calomel reference electrode are used to measure pAg potentiometrically. The potentiometer reading is $+0.419$ V; the potential of the SCE is $+0.246$ V. Calculate pAg and [Ag^+] to the correct number of significant figures.

Solution: Use **16-4** to calculate the potential at the indicating electrode.

$$E_{\text{ind el}} = +0.419 \text{ V} + 0.246 \text{ V} = +0.665 \text{ V}$$

Next, use the $E°$ of $+0.800$ V for the reduction of Ag^+ to $Ag°(s)$, in conjunction with the Nernst equation **(16-1)**, to calculate $[Ag^+]$.

$$E_{\text{ind el}} = +0.665 \text{ V} = +0.800 \text{ V} + 0.0590 \text{ V} \log a_{Ag^+}$$

(Since the activity of silver metal is unity, its activity does not affect the above expression.) Combining the indicating electrode potential and the $E°$ gives

$$+0.665 \text{ V} - 0.800 \text{ V} = -0.135 \text{ V} = 0.0590 \text{ V} \log a_{Ag^+}$$

Solving for the activity of the Ag^+ ion, we obtain

$$\log a_{Ag^+} = \frac{-0.135 \text{ V}}{0.0590 \text{ V}} = -2.28_8 = +0.71_2 - 3$$

$$a_{Ag^+} = 5.1_5 \times 10^{-3} M \text{ (2 sig. figs.)}$$

Since pAg is the negative log of the activity of the silver(I) ion,

$$pAg = 2.28_8, \text{ or } 2.29 \text{ (3 sig. figs.)}$$

Note that the activity of the silver(I) ion has one fewer significant figure than the log value, because the first significant figure of the log represents the exponent.

At this concentration, the activity of the silver(I) ion is about 5% lower than the molarity, so $[Ag^+] = 5.4 \times 10^{-3} M$.

16-2 | POTENTIOMETRIC MEASUREMENT OF pH

The pH of a solution can be measured by using an indicating electrode, which responds to the $[H^+]$, and a reference electrode, which will maintain a constant potential in the solutions to be measured. The classical example of an indicating electrode is the hydrogen electrode. This is a platinized platinum electrode over which hydrogen gas is bubbled. The oxidation-reduction half reaction for this electrode is

$$2H^+ + 2e^- = H_2(g)$$

If the pressure of the hydrogen gas is one atomosphere, the Nernst equation **(11-20)** for the potential measured at this electrode is

$$E = E° + 0.059 \text{ V} \log a_{H^+}$$

Because $E°$ for a standard hydrogen electrode is zero,

$$-E = -0.059 \text{ V} \log a_{H^+} = 0.059 \text{ V pH}$$

Note that this measurement is based on the exact definition of pH as the negative log of the *activity* of H^+ (see Sec. 9-1). A reference electrode, such as a standard calomel electrode (Fig. 16-3), is used to complete the circuit.

Because the hydrogen electrode is very awkward to use and not always resistant to chemical attack, a glass electrode (Fig. 16-3) is usually used as the indicator electrode for measuring pH (see below).

Modern pH Measurement

The classical method for measurement of pH, using the hydrogen indicating electrode, was discussed above. Now we will describe the modern method for measurement of pH, using the glass indicating electrode. Recall that pH is defined as

$$pH = -\log a_{H^+}$$

and that the activity, a_{H^+}, of the hydrogen ion is the *product* of the activity coefficient and $[H^+]$. At an ionic strength of 0.001, the activity coefficient of hydrogen ion is 0.97; so at concentrations below $10^{-3} M$ H^+, the activity and molarity of the hydrogen ion are about the same (as long as the ionic strength is $< 10^{-3}$).

The usual reference electrode is the saturated calomel electrode (Fig. 16-3). It has a constant potential of $+0.2458$ V and does not interact with the sample solution. It is connected to the back of the pH meter (Fig. 16-4) and held upright in a rigid holder.

The glass electrode shown in Figure 16-3 is used as the indicating electrode for most of the pH measurements made presently. The H^+ in a sample solution equilibrates

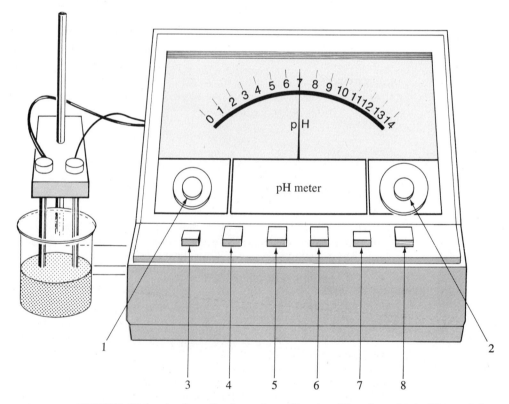

FIGURE 16-4. A schematic of a modern pH meter with scale readout. The controls are: (1) standardization control for measurements; (2) temperature compensator control; (3) manual pushbutton to connect temperature compensator (usually kept "on"); (4) automatic pushbutton to disconnect temperature compensator for automatic temperature compensation; (5) ± 700 mV pushbutton; (6) 1400 mV pushbutton; (7) standby pushbutton for removing electrodes from solution; (8) read pushbutton for reading pH from scale.

with the —SiO$^-$Na$^+$ groups on the electrode tip's surface, and affects the potential of the electrode. Strictly speaking, the activity of the sodium(I) ion in the sample solution also affects the potential of the glass electrode, because this ion is involved in the equilibria in **16-2**.

The relation between $E_{ind\ el}$, the potential at the glass electrode, and the activity of the hydrogen ion in a sample containing sodium(I) ion is

$$E_{ind\ el} = C + 0.0590 \text{ V} \log[a_{H^+} + (K_{H^+,Na})a_{Na^+}] \qquad \textbf{(16-5)}$$

where C is a constant that includes a number of sources of potential—such as the potential of the SCE, a liquid junction potential, and an asymmetry potential—and K_{H^+,Na^+} is a type of selectivity coefficient of the glass electrode for H^+ in the presence of Na^+.

Equation **16-5** can be simplified for lithium-oxide glass electrodes used at pHs below 13. Below this pH, the second term in the brackets is negligible compared to the a_{H^+}, so that **16-5** becomes the following:

$$E_{ind\ el} = C + 0.0590 \text{ V} \log a_{H^+} = C - 0.0590 \text{ V pH} \qquad \textbf{(16-6)}$$

Unfortunately, the value of C in **16-6** can never be known accurately. The value of the asymmetry potential varies from day to day, and the value of the liquid junction potential is uncertain. Therefore, most laboratory pH measurements involve calibration of the pH meter against a standard buffer solution, which is assumed to have the same value of C. The National Bureau of Standards has determined the pH values of the buffers listed below very accurately.

Buffer	pH(25°C)
0.05M potassium tetroxalate	1.679
Saturated potassium hydrogen tartrate	3.557
0.05M potassium acid phthalate	4.008
0.025M H$_2$PO$_4^-$, 0.025M HPO$_4^{-2}$	6.865
0.01M borax	9.180

It is recommended that two buffers whose pHs bracket the pH of the sample be used to standardize the pH meter to reduce the error from variation in liquid junction potentials.

Sodium Error

In alkaline media, glass membrane electrodes exhibit a sodium error, usually referred to as an alkaline error. The error arises from high activities of sodium ion interacting with —AlOSi$^-$H$^+$ sites via equilibria like that in **16-2**, causing a change in potential. The magnitude of this effect depends on the type of glass in the electrode and on the ratio of the sodium ion to the hydrogen ion. The —AlOSi$^-$H$^+$ sites in the older soda-lime-aluminum glass electrodes have a relatively small selectivity coefficient for H^+ over Na^+, compared to that of the sites in the new lithium-oxide glass electrodes.

With the older soda-lime-aluminum glass electrodes, the presence of 1M sodium ion caused pH values to be low by 0.2 pH unit at pH 10 and by 1.0 unit at pH 12. When the newer lithium-oxide glass electrodes are used, the presence of 1M sodium ion causes an error of less than 0.2 pH unit at pH 13. Almost all commercially available glass pH electrodes are of the latter type.

Glass electrodes should be stored upright in distilled water when not in use. New glass electrodes should be presoaked in distilled water before use. The reason for both of these requirements is that the outer layers of the glass electrode must become hydrated before their ion-exchange sites can function effectively. During the soaking, several layers of the glass surface of an electrode become hydrated and the ion-exchange sites exchange metal ions for hydrogen ions. Without this soaking, a steady potential reading is not attained. Of course, if a glass electrode is not to be used for a long time, it is best stored dry since the glass, like any glass, slowly dissolves in water.

The Use of the pH Meter

There are three basic types of pH meters, two of which are based on the potentiometric principle of opposing the voltage of the sample solution. These two are the classic potentiometric null-point indicator type and the new, direct-reading, feedback circuit type (Fig. 16-4). With the former, time and care must be taken in the calibration and adjustment of the null-point indicator before reading. The reading obtained, however, is generally very accurate and precise. With the latter, the direct-reading feature gives a more rapid, but possibly less accurate and precise, reading of the pH. The third type of pH meter is simply an electronic voltmeter.

Because of the high electrical resistance of the glass electrode, all types require special design. Electronic amplification of the small current flowing into the galvanometer is required in the potentiometric type. The direct-reading types require a high input resistance or a corrector amplifier–capacitor combination to correct for zero drift (Beckmand Zeromatic® pH meters).

Operation

To standardize a particular pH meter with a particular set of electrodes, two standard buffer solutions are used. The meter is put in the standby mode (control 7, Fig. 16-4) to protect the circuits when electrodes are being removed from a solution. The electrodes are then removed from solution, rinsed of the previous sample solution, and excess water is blotted off. They then are immersed in the first standard buffer solution, and the circuits are reconnected (control 8, Fig. 16-4). The temperature compensator (control 2, Fig. 16-4) is adjusted to the temperature of the buffer solution, and the standardization control (control 1, Fig. 16-4) is turned until the meter reading corresponds to the pH of the first buffer. The above is repeated for the second standard buffer.

After the electrodes have been removed from the second buffer, they are dipped into the sample solution. A temperature adjustment is made if necessary (control 2, Fig. 16-14), and the pH of the sample is read. Generally one standardization is enough for several samples.

pH Readout

There are two types of readout for pH meters—scale readout (Fig. 16-4) and digital readout. In the usual scale readout, only two significant figures to the right of the decimal point are justified. The first is read directly from the scale, and the second is interpolated. Since pH is a log term, *all* zeroes to the right of the decimal are significant. Thus pH readings of 0.05 and 0.00 both have two significant figures to the right of the decimal point.

> **EXAMPLE:** A scale readout pH meter gives a pH reading of 5.96 for a sample of water, which has been purified in a water plant. Calculate the activity of the hydrogen ion and $[H^+]$.
> *Solution:* Insert the pH value into the definition of pH.
>
> $$pH = -\log a_{H^+} = 5.96$$
>
> $$\log a_{H^+} = -5.96 = +0.04 - 6$$
>
> $$a_{H^+} = 1.1 \times 10^{-6}$$

Note that the 0.04 term represents only two significant figures, since it is a log term (Sec. 3-1), so the activity should be expressed with two significant figures.

Since this is a purified water sample, we may tentatively assume that the activity and the molarity are the same; thus, $[H^+] = 1.1 \times 10^{-6}M$.

Digital readout pH meters give a direct readout of pH to two or three digits to the right of the decimal point. There is some question as to whether the third digit is significant because of stability problems and activity coefficients. Some expensive pH meters do possess the stability necessary for a third digit when they are used with special low resistance glass electrodes. In general, it is best to check the manufacturer's literature and the performance of the pH meter before assuming an uncertainty of ± 0.001 for pH readings.

16-3 | POTENTIOMETRIC MEASUREMENT OF SPECIFIC IONS

We have already described indicating electrodes useful for direct potentiometric measurement of other ions (Sec. 16-1). We will now discuss the measurement of a few important ions using specific ion-indicating electrodes. As in pH measurement, such measurements give the *activity* of the ion, not its molarity. Thus pNa is

$$pNa = -\log a_{Na^+}$$

where a_{Na^+} is the product of the activity coefficient and $[Na^+]$. There is quite a difference between the molarity of Na^+ and the activity of Na^+ in body fluids. In normal human serum, the activity coefficient of sodium(I) ion is 0.78, for example. If analytical results in terms of concentration are desired, a plot of potential in millivolts against molarity, or ppm, is constructed.

Measurement of Chloride Ion

The measurement of chloride is very important. For example, in food processing the chloride content of milk and butter is measured potentiometrically using a solid-state electrode (Sec. 16-1). An increase in the amount of chloride in milk is an indication of mastitis in cows; this is easily detected with the chloride electrode.

In biomedical research, increased or dereased chloride levels can be quickly measured by potentiometric measurement. Measurement of chloride is important in this field because of chloride's contribution to the electrolyte balance of the blood (Sec. 22-1); in addition, chloride is necessary for the manufacture of hydrochloric acid in the stomach.

Sweat Chloride

The measurement of chloride in body perspiration is a useful clinical measurement. Formerly, a thirty-minute sweat collection procedure was needed before measurement. Now, sweating is induced by rubbing the palm or wrist with a compound called pilocarpine and applying a weak electric current to induce passage of chloride ions through the skin (iontophoresis). (Pilocarpine is an alkaloid, $C_{11}H_{16}O_2N_2$. The two nitrogens are part of a five-membered heterocyclic ring. One nitrogen is basic enough to form a hydrochloride salt.)

The chloride in the sweat is measured directly on the unbroken skin using a *flat-headed* chloride specific ion electrode (Fig. 16-5). The entire procedure takes five minutes when a commercial chloride screening system is used.

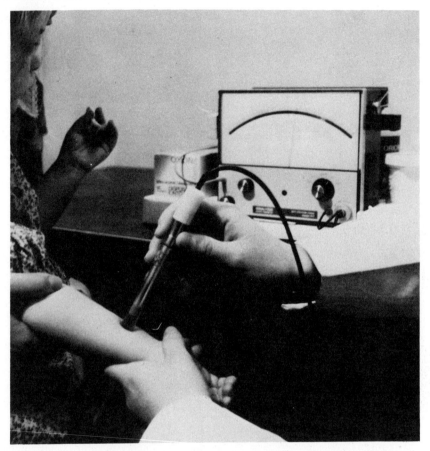

FIGURE 16-5. Orion cystic fibrosis screening system with a flat-headed chloride specific ion electrode for pCl measurement. (Courtesy of Orion Research, Inc.)

Since most chloride abnormalities result in decreased chloride levels, the occurrence of increased chloride levels is of great significance in diagnosing *cystic fibrosis*, a childhood lung disease. Cystic fibrosis affects one in every 600 to 2000 babies born and is characterized by high sweat levels of chloride and other ions. The flat-headed chloride specific ion electrode (Fig. 16-5) is used for rapid screening of babies in hospitals. The results are interpreted with reference to a mean chloride level for infants of 20 meq/liter $(2.0 \times 10^{-2}M)$ and a safe upper level for infants of 40 meq/liter $(4.0 \times 10^{-2}M)$. If the level of chloride exceeds 60 meq/liter $(6.0 \times 10^{-2}M)$, then cystic fibrosis is indicated. The chloride level has been known to run as high as 140 meq/liter [1]. If the chloride level is between 40 and 60 meq/liter, cystic fibrosis is not decisively indicated. Further observation is necessary in such cases.

Measure-ment of Fluoride Ion   The measurement of fluoride ion is also very important in many areas related to health. Fluoride is measured with a solid-state, single crystal, specific ion electrode (Sec. 16-1). Probably its most important application is in the analysis of drinking water to which

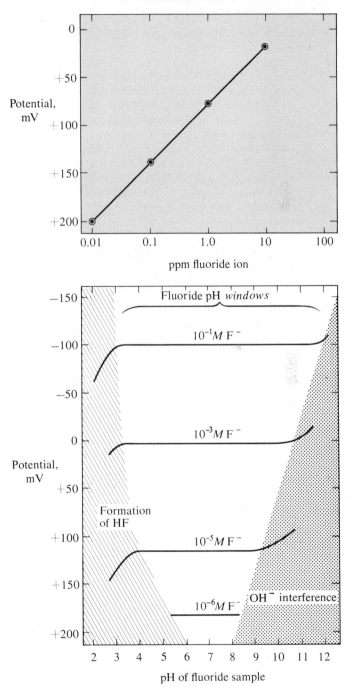

FIGURE 16-6. Above: typical calibration plot for fluoride analysis. Below: behavior of fluoride electrode as a function of pH.

fluoride ion has been added for the protection of teeth. The single crystal used in the electrode is a lanthanum fluoride crystal doped with divalent europium to increase the electrical conductivity.

The fluoride ion electrode is highly specific for fluoride ion down to an activity of about 10^{-6}. The only known anion interference is from the hydroxide ion. In spite of this, there are definite pH limitations to the method. As the level of fluoride decreases, the pH *window*, or *range*, over which it may be accurately measured shrinks. Below pH 3–5 (Fig. 16-6), fluoride forms hydrofluoric acid, which is not measured by the electrode.

To analyze drinking water for 1 ppm ($5 \times 10^{-5}M$) of fluoride, the pH must be adjusted to be within a pH window of 5 to 8. In addition, iron(III) in the water must be prevented from forming fluoride complex ions, since these are not measured by the electrode. A buffer is added to keep the solution slightly acidic; a solution of sodium citrate is also added to complex the iron(III) and prevent it from complexing the fluoride ion. Once this is done, the potential measured is compared to a calibration plot (Fig. 16-6) to obtain a value of ppm fluoride.

Measure-
ment of
Cations

Modified glass electrodes have been designed for the measurement of certain monovalent cations such as Na^+, K^+, Ag^+, Li^+, and even NH_4^+. Since the hydrogen ion also interacts with the ion-exchange sites on such electrodes, most measurements using them are done in neutral or basic solution. In neutral solution, for example, a typical sodium electrode has a selectivity coefficient **(16-5)** for sodium over potassium of about 300. This means sodium(I) can be measured in nearly a hundred times as much potassium(I) without appreciable error (10%).

A glass electrode has been proposed for the measurement of sodium(I) in body fluids, such as blood, but to use it a preliminary separation of interfering materials is necessary. Flame emission (photometry) analysis for sodium is still preferred by many clinical and hospital laboratories [2].

| QUESTIONS AND PROBLEMS

(Answers to even-numbered problems are in Appendix 5.)

Concepts and Definitions

1. Name the items of general equipment that are necessary for the direct potentiometric measurement of a particular ion in solution.
2. Explain the difference between an indicating electrode and a reference electrode.
3. Explain why a measured cell voltage is a form of potential energy, not kinetic energy.
4. Describe a calomel electrode.
5. List the four types of indicating electrodes and state how the ion-sensing tip of each differs.
6. Explain what is meant by a sodium error for the glass electrode.

7. Modern glass electrodes (lithium oxide) have a very low sodium error. Would you predict that they also have a very low lithium error? If so, why is this not as serious a problem as the sodium error for most measurement?
8. Explain how a potentiometer differs from a voltmeter.
9. Why does a pH measurement give only the approximate $[H^+]$?
10. Explain how a sweat chloride measurement is made to detect cystic fibrosis.

pH Measurements

11. Calculate the activity of the hydrogen ion corresponding to each of the following pH

values (assume each represents the correct
number of significant figures):

a. pH = 1.96
b. pH = 0.939
c. pH = −0.04
d. pH = 0.00
e. pH = 4.996
f. pH = 6.9

12. Assuming that no additional ionic species are
present (other than the acid ionizing to
hydrogen ion) in Problem 11, state whether
the $[H^+]$ will be larger than, smaller than, or
the same as the activity in each case.

13. If K_w for water is 1.00×10^{-14}, calculate the
pOH corresponding to the following pH
values, using the correct number of significant
figures.

a. pH = 1.96
b. pH = 0.939
c. pH = −0.51
d. pH = 14.00

14. The following pH measurements were made
with the older soda-lime-aluminum glass
electrode. State whether there will be an error
in the pH and, if so, how large the error
would be.

a. pH = 12.1 $(1M\ Na^+)$
b. pH = 10.2 $(1M\ Na^+)$

15. Choose the correct National Bureau of
Standards–approved buffer(s) for each of the

following samples whose $[H^+]$ falls in the
range of $[H^+]$ given.

a. $5 \times 10^{-4}M$–$9 \times 10^{-4}M\ H^+$
b. $1.1 \times 10^{-4}M$–$2.3 \times 10^{-4}M\ H^+$
c. $8.1 \times 10^{-6}M$–$2.9 \times 10^{-5}M\ H^+$

Specific Ion Electrodes

16. Calculate the activity of each ion below from
the pM reading obtained from potentiometric
measurement using a specific ion indicating
electrode.

a. pNa = 1.23
b. pK = 10.96
c. pF = 0.20
d. pCa = −0.301

17. An unknown sample of calcium(II) ion has a
potentiometer reading of +10 millivolts,
using a calcium specific ion electrode. By
making a semilog plot of millivolt readings vs.
concentration, using the following calcium(II)
ion standards, calculate the concentration in
molarity of the calcium(II) ion in the
unknown.

M of Ca^{+2}	mV
1.0×10^{-4}	−26
8.0×10^{-4}	0.0
6.0×10^{-3}	+25
3.0×10^{-2}	+47

| **NOTES**

[1] R. L. Searcy, *Diagnostic Biochemistry*, McGraw-Hill, New York, 1969, p. 159.
[2] B. Zak, Wayne State University, Detroit, personal communication, 1976.

17 | Electrochemical Methods of Analysis

"You know my methods. Apply them," said Holmes.

ARTHUR CONAN DOYLE
The Hound of the Baskervilles, Chapter 1

The previous chapter dealt entirely with one electrical method of analysis—the direct measurement of the potential energy (or available voltage) of a particular reaction. No actual electrochemical reaction occurs to an appreciable extent during such measurement. In this chapter we deal with methods of analysis that are truly *electrochemical*; that is, there is a conversion of electrical energy into chemical energy, or vice versa. It should be your goal for this chapter to "know" these methods and be able to "apply them."

We will start this chapter with a short review of the principles of electrolytic and galvanic cells. These principles will then be applied to the succeeding topics of electrogravimetry (electrodeposition), coulometry, and voltammetry, including polarography. Don't become discouraged as you realize that these methods are complicated. Try to grasp the essentials first—later, master the details.

These methods are important because they have several important health science-oriented applications. For one, coulometry is used for very accurate measurement of the chloride ion, an important constituent of body fluids and an ion not easily measured by spectrophotometry. For another, the amount of oxygen in body fluids, such as the blood, can be measured by the use of two different types of electrodes. One of these is based on a glavanic type of electrical measurement, and the other is based on a polarographic type of measurement.

17-1 | A REVIEW OF GALVANIC AND ELECTROLYTIC CELLS

Electrolytic and Galvanic Cell Reactions

There are two types of electrochemical cells—the galvanic cell and the electrolytic cell. In the galvanic cell a chemical reaction occurs *spontaneously* and produces electrical energy. Such chemical reactions were discussed in Chapter 11, and the calculation of the standard potential was outlined in Section 11-3. In the electrolytic cell there is no spontaneous reaction; electrical energy is needed to cause a cell reaction, which process is the opposite of the spontaneous reaction.

To illustrate the difference between the two types of cells, consider the spontaneous reaction and the nonspontaneous reaction that can occur from the combination of the same two half reactions. The half reactions are

$$Cu^{+2} + 2e^- = Cu^\circ(s) \qquad E^\circ = +0.337 \text{ V}$$

$$\frac{1}{2}O_2(g) + 2H^+ + 2e^- = H_2O \qquad E^\circ = +1.229 \text{ V}$$

The spontaneous reaction is that which results from the subtraction of the first half reaction and its E° from the second half reaction and its E° to give the cell emf, E°_{cell}:

$$\frac{1}{2}O_2(g) + 2H^+ + Cu^\circ(s) = H_2O + Cu^{+2} \qquad E^\circ_{cell} = +0.892 \text{ V} \qquad \textbf{(17-1)}$$

We see that the reaction is spontaneous because the cell emf is positive. The nonspontaneous reaction is that which results from the subtraction of the second half reaction and its E° from the first half reaction and its E° as follows:

$$H_2O + Cu^{+2} = \tfrac{1}{2}O_2(g) + 2H^+ + Cu^\circ(s) \qquad E^\circ_{cell} = -0.892 \text{ V} \qquad \textbf{(17-2)}$$

This reaction does not occur spontaneously because E°_{cell} is negative, but it will occur if enough electrical energy is applied to overcome the voltage of the spontaneous reaction and thus reverse the direction of the equilibrium.

In the galvanic cell, measurement of the potential, or the flow of current, is often used for analytical purposes. But how can an electrolytic cell be used for analytical purposes? In an electrolytic cell involving a reaction such as that in **17-2**, the applied voltage forces electrons from the anode to the cathode. At the anode, electrons are taken up from the species that is oxidized (in **17-2** water is oxidized to oxygen gas). At the cathode, electrons combine with metal ions to form the free metal, which deposits on the cathode. If the deposition is quantitative, the gain in weight of the cathode can be measured, indirectly giving the amount of the metal ion originally in the sample solution. This process is called *electrogravimetry*, or *electrodeposition*. Because not all metal ions deposit quantitatively, a more general procedure is to measure the quantity of electrical current required to reduce or oxidize an ion in solution. This is known as *coulometry*. Frequently, a so-called coulometric titration is performed, which is a variation of direct coulometry. In addition to electrogravimetry and coulometry, a third method is the *polarographic* reduction of metal ions at a dropping mercury electrode, which will be discussed below. Measurement of the so-called diffusion current gives a measure of the ionic concentration.

Electrode Reactions

Electrode reactions in a galvanic cell are simply the spontaneous half reactions that ordinarily occur as parts of a complete oxidation-reduction reaction. The oxidizing agent is reduced at the cathode, and the reducing agent is oxidized at the anode. In an electrolytic cell, a number of reactions are possible, depending on the applied voltage. Whatever reaction occurs first obviously can be used most easily for analysis. The half reaction that occurs first at the cathode is that which has the most positive, or least negative, potential. For example, in a mixture of Ag^+ ($E° = +0.800$ V) and Cu^{+2} ($E° = +0.337$ V), silver(I) will be deposited first, as silver metal.

Since most metal ions are soluble only in acid solution, electrolysis in acid solution involves a separation of the metal ion from the hydrogen ion.

The only metal ions commonly reduced before hydrogen ion in acid solution are Ag^+, Cu^{+2}, Hg^{+2}, and BiO^+; the $E°$ values of these are all more positive than the $E°$ of 0.0 V for the reduction of $1M$ hydrogen ion. If the pH is raised to 7, the $E°$ for the reduction of hydrogen ion is lowered to -0.43 V. Any metal ions in solution at that pH which are reduced at a potential more positive than that of hydrogen ion, such as $Zn(NH_3)_4^{+2}$, are deposited at the cathode.

Another way of preventing the interference of hydrogen ion is through the polarographic use of a mercury cathode, rather than a solid inert cathode, for deposition. This has the advantage of shifting the point of equilibrium to the right for the reduction process as follows:

$$M^{+n} + ne^- + Hg(l) = M[Hg(l)]$$

Because the metal is dissolved in the mercury, it has less tendency to dissolve into the aqueous sample solution. This has the effect of making the $E°$ value for the reduction of the metal less negative, or more positive. Many metal ions that are reduced after hydrogen ion using an inert cathode are reduced before hydrogen ion using a mercury electrode. Thus polarography can be used for the determination of about one-third of all the elements.

Overvoltage
An effect that is of practical importance, but cannot be predicted theoretically, is the overvoltage effect for a cell. This is most significant in cases where a gas is liberated at any electrode; in such a case, additional voltage is necessary to overcome the slow rate of some step. Typical overvoltages are listed in Table 17-1. Note that the overvoltage for a specific gas, such as hydrogen, varies with the kind of electrodes used. There is a wide variation between the overvoltage necessary for a smooth platinum electrode, which is inert, and for a liquid mercury electrode, which forms an amalgam with metals. The high overvoltage for the mercury electrode is an important factor in that it allows many metal ions with $E°$ values more negative than that of hydrogen ion to be reduced and amalgamated into the mercury electrode before hydrogen ion is reduced to hydrogen gas.

TABLE 17-1. Overvoltages for Formation of $O_2(g)$ and $H_2(g)$

Current density A/cm^2	Hydrogen overvoltage			Oxygen overvoltage	
	Smooth Pt	Platinized Pt	Hg	Smooth Pt	Platinized Pt
0.001	0.024	0.015	0.9	0.72	0.40
0.01	0.068	0.030	1.04	0.85	0.52
0.1	0.29	0.041	1.07	1.28	0.64
1.0	0.68	0.048	1.11	1.49	0.77

In addition to cell overvoltage, there is also an *electrode overpotential*, or *concentration overpotential*. This effect arises when metal ions (or anions) at the surface of the electrode are removed from solution at a faster rate than they diffuse to the surface, resulting in a lower concentration at the surface than in the bulk of the solution. It is the concentration at the surface, rather than that in the bulk of the solution, that should be used in the theoretical (Nernst equation) calculation of the potential required for an electrochemical reaction. Since this surface concentration is not known, it cannot be used; the difference between the theoretical potential and the observed potential (excluding other effects such as overvoltage) is simply said to be the electrode overpotential.

17-2 | ELECTROGRAVIMETRY

**Electro-
gravimetric
(Electro-
deposition)
Apparatus**

In electrogravimetry, or electrodeposition, the species that can be determined are limited to a few metal ions, which are deposited on the cathode. Usually only one metal ion from a sample is electrodeposited. The metal ion to be determined is reduced at a preweighed platinum cathode (Fig. 17-1) and deposited quantitatively on that cathode. After completion of the electrodeposition, the cathode is dried and weighed. The weight of the metal is found by subtracting the weight of the cathode before deposition from that after deposition. This weight is then used to calculate the concentration of the metal ion in the sample solution.

In the electrogravimetric apparatus (Fig. 17-1), the current for the electrolysis is supplied by a battery or other source of direct current. The voltmeter, V, measures the applied voltage, and an ammeter, A, measures the current flow. Since the voltage must be controlled to deposit a particular metal, the variable resistor, R, is used to adjust

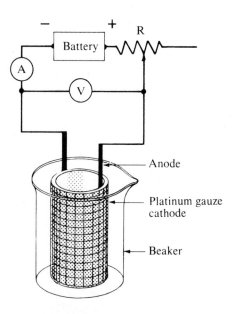

FIGURE 17-1. Apparatus for electrodeposition.

the applied voltage. (As the amount of metal ion in solution decreases near the end of the electrodeposition, the variable resistor is adjusted to supply the voltage required at each point.)

Metal Ions Commonly Determined

The copper(II) ion is the metal ion most commonly determined by electrogravimetry. The reaction at the cathode is

$$Cu^{+2} + 2e^- = Cu^\circ(s)$$

and the reaction at the anode is

$$H_2O = \frac{1}{2}O_2(g) + 2H^+ + 2e^-$$

The overall cell reaction is the sum of these **(17-2)**. The silver(I) ion is the most likely interference and should be removed by precipitation before analysis. Even small amounts of silver(I) will cause problems. More than 0.2% of silver(I) causes a rough deposit of copper metal, making handling difficult; more than 0.4% affects the accuracy of the method. Some nitric acid must be added to the electrolyte to prevent the reduction of hydrogen ions to hydrogen gas during the latter stages of the deposition. If too much nitric acid is used, and more than 0.5% of iron(III) is present, quantitative deposition of copper may be inhibited. The separation of copper(II) from other metal ions by this method is possible.

The sequential determination of copper(II) and then nickel(II) is possible in mixtures of the two metal ions. The copper(II) is deposited from acid solution as above, leaving nickel(II) in solution since its E° is -0.25 V. The electrodeposition is temporarily stopped after electrodeposition of copper is complete, and the cathode is weighed to determine the weight of the copper deposit. The cathode is then reinserted into the solution, and the solution is neutralized with ammonia to form the $Ni(NH_3)_6^{+2}$ ion. The pH at this point is such that hydrogen ion cannot be reduced to hydrogen gas before $Ni(NH_3)_6^{+2}$ is reduced to nickel on the cathode. The electrodeposition of nickel is then performed, and the cathode plated with both copper and nickel is again weighed. The increase in weight over the previous weighing gives the weight of nickel.

The silver(I) ion has the most positive E° in acid solution and could theoretically be determined in the presence of any other metal ion. However, it is not deposited quantitatively in acid solution, but is deposited quantitatively from an ammonical solution or from an alkaline cyanide solution. The reaction at the cathode for the former complex ion is

$$Ag(NH_3)_2^+ + e^- = Ag^\circ(s) + 2NH_3 \qquad E^\circ = +0.375 \text{ V}$$

Since the E° for the electrodeposition of the $Cu(NH_3)_4^{+2}$ ion is -0.06 V, the diammine silver(I) ion can be electrodeposited in the presence of the tetrammine copper(II) ion in ammonia solution.

Lead(II) ion is always oxidized at the anode to lead(IV) ion and is deposited as lead(IV) oxide. The reaction at the anode is

$$Pb^{+2} + 2H_2O = PbO_2(s) + 4H^+ + 2e^- \qquad E = -(+1.455 \text{ V})$$

(The $+1.455$ V is for the *reduction* of lead(IV) oxide.) Lead(II) and copper(II) can be deposited simultaneously, the lead(II) at the anode and the copper(II) at the cathode, even in the presence of other ions, except, of course, silver(I).

Voltage Requirements for Electrolysis
The voltage, E, required to start electrodeposition depends on the potential at the cathode, E_c, the potential at the anode, E_a, the overpotential at the cathode, E_{oc}, the overpotential at the anode, E_{oa}, and the cell resistance, R_c. The equation relating these is the following:

$$E = (E_c + E_{oc}) - (E_a + E_{oa}) + R_c$$

Since metal ions are generally determined by electrolysis, the potential at the cathode, E_c, is the most important variable in the equation above. This variable depends on the concentration of the metal ion to be electrodeposited, and it can be calculated using the Nernst equation:

$$E_c = E^\circ + 0.059 \text{ V}/n \log[M^{+n}] \qquad (17\text{-}3)$$

where E° is the standard reduction potential of metal ion M^{+n} and n is the number of electrons required to reduce it to metal $M^\circ(s)$.

For example, if $0.010M$ copper(II) ion is to be electrodeposited at the cathode as copper metal, then E_c is calculated, using **17-3**, as follows:

$$E_c = +0.337 \text{ V} + \frac{0.059 \text{ V}}{2} \log[0.010M \text{ Cu}^{+2}]$$

$$E_c = +0.337 \text{ V} - 0.059 \text{ V} = +0.278 \text{ V}$$

Of all the parameters contributing to E, E_c varies most significantly during electrodeposition. From the calculation above, it is apparent that E_c will become less positive as the concentration of Cu^{+2} decreases during deposition. For example, if $[Cu^{+2}]$ were $1.0 \times 10^{-6}M$ at the end of the electrodeposition, then E_c would be $+0.160$ V.

17-3 | COULOMETRIC ANALYSIS

Faraday's Law and the Coulomb
Coulometry, or coulometric analysis, depends on the measurement of the amount of electrical current required to reduce or oxidize an ion in solution. It is a more general method than electrogravimetry, because not many ions are quantitatively deposited on an electrode, and because it can be used for reactions resulting in soluble products or gases as well as metals.

The key to understanding coulometry is Faraday's law. This law is based on two electrical units—the *coulomb* (C) and the faraday (F). The coulomb is defined in terms of amperes of electrical current and the total time it flows in seconds.

$$C = A \cdot s \qquad (17\text{-}4)$$

The faraday is defined in terms of the coulomb or in terms of moles of electrons. The faraday is frequently defined to three significant figures as 9.65×10^4 C, but for analytical work requiring more than three significant figures, it is defined as

$$F = 96{,}487 \pm 1 \text{ C} \qquad (17\text{-}5)$$

In terms of moles of electrons or equivalents, it is defined as

$$F = 1 \text{ mole of electrons} = 1 \text{ eq of a chemical species} \qquad (17\text{-}6)$$

(Recall that one mole of electrons will reduce one equivalent of a species, by definition.) The relation between the number of equivalents, the weight in grams of a species that is oxidized or reduced, and n, the number of electrons involved per formula weight, is

$$eq = \frac{g}{form\ wt/n} \qquad (17\text{-}7)$$

The number of equivalents of any species reduced or oxidized at a constant current for a given time in seconds may be obtained by combining **17-4** through **17-6**:

$$eq = \frac{C}{96,487\ C/F} = \frac{A \cdot s}{96,487\ C/F} \qquad (17\text{-}8)$$

Once the number of equivalents have been calculated, **17-7** may be used to calculate the number of grams involved.

> **EXAMPLE:** A 100.0-ml volume of a copper(II) sulfate solution of unknown concentration is analyzed coulometrically by reducing the copper(II) to copper metal, using a constant current of 2.000 A. (a) If 20.00 minutes are required for the electrolysis calculate the equivalents of copper(II) in the sample, and the molarity. (b) Calculate the weight of the copper deposited at the cathode.
>
> *Solution to a:* Using **17-8**, calculate the eq of copper(II).
>
> $$eq = \frac{(2.000\ A)(1200\ s)}{96,487\ C/F} = 0.02487_4\ eq$$
>
> The molarity of the copper(II) in the sample is
>
> $$0.02487_4\ eq/2\ eq\ per\ mole = 0.01243_7\ mole$$
>
> $$\frac{0.01243_7\ mole}{0.1000\ liter} = 0.1243_7 M$$
>
> *Solution to b:* The weight of the copper deposited in grams is found by rearranging **17-7**.
>
> $$g = (eq)(form\ wt/n)$$
>
> $$g = (0.02487_4\ eq)(63.54/2) = 0.7902_5\ g$$

Coulometric Methods and Apparatus

There are two experimental approaches to coulometric measurement—constant-potential (applied voltage) coulometry and constant-current coulometry.

In constant-potential coulometry, the potential is set at the desired value and the electrolysis is allowed to run until the reaction is complete, at which point the current is essentially zero. The coulombs of current used can be calculated in one of two ways. Classically, a separate electrolysis cell, or *coulometer*, is connected in series with the working cell. For example, silver(I) may be deposited in the coulometer, and the silver metal weighed to give the coulombs of current used. The modern way of measuring coulombs is to record a plot of current in amperes against time in seconds and to calculate the area under this curve to give the total coulombs used. Various mechanical and electronic devices have been designed to integrate the area under the curve automatically [1].

In the constant-current approach to coulometry, the time necessary to complete the electrolysis, rather than the current, is measured. A digital electric timer that can be turned off automatically is used to record the exact time. The coulombs are calculated from the magnitude of the current and the time.

$$C = A \cdot s$$

The number of coulombs is then converted to equivalents using Faraday's law **(17-8)**. In order to tell when the reaction has ceased, one of the various indicating techniques used in conventional titrations is employed. Visual indicators or potentiometric detection of the end point can be used. There is also an amperometric method employing two electrodes; a diffusion current that results from excess titrant or excess sample ion is used to signal the end point (Sec. 17-4).

Coulometric methods may be *direct* or *indirect*. In a direct method, the species being determined is reduced at the cathode or oxidized at the anode. In an indirect method, the species being determined does not react at either electrode but reacts with a second species generated at one of the electrodes. This second species is called a *titrant* even though it is not delivered from a buret, and the indirect method is frequently called a *coulometric titration*.

A schematic of a coulometric apparatus for an indirect coulometric method is shown in Figure 17-2. The generating electrode is shown at the left. If it is to generate a metal ion, it will be an anode, and the other electrode will be the cathode. (No method of indicating the end point is shown.)

The constant-current source is either a battery in series with a large resistance or a regulated constant-current device automatically controlled with an electromechanical or electronic regulator.

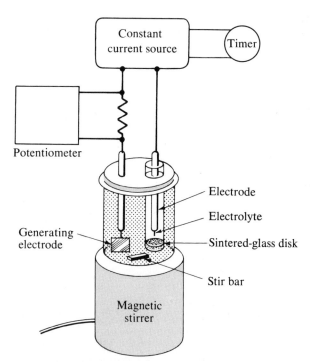

FIGURE 17-2. Schematic of a constant-current coulometer.

A precision electric stopwatch that can be turned on and off in perfect synchronization with the current can be used as a timer. A better, but more expensive, timer is a clock with a solenoid-operated brake, which has a deviation of ± 0.01 seconds per complete cycle.

Ions Commonly Determined

Coulometric methods are especially useful for the determination of small amounts of reducible or oxidizable species. This is because an amount of electricity as small as 0.1000 C can be measured with a relative deviation of 0.1 pph. From **17-8**, this is equal to:

$$0.1000 \text{ C} = 0.0000010364 \text{ eq} = 1.036_4 \times 10^{-6} \text{ eq}$$

If this amount is dissolved in 10 ml of solution, it will be roughly a $10^{-7} M$ solution, a fairly low concentration.

Direct coulometric methods are not used for many determinations because side reactions occur in many cases near the end of the electrolysis. Indirect coulometry, or coulometric titration, is used much more because the *titrant* generated at one of the electrodes is used up immediately in a chemical reaction before any side reactions can occur. Coulometric titrations also have the advantage of being able to utilize unstable reagents, such as silver(II) and copper(I), which are difficult to use in conventional titrations.

Determination of Chloride

From a medical viewpoint, the most important coulometric method is the determination of chloride ion in body fluids. The chloride ion is colorless and is not readily determined by colorimetry, so the determination of small amounts of chloride by coulometric titration is widely used.

The basic coulometric reactions for the determination of chloride include, first of all, coulometric generation of silver(I) from a silver wire *anode* immersed in the sample.

$$\text{Ag}^\circ(\text{anode}) \longrightarrow \text{Ag}^+ + e^-$$

As it is generated, the silver(I) ion reacts with the chloride ion.

$$\text{Ag}^+ + \text{Cl}^- \longrightarrow \text{AgCl(s)}$$

At the point at which the silver(I) is no longer able to react with the chloride ion (equivalence point), it is then reduced at the *cathode* as follows:

$$\text{Ag}^+ + e^- \longrightarrow \text{Ag}^\circ(\text{cathode})$$

This reduction is detected amperometrically, wherein the sudden increase in diffusion current (Sec. 17-4) signals the equivalence point.

A commercially available instrument called the Cotlove Chloridometer has been designed for the coulometric determination of chloride. It is an automatic constant-current coulometer and ultilizes the above reactions. The titration is performed automatically and uses the amperometric end point. The sudden rise in the amperometric current at the end point activates a microswitch, which automatically terminates the generation of silver(I) ion from the cathode and stops the timer. The chloridometer is standardized against a known chloride solution before analysis, so the time needed for the titration is related to the concentration of the chloride directly. The results can be printed out on paper tape by the instrument. Protein does not interfere with the titration because no excess silver(I) ion is generated. (If an excess of silver(I) were present, it would react with the protein and give high results.) This is an advantage, because protein must be precipitated in many clinical methods before the determination.

Determination of Other Ions

Almost any ion that can be titrated by conventional methods can be determined by coulometric titration. As an example, consider the coulometric titration of arsenic(III). As a titrant, iodine is generated at a platinum anode by oxidation of the iodide ion.

$$2I^- \longrightarrow I_2 + 2e^-$$

As it is generated, the iodine reacts with arsenic(III).

$$I_2 + As(III) \longrightarrow 2I^- + As(V)$$

As soon as the first excess of iodine is generated past the equivalence point, it reacts with the starch indicator, giving a blue-purple color.

$$I_2 + starch + I^- \longrightarrow I_3(starch)^-$$

Iron(II) can be similarly determined by generating cerium(IV) titrant from a solution of cerium(III). The same approach can be used for many other ions.

17-4 | POLAROGRAPHY AND AMPEROMETRIC TITRATIONS

Polaro- graphy: Voltam- metry at a Mercury Cathode

Voltammetry involves the measurement of small amounts of a chemical species by recording the current at various applied voltages. A voltammogram, or a curve of current plotted against applied voltage, is used for the determination of the species. The *working electrode*, or electrode at which the species to be determined reacts, can be either the anode or the cathode. The first reproducible voltammetric apparatus was devised by the Nobel laureate Jaroslav Heyrovsky in 1922 [2]. He used a dropping mercury cathode as his working electrode. He called his technique polarography, and the plot of current vs. voltage a *polarogram*. Although polarography is really just one specialized form of voltammetry, it is used so much for analysis that we will confine our discussion to it.

The polarogram is really a plot of the *diffusion current*, i_d, against the applied voltage, E. One reason for this is that the solution is not stirred during polarography, so the current passing through the polarographic cell depends on the rate of diffusion of ions to the dropping mercury cathode.

Since polarography is used for the *reduction* of chemical species, consider the hypothetical reduction of two metal ions, M_1 and M_2, in a polarographic cell. We will stipulate that M_1 is more easily reduced (has a larger positive $E°$) than M_2. Figure 17-3 is a *hypothetical* polarogram for this hypothetical reduction.

We start the polarographic reduction with the applied voltage at zero; no current flows at this voltage ($i_d = 0$). We then increase the voltage until it reaches a value corresponding to E_1, the voltage at which M_1 is reduced. At this point a current begins to flow. The current increases until it reaches a limit governed by the rate at which ions of M_1 diffuse to the mercury cathode. The greater the concentration of M_1, the greater the current. The distance i_d for M_1 is called the diffusion current for M_1. The relation between i_d and the concentration of M_1 is expressed as follows:

$$i_d = k_1[M_1] \tag{17-9}$$

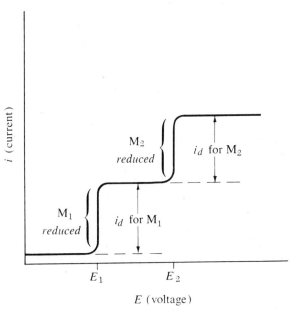

FIGURE 17-3. **Hypothetical current-voltage curve obtained for stepwise reduction of metal ion M_1 and metal ion M_2.**

where k_1 is a proportionality constant and $[M_1]$ is the molarity of M_1 in the solution. It should be noted that the amount of M_1 reduced during polarographic measurement is negligible, so the $[M_1]$ term is constant during the recording of the diffusion current.

If the voltage is increased above E_1, the current remains constant for a time because M_1 is still being reduced. When the voltage reaches a value corresponding to E_2, the current begins to increase again until it reaches a limit governed by the rate at which ions of M_2 diffuse to the mercury cathode. The total diffusion current at this point is the sum of that from M_1 and that from M_2. However, the diffusion current from M_1 can be subtracted graphically (Fig. 17-3) from the total to give the i_d for M_2. The relation between i_d and the concentration of M_2 is similar to **17-9**.

$$i_d = k_2[M_2] \qquad (17\text{-}10)$$

If we compare the relative sizes of i_d for M_1 and i_d for M_2 in Figure 17-3, we see that i_d for M_2 is somewhat smaller. Because of this, the concentration of M_2 is very likely to be smaller than that of M_1. However, the values of k_1 and k_2 would have to be determined before we could be sure that this is so.

The hypothetical polarogram in Figure 17-3 is said to exhibit two waves or steps, the first starting at point E_1 and continuing to point E_2 with a *height* corresponding to i_d for M_1. The second wave starts at point E_2 and continues to the end of the curve; it has a *height* corresponding to i_d for M_2.

An Experimental Polarogram

Now we will consider an actual experimental polarographic reduction of metal ions M_1 and M_2, Since many mercury drops form and fall off the dropping mercury cathode during the plotting of one polarogram, it is necessary to first visualize the stages of growth of a mercury drop and their effect on the diffusion current. As shown in Figure

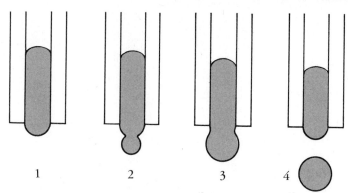

1   2   3   4

FIGURE 17-4. **Stages of formation of a mercury drop from capillary in dropping mercury cathode. (1) Drop starts to form and current is at a minimum. (2) Drop grows and current increases slightly. (3) Drop reaches maximum size and current increases slightly to maximum. (4) Drop falls off, new drop starts to form, and current decreases to same magnitude as in stage 1.**

17-4, the drop starts out as a small *bubble* (stage 1), with only a moderate surface available for reduction of the ions diffusing to it. The diffusion current is at a minimum at this stage. The drop grows through stages 2 and 3 to a maximum, and so does the surface available for reduction of the ions diffusing to it. The diffusion current also increases to a maximum. Then the drop falls off (stage 4), and the surface area and diffusion current decrease to a minimum again. This whole process repeats itself many times during the course of one polarogram.

We can now proceed to the actual polarogram of the reduction of ions M_1 and M_2 (Fig. 17-5). As the applied voltage is increased from zero to E_1, we observe a *residual current* rather than none at all. This is partly the result of the so-called *double layer* of negative and positive charges around the mercury drop. We note that here, as well as elsewhere, the curve is not a straight line, but has a sawtooth shape. This is the result of the changing size of the mercury drop.

The curve does not exhibit a sharp perpendicular break for the reduction of M_1, but instead it increases at an angle of less than 90°. Instead of all of the ions of M_1 being reduced at one voltage, they are reduced at slightly different voltages. This is because the ions themselves have slightly different energies. The average voltage at which the M_1 ions are reduced is symbolized as $E_{1/2}$, the *half-wave potential*. It is the value of the applied voltage at which the diffusion current is one-half the value of i_d for M_1. The half-wave potential, like the standard potential, is a constant that is characteristic of a redox half reaction such as $M^{+n} + ne^- = M°(s)$. It does not depend on concentration, and it can be used for identifying a particular ion in a simple mixture. (Too many species have similar values of $E_{1/2}$ for this approach to be used for complex mixtures.)

The wave for the reduction of M_2 is similar to that for M_1, except that it exhibits a *maximum* at the beginning. The reasons for such a maximum are not completely understood, but it is thought that the maximum is an effect arising from adsorption onto the mercury surface. Surface-active agents, such as gelatin, can be added to eliminate such a maximum.

The polarogram in Figure 17-5 is only obtainable if the solution is kept free of oxygen by bubbling nitrogen gas through it for at least five minutes. If oxygen is not removed, it

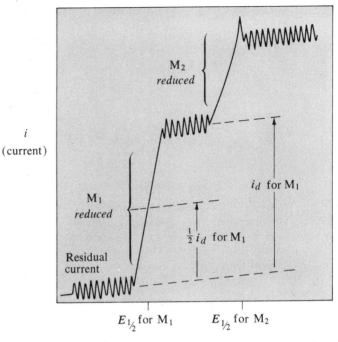

FIGURE 17-5. Actual current-voltage curve obtained for stepwise reduction of metal ion M_1 and metal ion M_2.

is easily reduced at the dropping mercury cathode in two steps (giving two waves) as follows:

$$O_2 + 2H^+ + 2e^- = H_2O_2 \qquad E_{1/2} = -0.05 \text{ V (v. SCE)} \qquad \textbf{(17-11)}$$

$$H_2O_2 + 2H^+ + 2e^- = 2H_2O \qquad E_{1/2} = -0.9 \text{ V (v. SCE)} \qquad \textbf{(17-12)}$$

The half-wave potentials given for these reactions are with reference to the saturated calomel electrode (SCE).

Polaro-
graphic
Apparatus
and
Methods
A schematic of a polarographic cell is given in Figure 17-6. Nitrogen gas is bubbled into the solution through a separate arm at the left side to remove oxygen to prevent the reactions given in **17-11** and **17-12**. The anode is usually a saturated calomel electrode; it completes the electrical cell as well as acts as the reference electrode. The anode is kept in electrical contact with the sample solution by a bridge plugged with agar.

The mercury is kept in a mercury reservoir and is forced through the capillary tip (Fig. 17-4) into the sample solution. Let us assume that we are going to reduce a solution of Cd^{+2} in a $0.1M$ potassium chloride *supporting electrolyte* (such an electrolyte eliminates electrical migration of Cd^{+2} so that it reaches the cathode only by diffusion). The cadmium(II) is reduced at the cathode as follows:

$$Cd^{+2} + 2e^- + Hg(l) = Cd[Hg](l) \qquad E_{1/2} = -0.60 \text{ V (v. SCE)}$$

As cadmium(II) is reduced, mercury in the calomel electrode is oxidized to mercury(I) and forms mercury(I) chloride.

$$2Hg(l) + 2Cl^- = Hg_2Cl_2(s)$$

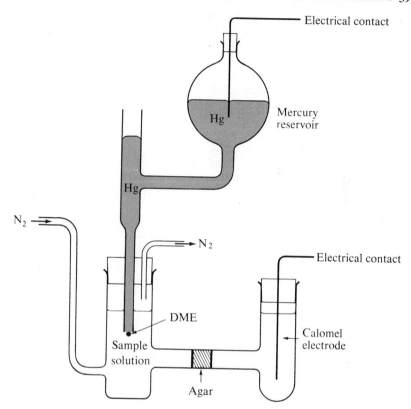

FIGURE 17-6.   Schematic of a polarographic cell, showing the dropping mercury electrode (DME).   The anode is the calomel electrode.

The overall cell reaction is thus the following:

$$Cd^{+2} + 2\,Hg(l) + 2\,Cl^- = Cd[Hg](l) + Hg_2Cl_2(s)$$

Fortunately, the amount of cadmium(II) ion that is reduced is very small, so there is not much change in concentration of the chloride ions or much loss of mercury metal in the calomel anode.

The Ilkovic Equation   Under controlled conditions, the relation between the diffusion current, i_d, in microamps and the concentration, C, in mmole/liter of a reducible species is given by the Ilkovic equation:

$$i_d = 607nCD^{1/2}m^{2/3}t^{1/6} \tag{17-13}$$

where: 607 = a constant if i_d = *ave.* value of current ($= 708$, if i_d = *max.* value)
n = number of electrons in the half reaction at the mercury cathode
D = diffusion coefficient of the species reduced in cm^2/s
m = mass of mercury passing through the capillary in mg/s
t = mercury drop time in s

Temperature does not appear in **17-13**, but D, the diffusion coefficient, genrally undergoes a 1 to 2% change per degree Celsius at room temperature, so temperature should be controlled for careful work.

Quantitative Methods

There are several general methods for quantitative analysis using polarography. The *absolute method* involves measuring all the parameters except C in **17-13** and then solving for C. This is difficult experimentally because of the effects of temperature and of the nature of the solution on D, m, and t.

The *calibration plot method* is the best method for the analysis of large numbers of samples. The values of i_d for four or five solutions of varying concentrations of the sample species are measured, and a plot of i_d against C of the species is made. This is usually a straight line, much like a Beer–Lambert law plot. The i_d of an unknown sample is then measured, entered on the plot, and the concentration of the unknown read from the plot.

The *standard addition method* is recommended when only one to three samples are to be analyzed, because it is faster, though somewhat less accurate. A polarogram of the unknown sample is recorded, and i_d of the sample species is measured. Then a known amount of a standard solution of the same species is added to the unknown sample, and a second polarogram is recorded, giving a value of i_d corresponding to the total of the sample species in the unknown and in the standard solution. If the volume of the standard solution is small enough compared to the volume of the unknown sample to be neglected, then the increase in the diffusion current, $i_{d(s)}$, the standard concentration, C_s, and the diffusion current from the unknown alone, $i_{d(u)}$, can be used to calculate the concentration of the sample species in the unknown, C_u, as follows:

$$C_u = C_s \frac{i_{d(u)}}{i_{d(s)}} \qquad (17\text{-}14)$$

Species Commonly Determined

The useful applied voltage varies from $+0.2$ to -1.9 V against SCE on a polarogram with a DME. Any soluble species that can be reproducibly reduced within this range can be determined, in theory, by polarography. This includes many inorganic metal ions, such as cadmium(II), lead(II), and others that can form mercury amalgams. Alkali metal ions such as sodium(I) and potassium(I) are not determined polarographically, unless a special electrolyte containing tetralkylammonium hydroxide is used. Inorganic anions, as long as they are reducible, can also be determined. For example, sulfate is determinable, but sulfide is not, because sulfur is in its lowest oxidation state in the sulfide anion.

Reducible organic and biochemical compounds can be determined if they can be dissolved in a suitable polarographic solvent. For example, aldehydes and ketones are reduced to alcohols.

$$RCHO + 2e^- + 2H^+ \longrightarrow RCH_2OH$$

In addition, sugars that contain aldehyde or ketone functional groups should also be amenable to polarographic determination. Glucose contains an aldehyde functional group and undergoes a two-electron reduction similar to that above.

$$C_5H_{11}O_5CHO + 2e^- + 2H^+ \longrightarrow C_5H_{11}O_5CH_2OH$$

(glucose)   (glucitol, or sorbitol)

Ampero-
metric
Methods
An amperometric method differs from a polarographic method primarily in that the applied voltage is kept *constant* in the amperometric method; in the polarographic method the voltage is *continuously varied*. There are at least two general amperometric methods. In one, the determination is made by comparing the diffusion current of the unknown sample to that of a standard sample and then calculating the unknown concentration, using an equation such as **17-14**. This approach is best for the determination of the same species in a large number of samples—for example, the determination of oxygen in aqueous solutions, such as body fluids. In the second general method, an amperometric detection of the end point is used.

Determi-
nation of
Oxygen
Recall that oxygen is readily reduced to hydrogen peroxide **(17-11)** and then to water **(17-12)**, using a mercury dropping electrode. It is also possible to reduce oxygen with an inert working electrode, such as a gold cathode. The determination of oxygen using the commercially available oxygen meter is done with such a gold cathode.

The probe holding both the gold cathode and a silver anode is inserted into the aqueous solution to be measured (Fig. 17-7). A constant voltage is applied to cause the reduction of oxygen. The oxygen diffuses through a membrane and is reduced at the gold cathode. The diffusion current that results depends on the rate of diffusion of oxygen through the membrane, which in turn depends on the pressure of the oxygen dissolved in the aqueous solution.

The magnitude of the diffusion current reading on the oxygen meter (not shown) is calibrated by using an oxygen standard, such as fresh water for which the oxygen content is known at the given temperature.

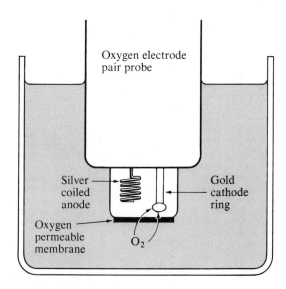

FIGURE 17-7. Oxygen electrode pair used in connection with oxygen meter (not shown) to reduce oxygen and give a diffusion current for amperometric measurement of oxygen.

Amperometric
Titrations

An amperometric titration is one in which the equivalence point is located by measuring the change in the diffusion current throughout the titration, while holding the applied voltage constant. A plot of diffusion current against the volume of titrant will show a change in slope at the equivalence point (Fig. 17-8).

In an amperometric titration, the applied voltage is selected to cause reduction of the titrant, the sample species, or both. One example is the titration of lead(II) by sodium sulfide, forming insoluble lead(II) sulfide. The potential is set so the lead(II) is reduced; sulfide ion, of course, is in its most reduced form and cannot be further reduced. The shape of this titration curve is shown in the top half of Figure 17-8. During the titration, the concentration of lead(II) is continually decreasing, and so is the diffusion current. At the equivalence point, the diffusion current is zero; it remains zero when an excess of the titrant is added.

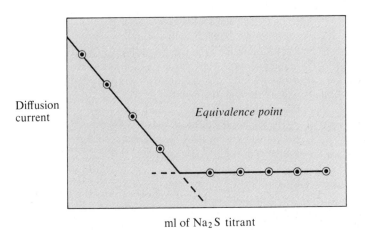

ml of Na_2S titrant

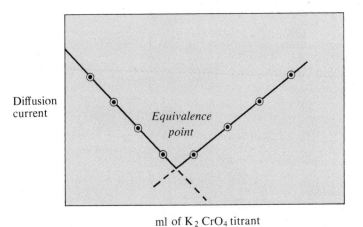

ml of K_2CrO_4 titrant

FIGURE 17-8. Two amperometric titration curves of lead(II) ion. Above: titration of lead(II) with sodium sulfide titrant. Below: titration of lead(II) with potassium chromate titrant.

Another example.is the titration of lead(II) by potassium chromate, forming insoluble lead(II) chromate as follows:

$$Pb^{+2} + K_2CrO_4 \longrightarrow PbCrO_4(s) + 2K^+$$

Both of the reacting species are reduced, but the chromate ion is not present to be reduced until after the equivalence point. As shown in the bottom half of Figure 17-8, the diffusion current increases beyond the equivalence point as the result of the excess of chromate ion added to the solution.

The scope of amperometric titrations is obviously wider than these examples show; any ion that is polarographically reducible can in theory be used in an amperometric titration.

| QUESTIONS AND PROBLEMS

(Answers to even-numbered problems are in Appendix 5.)

Concepts and Definitions

1. Define what is meant by an electrogravimetric method of analysis. How does it differ from a gravimetric method of analysis?
2. Explain the difference between:
   a. An electrogravimetric method and a direct coulometric method of analysis.
   b. Electrogravimetry and polarography.
   c. A polarographic method and an amperometric titration method.
   d. Direct coulometry and indirect coulometry.
3. Explain why coulometry can be used to determine more elements than electrogravimetry.
4. Explain how oxygen can be determined amperometrically. Of what importance might this be for blood analysis?
5. If a metal does not form a mercury amalgam, can it be determined polarographically?
6. Indicate whether each of the anions below can be determined polarographically. (*Hint:* Are they reducible?)
   a. Sulfate ion
   b. Sulfide ion
   c. Iodate ($IO_3{}^-$) ion
   d. Iodide ion
   e. Triiodide ($I_3{}^-$) ion
   f. Iodine (I_2)
7. Define diffusion current and tell how it can be used for quantitative polarographic analysis.
8. What is the nature of the titrant in:

a. An amperometric titration?
b. A coulometric titration?

Quantitative Analysis Calculations

9. A solution containing an unknown concentration of lead(II) chloride gives a polarographic diffusion current of 1.74 microamperes. To 50 ml of this solution is added a 1.0-ml volume of lead(II) chloride such that the concentration of lead(II) chloride in the 50 ml of solution from the 1.0-ml addition alone is $2.0 \times 10^{-4}M$. The polarographic diffusion current of the resulting mixture is 2.61 microamperes. Calculate the concentration of lead(II) chloride originally present in the 50 ml of solution.
10. A solution of copper(II) nitrate is analyzed by electrogravimetry. Calculate the molarity of this solution if 0.3000 g of pure copper is deposited from:
    a. A 25.00-ml aliquot.
    b. A 75.0-ml aliquot.
11. A solution contains $0.010M$ Cu^{+2} and $0.010M$ Ag^+, and it is planned to electrodeposit all of the Ag^+ before the Cu^{+2} begins to deposit. Assuming that all potential requirements for electrolysis at the cathode are the same except for the potentials calculated from the Nernst equation, calculate the percentage of Ag^+ remaining when Cu^{+2} begins to deposit.

12. A 25.00-ml aliquot of a 100-ml sample containing 1.100 g of an arsenic(III) sample is analyzed by coulometric titration, generating iodine from iodide. At constant current of 2.50 A it requires 125 seconds to reach the starch end point, at which point all of the arsenic(III) is oxidized to arsenic(V). Calculate the percentage of arsenic in the sample.

13. In a coulometric determination of permanganate, iron(III) is reduced to iron(II), which then reduces the permanganate to manganese(II) ion. Titration is complete when a constant current of 0.00200 A runs for 11.0 minutes. Calculate either the molarity or the normality of the permanganate solution, assuming a 25.00-ml volume.

| NOTES

[1] J. J. Lingane, *Electroanalytical Chemistry*, 2nd ed., Wiley Interscience, New York, 1958, pp. 340–50.

[2] J. Heyrovsky, *Chem. Listy* **16**, 256 (1922).

18 | Introduction to Radiochemistry and Radioimmunoassay

"Are they blood-stains, or mud stains, or rust stains, or fruit stains, or what are they? That is a question which has puzzled many an expert, and why? *Because there was no reliable test.*"
[Holmes said.]

ARTHUR CONAN DOYLE
A Study in Scarlet, Chapter 1

Another instrumental method of analysis that gives reliable results for traces of ions and molecules is radiochemical analysis. This method requires a radiochemistry background and is so broad that it embraces many techniques such as radioimmunoassay, isotope dilution, etc. In this brief chapter, we have space only for an introduction to radiochemistry and one technique, that of radioimmunoassay.

18-1 | INTRODUCTION TO RADIOACTIVE EMISSION

Radioactive isotopes of a number of elements decompose to yield one or more of the three types of radioactive emission—alpha (α), beta (β), and gamma (γ). The alpha and beta emissions are particles; we will use both "particle" and "emission" in our discussion. The gamma emission is electromagnetic in nature and is also referred to as gamma rays.

Alpha Particles (Emission)

Alpha (α) particles are helium (He^{+2}) nuclei, consisting of two protons and two neutrons. Since the mass number of an alpha particle is four, its symbol is given as ^4_2He, or as $_2He^4$. We will use the former symbol. Thus for the radioactive decay of the uranium-238 isotope to the thorium-234 isotope, we write the following equation:

$$^{238}_{92}U \longrightarrow {}^{234}_{90}Th + {}^4_2He \qquad (18\text{-}1)$$

Radioactive particles or emissions are usually characterized by the energy of the particles or rays, rather than by wavelength or wavenumber. Alpha particles range in energy from three to nine million electron volts (MeV). One eV equals 3.8×10^{-20} cal; 1 eV of energy per molecule corresponds to 23.06 kcal of energy per mole.

Alpha decay produces alpha particles with *discrete* energies. Thus if we measure the energies of the alpha particles produced in the decay of $^{238}_{92}U$ **(18-1)**, we observe that 77% of the uranium emits 4.18 MeV alpha particles (and produces ground-state thorium) and 23% of the uranium emits 4.13 MeV alpha particles (and produces excited-state thorium with 0.05 MeV excess energy). Thus each of the two types of alpha particles has its own discrete energy and no other energy.

Beta Particles (Emission)

Beta (β) particles are electrons ejected at high velocity from the nucleus of a decaying isotope. Since a beta particle has almost no mass, it is symbolized as $_{-1}^{0}\beta$, as well as just β. We will use the latter symbol. (Positrons, $_{+1}^{0}\beta$, may be emitted in certain radioactive decay processes. This is not a common mode of decay, and we shall discuss only the negatively charged β particles.) A good example of an isotope's emitting a beta particle is the radioactive decay of tritium, $_{1}^{3}H$:

$$_{1}^{3}H \longrightarrow \quad _{2}^{3}He + \beta \qquad\qquad \textbf{(18-2)}$$

This decay requires 12.26 years for half of the tritium to decompose; thus its half-life is said to be 12.26 years, as will be discussed in the next section.

Like alpha particles, beta particles are characterized by their energies. Unlike alpha particles, however, they are not emitted with discrete energies. Instead there is a broad continuous distribution of energies, ranging from close to zero to a maximum value. For the decay of tritium in **18-2**, the maximum energy is 0.018 MeV. The average energy is much lower than that value.

Gamma Emission (Rays)

Gamma (γ) rays are electromagnetic radiation like light or X-rays. They carry away a significant amount of energy from a radioactive isotope, but have no rest mass themselves. Hence they do not cause *unit changes* in the atomic number or mass of an isotope; however, they do cause a *fractional* decrease in mass corresponding to the energy of the gamma ray(s) emitted. (Recall that mass and energy are interconvertible.) Emission of a gamma ray often occurs just after the ejection of an alpha or a beta particle. A good example of this is the radioactive decay of iodine, $_{53}^{131}I$, to xenon, $_{54}^{131}Xe$:

$$_{53}^{131}I \longrightarrow \beta + {_{54}^{131}Xe^{*}} \longrightarrow {_{54}^{131}Xe} + \gamma \qquad\qquad \textbf{(18-3)}$$

Note that ejection of a beta particle from iodine produces $_{54}^{131}Xe^{*}$, an excited nuclear state of xenon. This excited state then returns to the ground state by emission of a gamma ray.

Like alpha particles, gamma rays are produced with discrete energies. Thus if we measure the energies of the gamma rays produced in the decay of $_{53}^{131}I$ **(18-3)**, we will always find that a certain number of gamma rays are emitted with an exact energy of 0.364 MeV and a certain number are emitted with an exact energy of 0.637 MeV. There is no broad range of energies as in the case of beta particles.

18-2 | RATE OF RADIOACTIVE DECAY

It is important to recognize that radioactive isotopes decay at vastly different rates. For certain radiochemical methods, it is important to choose an isotope with a certain rate of decay. To understand this requirement, we first will consider the kinetic rate law obeyed by all radioactive decay and then the *half-life* constant that is used to classify fast and slow decay rates.

Rate Law for Radioactive Decay

The decay of any radioactive isotope obeys the so-called first-order kinetic rate law. Consider the general case for decay of a given radioisotope A to product P_a:

$$A \longrightarrow P_a + \alpha, \beta, \text{or } \gamma$$

The rate of this decay is given by the expression (rate law):

$$\text{Rate} = k[A]$$

where [A] is the number of atoms of A or the amount of A in grams or moles. The symbol k is called the specific reaction rate constant and has the dimension of time^{-1}, as sec^{-1}, year^{-1}, etc.

The rate law indicates that the rate of the reaction depends only on the amount of A present. Since the amount of A is constantly decreasing during the time of the radioactive decay, the rate is also constantly decreasing.

Mathematically, the rate of the reaction can also be expressed as

$$\frac{-d[A]}{dt} = k[A]$$

This is known as the *differential form* of the first-order rate law. The minus sign is used because A is disappearing as the reaction proceeds; the decrease in [A] with respect to time depends only on [A].

It is often desirable to know the *total amount* of A remaining after a given time. This can be calculated from the *integrated form* of the differential equation; the integrated form is

$$kt = \ln \frac{[A_0]}{[A]} = 2.303 \log \frac{[A_0]}{[A]} \tag{18-4}$$

where $[A_0]$ is the amount of A at zero reaction time and [A] is the instantaneous amount of A at a given time t.

Half-Life of Radioactive Isotopes

Equation 18-4 can be used to show that the time required for one-half of any number of radioactive atoms to disintegrate is constant and is inversely proportional to the value of k. Substituting $[A_0] = 100\%$ and $[A] = 50\%$ into the right-hand term of 18-4 and solving gives·

$$kt_{1/2} = 2.303 \log \frac{[100\%]}{[50\%]}$$

$$kt_{1/2} = 2.303 (\log 2) = 0.693$$

$$t_{1/2} = 0.693/k \tag{18-5}$$

$$k = 0.693/t_{1/2} \tag{18-6}$$

Equation **18-5** implies that $t_{1/2}$, the half-life, is independent of the amount of A and inversely proportional only to the first-order rate constant. Most radiochemical texts list values of $t_{1/2}$ rather than values of k for radioisotopes. However, k can be calculated from $t_{1/2}$ by using **18-6**.

Knowledge of the length of the half-life for a particular radioisotope has some useful applications in the field of radioimmunoassay (Sec. 18-4). For example, some radio-immunoassays involve an isotope with a fairly long half-life, so that it can be purchased in advance and be available at any time for immediate use. A common radioisotope used is tritium, which decays according to **18-2**. The half-life of tritium is 12.26 years. This means that tritium decays slowly enough that its activity will remain constant long enough for it to be purchased in advance of analysis. It obviously would take 12.26 years for 50% of it to decay, but even the time required for a decay of 1% is quite long. That time span can be calculated by first finding the value of k from **18-6**:

$$k = 0.693/12.26 \text{ year} = 0.0565 \text{ year}^{-1}$$

Then we substitute $[A_0] = 100\%$ and $[A] = 99\%$ into **18-4** to calculate $t_{1\%}$, the time for 1% reaction:

$$(0.0565 \text{ year}^{-1})t_{1\%} = 2.303 \log \frac{[100\%]}{[99\%]} = 0.010039$$

$$t_{1\%} = 0.010039/0.0565 \text{ year}^{-1} = 0.17 \text{ year} = 2.1 \text{ month}$$

Thus the tritium could be kept for 2.1 months before it lost even 1% of its activity.

Another radioactive isotope once widely used in radiochemical analysis is the $^{131}_{53}\text{I}$ isotope. It has a fairly short half-life of 8.08 days **(18-3)**. Thus this isotope must be purchased at or near the time of assay to avoid significant loss. For example, if 600 μg of 100%-pure iodine-131 were allowed to stand for 8.08 days, only 300 μg of iodine-131 would be left. Furthermore, the half-life principle can also be applied to this 300 μg; it would take only another 8.08 days for 150 μg of this remaining 300 μg to decay. Thus $t_{3/4} = 2t_{1/2}$; $t_{7/8} = 3t_{1/2}$, etc. The $^{125}_{53}\text{I}$ isotope, which has a half-life of 56 days and a shelf life of almost a year, is now generally used in place of the former isotope.

18-3 | INSTRUMENTAL MEASUREMENT OF RADIATION

Most instrumental measurements of radiation are based on the excitation and/or ionization caused by the radiation in solids, liquids, or gases. The oldest technique is the use of photographic emulsions, in which radioactive particles or gamma rays darken photographic film just as light does. As this technique is not widely used, we will instead discuss the use of Geiger–Müller counters, ionization counters, proportional counters, and scintillation counters. It should be noted that the scintillation counter does not depend on the ionization of any material.

Geiger–Müller Counters   The Geiger counter is properly called the Geiger–Müller counter. It became well known to those prospecting for uranium during the early days of the atomic bomb in World War II. It consists of a tube with a thin mica window at one end to allow radiation to pass through. A typical Geiger tube is filled with about 90% inert gas (helium or argon) and 10% quenching gas. The tube also contains an anode and a cathode.

As radioactive particles pass through the mica window, they collide with gas molecules and cause them to ionize, producing primary electrons and positive gaseous ions. The

Geiger tube is operated at such high voltages that the primary electrons cause the formation of many secondary electrons, giving rise to an avalanche effect. This induces a current at the wire anode of the Geiger tube. Each individual radioactive particle produces a pulse of current; the total current is a measure of the number of radioactive particles entering the tube.

Geiger tubes are most useful for counting beta particles of high energy; they can also be used under special conditions for beta particles with energies as low as 0.155 MeV and for alpha particles. They cannot be used to measure the very low energy beta particles that are emitted by tritium **(18-2)**. Usually alpha particles are not counted by a Geiger counter, since they cannot penetrate the window of the Geiger tube. Gamma rays do penetrate the window but are counted at low efficiencies (2–5%).

Ionization Chambers and Proportional Counters

Ionization chambers and proportional counters are similar to Geiger counters in that each of the former is a gas-filled detector with an anode and a cathode. The ionization chamber operates at much lower voltages than the Geiger tube does. The result is a much smaller current, so that in many cases it must be amplified greatly; this increases the cost and results in a much more complicated electronic system.

Proportional counters, as the name implies, give a current that is directly proportional to the number of primary electrons created by the radioactive particles entering the detector. A proportional counter is operated at higher voltages than the ionization chamber, so that secondary electrons are formed, multiplying the current to 10^3 to 10^5 times that of an ionization chamber. The current still requires amplification, but the construction of the proportional counter is such that it is more precise and versatile than the ionization chamber or the Geiger tube. For example, it can distinguish between alpha and beta particles, although it cannot measure gamma rays.

In both of these detectors, the sample must be introduced directly into the measuring chamber containing the gases and electrodes; hence liquids and powdered samples cannot be measured, as they contaminate the chamber by mixing with the gases.

Scintillation Counters

One of the prototypical scintillation counters is shown in Figure 18-1. It consists of a scintillator (a crystal or liquid that fluoresces or phosphoresces) and a photomultiplier (Fig. 12-12). When a gamma ray strikes a scintillator, photons are emitted, which are

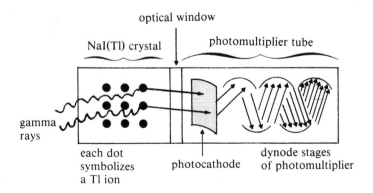

FIGURE 18-1. **A scintillation counter for measuring gamma rays.**

converted to electrical energy that can be measured. Most of the scintillators used are fluorescent (Ch. 14). However, instead of being excited by ultraviolet radiation to emit photons, they are excited by gamma rays to emit photons. One such scintillator is the thallium ion (Tl^+) in a crystal of sodium iodide. It is thought that when a gamma ray strikes the iodide ion, an electron on iodide is transferred to the thallium ion. The thallium ion then emits a photon in the ultraviolet-violet region of the spectrum. Although this emission is probably fluorescence, the general term "phosphor" (instead of fluorophor) is applied to this scintillator, as well as to others.

The photon that is emitted by a thallium ion passes through an optical window (quartz) into the photomultiplier tube and is converted to electricity (electrons) and then amplified many times. The scintillation counter is so constructed that it can measure the energies of all of the gamma rays striking the detector, just as a spectrophotometer can measure the wavelengths of all of the light striking it.

In modern liquid scintillation counters, the sample is dissolved or suspended in an organic liquid containing molecules that scintillate (emit photons) when they interact with beta particles. The sample is completely surrounded by the scintillating liquid; hence all emissions interact and are counted. Weak beta particles, which cannot penetrate the window of a Geiger counter, are counted with 100% efficiency. Because of these advantages, liquid scintillation counting is preferred almost exclusively to other scintillation methods.

18-4 | TRACERS AND RADIOIMMUNOASSAY

Covering the many uses of radiochemistry in analysis is beyond the scope of this introductory chapter. However, we will mention the use of tracers in chemistry and the newer technique of radioimmunoassay.

Use of Tracers By substituting a radioisotope for a stable isotope of an element in a compound, it is possible to "trace" the progress of the compound through a physical or chemical process; hence the name "tracer." The basic assumption is that the stable isotope and the radioisotope will behave exactly the same way and that the substitution will not alter the chemical or physical properties of the compound. There is, however, an "isotope effect," which must be evaluated to gauge its significance. For example, substituting tritium (^3_1H) for hydrogen could cause a significant isotope effect because of the large difference in masses.

In many cases, compounds already "labeled" or "tagged" with a radioisotope, such as carbon-14 or tritium, are commercially available. In other cases, a chemist synthesizes the compound. For example, a weakly ionized carboxylic acid may be labeled with tritium by mixing with a solution of strongly ionized ^3_1HCl, producing $RCOO^3_1H$. The exact amount of the radioisotope present is then measured, using a scintillation counter or a proportional counter. In this way, before any experiments are performed, the exact amount of radioactivity per volume of the solution of a compound is established.

In most experiments, a small amount of labeled, or "hot," compound is mixed with a large amount of nonradioactive compound, called the carrier, or "cold," compound. The concentration of the labeled compound is usually not significant compared to the concentration of the carrier.

One example of the use of tracers in chemistry is in checking the effectiveness of a separation of calcium ions from magnesium ions. To do this, we add a measured portion of "cold" calcium ions containing a small known amount of "hot" calcium ions to "cold" magnesium ions. After the first trial separation of the two ions, we measure the solution containing the magnesium to determine the amount of radioactive calcium still present. We also measure the separated fraction (solution) containing the bulk of the calcium to determine the amount of radioactive calcium present. In each case, the radioactivity of the solution is proportional to the total amount of the calcium present.

In medicine, for another example, tracers are used to determine the uptake of a particular element in the liver, heart, kidneys, etc.

Radio-immuno-assay

Radioimmunoassay (RIA) is an extremely useful method for the specific measurement of traces of substances such as hormones, vitamins, or drugs in body fluids, etc. Although there are good chemical and instrumental methods available for measuring micro amounts ($10^{-4}M$) of these substances, it is often difficult to measure them at trace levels. The principles of RIA were first established in 1962 for the determination of the hormone insulin. A good discussion of this method is given by Skelly, Brown, and Besch [1]. More detailed discussions have also been published by Walther [2] and by Gochman and Bowie [3].

Terminology and Reaction

Since RIA is associated with the science of immunology, we will first define the two key terms used in this field, *antibody* and *antigen*. An antibody is a protein, usually a gamma globulin or immunoglobulin, that will react specifically with an antigen or hapten. It is generally symbolized as Ab. An antigen is a foreign substance that causes the formation of an antibody in the body and that reacts specifically with that antibody to give a kind of complex.

Radiolabeled antigen is symbolized as Ag*, while unlabeled (cold) antigen is symbolized as Ag. The specific reactions, or complex formations, of an antibody with the above antigens are written as follows:

$$Ag^* + Ab \rightleftharpoons Ag^*\text{---}Ab \qquad \textbf{(18-7)}$$

$$Ag + Ab \rightleftharpoons Ag\text{---}Ab \qquad \textbf{(18-8)}$$

To determine cold antigen in a body fluid, it is necessary to first prepare a hot portion of the antigen. Radiolabeling of hormones, for example, is done using iodine-125 or tritium. Since most polypeptide hormones contain tyrosine, they are labeled with either of the iodine isotopes through substitution of the iodine on the 3 and 5 positions of the benzene ring of tyrosine.

The use of iodine radioisotopes has certain advantages and disadvantages compared to use of tritium. Iodine-125 is a gamma emitter and thus is readily counted on a scintillation counter. This isotope has a moderate half-life of 56 days (Sec. 18-2); a supply once purchased can be used for almost a year before decay reduces its radioactivity to insignificant levels.

Determination of Unlabeled Antigen

The first requirement for the determination of an antigen (unlabeled) in the body is to produce the required antibody (also called antiserum). This may be done by injecting the antigen into the blood serum of an animal. Once the antibody has been produced, a sample of it is obtained by removing a sample of the animal's blood.

Next, the antigen-antibody reaction is carried out by incubating the antibody with both the unlabeled antigen (to be determined) and the radiolabeled antigen. Both reactions **18-7** and **18-8** then occur. An excess of unlabeled antigen (Ag) over antibody is used, so that only a portion of both Ag and Ag* is complexed to the antibody. After incubation, the solution contains:

$$Ag^*—Ab, \ Ag—Ab, \ free \ Ag^*, \ and \ free \ Ag$$

(18-7) (18-8)

The next analytical step involves the separation of the complexes from the free Ag* and the free Ag. This is done using such separation techniques as precipitation followed by centrifugation, electrophoresis, or chromatoelectrophoresis (a combination of chromatography and electrophoresis).

Measurement of Radioactivity. The initial conditions are always adjusted so that when the antibody and the radiolabeled antigen are mixed without the cold antigen, only about 50% of the Ag* will be bound to the Ab. Then a calibration curve (Fig. 18-2) is established by adding various amounts of cold antigen (Ag) and calculating the percentage of Ag* bound for each concentration of Ag. Finally, one or more samples containing an unknown amount of Ag are measured.

A hypothetical RIA might involve calculations as follows. Suppose the initial radioactivity of labeled antigen in the final solution is 10,000 cpm/ml (counts per minute per ml final solution). Then the amount of antibody to be added is adjusted so that after separation, the radioactivity of the solution containing only Ag*—Ab is 5,000 cpm/ml. The percentage of Ag* bound is then:

$$\frac{5,000 \text{ cpm/ml}}{10,000 \text{ cpm/ml}} (100) = 50.0\%$$

A calibration curve, such as Figure 18-2, is then constructed using measurements of the radioactivity of Ag*—Ab after separation and plotting the calculated percentage of Ag* bound vs. the concentration of Ag added in ng/ml. The counts per minute will always be less than 5,000 because of the added Ag. A sample containing an unknown amount of Ag is then incubated, and after separation, the radioactivity of the solution containing only Ag*—Ab is 2,000 cpm/ml. The percentage of Ag* bound is then:

$$\frac{2,000 \text{ cpm/ml}}{10,000 \text{ cpm/ml}} (100) = 20.0\%$$

From Figure 18-2, the concentration of Ag at 20% Ag* bound is 100 ng Ag per ml.

Applications of RIA

Most radioimmunoassays are done using commercially available screening kits that make such assays very convenient. Using iodine-125 screening kits, various digestive hormones, thyroid hormones, and sex hormones can be determined. Other compounds

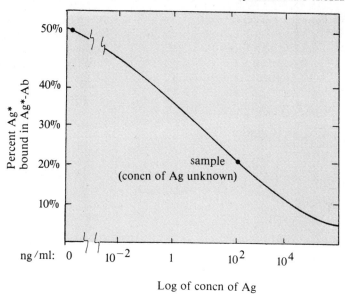

FIGURE 18-2. **Radioimmunoassay calibration curve.**

regularly determined with iodine-125 kits include digitoxin, digoxin, gentamicin, morphine (p. 338), phenobarbital (p. 92), phenytoin, theophylline, and tobramycin. Compounds determined with tritium or carbon-14 kits include morphine (p. 338), barbiturates
(p. 309), marijuana, LSD, vitamin D_3, and many hormones.

⌷ QUESTIONS AND PROBLEMS

(Answers to even-numbered problems are in Appendix 5.)

1. Define the following terms:
   a. Alpha particle.
   b. Beta particle.
   c. Gamma ray.
2. Define the term "half-life" and show
   mathematically that $t_{3/4} = 2t_{1/2}$; that is, that
   the time required for three-fourths of a species
   to decay radioactively is twice the half-life.
3. For each isotope below, calculate the first-order
   rate constant and the time required for
   three-fourths of the isotope to decay.
   a. Iodine-125 ($t_{1/2} = 56$ days)
   b. Iodine-131 ($t_{1/2} = 8.08$ days)
4. Calculate the time required for only 1% of
   each of the isotopes below to undergo

radioactive decay, assuming each was 100%
pure. (See previous problem.)
   a. Iodine-125
   b. Iodine-131
5. What instrument(s) can be used to measure the
   following:
   a. Alpha particle?
   b. Beta particle?
   c. Gamma ray?
6. Construct a radioimmunoassay curve of the
   percentage of Ag* bound in the Ag*—Ab
   complex from the following data points,
   assuming that the initial radioactivity of
   labeled antigen in the final solution is 8,000
   cpm/ml.

a. No Ag added; cpm/ml = 4,000
b. 10^{-11} g/ml Ag added; cpm/ml = 3,500
c. 10^{-10} g/ml Ag added; cpm/ml = 3,000
d. 10^{-8} g/ml Ag added; cpm/ml = 2,000
e. 10^{-7} g/ml Ag added; cpm/ml = 1,600
g. 10^{-6} g/ml Ag added; cpm/ml = 1,400

7. An unknown containing Ag measures 2,500 cpm/ml under the same conditions as in Problem 6. Read the concentration of Ag from the graph from that problem.

| NOTES

[1] D. S. Skelley, L. P. Brown, and P. K. Besch, *Clin. Chem.* **19**, 146 (1973).
[2] W. H. C. Walther, *Clin. Chem.* **23**, 384 (1977).
[3] N. Gochman and L. J. Bowie, *Anal. Chem.* **49**, 1183a (Nov., 1977).

PART THREE

Separations and Clinical Analysis

This part of the book includes an elementary treatment of some selected separation techniques along with an introduction to clinical analysis. We do not attempt to cover every separation method because little useful information could then be included. This part can be used in two ways—as a brief survey for short courses and as a more advanced treatment for long courses.

For courses requiring only an introduction to separation techniques, we recommend a careful reading of Chapter 19. This chapter surveys both single-stage and multiple-stage separation techniques, so the student becomes familiar with general techniques such as thin-layer chromatography and gas chromatography. The chapter also discusses liquid-liquid extraction at length, so the student receives an in-depth exposure to at least one separation technique that can be both single-stage and multiple-stage.

For longer courses we recommend a careful reading of Chapters 19, 20, and 21. After being exposed to the survey of separations in Chapter 19, the student can then master the theory of chromatography in Chapter 20 and the applications of that theory to separations in Chapter 21.

For both types of courses we recommend that students read Chapter 22 to see how clinical analysis is done.

19

Introduction to Separations: Liquid-Liquid Extraction

"Sherlock Holmes put the sopping bundle upon the table. . . . From within he *extracted a dumb-bell*, which he tossed down to its fellow in the corner."

ARTHUR CONAN DOYLE
The Valley of Fear, Chapter 7

A separation involves physically removing interferences from a species as completely as possible before measuring it. In the above quote Sherlock Holmes was actually performing a separation of sorts by "extracting the dumb-bell" by hand. Although the separation of solid crystals is occasionally done by hand, the chemist generally performs separations with the help of an instrument and/or specialized lab apparatus.

Some separations, such as liquid-liquid extraction, involve a simple transfer of a species from one phase to an immiscible phase of the same type. For a mixture of species A, B, and C in liquid phase 1, the separation of species A by removal into phase 2 is symbolized as follows:

$$(A, B, C)_{\text{ph 1(l)}} + (\quad)_{\text{ph 2(l)}} \xrightarrow{\text{extraction}} (A)_{\text{ph 2(l)}} + (B, C)_{\text{ph 1(l)}}$$

Many more separations, however, involve transferring a species from one phase to an immiscible phase of a different type. For a mixture of species A, B, and C in phase 1 (gas, liquid, or solid), the separation of species A by removal into phase 2 (a different phase from phase 1) is symbolized as follows:

$$(A, B, C)_{\text{ph 1}} + (\quad)_{\text{ph 2}} \xrightarrow{\text{separation}} (A)_{\text{ph 2}} + (B, C)_{\text{ph 1}}$$

We will first survey the various types of separation methods with two purposes in mind—to provide a brief survey of separation methods for shorter courses and to provide an introduction to separations as preparation for Chapters 20 and 21. Then we will discuss liquid-liquid extraction in detail as an example of a useful separation technique.

415

19-1 | SURVEY OF SEPARATION TECHNIQUES

Introduction   In this section we will survey various separation techniques rather than discuss them in detail. The key to understanding these techniques is recognizing that each involves a separation of two phases. Some separations involve only a single stage, or step, but since quantitative separation in a single stage is not usually achievable, most separations are based on multiple stages, or steps. Some examples of each type of separation follow:

Single-stage separation techniques
Single-step extraction
Precipitation
Electrodeposition

Multiple-stage separation techniques
Multiple-step extraction
Liquid chromatography: liquid-liquid and liquid-solid
　Column chromatography (atmospheric pressure)
　High-performance (pressure) liquid chromatography (HPLC)
　Thin-layer chromatography and paper chromatography
　Exclusion chromatography
Ion-exchange chromatography
Gas-liquid and gas-solid chromatography

We will begin with a discussion of the single-stage separation techniques to establish the basic principles. Then we will introduce some of the multiple-stage separation techniques.

Single-Stage   Since liquid-liquid extraction is to be discussed separately in the next section, we will
Separation   confine our discussion here to separations by precipitation and by electrodeposition. In
Techniques   both of these techniques there is a separation of phases that requires some time to accomplish.

Separation by   In a separation done by precipitation, quantitative separation in a single step is often
Precipitation   achievable, because an excess of the precipitating reagent can be added without causing serious coprecipitation. Since a separation by precipitation is a necessary part of any gravimetric method of analysis, all of the gravimetric methods discussed in Chapter 5 can be considered to be examples of this type of separation.

A separation by precipitation is, of course, often used in connection with methods other than gravimetric analysis. Frequently, the desired species is precipitated to separate it from interferences, and then it is measured by a rapid method of analysis, such as spectrophotometric measurement. An example of this is the separation of cholesterol from interferences before measurement. In body fluids cholesterol occurs in the free form and in the esterified form.

$$C_{27}H_{45}OH \qquad\qquad C_{27}H_{45}O-\overset{\displaystyle O}{\overset{\displaystyle \|}{C}}-R$$

Free cholesterol　　　　　　Esterified cholesterol
(a steroid alcohol)　　　　　　(a steroid ester)

The Zlatkis-Zak [1] colorimetric method for cholesterol measures the sum of both the free cholesterol and the esterified cholesterol. If free cholesterol alone is to be measured, then it is commonly separated by precipitating it with digitonin, $C_{55}H_{90}O_{29}$, as follows:

$$C_{27}H_{45}OH + C_{55}H_{90}O_{29} + H_2O \longrightarrow C_{27}H_{45}OH[C_{55}H_{90}O_{29} \cdot H_2O](s)$$

The digitonin precipitation is quite specific for cholesterol; it will not precipitate esterified cholesterol or other sterols (steroid alcohols). Unfortunately, the precipitation is slow, requiring an overnight reaction. After the precipitate has been isolated, it is washed, dried, and dissolved before the free cholesterol is measured colorimetrically.

Separation by Electro- deposition

In a separation done by electrodeposition, quantitative separation in a single operation can be achieved as long as the voltage applied during the deposition does not reach a value where other metal ions are deposited at the cathode. Recall from Section 17-2 that electrodeposition, or electrogravimetry, involves the quantitative deposition of a *single metal ion* at the cathode. (See Fig. 17-1 for the electric cell and apparatus used for electrodeposition.) In this chapter, then, we are concerned with a quantitative separation of one metal ion from others, as well as a quantitative deposition of the ion on the cathode.

The key to separating one metal ion from others is controlling the voltage applied to the cathode. By applying a voltage just large enough to start electrodeposition of the metal ion with the most positive $E°$ value, we can avoid deposition of metal ions with less positive $E°$s. Of the soluble ions listed in Table 19-1, the metal ion most readily separated by electrodeposition at a cathode is silver(I), because it has the most positive $E°$. (Lead(IV) dioxide, of course, does not exist in solution.) The metal ion *most commonly* separated by electrodeposition at the cathode is the copper(II) ion. This is because copper occurs in so many substances.

By looking at Table 19-1, you can see that copper(II) can easily be separated from alkali metal ions, such as sodium(I), etc. However, unless the BiO^+ ion is absent, bismuth will be electrodeposited along with copper. Whether ions with an $E°$ similar to zinc(II) will interfere depends on the voltage at the *end* of the electrodeposition of copper. In practice, this can be as low as -0.4 V; if electrodeposition is terminated at this point, neither zinc(II) nor alkali metals, such as sodium(I), interfere. They are left in solution. (Silver, of course, would interfere.)

Another metal that can be readily separated is lead(II), which is oxidized *at the anode* to lead(IV) dioxide. It can be deposited at the anode at the same time that copper is

TABLE 19-1. Potentials of Reducible Metal Ions

Reduction half reaction	$E°$, volts
$PbO_2(s) + 2e^- = Pb^{+2}$	$+1.455$
$Ag^+ + e^- = Ag°(s)$	$+0.800$
$Cu^{+2} + 2e^- = Cu°(s)$	$+0.337$
$BiO^+ + 3e^- = Bi°(s)$	$+0.32$
$2H^+ + 2e^- = H_2(g)$	0.00
$Zn^{+2} + 2e^- = Zn°(s)$	-0.76
$Na^+ + e^- = Na°(s)$	-2.71

deposited at the cathode. Silver(I) can be separated from any of the metal ions listed in Table 19-1; the only interference would be some species that would have a similar or more positive $E°$ value.

Multiple-Stage Separation Techniques

Since multiple-step solvent extraction is to be covered in the next section, our discussion here will be concerned only with various types of chromatography. The term chromatography has come to include so many physically different separation techniques that it is difficult to define. The following are four probable, common characteristics of a chromatographic technique:

1. A chromatographic separation involves a distribution of the sample components between a stationary phase and an eluent (carrier) phase, or mobile phase.
2. The stationary phase is a bed of large surface area.
3. The eluent (carrier) is a fluid (liquid or gas) that percolates through or along the bed.
4. The sample components are eluted at different rates through the bed and emerge separated at different times.

A schematic of a chromatographic separation is shown in Figure 19-1. A two-component mixture of A and B (soln.) is added to the bed contained in a column, and initially A and B remain evenly distributed at the top of the column (left). As eluent continues to be added, the faster-moving A migrates ahead of B to some extent, but still remains partially mixed with B (middle). Finally, a complete separation of A and B is achieved, as essentially all of A migrates ahead of B (right). Ultimately, A is eluted out of the column before B reaches the bottom of the bed.

Liquid Chroma-tography

Liquid chromatography includes many different techniques. In addition, it is divided into two *mechanistic areas*—liquid-liquid (partition) chromatography and liquid-solid (adsorption) chromatography. In *partition chromatography*, a stationary bed is coated with a stationary aqueous or nonaqueous liquid. The eluent used is an immiscible liquid of the opposite type (nonaqueous or aqueous). The separation works much on the principle of a multiple-step extraction, with sample components dissolving in the stationary liquid to different degrees. In *adsorption chromatography*, the stationary bed itself interacts with the sample components; each component is adsorbed from the liquid eluent to a certain degree onto the stationary bed. In order for the bed material to interact with the sample, it must be a finely divided pure solid, such as activated alumina, silica, nylon, or starch.

Since we are not as interested here in the mechanism of liquid chromatography as in the various techniques, we will now proceed to a discussion of column chromatography at atmospheric pressure, high-performance (pressure) liquid chromatography, exclusion column chromatography, thin-layer chromatography, and paper chromatography.

Column chromatography is carried out at atmospheric pressure using fairly wide diameter columns (1–5 cm) packed with an adsorbent. The separation is done by hand; the sample is poured into the column, and elution proceeds as diagrammed in Figure 19-1. Because the adsorbent is finely divided, elution is slow, requiring from thirty minutes to several hours. The components can be detected automatically as they emerge from the column if they absorb visible or ultraviolet radiation. Similarly, an automatic fraction collector can be set up so that uniform-size fractions are collected. This type of column chromatography is suitable for some qualitative and quantitative analyses, but is *not efficient* enough for large numbers of samples involving more than a few components.

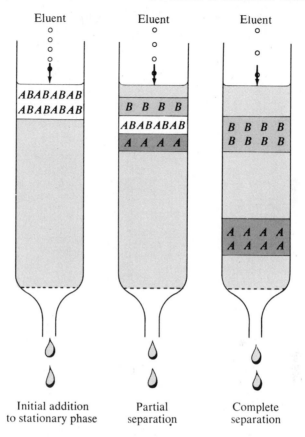

FIGURE 19-1. Schematic of a chromatographic separation of species *A*, a fast-moving species, and species *B*, a slow-moving species, down a chromatographic column.

High-performance (*pressure*) *liquid chromatography* (*HPLC*) permits efficient, usually fast, column separation of complex mixtures (Fig. 19-2). It employs a pump and operates at pressures high enough to achieve separations in minutes using micro-sized stationary bed particles. The column is closed to the atmosphere and constructed of metal to withstand the operating pressures of up to 10,000 psi. The column is fairly thin, with an inside diameter of 2 to 10 mm.

After a sample is introduced through a valve or inlet port onto such a column, liquid eluent is added to the column. The high pressure forces the eluent through the adsorbent at a rapid rate. The component that interacts the least with the adsorbent or stationary liquid phase is eluted first; its exit from the column is indicated by the detector system located just below the exit.

As each component leaves the column, it enters a very narrow tube that is part of the detector cell. Some type of radiation, such as UV radiation, passes through this cell and is partially absorbed by the component. (Alternatively, refractive index changes

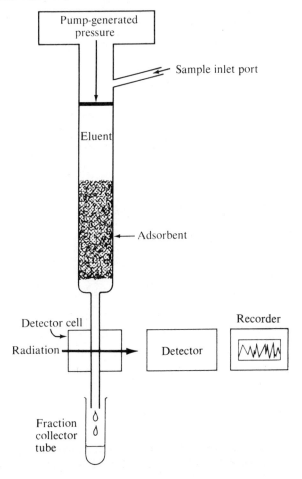

FIGURE 19-2. Schematic of a high-performance (pressure) liquid chromatographic column. Pump-generated pressure forces the eluent and sample through the bed of adsorbent, into the detector cell, and finally into a fraction collector tube.

are measured.) The absorption is measured by the detector; its output is displayed on the record graph. The component may then be collected in a fraction collector tube, as part of an automatic fraction collector system. A fresh collector tube is attached to the detector cell to collect the next increment of the eluent.

Exclusion column chromatography involves the use of a column packed with porous beads. The pores, or interiors, of the beads are limited to within a certain size range; these pores hold a majority of the liquid mobile phase. An exclusion chromatographic separation is based on a slight, partial, or complete exclusion of certain molecules from the liquid mobile phase inside the stationary porous beads. As shown in Figure 19-3, the largest molecules are too large to penetrate the pores of even the largest beads. Intermediate-size molecules can penetrate some of the pores of the beads, and small molecules can penetrate all of the pores of all beads. As a result, the largest molecules

Eluent

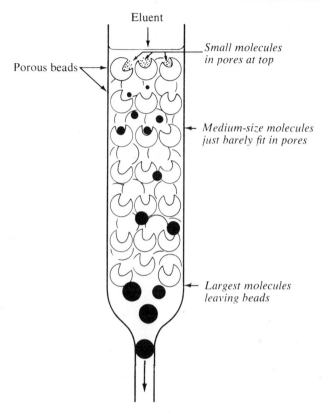

Porous beads

Small molecules
in pores at top

Medium-size molecules
just barely fit in pores

Largest molecules
leaving beads

FIGURE 19-3. Schematic of an exclusion chromatographic column separation showing the largest molecules leaving the column first and the smallest molecules traveling the slowest.

pass through the column first, as they are not held back at all, followed by intermediate-size molecules that are held back somewhat, and finally small molecules that are held back the most.

Exclusion chromatography is termed *gel permeation chromatography* when the system is composed of an organic solvent and a gel, such as cross-linked styrene or porous silica beads. Because the amount of swelling and the resultant pore size of the gel are a function of the specific organic solvent used, the separation required can sometimes be achieved by picking the correct organic solvent.

Exclusion chromatography is termed *gel filtration chromatography* when the system is composed of an aqueous solvent and a hydrophilic solid, such as dextran, acrylamide, or agarose gel. The aqueous solvent may be an unbuffered saline solution, a buffered saline solution, or a solution of a salt other than sodium chloride. Gel materials with as many as ten different ranges of exclusion size are available.

Thin-layer chromatography (*TLC*) is chromatography carried out on a thin (0.1–2 mm thick) layer of adsorbent cemented to a rigid plate of thin glass or plastic. The adsorbent

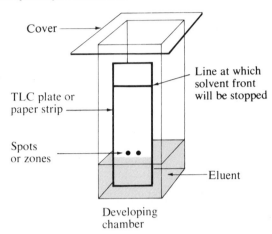

FIGURE 19-4. Diagram of a developing chamber for a thin-layer chromatogram or a paper chromatogram.

is usually a finely divided material, such as silica gel or alumina. The samples are applied as uniform spots at a short distance from one end of the plate. Then the plate is placed vertically in a closed glass developing chamber filled to about a 1-cm depth with the eluting solvent (Fig. 19-4). The eluting solvent migrates up the plate by capillary action, eluting first the component least attracted to the adsorbent. Eventually all the components are eluted to some extent as the solvent front is allowed to reach a premarked point on the plate. The entire process is fairly rapid, requiring from fifteen minutes to an hour.

After separation, each component exists as an invisible or a colored spot on the plate. If it is invisible, it is visualized as a dark spot on a fluorescent plate or by spraying with a color-forming reagent, such as iodine in methyl alcohol or sulfuric acid. The iodine forms brown complexes with many organic compounds, and sulfuric acid forms a black carbon spot after heating. The visualization of the spot is adequate for qualitative analysis. Quantitative analysis requires special techniques, which will be discussed in Chapter 21.

Paper chromatography is chromatography on a fairly thick sheet of uniform, pure paper, rather than on the thin, adsorbent layer used in TLC. (The paper is similar to that used for some filter papers.) The samples are applied as spots a short distance from one end of the paper, and the sheet is suspended in a glass developing chamber filled to a 1-cm depth with the eluting solvent (Fig. 19-4). Elution occurs as in TLC, except that components partition themselves between the layer of solvent (usually water) adsorbed on the paper and the eluting solvent. Elution is usually slower than in TLC. After separation, visualization is done by many of the same methods as in TLC, except that fluorescent paper is not available for use. If the compounds themselves fluoresce, this can be observed on the paper.

Ion-Exchange Chroma-tography   Ion-exchange chromatography differs from the previous types of chromatography because it is restricted to the determination of ions rather than molecules and because the separation of the sample components is based on a chemical *reaction* rather than on a chemical or physical *interaction*, or attraction.

Ion-exchange resins are of two types—cation-exchange or anion-exchange resins. The most common type of cation-exchange resin is a sulfonated polymer, which contains sulfonic acid groups, $-SO_3^-H^+$. The proton in the sulfonic acid group can be exchanged for $+1$, $+2$, or $+3$ cations; for example,

$$2\,Res-SO_3^-H^+ + Ca^{+2} \longrightarrow (Res-SO_3^-)_2Ca^{+2} + 2H^+ \qquad \text{(19-1)}$$

The most common type of anion-exchange resin is that which contains a quaternary ammonium group, $-NR_3^+Cl^-$. The chloride ion in this group can be exchanged for -1, -2, or -3 anions; for example,

$$2\,Res-NR_3^+Cl^- + CO_3^{-2} \longrightarrow (Res-NR_3^+)_2CO_3^{-2} + 2Cl^-$$

For an ion-exchange chromatographic separation of two cations, a column similar to that in Figure 19-1 is packed with a cation-exchange resin mixed with a solvent, frequently the eluting solvent. Then a solution of the sample containing cations, such as Ca^{+2} and Mg^{+2}, is poured onto the column. Both cations react with the resin (19-1) and are held on the resin. An eluent such as hydrochloric acid is then poured down the column, gradually displacing first the Ca^{+2} and then the Mg^{+2}. Each ion passes down the column and is detected as it leaves. An automatic fraction collector may be used to collect increments of the eluting solvent containing each ion.

Gas Chromatography

Gas chromatography includes both gas-solid and gas-liquid chromatography. We will discuss the latter in this chapter. Gas-liquid chromatography takes place on a solid support, or packing, in a coiled column. The solid support is coated with a thin film of a nonvolatile organic liquid, called the stationary phase. The column is heated to maintain the sample in the gaseous state.

To separate the sample into its components, it is injected through the heated sample injection port (Fig. 19-5) into the column through which the carrier gas is flowing at a constant rate. (Helium is a typical carrier gas.) The temperature of the injection port causes the sample to volatilize if it is not already a gas. The carrier gas transports the sample through the column packing, where it interacts with the liquid coating and is partially absorbed by it. The sample components are partitioned between the moving carrier gas and the liquid coating. The component that interacts least with the liquid coating is the first to be eluted, followed by other components, all mixed with the carrier gas.

As each component leaves the column, it enters the detector cell (Fig. 19-5). Various types of detectors are used to sense the emergence of each component; a thermal conductivity detector is shown in Figure 19-5. In this type of detector, the thermal conductivity of the carrier gas that is mixed with a gaseous sample component is different from that of the pure carrier gas. This difference is measured electrically and displayed on a recorder to give an elution curve.

The detector indicates the presence of, and measures the quantity of, each component in the sample. It does indicate the number of components in a sample; however, it does not automatically indicate what the chemical identity of each is. Identification is done by comparison with known compounds and their *retention times*. Retention time is the number of minutes between the injection of the sample and its emergence from the column. Quantitative analysis is done on the basis of measurement of peak areas, or sometimes peak heights.

Individual components can be collected after being separated on some chromatographic columns; further analysis can then be performed to help identify the components.

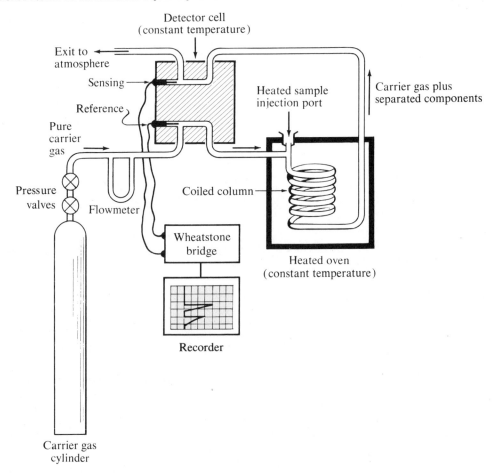

FIGURE 19-5. **Schematic of a gas chromatograph having a thermal conductivity detector cell. (Note that the detector measures the difference in the thermal conductivities of pure carrier gas and of gas plus components.)**

19-2 | LIQUID-LIQUID (SOLVENT) EXTRACTION

Principles Liquid-liquid extraction is the separation of two or more components in a particular solvent by preferentially dissolving one of them in a second solvent immiscible with the first. Since water is generally one of the solvents, solvent extraction is based on the facts that many organic liquids are not miscible with water and are either heavier or lighter than water.

A *one-step* liquid-liquid extraction can be explained in terms of the following discussion. Suppose that an immiscible organic liquid is added to an aqueous solution containing components O (organic-soluble) and W (organic-insoluble) in a separatory funnel (Fig. 19-6). Assume that the organic liquid has a smaller specific gravity (density) than water (sp gr = 1.0) so that it remains in a layer on top of the aqueous layer. We

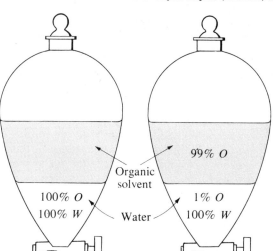

Before extraction   After extraction

FIGURE 19-6. **Solvent extraction of organic-soluble substance O from organic-insoluble substance W into an organic solvent lighter than water.**

then shake the stoppered separatory funnel vigorously for a sufficient time for equilibrium to be attained. Since O is soluble in the organic layer, most of it passes into the upper layer. However, since O is also soluble in water, a small amount of it remains in the lower aqueous layer (right, Fig. 19-6). This amount is shown as 1% in Figure 19-6, but it may actually be more or less than that. Since W is insoluble in the organic solvent, it remains in the aqueous phase. In some cases, a small amount of W will also be extracted into the upper layer, making the separation poorer than that shown in Figure 19-6.

If substance O is not extracted at least 99% into the organic layer, then the two layers may be separated by drawing off the heavier layer and extracting the aqueous layer with fresh organic solvent. This is a *multiple-step* extraction. More elaborate procedures employing the principle of countercurrent extraction and the Craig Countercurrent Distribution Apparatus can be used for less favorable cases.

Choosing the Organic Solvent Not just any organic solvent will efficiently extract chemical substances from water. Obviously it must be a good solvent for the substance being extracted, but it should also coalesce quickly and not form emulsions with water. Finally, it should be significantly heavier or lighter than water so that two definite layers are formed. Some organic solvents that are suitable for extraction are, in decreasing order of specific gravity, as follows:

Chlorofrom, $CHCl_3$	1.49
Dichloromethane, CH_2Cl_2	1.31
Tributyl phosphate, $(BuO)_3PO$	0.98
Benzene, C_6H_6	0.88
Methyl isobutyl ketone, $CH_3COCH_2CH(CH_3)_2$	0.80
Ether, $(C_2H_5)_2O$	0.71

All of these differ significantly from water with respect to specific gravity, with the exception of tributyl phosphate. A lighter solvent, such as benzene, is frequently added to it to reduce the specific gravity and to help the layers separate more readily.

It is difficult to predict just which chemical species will be soluble in water and which will be soluble in organic solvents. In general, organic compounds with an oxygen atom or an —OH group are highly soluble in water if they have fewer than six carbon atoms to one oxygen or —OH. When the ratio of carbon atoms to oxygens is more than six to one, the solubility decreases steadily as the ratio increases. Ionic substances, of course, tend to dissolve in water; if they have a large carbon chain on them, it is possible that they will dissolve in organic solvents if they can form an ion pair. The best way to approach such a problem is to test the solubility in the laboratory.

Percent Extraction

As analytical chemists, we are interested in the *percent extraction* of any component extracted into an organic solvent. This, however, varies with the relative volumes of the two solvents used and the initial amount of the extractable component present. We must therefore define some type of equilibrium constant to enable us to calculate the percent extraction under any conditions.

If we are dealing with a species that exists in only one form (does not ionize, etc.), we can define a partition coefficient, K_p:

$$K_p = \frac{[S]_o}{[S]_w}$$

where $[S]_o$ is the concentration of the species in the organic liquid phase and $[S]_w$ is the concentration of the species in the aqueous phase. However, since many species exist in more than one form, most authors prefer to define a *distribution ratio*, D, as follows:

$$D = \frac{[C_s]_o}{[C_s]_w} = \frac{(\text{moles of all forms of } S)_o / V_o}{(\text{moles of all forms of } S)_w / V_w} \qquad (19\text{-}2)$$

where $[C_s]_o$ is the concentration of all forms of species S in the organic phase, $[C_s]_w$ is the concentration of all forms of species S in the aqueous phase, and V_o and V_w are the volumes of the organic and aqueous phases.

As an example of a real situation, the distribution ratio for the extraction of benzoic acid from water into the organic solvent benzene is defined as follows:

$$D = \frac{[C_{bz}]_o}{[C_{bz}]_w} = \frac{[HBz]_o + [HBz \text{ dimer}]_o}{[HBz]_w + [Bz^-]_w}$$

where $[C_{bz}]_o$ and $[C_{bz}]_w$ are total concentrations of benzoic acid in each phase. $[C_{bz}]_o$ is the sum of un-ionized benzoic acid $[HBz]_o$ and benzoic acid dimer $[HBz \text{ dimer}]_o$; $[C_{bz}]_w$ is the sum of un-ionized benzoic acid $[HBz]_w$ and ionized benzoic acid $[Bz^-]_w$. It is clear that defining a partition coefficient for such a situation is much more difficult than defining a distribution ratio and dealing with the sums of the species. The disadvantage of using D is that it varies with pH, so the pH has to be constant for experiments where D is used.

Percent Extraction and D

The percent extraction can be related to D, so it is possible to calculate the percent extraction corresponding to various theoretical or real values of D. Assume we are extracting a chemical species from water to an organic phase. Then % Ex, the percentage of species S extracted into the organic phase, is

$$\% \text{ Ex} = \left[\frac{(\text{moles of all forms of } S)_o}{(\text{moles of all forms of } S)_o + (\text{moles of all forms of } S)_w} \right](100) \quad \textbf{(19-3)}$$

By manipulation of this equation and substitution from **19-2**, we obtain the relation between the percent extraction and D as follows:

$$\% \text{ Ex} = \left[\frac{D}{D + (V_w/V_o)} \right] 100 \quad \textbf{(19-4)}$$

where again V_w and V_o are the volumes of the aqueous phase and the organic phase. In the special case where V_w and V_o are equal, **19-4** simplifies to the following:

$$\% \text{ Ex} = \left[\frac{D}{D + 1} \right] 100 \quad \textbf{(19-5)}$$

Equations **19-3** and **19-4** apply only to a *single* extraction of an aqueous phase with an organic phase. The percent extraction that is obtained after n repeated extractions with a fresh volume of organic solvent always equal to that of the aqueous phase is expressed as follows:

$$\% \text{ Ex} = 100\% - \frac{100\%}{(D + 1)^n} \quad \textbf{(19-6)}$$

In general, the percent extraction for a given value of D is larger for repeated extractions than if the volume of the organic phase is increased, as the following example will demonstrate.

EXAMPLE: A hypothetical organic compound B has a partition coefficient of 10.0 for extraction into ether from water.   (a) It is desired to extract B at least 99.0% into the ether. Will one extraction using equal volumes of ether and water be adequate? (b) If not, would it be better to increase the volume of ether relative to that of water or to use two or more repeated extractions?

Solution to a: Use **19-5** to calculate the percent extraction.

$$\% \text{ Ex} = \left[\frac{10.0}{10.0 + 1} \right] 100 = 90.9\%$$

Obviously it is not adequate to use one extraction with $V_o = V_w$. We must either increase the volume of ether or use more than one extraction.

Solution to b: First, consider using twice the volume of ether as water, and calculate the percent extracted using **19-4**.

$$\% \text{ Ex} = \left[\frac{10}{10 + 1/2} \right] 100 = 95.2\%$$

The percent extraction has increased from 90.9 to 95.2%, but it is still not a 99.0% extraction. Before considering further increases in the ratio of organic solvent to water for one extraction, let us consider two extractions with the same volume of ether as water. Using **19-6**,

$$\% \text{ Ex} = 100\% - \frac{100\%}{(10.0 + 1)^2} = 99.2\%$$

Obviously two extractions with volumes equal are more satisfactory than increasing the volume of the organic solvent.

TABLE 19-2. Relation of Percent Extraction to D

	$V_o = V_w$		$V_o = 2V_w$
D	% Ex for 1 extraction	% Ex for 2 extractions	% Ex for 1 extraction
1.00	50.0	75.0	66.7
10.0	90.9	99.2	95.2
20.0	95.2	99.8	97.6
50.0	98.0	99.9_6	99.0
90.0	98.9	99.9_9	99.4
500.0	99.8	99.9_{994}	99.9_0
900.0	99.9	99.9_{999}	99.9_4

The results of the preceding example are general for other values of D. Table 19-2 contains the results of calculations of the percent extraction for all three of the above situations for various values of D. Several conclusions can be drawn from the table, depending on the definition of a quantitative extraction. If we accept a 99.0% extraction as quantitative, we see that for a one-step extraction involving equal volumes of solvent and water, D must be above 90.0. If we double the volume of organic solvent used to extract, then D can be as low as 50.0. Further, if we use two extractions with equal volumes instead of one, D can be as low as 10.0 for a quantitative extraction. If we want a 99.9% extraction for quantitative purposes, the minimum values of D are much larger. For one step with equal solvent volumes, D must be above 900, a rather uncommon value. Even if two extractions with equal volumes are used, D must be 500.0 or larger, also a rather large value. In such cases, manual methods involving one or more steps are often discarded in favor of a continuous automatic extraction procedure, such as that involving the Craig Countercurrent Distribution Apparatus, which we will discuss on p. 429.

Analytical Applications of Manual Extractions

Manual extractions are important in the analytical laboratory for isolating a chemical species to be determined, or a species that might interfere in the determination of another species.

The classic inorganic example of extraction is the extraction of iron(III) from an aqueous $6M$ hydrochloric acid solution into ether, or diethyl ether, as it is more accurately called. This extraction can be used to isolate iron(III) before its determination, but more often it is used to remove iron(III) as an interfering species. In $6M$ hydrochloric acid, iron(III) forms the tetrachloroferrate(III) ion, $FeCl_4^-$, and it is extracted as the $H^+FeCl_4^-$ ion pair into ether. The distribution ratio for this extraction, using $6M$ hydrochloric acid as the aqueous phase, is 140. (It has a different value for other concentrations of hydrochloric acid.) The percentage of iron(III) extracted using equal volumes of $6M$ acid and ether is calculated from **19-5** as follows:

$$\% \ Ex = \left[\frac{140}{140 + 1} \right] 100 = 99.3\%$$

If a more quantitative extraction, such as 99.9%, is desired, a larger ratio of ether to $6M$ acid can be used. The ratio needed can be calculated by rearranging **19-4** and solving for V_w/V_o.

$$\frac{V_w}{V_o} = 140\left[\frac{100}{99.9\%}\right] - 140 = 0.140 = \frac{1}{7.14}$$

Thus about 7.14 volumes of ether per volume of $6M$ acid gives a theoretical value of 99.9% extraction.

Certain other species are also extracted into ether. Arsenic(III) chloride has a D value of 2 and is about 68% extracted into an equal volume of ether in one extraction step. In contrast, arsenic(V) chloride has a D value of 0.02 and is only about 2% extracted under the same conditions. Thus, the interference of arsenic(III) chloride can be effectively eliminated by oxidizing the sample to produce arsenic(V) chloride before extraction. Antimony(V) interferes greatly, however, as it is 81% extracted, so a simple one-step manual extraction is not always feasible.

A classic organic example of extraction is the extraction of phenols from samples such as oil refinery waste water [2]. Phenols are weakly acidic and are converted to the phenoxide anion only in strongly basic solutions above pH 11.

$$C_6H_5OH + OH^- \longrightarrow C_6H_6O^- + H_2O$$

If the refinery waste water is extracted at its normal pH, a number of other types of organic compounds are extracted with phenol into carbon tetrachloride, the organic solvent used. However, the waste water is first adjusted to pH 12, which converts the phenols to anions that are only soluble in water and not in carbon tetrachloride. The other types of organic compounds are then extracted away from the phenoxide anions into the carbon tetrachloride. The pH of the waste water is then adjusted back to 5, and the resulting phenols are extracted into carbon tetrachloride. The phenols can then be measured using spectrophotometry without interference from the other organic compounds.

A good clinical example of extraction is the extraction of cholesterol and cholesterol esters away from the protein in blood serum. Since the blood serum is primarily an aqueous medium, organic solvents such as ether-alcohol, or acetone, can be added to extract the cholesterol and its esters away from the protein that might interfere with the colorimetric determination of cholesterol. In Section 19-1 the further separation of cholesterol and cholesterol esters by precipitation was discussed.

Counter-current Extraction: The Craig Apparatus

So far we have discussed separations where the differences in the distribution ratio are fairly large and separations can be achieved by one to three manual extractions. Where the differences are small, countercurrent extraction may be used; the apparatus for this is known as the Craig Countercurrent Distribution Apparatus. The apparatus itself is rather complicated, and an accurate description of it is rather involved. Essentially, it consists of a series of glass tubes mounted in a rack so that a whole set of stepwise extractions can be conducted simultaneously. When the solvent layers have separated, the rack is tilted so that the upper, lighter phase in every tube is completely transferred to the upper part of the next tube.

We will next consider a simple example of a separation between two species, S and T. They are present in equal quantities in a solvent that is heavier than the extracting solvent. Further, assume that T has a D value of 9.00 and that S has a D value of 1.00 for

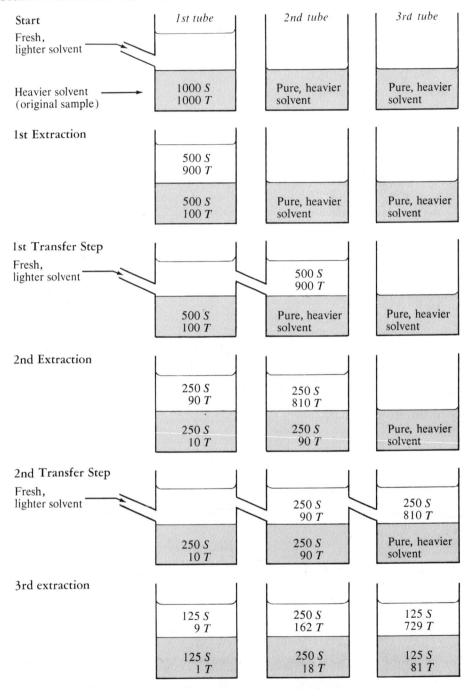

FIGURE 19-7. Schematic of the first three extraction steps of the Craig Countercurrent extraction process. For species T, $D = 9.0$; for species S, $D = 1.0$.

the extraction of each into the lighter extracting solvent. At the start, each tube, except the first, in the Craig Apparatus is filled with an equal volume of the same solvent as is in the sample. The first tube is filled with the sample. Then an equal volume of the lighter extracting solvent is poured into the first tube (see *Start*, Fig. 19-7).

To simplify the example, assume that 1000 molecules of S and 1000 molecules of T are present in the sample. After the first extraction (Fig. 19-7), half of the molecules of S are found in each phase, since D is 1.00. Exactly 90% of the original 1000 molecules of T, or 900 molecules, are extracted into the upper phase. Next, the first transfer step is carried out. The upper layer containing 500 S molecules and 900 T molecules is transferred into the second tube, and fresh extracting solvent is poured into the first tube.

In the second extraction, two extractions occur—one in the first tube and one in the second tube. In each tube 250 molecules of S end up in each phase, because its distribution ratio is 1.00. Note, however, that only 10 molecules of T remain in the lower phase in the first tube. This is because its larger D of 9.00 enables 90% of the T molecules in each lower phase to be extracted into the upper phase. Then the second transfer step is executed. The upper layer in each of the two tubes is transferred to the next tube, and fresh extracting solvent is again poured into the first tube.

In the third extraction, three extractions occur—one in each of the three tubes shown in Figure 19-7 (in the actual Craig Apparatus there are many more than three tubes).

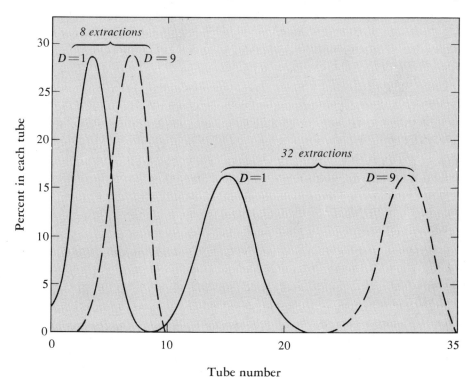

FIGURE 19-8. Theoretical distribution of molecules having D values of 1.00 and 9.00 after different numbers of transfers in a Craig Countercurrent Distribution Apparatus.

Note that all but one molecule of T has been extracted from the original sample and that most of T is now in the third tube, whereas most of S is in the second tube. It may seem odd that there is yet no change in the number of molecules of S in each phase in any tube, but this will change as soon as the third transfer occurs.

Of course, there is no separation at this point, but you should now have grasped the concept of the Craig Countercurrent extraction. Eventually, all of the T molecules are to be found in the tubes toward the end of the apparatus, whereas the S molecules tend to "bunch up" toward the center. No matter how many tubes are used, all of T will be found farther along than all of S, because of the greater value of D for T. This is illustrated in Figure 19-8 for two different numbers of extractions. After eight extractions, there is not a complete separation of T and S. When the number of extractions is increased to 32, there is a fairly good separation of the two species.

| QUESTIONS AND PROBLEMS

(Answers to most even-numbered problems are in Appendix 5.)

Concepts and Definitions
1. Differentiate between the following pairs of separation techniques with respect to the number of stages or steps involved:
   a. Liquid chromatography and single-step extraction.
   b. Precipitation and paper chromatography.
   c. Gas chromatography and electrodeposition.
2. What is the difference between a separation by precipitation and a gravimetric method of analysis?
3. Give one point of similarity and one point of difference between a separation by precipitation and a separation by electrodeposition.
4. Why must cholesterol be separated by precipitation from cholesterol esters if the Zlatkis-Zak colorimetric method is to be used to measure cholesterol in body fluids?
5. Indicate whether the first metal ion in each of the following pairs can be separated without interference from the second metal ion by electrodeposition:
   a. Cu^{+2} and Ag^+.
   b. Cu^{+2} and Na^+.
   c. Ag^+ and Cu^{+2}.
   d. Cu^{+2} and BiO^+.
6. List the four common characteristics of a chromatographic technique.
7. Differentiate between liquid-liquid chromatography and liquid-solid chromatography.
8. Differentiate between thin-layer chromatography and paper chromatography.
9. List three advantages of thin-layer chromatography over paper chromatography, including at least one advantage regarding visualization.
10. Describe the carrier phase and the stationary phase used in gas chromatography.
11. What is the difference between ion-exchange chromatography and liquid chromatography of organic compounds?

Solvent Extraction Concepts
12. Describe the nature of the two phases in a solvent extraction.
13. Indicate why organic solvents such as ethyl alcohol and acetic acid are not used to extract chemical species from water.
14. What is the difference between the partition coefficient and the distribution ratio?
15. Derive the relation between the percent extraction and D, as expressed in **19-4**. (*Hint:* Start with **19-3**.)
16. Why should two extractions using equal volumes be more efficient than a single extraction using a volume of the extracting solvent that is double that of the solvent containing the sample?
17. The distribution ratio for extracting iron(III)

into ether from $4M$ hydrochloric acid is less than that for $6M$ hydrochloric acid. Explain.
18. Differentiate between a manual extraction involving thirty repeated extractions and an extraction involving a thirty-tube Craig Countercurrent Distribution Apparatus.

Solvent Extraction Problems
19. Calculate the percent extraction for a single extraction into an organic phase from an aqueous sample phase of equal volume for each following value of D:
   a. $D = 0.100$.
   b. $D = 5.00$.
   c. $D = 30.0$.
   d. $D = 200.0$.
20. Calculate the percent extraction for a single extraction into an organic phase that is

double (2.00 : 1.00) the volume of the aqueous sample phase for each value of D given in Problem 19.
21. Calculate the percent extraction for two extractions using each time a volume of organic phase that is the same as that of the aqueous sample phase for each value of D given in Problem 19.
22. A certain species has $D = 5.00$ for extraction into an organic solvent. Which will give the higher percent extraction: A single extraction using a volume of organic solvent ten times greater than that of the aqueous phase? Or two extractions of the aqueous phase with the same volume of fresh organic solvent each time?
23. Calculate the number of molecules of S and T in both phases after the third transfer step, and fourth extraction step, of the separation in Figure 19-7.

| **NOTES**

[1] A. Zlatkis, B. Zak, and A. J. Boyle, *J. Lab. Clin. Med.* **41**, 486 (1953).
[2] J. Schmauch and H. M. Grubb, *Anal. Chem.* **26**, 308 (1954).

20 | Theory of Chromatography

"You have a *theory*?" "Yes, a *provisional one*. But I shall be surprised if it does not turn out to be correct." [said Holmes.]

ARTHUR CONAN DOYLE
"The Yellow Face"

Theories are proposed in order to be confirmed or disproved at some future time. Even Sherlock Holmes's theories were not always correct, as he found to his chagrin in the above story. Chromatographic theory is still in a partially unsettled state, although it does not appear that any of the present theory will be disproved in the near future. Two theoretical approaches, plate theory and rate theory, have been used by classical chromatographers to guide their research work. This classical treatment is a little like looking at the same situation twice through different glasses.

In the following treatment we want to approach chromatography from as *unified* a viewpoint as possible. We have avoided classical phrases such as the "height equivalent of a theoretical plate" and instead use the concept of plate height in a practical, real sense. We start our treatment with the concept of resolution, since, after all, the resolution of two or more sample constituents is the desired result. We then follow this with equilibrium and rate considerations, but without emphasizing two different theories, thereby setting up artificial barriers to a unified understanding of chromatography. Thus the words *plate theory* and *rate theory* do not appear in this chapter. We hope that our treatment is successful and that you acquire enough chromatographic theory to enhance your reading of the following chapter with a greater depth of meaning.

20-1 | CHROMATOGRAPHIC RESOLUTION

In this section we will focus on chromatographic separations in which, one by one, sample constituents on the stationary phase dissolve in or are eluted by the mobile phase as it moves by them. (This is in contrast to displacement chromatography, where the mobile phase *displaces* sample constituents from the stationary phase.) If the stationary phase is packed in a column, the term *elution* is used to describe the process. As shown in Figure 19-1, the mobile phase passes over the stationary phase continuously until all the sample constituents are eluted from the column. If the stationary phase is an open sheet (thin layer or paper), the term *development* is used to describe the elution chromatography [1]. Because the mobile phase is stopped at a line short of the end of the sheet (Fig. 19-4), the sample constituents cannot be eluted from the stationary phase.

Because elution chromatography on a column *cannot be visualized*, it is *conceptualized* in most texts in terms of the detector response curve. To help you understand this, we will constantly compare it with developmental elution chromatography on thin-layer plates, which *can be visualized*.

Resolution   The quality of a chromatographic separation depends on both efficiency and selectivity. We will begin by considering efficiency, first in terms of a single sample constituent, and then in terms of two constituents.

Efficiency   Each sample constituent can be viewed as existing and traveling in a *zone* on a thin-layer plate or in a column. On a thin-layer plate, the constituent can actually be seen as a circular zone, or spot, on the stationary phase (Fig. 20-1, upper). In a column of the

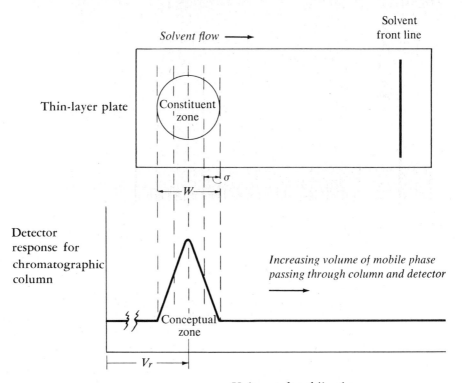

FIGURE 20-1.   Constituent zone on a thin-layer plate (upper) and conceptual zone of a sample constituent eluted from a column (lower). The W term is the diameter of the thin-layer zone or the width of the conceptual zone at its base.

stationary phase, the zone must be conceptualized in terms of the detector response curve plotted by a recorder (Fig. 20-1, lower). As the sample constituent leaves the column, its recorded concentration increases gradually to some maximum value, at V_r, the *retention volume*. (Note that the curve in the lower half of Fig. 20-1 is a plot of response against *volume* of mobile phase. The retention volume is the volume at which the constituent's concentration is at a maximum and at which half of the constituent ideally has been eluted.) Past the retention volume, the concentration of the constituent then decreases until it is completely eluted from the column.

In qualitative terms, the efficiency of a stationary phase can be evaluated in terms of how well it functions in keeping the constituent zone or the conceptual zone from spreading apart. In quantitative terms, efficiency is evaluated by first defining the quantity W (Fig. 20-1) as follows:

$$W = \text{diameter of constituent zone (TLC) or}$$
$$= \text{width of conceptual zone at its base}$$

For symmetrical zones, the usual practice is to use the quantity σ (Fig. 20-1):

$$\sigma = \frac{W}{4} \tag{20-1}$$

The σ term is sometimes called the quarter-width of the zone. Column efficiency is then defined in terms of L, the length of the column, and either W or σ.

$$\text{Efficiency} = \frac{W^2}{16L} \quad \text{or} \tag{20-2}$$

$$= \frac{\sigma^2}{L} \tag{20-3}$$

In principle, the same definitions can be applied to thin-layer plates.

Selectivity In addition to efficiency, the selectivity of a stationary phase is important in influencing the resolution of two or more chromatographic zones. In qualitative terms, selectivity is the separation of the *centers* of each of the constituent zones. Selectivity is expressed quantitatively by symbolizing the distance between the zone centers as d. In the upper half of Figure 20-2, it is apparent that d is the distance between the zone centers on a thin-layer plate. However, for a chromatographic column, d is conceptually the distance between the maxima of the sample conceptual zones (Fig. 20-2, lower). Since the maxima *represent* the centers of the zones leaving the column, d is also the distance between zone centers on a column.

Note in the lower half of Figure 20-2 that if the "inner" line of each conceptual zone is continued along the dotted lines shown, the dotted lines intersect close to a point at the baseline. This represents a very small amount of overlap of zones A and B on a column.

Resolution We are now ready to discuss the concept of resolution in terms of the efficiency and selectivity of the separation shown in Figure 20-2. It is easily seen that *resolution is directly proportional to selectivity* (distance between zone centers) and *inversely proportional to efficiency* (width of each zone). In other words, resolution is a measure of the overlap of any two adjacent zones.

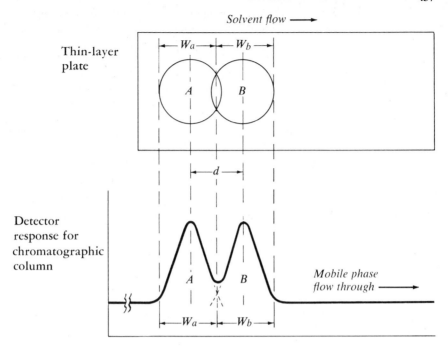

FIGURE 20-2. Two nearly resolved sample constituent zones.

Qualitatively speaking, we define resolution, R, as follows:

$$R = \frac{\text{zone separation}}{\text{zone width}}$$

This equation clearly indicates that resolution is dependent on the separation of the zones and that this separation is limited by how wide the zones are.

In quantitative terms, R is defined in two slightly different ways, using the symbols in Figure 20-2, as follows:

$$R = \frac{d}{W_a/2 + W_b/2} = \frac{2d}{W_a + W_b} \qquad (20\text{-}4)$$

If the A and B zones have the same width or diameter, **20-4** becomes the following:

$$R = \frac{d}{4\sigma} \qquad (20\text{-}5)$$

(Recall from **20-1** that $W = 4\sigma$.)

Values for R can be calculated for various situations by assuming that σ for zones A and B is the same as the standard deviation from the normal distribution curve (Fig. 3-2). Note that a distance of $\pm 2\sigma$ from the center of the curve in Figure 3-2 includes 95.46% of the population. By subtraction, then, 4.54%/2 of the population is on one side or the other of that portion of the curve having a total base length of 4σ.

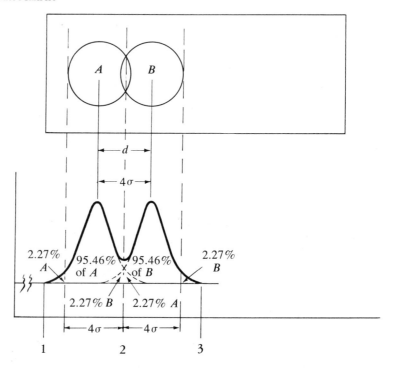

FIGURE 20-3. The percentages of A and B in two nearly resolved constituent zones. Constituent A is collected from point 1 to point 2; constituent B is collected from point 2 to point 3.

Let us now consider the separation of constituents A and B with zones as shown in Figure 20-3. At point 1 we begin to collect constituent A, at point 2 we begin to collect constituent B, and at point 3 we stop collecting. Assume that the distance from point 2 to the maxima of either zone is the same, 2σ. Then the distance between the maxima is 4σ. The width of each zone at the base is then 4σ also. Using **20-4** or **20-5**, R is calculated as follows:

$$R = \frac{4\sigma}{4\sigma} = 1.0$$

Next, we develop an interpretation of this value of unity for R, always assuming two zones of equal width. To collect constituent A, we begin at point 1, at a distance greater than 2σ from the center of zone A, because there are no other constituents on that side of the zone. We must stop collecting A at point 2, because the contamination from B becomes too large. The percentages of A and B that have been collected at point 2 are

$$\%A \text{ collected} = 2.27\% + 95.46\% = 97.73\%A$$

$$\%B \text{ collected} = 0\% + 2.27\% = 2.27\%B \text{ (contamination)}$$

To collect constituent B, we begin at point 2, where the collection of A was terminated. Since there is no other constituent on the other side of the zone of B, we collect B until point 3, where it has stopped coming off the column. The percentages of B and A collected from point 2 to point 3 are

$$\%B \text{ collected} = 95.46\% + 2.27\% = 97.73\%B$$

$$\%A \text{ collected} = 2.27\% + 0\% = 2.27\%A \text{ (contamination)}$$

We see that a numerical value of 1.0 for R implies that the separation is a borderline one; the amount of contamination of A or B by the other is just beginning to be significant. Similar calculations for other values of R for two zones of equal width give the following interpretations:

d	R	$\%A$ collected	$\%$ contamination by B
4σ	1.0	97.73	2.27
5σ	1.25	98.76	0.62 ($= 50.00 - 49.38$, Appendix 4)
6σ	1.5	99.87	0.13 (Appendix 4)

Since a quantitative process usually implies 99.9%, a strictly quantitative separation is here achieved with an R value of 1.5.

20-2 | EQUILIBRIUM CONSIDERATIONS

As noted in Section 19-1, chromatography is a continuous multiple-stage separation technique. You might wonder whether it is possible for equilibrium to be achieved anywhere in such a multiple-stage process. An understanding of the mechanism of a sample zone would seem to make it likely that equilibrium would be achieved in the *center* of the zone. There molecules of a particular constituent are effectively in complete equilibrium between the stationary phase and the mobile phase. This equilibrium can be described by the same D, or distribution ratio, as is used in liquid-liquid extraction (Sec. 19-2).

For example, in a liquid chromatographic system involving water as the mobile phase and benzene as the stationary phase, the distribution ratio for the chromatographic process has the same value as D for the liquid-liquid extraction process **(19-2)**. Although this is gratifying information, values of D are unfortunately not among the fundamental parameters measured in chromatography. After discussing some of these parameters, we will describe how D is related to them.

Measured Chromatographic Parameters Both development (planar) chromatography and column elution chromatography are characterized by a retardation parameter, R_r, called the retention ratio or retardation factor. First, we will consider R_r in general terms.

General Definition of R_r R_r is defined in terms of time as follows:

$$R_r = \frac{t_m}{t_m + t_s} \tag{20-6}$$

where t_m is the time the average sample constituent molecule spends in the mobile phase and t_s is the time that the same molecule spends in the stationary phase. R_r thus gives the fraction of time that the average molecule spends in the mobile phase.

R_r is also defined in terms of $\bar{u}$, the average velocity of the mobile phase, and $\bar{v}$, the average velocity of the sample constituent molecule, as follows:

$$R_r = \frac{\bar{v}}{\bar{u}} \tag{20-7}$$

If R_r is 0.40, for example, this means that the average constituent molecule spends 40% of its time in the mobile phase or travels 40% as fast as the mobile phase. Similarly, if R_r is 0.0, this indicates that the constituent is not eluted or moved by the mobile phase at all. A R_r of 1.0 indicates that the constituent moves along as fast as the mobile phase and is not retained at all by the stationary phase.

Retardation
Factor in
Development
Chroma-
tography

In thin-layer and paper chromatography, both forms of development chromatography, R_r as defined above is not used. Instead a retardation factor, R_f, measured in terms of distance, is employed. Recall that the sample is placed in a spot, or zone, at the bottom of a vertically aligned plate or sheet (Fig. 19-4), and each constituent is eluted from the spot, or zone, by the mobile phase traveling up the plate or sheet.

As shown in Figure 20-4, an R_f is calculated for each constituent zone in terms of the distance traveled from the original sample zone.

$$R_f = \frac{d_s}{d_m}$$

In this equation, d_s is the distance the average constituent molecule travels along the planar stationary phase during the time the front of the mobile phase (solvent) travels the distance d_m. Note that d_s is measured from the *center* of the original sample zone to the *center* of the sample constituent zone. The measurement of d_m should theoretically utilize the location of the bulk of the mobile phase, rather than the front, but this is a practical impossibility.

Note in Figure 20-4 that R_f values fall in the range between zero and one. Slower moving constituents, such as constituent A, have R_f values closer to zero, and faster moving constituents, such as constituent B, have R_f values closer to one.

Since the average velocity, $\bar{v}$, of each constituent should be proportional to d_s, and since it *appears* that $\bar{u}$, the average velocity of the mobile phase, should be proportional to d_m, it appears that R_f can be exactly defined in an equation similar to **20-7**. However, this is only approximately true, since the *front* of the mobile phase or solvent creeps ahead a bit faster that the *bulk* of the mobile phase. Thus since d_m is only approximately proportional to $\bar{u}$, we can say that

$$R_f \cong \frac{\bar{v}}{\bar{u}} \cong R_r \tag{20-8}$$

Strictly speaking, R_r is about 10% larger than R_f on the average, but we will neglect this difference in the rest of the discussion.

Column
Elution
Chroma-
tography

In column elution chromatography, we measure elution or retention volumes rather than times (or distances). These volumes must be related to V_m, the volume of the mobile phase held in the column outside the stationary phase. (We have already defined the retention volume, V_r, as the volume of the mobile phase at which the concentration of the constituent has its maximum value.) Before any constituent can appear at the

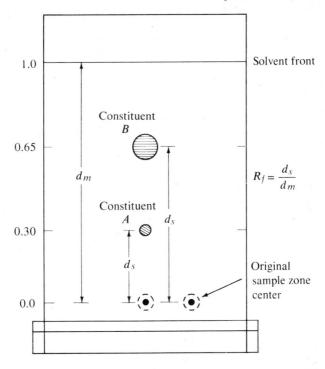

FIGURE 20-4. Schematic of the measurement of R_f on a thin-layer plate or paper chromatogram.

bottom of a column, a volume of mobile phase equivalent to V_m must pass through the column.

Now suppose that for two constituents being separated on a column, a volume of mobile phase equivalent to $2V_m$ is needed to elute constituent A to its maximum concentration at point V_r for A (Fig. 20-5). Further suppose that a volume of mobile phase equivalent to $4V_m$ is necessary for elution of constituent B to its maximum concentration at point V_r for B (Fig. 20-5). From this information it is possible to calculate the retardation ratio, R_r, as follows:

$$R_r = \frac{t_m}{t_r} = \frac{V_m}{V_r} \tag{20-9}$$

Using this equation, R_r for A is 0.5 and R_r for B is 0.25.

Equation 20-9 is derived as follows. Recall that t_m is the time the average molecule spends in the mobile phase. We then define t_r as the longer time it takes for the average molecule to be eluted to its maximum concentration at volume V_r (Fig. 20-5). Going back to 20-6, we see that t_s, the time the average molecule spends in the stationary phase, must be the difference between t_r and t_m.

$$t_s = t_r - t_m \tag{20-10}$$

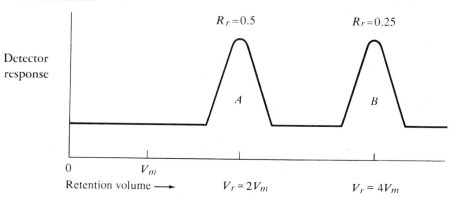

FIGURE 20-5. Column elution separation of two constituents with different retention volumes.

Substituting **20-10** into **20-6**, we obtain

$$R_r = \frac{t_m}{t_m + (t_r - t_m)} = \frac{t_m}{t_r} \tag{20-11}$$

This gives the first equality of **20-9**. The second is obtained by assuming that at constant flow rate, F, the following equations are true.

$$t_m = \frac{V_m}{F} \tag{20-12}$$

$$t_r = \frac{V_r}{F} \tag{20-13}$$

Substituting **20-12** and **20-13** into **20-11** gives

$$R_r = \frac{V_m/F}{V_r/F} = \frac{V_m}{V_r} \tag{20-14}$$

Relation Between R_r and D Recall that in Section 19-2 we defined a partition coefficient, K_p, and a distribution ratio, D (**19-2**), for liquid-liquid extraction. We similarly define such constants for chromatography as follows:

$$K = \frac{[S]_s}{[S]_m} \tag{20-15}$$

$$D = \frac{[C_s]_s}{[C_s]_m} \tag{20-16}$$

where $[S]$ is the concentration of a particular form of a solute in either the mobile or the stationary phase and $[C_s]$ is the concentration of all forms of the solute in either the mobile or the stationary phase.

We can also define a mass ratio constant, k', as follows:

$$k' = \frac{(mole)_s}{(mole)_m}$$

For a constituent that exists in *only one equilibrium form,*

$$k' = K\left(\frac{V_s}{V_m}\right) = D\left(\frac{V_s}{V_m}\right) \tag{20-17}$$

where V_s is the volume of the stationary phase and V_m is the volume of the mobile phase. Equation **20-17** indicates that the partition coefficient and distribution ratio for a given separation system, such as liquid-liquid extraction or liquid-liquid chromatography, are directly proportional to the ratio of the masses on the stationary and mobile phases.

Now how do we relate K or D to R_r, the retention ratio? First we define the mass ratio constant in terms of times, as follows:

$$k' = \frac{t_s}{t_m} \tag{20-18}$$

Rearranging,

$$t_s = (k')(t_m) \tag{20-19}$$

Substituting **20-19** into **20-6**, we obtain

$$R_r = \frac{t_m}{t_m + (k')(t_m)} = \frac{1}{1 + k'} \tag{20-20}$$

Since R_r has the inverse relationship to k' given in **20-20**, it also has the same relationship to K or D, since they are proportional to k'.

Equation **20-20** is relevant to development chromatography, since **20-8** states that R_f is approximately the same as R_r. Thus for thin-layer or paper chromatography,

$$R_f \cong \frac{1}{1 + k'} \tag{20-21}$$

Equation **20-20** is also used to relate volumes in column elution chromatography. Combining the right-hand side of **20-20** and the right-hand side of **20-11**, we obtain

$$R_r = \frac{t_m}{t_r} = \frac{1}{1 + k'}$$

Rearranging the above gives

$$t_r = t_m(1 + k') \tag{20-22}$$

We now substitute volumes for times, using **20-12** and **20-13** at a constant flow rate, F.

$$\frac{V_r}{F} = \frac{V_m}{F}(1 + k')$$

We then eliminate F and rearrange to obtain

$$V_r = V_m + k'V_m \tag{20-23}$$

By substituting the middle term of **20-17** for k', we can also obtain

$$V_r = V_m + \left(K \frac{V_s}{V_m} \right) V_m$$

which simplifies to

$$V_r = V_m + K V_s \tag{20-24}$$

again at constant flow rate of the mobile phase.

Applications
: It is possible to apply some of the above equations to improve chromatographic separations. For example, in thin-layer chromatography, using a certain mobile phase may give R_f values in the 0.85–0.95 range, which means the constituents are moving too fast to be resolved effectively. By inspecting **20-21**, we see that R_f is affected only by the reciprocal of $(1 + k')$. If values of k' are known or can be deduced through **20-17**, then a mobile phase with a *larger* value of k' should be substituted for the original mobile phase. Since k' for the original mobile phase must have been of the order of 0.1 to give an R_f of the order of 0.9, it is clear that a mobile phase with a k' of the order of 2 should be chosen to achieve an R_f value in the 0.5 range. If values of k' are not available, it may be possible to estimate k' from values of K_p or D from a liquid-liquid extraction system involving the same phases as are being used in the thin-layer separation. (This assumes that the thin-layer separation occurs via partition chromatography rather than adsorption chromatography.)

As another example, assume that a second sample constituent is being eluted from a column too soon to permit a good separation from the first constituent eluted from the column. If the separation is being effected by partition chromatography, the liquid stationary phase should be changed so that its V_r is increased to a greater extent than V_r for the first constituent. By means of either **20-23** or **20-24**, we are guided to choose a mobile phase with a higher k', or K, for the sample constituent. In effect, we are changing to a mobile phase that interacts *to a greater degree* with the constituent.

A final example is taken from the field of gas chromatography. Suppose that all the sample constituents are eluted from the column too rapidly to be differentiated. In other words, V_r for all constituents is too small. This time, however, we will assume that the values of K are satisfactory. Equation **20-24** predicts that we should then increase V_s, the amount of the stationary phase. In effect, we are giving the sample more reactive solvent on which to undergo a proper separation.

Application of the Separation Quotient (Factor)
: It is often useful to consider the separation of any two sample constituents, A and B, in terms of α, the separation quotient or separation factor. The α term is usually defined as a ratio numerically greater rather than less than one. For liquid-liquid extraction of A and B (both of which, we assume, exist in only one form), it is defined as follows:

$$\alpha = \frac{(K_p)_B}{(K_p)_A}$$

where constituent B is chosen to have the larger K_p. Similarly, for development or column elution chromatography, it is defined as follows:

$$\alpha = \frac{(k')_B}{(k')_A} \tag{20-25}$$

where B again has the larger k' value. For column elution chromatography, **20-25** and **20-23** can be used to show that α gives a ratio of the adjusted retention volume as follows:

$$\alpha = \frac{(V_r - V_m)_B}{(V_r - V_m)_A} \tag{20-26}$$

By reference to Figure 20-5, you can deduce that the adjusted retention volume is corrected for variation in V_m, the volume of the mobile phase held in the column.

By means of α, we can transfer separation information, for example, from thin-layer chromatography to column elution chromatography. For example, for a thin-layer separation of A and B performed by partition chromatography, suppose that A and B have R_f values of 0.50 and 0.30, respectively. Then **20-21** can be used to solve for the respective mass ratio constants.

$$(R_f)_A = 0.50 = \frac{1}{1 + (k')_A} \qquad\qquad (R_f)_B = 0.30 = \frac{1}{1 + (k')_B}$$

$$(k')_A = 1.0 \qquad\qquad\qquad\qquad (k')_B = 2.3$$

Then α can be calculated, using **20-25**.

$$\alpha = \frac{(k')_B}{(k')_A} = \frac{2.3}{1.0} = 2.3$$

This predicts that the ratio of the mass ratio constants for the column elution separation will also be 2.3 to 1. In addition, **20-26** predicts that the adjusted retention volumes for B and A will have the same ratio.

A Final Word on Column Length

The above consideration of α ignores the effect of the length of a column on the resolution, or separation, of two or more constituents. The resolution of two or more constituents can be improved by increasing the length of the column used. It can be shown that resolution is proportional to the square root of the length of the column. For example, if we quadruple column length, we will increase the resolution **(20-4)** by a factor of two. The limitations to this are that the time required for the separation may as a result be made inconveniently long and that the constituents may become too dilute to be isolated or to be measured accurately.

20-3 | KINETIC CONSIDERATIONS

The discussion thus far has not included any information on the relationship of the length of a column and the rate of the separation (kinetics). It is possible to establish such a relationship by defining specific parameters that control the length of a column and the separation rate, and then relating the parameters. We will begin with the length of a column.

A chromatographic column or thin-layer plate may be thought of as consisting of a number, N, of theoretical plates or stages at which equilibrium is *postulated* to occur. (The term *theoretical plates* originates from separations carried out by distillation, in which the distillation column is conceptually divided into theoretical plates.) The number of theoretical plates can be estimated empirically by the following equation:

$$N = 16(d/W)^2 \tag{20-27}$$

where d and W are defined as in Figure 20-2.

The relationship between N and L, which is the length of the column or the migration distance in thin-layer chromatography, is

$$L = HN \tag{20-28}$$

where H is the height of each theoretical plate, or simply the *plate height*. It turns out that H is also an efficiency parameter which characterizes zone spreading and is defined as follows:

$$H = \frac{W^2}{16L} \tag{20-29}$$

It is H, not L, that is directly related to the rate of a chromatographic separation. Next, we will develop that very point.

Rate Factors Influencing H

The basic rate, or kinetic, consideration involves W, the width of a sample constituent zone (Fig. 20-1). As a constituent moves through the stationary phase, it invariably "spreads out" as it travels, so W is usually significantly larger at the end of the separation than at the start. It turns out that the rate at which the constituent moves through a column or up a plate affects the width or diameter of the constituent zone. It also affects both the efficiency and the plate height (20-29).

An equation that relates the plate height, H, and the velocity, v, of the mobile phase is the van Deemter equation:

$$H = A + B/v + Cv \tag{20-30}$$

where A, B, and C are constants whose values are described next.

A, or the eddy diffusion term, is *independent* of the velocity of the mobile phase. It is a measure of the nonideal chromatographic behavior of constituent molecules traveling a large number of pathways through the stationary phase. These pathways differ in length, so the molecules arrive at the end of the column or plate at different times.

B/v, or the longitudinal diffusion term, is *inversely proportional* to the velocity of the mobile phase. It is a measure of the molecular diffusion forces that cause migration from the center of the constituent zone toward the edges of the zone. This migration widens the zone, decreases the efficiency, and increases the plate height. As the velocity of the mobile phase is increased, longitudinal diffusion decreases.

Cv, or the mass transfer term, is *directly proportional* to the velocity of the mobile phase. It is a measure of the effect of the constituent's partitioning rate between the stationary and mobile phases. As the velocity of the mobile phase increases, the time available for equilibration (partitioning) between the two phases decreases, until at some point efficiency begins to decrease seriously and plate height increases significantly.

The cumulative effect, as velocity increases, of all these terms on H is shown in Figure 20-6 for gas chromatography and high-performance (pressure) liquid chromatography (HPLC). In gas chromatography, it is most desirable to achieve a minimum value of H, since **20-29** predicts that for a constant L, the number of theoretical plates increases as H decreases. At very low mobile phase velocities, the B/v term predominates and H is far too high for an efficient separation. At high velocities, the Cv term predominates and H is again too large. If the layer of stationary liquid phase can be adjusted to be fairly thin, column efficiency can be improved and the value of H lowered by a decrease in the value of C. Figure 20-6 does not predict what experimental adjustments are needed because of the A term, but from **20-30** it is obvious that the value of A should be reduced.

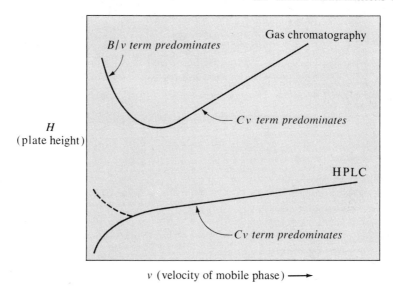

FIGURE 20-6. Typical curve of plate height, *H*, against velocity of the mobile phase for gas chromatography and high-performance (pressure) liquid chromatography (HPLC). The dotted portion at the left of the HPLC line indicates an alternative HPLC behavior.

In fact, eddy diffusion can be reduced by careful packing of the column with small, uniformly shaped particles to avoid open channels.

For high-performance liquid chromatography (HPLC), the curve in Figure 20-6 does not exhibit a minimum, because longitudinal diffusion and eddy diffusion are much slower than for gas chromatography. Therefore, *H* gradually increases as a function of the Cv term. Because the increase in *H* is so gradual, however, rather high flow rates can be used to compensate for the longer columns needed to maintain a large number of theoretical plates. (Recall from **20-28** that $N = L/H$.) To achieve rapid equilibrium with high flow rates, it is necessary to use a column packing consisting of small, uniformly shaped particles.

Summary We will now summarize very briefly the dependence of resolution on some key parameters. Increasing the column length, *L*, of course, increases resolution in proportion to the square root of the length. There are limitations to this with regard to time and constituent concentration.

Smaller diameter columns reduce the effects of eddy diffusion and of channeling of the sample constituents in directions that slow up the rate of longitudinal diffusion. Use of a smaller diameter gives rise to a need to use uniformly shaped, small particles as the stationary phase or as solid support for the stationary phase. Again, the use of this type of particle has made HPLC possible. Resolution is therefore increased by decreasing column diameter.

Since the sample constituent zones widen as the separation proceeds, the initial sample zone should be as narrow as possible to minimize the value of *W*, which in turn minimizes *d* (Figs. 20-1 and 20-2). Thus the sample should be as small as possible to minimize the size of the initial sample zone. Resolution is therefore increased by decreasing the sample size.

Finally, the velocity of the mobile phase affects the efficiency of the separation and H directly, rather than affecting resolution. However, since resolution is favored by greater efficiency, it is ultimately favored by minimizing the flow rate (Fig. 20-6).

| QUESTIONS AND PROBLEMS

(Answers to even-numbered problems are in Appendix 5.)

Definitions and Concepts

1. Define the following terms:
   a. Chromatographic efficiency.
   b. Chromatographic selectivity.
   c. Chromatographic resolution.
2. Explain how chromatographic resolution, R, the retardation ratio, R_r, and the retardation factor, R_f, differ.
3. Explain how the R_f value is measured in thin-layer or paper chromatography.
4. Explain from a theoretical point of view why R_f values are measured using the center of the zone rather than the front edge of the zone, as is used when measuring the solvent.
5. Explain why an R_f value is not exactly the same as an R_r value for the same chromatographic system.
6. Explain the difference between t_m, the time the average constituent molecule spends in the mobile phase, and t_r, the time required for the average constituent molecule to be eluted to volume V_r (Fig. 20-5).
7. Explain why t_s, the time spent in the stationary phase for an average constituent molecule, is equal to $t_r - t_m$, the difference between the time spent in the mobile phase and the time required for elution to volume V_r (Fig. 20-5).
8. Describe the effects of resolution and column length on the separation of two constituents on a chromatographic column.
9. What effect does the flow velocity have on the separation of two constituents on a chromatographic column?

Problems

10. Consider the separation of three constituents A, B, and C, rather than just the separation of A and B shown in Figure 20-3. Assume that the zone of C overlaps the zone of B to the same extent ($d = 4\sigma$) that B overlaps the zone of A. Calculate:
    a. The percentage of A and the percentage of B collected at point 2.
    b. The percentage of B and the percentage of all contaminants at point 3.
    c. The percentage of C and the percentage of B collected at the point where C is completely eluted.
11. Repeat the calculations in Problem 10 using $d = 6\sigma$ for all three zones (cf App. 4).
12. For a column elution chromatographic separation of constituents A and B, calculate R_r for each constituent, given that the volume of mobile phase held in the column is 3.8 ml and that:
    a. Constituent A has a retention volume of 7.6 ml.
    b. Constituent B has a retention volume of 11.4 ml.
13. For a certain column elution chromatographic system, it is known that constituent A has an R_r of 0.40 and constituent B has an R_r of 0.20. Given that the volume of the mobile phase held in the column is 4.7 ml, calculate:
    a. The retention volume for constituent A.
    b. The retention volume for constituent B.
14. For a certain column elution chromatographic system, it is known that the time the average sample constituent spends in the mobile phase is 1.6 minutes. Calculate the time for the average molecule to be eluted to its maximum concentration, given that:
    a. Constituent A has an R_r of 0.50.
    b. Constituent B has an R_r of 0.20.

15. On a thin-layer partition chromatographic plate, constituents A and B have R_f values of 0.60 and 0.40, respectively. For a column elution chromatographic separation employing the same stationary and mobile phases, calculate the ratio of the mass ratio constant of B to that of A.

| **NOTES**

[1] J. M. Miller, *Separation Methods in Chemical Analysis*, Wiley-Interscience, New York, 1975.

21 | Applications of Chromatography

"You know my methods. *Apply* them, and it
will be instructive to compare results," said
Holmes.

ARTHUR CONAN DOYLE
"The Sign of the Four"

Sherlock Holmes was always applying his methods and judging their effectiveness by
the success of the applications. In this chapter we give you the same opportunity—to
judge the effectiveness of chromatography by its applications. If necessary, you should
review the introductions to and the schematics of the various chromatographic techniques and equipment in Chapter 19. Although the theory of chromatography covered
in Chapter 20 will help you visualize how the separations occur, we will not stress the
theory in the applications presented. The order of the separations discussed is the same
as that in Chapter 19—liquid chromatography (column, thin-layer, paper), ion-exchange
chromatography, and gas chromatography.

21-1 | APPLICATIONS OF LIQUID CHROMATOGRAPHY

Liquid chromatography includes separations by column chromatography using either
atmospheric pressure or high-performance (pressure) liquid chromatography (HPLC),
by thin-layer chromatography, and by paper chromatography. Some of the above
separations are achieved by *adsorption* on active sites of a solid adsorbent; others
depend on a *partition* between a stationary liquid phase on a solid and an eluting
solvent. As we discuss the adsorbents and solvents that are common to all liquid
chromatographic techniques, we will refer often to these two mechanisms.

Adsorbents and Solid Supports

Common adsorbents used in column or thin-layer chromatography are listed in Table 21-1. The most common adsorbent is silica gel, which is a highly purified type of *porous* sand with a uniform particle size and a controlled percentage of water (usually 0–20%) achieved through activation (heating). As long as some water is present, Si—OH *active sites* form on the silica gel surface and hydrogen bond to many organic compounds.

The next most common adsorbent is alumina, an aluminum oxide that is highly pure and hydrated with from 0–15% water. The amount of water present can be controlled by the addition of water to a grade of alumina containing no water. Alumina is available in its natural basic form, in neutral form, or in an acidic form. These forms possess one or more types of *active sites*, such as Al^{+3}, Al—OH, Al—O$^-$, or Al—OH$^+$. Basic alumina is recommended for the separation of strong organic bases, such as amines; the other grades can be used for the separation of other compounds, such as aromatic hydrocarbons.

In general, we will focus on the *adsorption* of compounds on the active sites of silica gel and alumina, rather than on the *partitioning* of compounds between the eluent and the alumina or silica gel adsorbent. Some adsorbents, however, function primarily by retaining a layer of solvent on their surfaces; compounds to be separated are partitioned between the adsorbed solvent and the eluting solvent. These adsorbents include powdered cellulose and kieselguhr, as well as DEAE-cellulose (which also functions as an ion-exchanger).

Powdered cellulose is used mainly in thin-layer chromatography to separate polar compounds. Kieselguhr, a purified form of diatomaceous earth, is useful for the separation of various sugars, such as glucose, sucrose, etc. DEAE-cellulose is useful for the separation of large molecules, such as nucleotides.

Solvents

The effectiveness of a solvent used in most chromatographic elutions can be correlated to its degree of polarity. One measure of polarity is the dielectric constant, which is an electrical measurement of the effect of a solvent between two plates on the magnitude of electrical charge stored by a condenser connected to the plates.

TABLE 21-1. Common Adsorbents for Liquid Chromatography

Adsorbent	*Compounds separated*
Silica gel (Hydrated SiO_2)	Most compounds (Important exception: strongly basic compounds, such as amines.)
Alumina (Al_2O_3)	Bases, steroids, aromatic hydrocarbons, carbonyl compounds
Cellulose ($C_6H_{10}O_5$)$_x$	Polar compounds, ions
Kieselguhr	Sugars
Polyamides (Nylon-type polymer)	Anthocyanins, flavones, nitroanilines
DEAE-Cellulose (Diethylaminoethylcellulose)	Nucleotides

TABLE 21-2.   Polarity Series of Solvents

Solvent (increasing polarity)	Dielectric constant
n-Hexane	1.9
Cyclohexane	2.0
Carbon tetrachloride	2.2
Benzene	2.3
Trichloroethylene	3.4
Diethyl ether	4.3
Chloroform	4.8
Ethyl acetate	6.0
n-Propanol	20.1
Acetone	20.7
Ethanol	24.3
Methanol	32.6
Water	78.5 (25°C)

The solvents in Table 21-2 are arranged in order of increasing dielectric constant; water has the largest dielectric constant and is also the most polar solvent. Fairly polar solvents are frequently needed to elute compounds from silica gel and alumina; both adsorb polar compounds rather strongly.

In terms of adsorption chromatography, the solvents in Table 21-2 can move compounds down a column by *eluting* or *displacing* them. In elution, the compounds adsorbed on active sites dissolve in the solvent as it moves past them down the column. In displacement, the solvent itself competes with the adsorbed compounds for the active sites; because the solvent is much more concentrated, it gradually displaces the compounds from the active sites. Most of the nonprotonic solvents with low dielectric constants in Table 21-2 operate primarily by elution; the protonic solvents with higher dielectric constants, such as alcohols, operate by displacement.

Applications of Column Chromatography

Column chromatography is often done at atmospheric pressure using wide diameter columns and relatively little equipment. High-performance liquid chromatography (HPLC), in contrast, is carried out at high pressures in thin columns using an elaborate instrumental setup. Each technique has its specific uses.

Column Chromatography at Atmospheric Pressure

Column chromatography is used for the separation of a synthesized compound from its reaction mixture for some qualitative analysis purposes and for many quantitative purposes. (Thin-layer chromatography is preferred for most qualitative and many quantitative purposes, however.)

In general, less polar compounds are eluted or displaced from a column before the more polar compounds, using a solvent with as low a polarity as possible. (The order of polarity of compounds parallels the order of solvent polarity given in Table 21-2.) For example, if a mixture of an ester and an alcohol is to be separated on a column by means of adsorption on silica gel or alumina, a less polar solvent, such as chloroform or ethyl acetate, should be used to elute or displace the ester from the column first. Then a more polar solvent, such as acetone or ethanol, should be used to elute or displace the alcohol. It is not necessary in most cases to use a pure solvent, such as chloroform;

generally a mixture, such as hexane containing a certain percentage of chloroform, is satisfactory.

A good example of a separation by elution is the separation of vitamin A alcohol, $C_{19}H_{27}CH_2OH$, from vitamin A aldehyde, $C_{19}H_{27}CHO$ [1]. A column of alumina partially deactivated with water is used. Since the carbonyl group of the aldehyde is less polar than the hydroxyl group of the alcohol, the vitamin A aldehyde is the obvious choice to be eluted first. An eluting solvent of 2% acetone in hexane is found to elute vitamin A aldehyde, plus another compound, and leave vitamin A alcohol on the column. An eluting solvent of 8% ethanol (ethyl alcohol) is found to elute the vitamin A alcohol. The elution can be followed by ultraviolet spectrophotometry. A hypothetical elution curve of the separation is shown in Figure 21-1.

Separation of inorganic compounds or ions is generally not performed on alumina or silica gel adsorbents; however, a few important separations based on a partition between an adsorbed stationary liquid phase and a mobile eluting phase have been reported. Fritz and Schmitt [2] reported the separation of UO_2^{+2} from nearly all other metal ions on a silica gel column having an adsorbed $6M$ nitric acid aqueous phase. The $UO_2^{+2}(NO_3^-)_2$ ion pair is eluted from the column first, using an eluting solvent of methyl isobutyl ketone. The other metal ions are not dissolved by the methyl isobutyl ketone and remain on the column; they can be eluted later by washing the column with an aqueous eluting solvent.

Inorganic compounds are also separated by using inert adsorbents coated with an organic solvent; an aqueous eluting solvent is used to achieve a partition effect. An early example is the separation of UO_2^{+2}, Th^{+4}, and Pu^{+4} from other metal ions on an inert "Kel-F" resin coated with tributyl phosphate, an organic solvent [3]. The metal

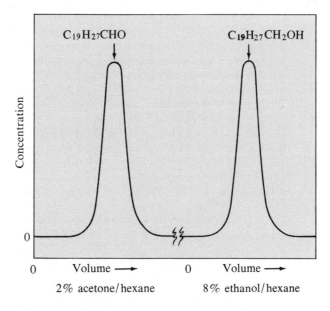

FIGURE 21-1. **Hypothetical elution curve of vitamin A aldehyde and vitamin A alcohol on an alumina column.**

ions are dissolved in $5.5M$ aqueous nitric acid; the UO_2^{+2}, Th^{+4}, and Pu^{+4} are selectively partitioned into the tributyl phosphate, but the other metal ions remain in the aqueous eluting solvent and pass through the column.

High-Performance Liquid Chromatography (HPLC)

HPLC partition or adsorption column chromatography can be used for the separation of preparative mixtures, but is mainly used for rapid qualitative and quantitative analysis. The high pressures employed (Sec. 19-1) enable analyses to be conducted in a few minutes.

The stationary phase in HPLC adsorption columns is either very fine particles of silica gel or alumina, or silica gel–coated micro-size glass particles. In HPLC partition columns, the stationary phase must consist of particles finely enough divided that they can be used in the small diameter (2–10 mm) columns used in HPLC column chromatography. In many cases, an organic coating is chemically bonded to the adsorbent particles using special techniques. A partition-like process takes place between the stationary liquid coating and the mobile phase (eluting solvent).

Either HPLC partition or adsorption column chromatography can be used for the analysis of mixtures of pharmaceuticals, drugs that are abused, vitamins, and other mixtures. HPLC is generally preferred to gas chromatography for the analysis of polymers and other heat-sensitive materials.

In general, the eluting solvent is not varied during elution in HPLC, as it is in atmospheric pressure column chromatography (Fig. 21-1). This facilitates the separation, permits rapid separation, and avoids problems with the refractive index detector (the refractive index of each solvent differs significantly). If the high pumping pressure does not permit a rapid enough separation, a technique known as gradient elution can be used. In this technique, the mixing of the components of the eluting solvent is controlled so that the proportion of the more polar component of the solvent is gradually increased. This elutes the more polar compound from the column faster.

An example of a separation by HPLC partitioning is the separation of the fat-soluble vitamin A acetate and vitamin D in vitamin tablets [4]. A preliminary treatment separates as much as possible of the inert tablet binder, or capsule gelatin, from the vitamins before chromatography. The separation is carried out on a chemically bonded nonpolar stationary phase consisting of n-$C_{18}H_{38}$ groups, using 95% methanol–5% water as the mobile phase. Note that the stationary phase is much less polar than the mobile phase and is chosen for that reason because of the large nonpolar groups on the vitamins. As shown in Figure 21-2, vitamin D (calciferol, $C_{28}H_{43}OH$) is eluted last, because its steroid-type structure interacts more strongly with the n-$C_{18}H_{38}$ groups than does the nonpolar part of vitamin A acetate ($C_{19}H_{27}CH_2O_2CCH_3$), which consists mostly of alternating single and double carbon-carbon bonds. As shown in Figure 21-2, vitamin A acetate is eluted quite soon after the solvent peak, reflecting injection of the hexane-methanol extract onto the column. The more polar vitamin D is eluted a *relatively* long time after the less polar vitamin A acetate.

The peaks (Fig. 21-2) can be quantitated by establishing calibration plots constructed from peak heights or peak areas of the pure individual components eluted under the same conditions as the mixture.

Applications of Exclusion Column Chromatography

Exclusion chromatography is a special kind of column chromatography that can be utilized both at room pressure and at high pressure. Separations are based on a partial or complete exclusion according to size of molecules from a mobile phase by a stationary phase consisting of porous beads. The largest molecules are almost completely excluded from entering the porous beads and so pass through the column first. Inter-

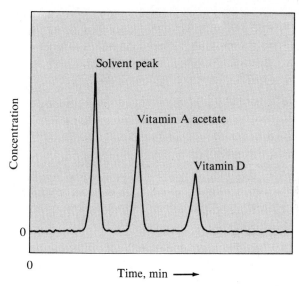

FIGURE 21-2. Hypothetical high-performance (pressure) liquid chromatographic separation of vitamins A and D on a Waters Micro Bondapak C_{18} column.

mediate-size molecules are able to enter only beads that have large pores, and thus pass through the column second. Small molecules can enter the pores of all of the beads and require the largest volume of solvent for elution, and therefore leave the column last.

When an aqueous mobile phase is used, exclusion chromatography is referred to as *gel filtration chromatography*; when an organic mobile phase is used, it is referred to as *gel permeation chromatography*. The latter can be performed with high-speed columns at high pressure, because the porous beads used with an organic mobile phase can withstand high pressures. The former must be done on columns with a low pressure drop, because the stationary phases used are very compressible.

Gel filtration chromatography is frequently used for *desalting* biochemical samples. After certain reactions are carried out, biochemists need to separate biochemical compounds from small inorganic salts or buffers. The mixture is poured onto a gel filtration column and eluted with pure water; the large biochemical compounds are excluded and pass through the column first, separated from the inorganic compounds. Gel permeation chromatography is frequently used for separating proteins of different molecular weights or for determining the range of molecular weights present in an organic polymer.

Applications of Thin-Layer Chroma-tography

Thin-layer chromatography is carried out using the same adsorbents employed in column chromatography (Table 21-1). The adsorbents are cemented to a plate (Fig. 19-4) with a calcium sulfate binder, but they still function either by adsorption or partition chromatography. Because thin-layer chromatography is easy to set up and requires about one-half hour on the average for a separation, it is frequently used for qualitative analysis. With a certain amount of care, it can also be used for quantitative estimation of certain compounds.

Qualitative
Analysis on
Thin-Layer
Plates

In general, selection of an adsorbent for thin-layer chromatography should be done so that the compounds to be separated exhibit the greatest differences in R_f values. As in Chapter 20, an R_f value is obtained by dividing the distance a sample spot moves by the distance the solvent front moves. For compounds of low polarity, the greatest differences in R_f values are obtained by using plates of silica gel or activated alumina and solvents of low polarity (Table 21-2). The separation is achieved by adsorption rather than partition chromatography. For compounds of high polarity, the greatest differences in R_f values are achieved by using plates with cellulose, nonactivated alumina, or silica gel [5].

Silica gel is used more extensively than any other adsorbent. It affords excellent separation of compounds of low polarity by adsorption; when coated with an appropriate stationary liquid phase, it also permits good separation of compounds of high polarity by partitioning. If acids are to be separated on silica gel, it is best to use a small amount of acetic acid in the eluting solvent. Bases are best separated on alumina, but if they are to be separated on silica gel, a small amount of ammonia or diethylamine should be added to the eluting solvent. Very polar molecules, such as amino acids and carbohydrates, should, of course, be separated on cellulose or kieselguhr rather than on silica gel.

A good example of a thin-layer separation for qualitative analysis is the separation of some of the B vitamins listed in Table 21-3 [5]. Two different solvents were used for elution—water and a mixture of 5% acetic acid, 5% acetone, and 20% methanol in benzene. Since the polarities of the B vitamins vary greatly, some good separations are possible. For example, thiamine (B_1) is a highly polar univalent cation, which is strongly retained by the adsorbent close to the original spot. Although cyanocobalamin (B_{12}) is also a univalent cation, its positive charge is somewhat shielded within a complex ion of cobalt(II), so it moves somewhat more than thiamine when water is used but the same when the benzene mixture is used. In contrast, pantothenic acid has a higher R_f value in the benzene mixture than in water. This is probably due to the combined effects of acetic acid, acetone, and methanol interacting with the many polar groups of this acid.

Quantitative
Analysis on
Thin-Layer
Plates

Quantitative estimation of compounds on a thin-layer plate can be carried out on the plate (in situ) or after the spots have been removed from the plate. In situ quantitation is performed using spectrophotometry, fluorometry, or radioisotope scanning. A special instrument called a photodensitometer, or densitometer, can be used for auto-

TABLE 21-3. Thin-Layer Chromatography of B Vitamins with Silica Gel Adsorbent [5]

B vitamin (Important polar group)	R_f value in Water	R_f value in Benzene mixture
B_1, Thiamine ($R-N^+Cl^-$)	0.05	0
B_{12}, Cyanocobalamin (Co^+PO^-)	0.22	0
B_2, Riboflavin ($3OH, 2C{=}O$)	0.42	0.35
Pantothenic acid ($COOH, -NH-C{=}O$)	0.60	0.89
Nicotonic acid ($COOH$)	0.76	0.75

matic scanning of the amount of light transmitted or reflected by colored substances on a plate. A special type of scanning fluorometer can be used for fluorescent compounds or for compounds on a fluorescent plate. Radioisotope scanning is used for carbon-14, phosphorus-32, or tritium (^3H) compounds. Quantitation after removal of spots from the plate can be done by cutting out an appropriate section of a plate, extracting the compound from the section, and measuring it spectrophotometrically.

An example of a thin-layer separation that probably involved both qualitative and quantitative analyses is the 1975 Florida oil spill investigation, in which oil samples from about 250 "suspect" ships were tested. Thin-layer chromatography, fluorometry, infrared spectrophotometry, and gas chromatography were used to analyze each oil sample [6]. The analyses conclusively proved that a Liberian tanker was the only ship that could have spilled the oil which blighted about fifty miles of Florida beaches.

Applications of Paper Chromatography

Paper chromatography is based on a partition of compounds between a stationary solvent held on the paper and the eluting solvent. It is not used as much as thin-layer chromatography because separations are much slower and spots are not as easy to detect. Paper chromatography finds its greatest uses in the separation of the same compounds as are best separated on cellulose thin-layer plates—amino acids, carbohydrates, and other biochemical compounds.

Selection of solvents for paper chromatography is made so that the greatest differences in R_f values are obtained for the compounds to be separated. If water is present, it is usually assumed that water is the stationary solvent held on the paper. Then polar compounds are adsorbed most strongly and have the lowest R_f values; compounds of low polarity have the highest R_f values.

A good example of paper chromatographic separation is the separation of amino acids on Whatman No. 1 paper with water-saturated phenol as the eluting solvent [7]. The R_f values for polar amino acids, such as aspartic acid and glutamic acid, would be predicted to be quite low, because they are diprotic carboxylic acids (such acids are highly soluble in water). As shown in Table 21-4, this is indeed the case. In contrast, amino acids with only one carboxylic acid group are not nearly as polar and thus are moved more readily by the water-saturated phenol eluting solvent. This is the case for alanine and glutamine, two amino acids that have amino groups of similar basicity. Evidently the amino group of proline interacts more strongly with the phenol eluting solvent than do the amino groups of alanine and glutamine, since proline is eluted before either of them.

TABLE 21-4. Paper Chromatography of Amino Acids

Amino acid	R_f value
Aspartic acid	0.14
Glutamic acid	0.24
Alanine	0.59
Glutamine	0.59
Proline	0.90

21-2 | APPLICATIONS OF ION-EXCHANGE CHROMATOGRAPHY

Ion-exchange chromatography is a separation based on a chemical reaction of ions with a stationary phase of ion-exchange resin. It can be performed in columns at atmospheric pressure or at high speeds using columns at high pressure (HPLC). It is used to separate various cations from one another and various anions from one another, using cation-exchange resin and anion-exchange resin, respectively.

Ion-Exchange Resins
Cation-exchange resins consist of a backbone of an organic polymer to which is attached an anionic functional group that will exchange cations. Some of the functional groups used on cation-exchange resins are the sulfonic acid group, $-SO_3^-$, the carboxylate group, $-CO_2^-$, and the phosphonate group, $-PO_3H^-$. The most commonly used resin is that which has the sulfonic acid group. It is used in the hydrogen or sodium form; the reaction of the hydrogen form with the sodium(I) ion is as follows:

$$Res-SO_3^-H^+ + Na^+ \longrightarrow Res-SO_3^-Na^+ + H^+ \qquad (21\text{-}1)$$

Anion-exchange resins consist of a backbone of an organic polymer to which is attached a cationic functional group that will exchange anions. Some of the functional groups used on anion-exchange resins are the quaternary ammonium group, $-NR_3^+$, and amine groups, such as $-NHR_2^+$. The quaternary ammonium resin is the most commonly used resin. It is used in the chloride form; the reaction of the chloride form with a nitrate anion is as follows:

$$Res-NR_3^+Cl^- + NO_3^- \longrightarrow Res-NR_3^+NO_3^- + Cl^- \qquad (21\text{-}2)$$

Eluting Solvents
The effectiveness of an eluting solvent in ion-exchange chromatography does not depend on polarity, as in conventional column chromatography, but on the ions dissolved in the eluting solvent. To separate various ions, they are first exchanged onto the ion-exchange resin. Then they must be displaced by an ion of the same charge at different rates, so that each ion in the mixture is completely eluted from the column before the next ion appears. The choice of the ion used in the eluting solvent is based on the *selectivity* of a particular resin for that ion in relation to the ions being separated. For example, suppose that it is desired to separate the Li^+, Na^+, and K^+ ions on a sulfonic acid resin. The best choice for an eluting ion is one that is preferred almost equally by the resin as compared to the ion least strongly held. Table 21-5 is a compilation of relative

**TABLE 21-5. Ion-Exchange Selectivities with 8%
Crosslinked Resin**

Ion	Relative equilibrium constant (25°C)
Li^+	1 (arbitrary standard)
H^+	1.27
Na^+	1.98
K^+	2.90
Mg^{+2}	3.29
Ca^{+2}	5.16
Sr^{+2}	6.51

equilibrium constants, which make "selectivity" more quantitative [8]. You can see that the best choice for the eluting ion is the hydrogen ion, in the form of a strong acid. This will elute lithium ion first, fairly rapidly, but will displace sodium ion and potassium ion more slowly, since these are more strongly held.

Applications of Ion-Exchange Chromatography

Ion-exchange chromatography can be used for qualitative analysis, but is mainly used for separations of ions prior to a quantitative analysis for each. The following generalizations are useful in predicting separations [9]:

1. In general, the more hydrated an ion is in solution, the smaller its selectivity on a given resin. For example, lithium ion is hydrated more than other univalent cations and has the lowest selectivity (Table 21-5).
2. Trivalent ions have higher selectivities than divalent or univalent ions, and divalent ions have higher selectivities than univalent ions.
3. The selectivity of a resin for an ion increases with increasing crosslinking of the organic polymer backbone of the resin.

The simplest example of ion-exchange chromatography is the separation of interfering anions from cations using an anion-exchange column, and vice versa. For example, iron(III) ions are seriously coprecipitated with barium sulfate during the precipitation for the determination of sulfate. To avoid this, the sample containing iron(III) ions and sulfate ions is passed through a cation-exchange column before precipitation, which removes the interfering iron(III) ions and allows the sulfate ions to pass through. Another simple example of ion-exchange chromatography is the

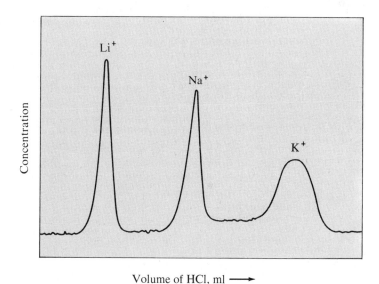

FIGURE 21-3. Hypothetical ion-exchange elution curve for the separation of three alkali metal ions using 0.7M hydrochloric acid on a sulfonic acid cation-exchange column.

separation of divalent or trivalent cations from univalent cations using a cation-exchange column. As can be seen from Table 21-5, cations such as strontium(II) and calcium(II) have much higher selectivities than any of the univalent cations listed. When a mixture of $+1$ and $+2$ ions is passed through a cation-exchange column, the univalent cations pass through rather quickly and the divalent cations are retained until all the univalent cations have been eluted.

A more difficult separation is the chromatographic separation of cations all having the same charge, such as the separation of the alkali metal cations. Fortunately, there are some fairly large differences in the hydration of these ions; lithium(I) ions are highly hydrated compared to sodium(I) and potassium(I) ions. The chromatographic separation of these three alkali metal ions can be done using $0.7M$ hydrochloric acid as the eluting solvent [10]. A hypothetical elution curve for the separation is shown in Figure 21-3. Since lithium ion has the lowest selectivity (Table 21-5), it is eluted first. The sodium ion is held about twice as strongly by the cation-exchange resin, so it is eluted late enough for the separation to be quantitative; the separation of potassium is also extremely good. If a divalent cation, such as magnesium(II), were present, it would be eluted much later than potassium ion without any interference.

21-3 | APPLICATIONS OF GAS CHROMATOGRAPHY

Gas chromatography is a separation technique usually based on the partitioning of a compound between a mobile vapor phase and a stationary liquid phase held on a solid support. (Strictly speaking, this is gas-liquid chromatography.) As you should recall from Figure 19-5, the gas flows through the support (coated with a stationary phase) in a heated column and finally through a detector. By choosing the proper stationary liquid phase it is possible to separate nearly all compounds with any significant volatility, excluding heat-sensitive compounds and most polymers.

Stationary Liquid Phases
Since the components in a sample must dissolve in the stationary phase to undergo partitioning, the effectiveness of a stationary phase depends on its ability to dissolve, and possibly interact with, each component. (It is, of course, necessary that the stationary phase be nonvolatile and stable at the temperatures used.) A useful rule of thumb is *like dissolves like*. If the stationary phase used is similar to the components being separated, then this allows the volatility of the components, as indicated by boiling point, to govern the order of the separation. Thus, a more volatile component should be eluted before a second component with a higher boiling point. Some recommended stationary phases [11] for various classes of organic compounds are listed in Table 21-6.

As a simple illustration of the use of Table 21-6, consider this situation. Suppose a mixture of benzene (b.p. $= 80.1°C$) and cyclohexane (b.p. $= 80.7°C$) is to be separated. A suitable stationary phase might be squalane, a long chain hydrocarbon that is saturated like cyclohexane rather than unsaturated like benzene. Cyclohexane would therefore be more soluble in squalane and would be eluted last (longer retention time) on the basis of its higher boiling point and higher solubility.

If dinonyl phthalate were chosen as a stationary phase instead of squalane, then benzene would be more soluble because the dinonyl phthalate is unsaturated like benzene. There is also undoubtedly a weak interaction between the electron-rich benzene

TABLE 21-6.   Recommended Stationary Phases [11]

Organic sample component	Recommended stationary phase
Aliphatic hydrocarbons	Squalane
Aromatic hydrocarbons	Dinonyl phthalate, Phenyl methyl silicone polymer
Alcohols	Dinonyl phthalate, Phenyl methyl silicone polymer
Aldehydes or ketones	Squalane
Acids	Dimethyl silicone polymer
Ethers, esters	Dinonyl phthalate
Amines	Polyethylene glycol
Alkyl halides	Dinonyl phthalate
Sulfur compounds	Dimethyl silicone polymer
Olefins	Silver nitrate + polyethylene glycol

and the electron-poor ester groups of the dinonyl phthalate that would further increase the solubility of benzene as compared to cyclohexane. Thus the cyclohexane would be eluted ahead of benzene, because the high relative solubility of benzene would outweigh its higher volatility. In other words, benzene would have a longer retention time.

Influence of Detector on Gas Chromatography Analysis

The type of detector (Table 21-7) available frequently places limitations on the analysis of complex samples. For example, the thermal conductivity detector measures the difference between the thermal conductivity of the carrier gas alone and that of the carrier gas containing the sample. It is suitable for general purposes but is not very sensitive to small amounts. Suppose that you must analyze for 10^{-3} μg amounts of an alcohol in a water solution. The thermal conductivity detector cannot detect such amounts, but the flame ionization detector can. Its output is proportional to the number of ions produced in burning the sample, which gives it a greater sensitivity than

TABLE 21-7.   Gas Chromatographic Detectors [11]

Characteristics of detectors	Thermal conductivity	Flame ionization	Electron capture
Minimum amount detected:	2–5 μg	10^{-5} μg	10^{-7} μg
Responds to:	All compounds	Only combustible compounds	Only compounds with electronegative atoms
Temperature limit:	450°C	400°C	225°C
Carrier gas:	Helium	Helium or nitrogen	Nitrogen or argon
Special applications:	Water	Water extracts	Halogen, phosphorus, sulfur, and $-NO_2$ compounds

the thermal conductivity detector. In addition, it has little or no response to water, so it detects the alcohol easily without "seeing" the relatively large amount of water present with the alcohol.

The electron capture detector has advantages over both the thermal conductivity and flame ionization detectors. Its response is based on measurement of a flow of "slow" electrons to an anode; reducible components such as halogen, phosphorus, sulfur, and nitro compounds react readily with these electrons and are detected at lower levels than with the other two types.

Qualitative Analysis Using Gas Chromatography

A gas chromatographic elution curve (Fig. 21-4) is a powerful tool for qualitative analysis because the number of peaks present generally indicates the number of components. However, the instrument does not identify the components giving rise to the peaks. This can be done by using the *retention time* of each component; the retention time is the number of minutes between injection of the sample and its entrance into the detector. Comparison of the retention time of an unknown component and that of several likely compounds on the same column under the same operating conditions frequently identifies the component. If this does not work, then the eluent from the column can be run through a spectrophotometer, such as an infrared spectrophotometer, for identification.

An example of a gas chromatographic qualitative analysis is the 1975 Florida oil spill investigation previously discussed [6]. Crude oil samples from about 250 "suspect" ships had to be matched to the crude oil that blighted Florida beaches. Since crude oil consists of a number of hydrocarbons, gas chromatographic elution curves, similar to that in Figure 21-4, of the oil from suspect ships were compared with the curve of the

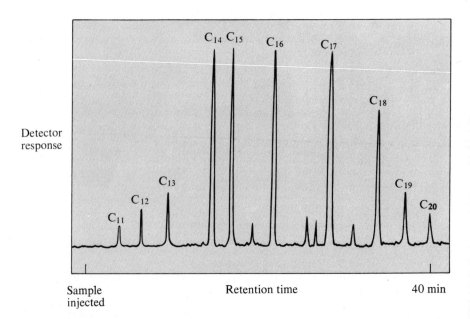

FIGURE 21-4. Gas chromatogram of a crude oil seepage sample using Apiezon L grease as stationary phase and a flame ionization detector.

crude oil on the beaches. The gass chromatographic evidence, along with thin-layer chromatographic evidence, was instrumental in proving that a Liberian tanker was guilty.

Quantitative Analysis   Under constant conditions (temperature, flow rate, etc.), the area under each gas chromatographic peak is proportional to the concentration of the component giving that peak. It is necessary to integrate the area under the peaks accurately. One method of doing this is to use electronic digital integration [12]. In this method, the peak is divided into a large number of segments or rectangles of equal width. Each segment (peak) height can be approximated by a voltage; in addition, the height of each segment is directly proportional to the area. Digital integration gives the sum of all the voltages, which is directly proportional to the peak area. Another way of integrating the area under chromatographic peaks involves using the ball and disc integrator [13]. The chromatographic output is fed into an electromechanical rotating disc. The disc rotations are plotted as up-and-down "sweeps" of a recorder pen. The number of such sweeps represents the peak area.

One method of quantitative analysis is to measure the peak areas of a number of standards of varying concentrations and to plot area vs. concentration; this is the so-called *calibration curve method*. Another method is the *internal standard method*, in which the calibration curve consists of a plot of the *ratios of the peak area* of a constant amount of a standard compound to the peak areas of different known amounts of the sample vs. the *weight ratios*. The unknown sample is then treated similarly and its weight read from the plot. This method avoids errors from unexpected experimental variations.

| QUESTIONS AND PROBLEMS

(Answers to most even-numbered problems are in Appendix 5.)

Concepts and Definitions

1. Explain the difference between adsorption and partition liquid chromatography.
2. Explain the difference(s) between:
   a. Silica gel and alumina adsorbents.
   b. Cellulose and silica gel adsorbents.
   c. Kieselguhr and alumina adsorbents.
3. In each solvent pair, decide which is the more polar solvent and which is the more nonpolar solvent.
   a. Ethanol and hexane
   b. Acetone and chloroform
   c. Propanol and water
   d. Cyclohexane and benzene
4. What advantage(s) does high-speed column chromatography have over column chromatography at atmospheric (room) pressure for:
   a. Qualitative analysis?
   b. Quantitative analysis?

5. In partition liquid chromatography, explain the difference between the stationary liquid phase and the mobile liquid phase.
6. Explain the three generalizations useful for predicting separations by ion-exchange chromatography.
7. Recommend a suitable stationary liquid phase for gas chromatographic separation of each pair of compounds below.
   a. Benzene and naphthalene
   b. Acetone and propionaldehyde
   c. Acetic acid and thiophenol
   d. Diethyl ether and ethyl acetate

Liquid Chromatographic Analysis

8. A mixture of salicylic acid and acetylsalicylic acid is separated by chromatography on a thin-layer plane. Under ultraviolet radiation, a blue fluorescent spot is detected at low R_f and a violet fluorescent spot is detected at high

R_f. Which compound has the low R_f and which the high R_f? (*Hint:* Look up their structures and see Table 21-2.)

9. Calculate the R_f value for each compound in parts a and b given that the solvent front has traveled 17.5 cm.
   a. Naphthalene spot at 7.9 cm
   b. Tetraphenylethylene spot at 3.5 cm
   c. Suppose that both compounds were chromatographed on a shorter plate where the solvent front travels 8.5 cm. Calculate where the spots of the two compounds should be located (use centimeters).

10. Suggest a suitable solvent for the elution of each of the following groups of compounds from a high-speed liquid chromatographic column.
    a. Phenol, *p*-bromophenol, and 2,4-dibromophenol
    b. Acetaldehyde, formaldehyde, and butyraldehyde
    c. Benzene, chlrorobenzene, and 2,4-dichlorobenzene

Ion-Exchange Chromatographic Analysis

11. In each of the following mixtures, predict the order of elution from a sulfonic acid cation-exchange column.
    a. Ca^{+2}, Mg^{+2}, Ba^{+2}
    b. Rb^+, Li^+, K^+, Na^+
    c. Mg^{+2}, K^+, Ca^{+2}, Li^+

12. Suggest a suitable eluting solvent for the separation of each of the following mixtures. (Do not use HCl for any of them because it elutes them too slowly.) Assume a sulfonic acid cation-exchange resin.
    a. K^+, Rb^+, Cs^+
    b. Ba^{+2}, Sr^{+2}, Ca^{+2}

Gas Chromatographic Analysis

13. The compounds below are not as volatile as desired for gas chromatographic separation. Suggest a derivative that might be used for each class of compounds to produce a more volatile, less hydrogen bonded type of compound.
    a. High molecular weight fatty acids
    b. High molecular weight alcohols
    c. Amino acids
    d. Carbohydrates

14. A sample is known to be a C_4H_9OH alcohol (a butanol). Write out the structures of the possible isomers (4) that the sample might be. Look up their boiling points and predict the order of their elution from a column with a dinonyl phthalate stationary liquid coating.

15. Suggest the appropriate detector for the analysis of the first component in the presence of the other components in each of the following mixtures:
    a. CH_3OH with H_2O and CH_3CCH_3.

$$\overset{\|}{O}$$

    b. CH_3CCH_3 $(10^{-2}\mu g)$ with CH_3CHO and

$$\overset{\|}{O}$$

       $CH_3CH(OH)CH_3$.
    c. CS_2 $(10^{-2}\mu g)$ with CH_3OH and C_4H_9OH.

| NOTES

[1] P. A. Plack, S. K. Kon, and S. Y. Thompson, *Biochem J.* **71**, 467 (1959).

[2] J. S. Fritz and D. H. Schmitt, *Talanta* **13**, 123 (1966).

[3] A. G. Hamlin, B. J. Roberts, W. Laughlin, and S. G. Walker, *Anal. Chem.* **33**, 1547 (1961).

[4] "Analysis of Pharmaceutical Products," Waters Associates Bulletin AN 138, Milford, Mass., Dec., 1973.

[5] V. W. Rodwell, *Thin-Layer Chromatography*, American Chemical Society (ACS Audio Course), Washington, D.C., 1975.

[6] *Chem. & Eng. News* **53**, 7 (Nov. 17, 1975).

[7] G. M. Price, *Biochem J.* **80**, 420 (1961).

[8] L. Meites, *Handbook of Analytical Chemistry*, McGraw-Hill, New York, 1963, pp. 10–159.

[9] H. F. Walton, *Principles and Methods of Chemical Analysis*, Prentice-Hall, Englewood Cliffs, N.J., 1964, pp. 138–59.

[10] W. Riemann, *Rec. Chem. Progr.* **15**, 85 (1954).

[11] J. M. Bobbitt, A. E. Schwarting, and R. J. Gritter, *Introduction to Chromatography*, Reinhold, New York, 1968, pp. 106–143.

[12] H. L. Pardue, M. F. Burke, and J. R. Barnes, *J. Chem. Educ.* **44**, 695 (1967).

[13] J. R. Barnes and H. L. Pardue, *Anal. Chem.* **38**, 156 (1966).

22 | Clinical Analysis

"*All knowledge* comes useful to the detective,"
remarked Holmes. "Even the trivial fact that in the year
1865 a picture of Greuze entitled La Jeune Fille
a l'Agneau fetched. . . more than forty thousand
pounds may start a train of reflection in your
mind."

ARTHUR CONAN DOYLE
The Valley of Fear, Chapter 2

Even as all knowledge is useful to the detective, so all chemical knowledge is useful to the clinical chemist. Throughout this chapter are a number of references to previous chapters, which verify this statement. Clinical chemists use all the methods of the analytical chemist plus some of their own to characterize and analyze medical samples.

Since most analytical chemists are more familiar with elements, we will begin the chapter by discussing the elements in the body. Naturally the elements exist as free or combined ions, so we will describe the major ions, or major electrolytes, as well as ions present in trace concentrations. We will also touch on the interpretation of electrolyte levels.

The main discussion centers on clinical analysis for selected components in the blood. Both the sampling of the blood and the determination of various compounds and substances in the blood will be discussed. Rather than discussing the determination of substances in other parts of the body, we concentrate on the analysis for common substances in the blood. We think that the lengthy discussion of the determination of glucose is of both value and interest. More extensive presentations of clinical analysis are available [1].

22-1 | ELEMENTS IN THE BODY

Nearly all the elements exist in ionic form in the body. The major ions in body fluids are termed the major electrolytes, whereas the minor ions are frequently labeled the trace elements, or trace metal ions. We will discuss the major electrolytes first.

Electrolytes in the Serum and Urine

The major serum electrolytes are Na^+, K^+, Ca^{+2}, Mg^{+2}, Cl^-, and HCO_3^-. Typical levels of the major electrolytes in the blood serum and in urine are listed in Table 22-1. These levels are given in both ppm [2] and molarity [3]. All of these ions are important for maintaining electrical neutrality in the blood, among other things. A loss of bicarbonate ion as carbon dioxide must be balanced by the movement of an equivalent amount of one of the cations out of that part of the bloodstream. Note from Table 22-1 that the molarity $(0.14M)$ of the sodium ion in blood is nearly balanced by the sum of the molarities $(0.13M)$ of the chloride ion and the bicarbonate ion. Unless pathological variations occur, the sum of the concentrations of the latter two ions is generally $0.01M$ less than that of the sodium ion [3].

Analysis for the major serum electrolytes is important because it indicates the nature of the electrolyte balance, rather than any toxic or lethal levels. (These levels will be important when the trace metals are considered.) Alkali (Group I) metal ions, such as sodium and potassium, are routinely determined by flame emission spectrometry

TABLE 22-1. Levels of the Elements in Blood Serum and Urine

Ionic form, or element	Blood serum level, ppm	Urine, mg/day
Alkali and alkaline earth metal ions [2, 3]		
Li^+	0.01 ppm	—
Na^+	3200 (0.14M)	1000–5000
K^+	120–214 (0.005M)	—
Mg^{+2}	43 (0.001M)	60–120
Ca^{+2}	90–110 (0.003M)	96–800
Sr^{+2}	—	0.4
Other metal ions [2]		
Al^{+3}	0.13–0.17 ppm	0.05
As(III) & (V)	0.04–0.2	$\geq$ 0.1 ppm
Cd^{+2}	0.0033	0.002–0.02 ppm
Cu^{+2}	1.05	—
Fe(II) & (III)	1.25 (men) 0.90 (women)	0.1–0.3
Pb^{+2}	0.3–0.4 (whole blood)	0.01–0.07
Anions [3]		
Cl^-	0.10M	—
HCO_3^-	0.03M	—
HPO_4^{-2}	0.002M	—
SO_4^{-2}	0.001M	—

(Sec. 15-2); they can also be measured by atomic absorption spectrometry (Sec. 15-3). On automated clinical analyzers (Sec. 13-4), sodium and potassium ion are nearly always measured by flame emission spectrometry.

There are many good methods for the determination of calcium and magnesium ions. They can be determined by titration with EDTA (Sec. 8-2), by atomic absorption spectrometry (Sec. 15-3), by flame emission spectrometry (Sec. 15-2), or by colorimetric analysis. On some automated clinical analyzers, an organic complexing dye is added to form a colored calcium(II) chelate which is measured colorimetrically.

The chloride ion can also be determined by a variety of methods—precipitation titration (Sec. 8-1), specific-ion electrode (Sec. 16-3), and coulometric measurement (Sec. 17-3). The bicarbonate ion is titrated as a base with standard acid or converted to carbon dioxide and measured as a gas.

One other measurement that is important in blood analysis is the measurement of blood pH. The pH of normal arterial blood is surprisingly constant, ranging between about 7.38 and 7.42 [2]. Blood pH is measured with the pH meter (Sec. 16-2) and is an important factor in indicating general disorders when used with electrolyte analyses. Table 22-2 gives a summary of four electrolyte patterns and their diagnostic indications.

Although the electrolyte analyses in Table 22-2 are interrelated, sometimes a single result rules out a certain physiological condition. For example, a normal serum pH value tends to rule out acidosis or alkalosis. On the other hand, more than an abnormal pH value must be observed to verify whether acidosis is metabolic or respiratory in nature. Metabolic acidosis is characterized by a low pH, a low chloride level, and a low level of carbon dioxide plus bicarbonate ion. In metabolic acidosis, a body malfunction causes an increase in the level of acids, such as ketonic acids, and the bicarbonate level falls because the bicarbonate ion reacts with part of the excess acid to attempt to restore a normal pH as follows:

$$HCO_3^- + \text{excess } H^+ \longrightarrow H_2CO_3 \text{ (exhaled as } CO_2) \qquad \textbf{(22-1)}$$

In repiratory acidosis, an increase in the carbon dioxide levels releases extra hydrogen ions through ionization as follows:

$$CO_2(g) + H_2O \longrightarrow H^+ + HCO_3^- \qquad \textbf{(22-2)}$$

In contrast, respiratory alkalosis is characterized by a high pH and a low level of carbon dioxide plus bicarbonate ion. In this condition a person is exhaling too much carbon

TABLE 22-2. Some Serum Electrolyte Patterns

	pH	$[Na^+]$	$[K^+]$	$[Cl^-]$	$[HCO_3^- + CO_2]$	Diagnostic indication
1.	N[a]	High	N	High	N	Probable dehydration
2.	Low	N	Any	Low	Low	Probable metabolic acidosis
3.	Low	N	N or high	Low	High	Probable respiratory acidosis
4.	High	N	N or low	N	Low	Probable respiratory tory alkalosis

[a] N = Normal.

dioxide; this plus a secondary drop in bicarbonate level causes the pH to rise above normal (the reverse of **22-1**).

(In the well-known science fiction novel, *The Andromeda Strain*, author Michael Crichton narrates a situation in which a deadly strain of bacteria from outer space is able to cause fatal coagulation of the blood within a narrow pH range of 7.39 to 7.43. The story hinges on the scientists' finally recognizing that an acidotic old man and an alkalotic, crying baby survived because their blood pHs were outside the pH range in which the strain was fatal.

Electrolytes in the Urine

The diagnostic significance of chloride, sodium, and potassium ion levels in the urine is a complicated problem, since electrolyte levels in the urine are influenced by a large number of factors [3]. The amounts excreted by the body over a twenty-four-hour period are more important than the concentrations of random samples; it is for this reason that Table 22-1 reports urinary electrolyte levels in milligrams per day rather than as concentrations.

Trace Elements

A number of trace elements in body fluids or in the tissues are *essential* to life at low levels [1]; at higher levels some of these elements are toxic or lethal. There are also a number of elements that may yet be shown to have an essential function at low levels, but at present are only known to be poisonous at higher levels. Finally, there are elements, such as iron, that perform a useful function at higher than trace levels. The essential elements function by activating or deactivating enzymes, or by controlling vitamin or hormone action. Iron, of course, is utilized in hemoglobin for binding oxygen. The so-called poisonous elements affect the brain or otherwise impair a vital body function.

Analysis for most trace metals is not done routinely in the clinical laboratory, but can be performed by flame emission or atomic absorption spectrometry. Analysis for mercury is best done by flameless atomic absorption spectrometry (Sec. 15-3). The analysis for poisonous elements is vital in situations where a poisoning is suspected. Lead, arsenic, mercury, and thallium are commonly implicated as poisons.

Lead poisoning is common among children who live in dwellings where lead-based paints are still present, and therefore analysis for lead(II) ion in as small a sample of blood as possible is important. Unfortunately, atomic absorption analysis requires a relatively large sample of blood so that lead(II) ion can be extracted away from inter-ferences before it is measured. A recent fluorometric method appears to be more suitable, because only a drop of blood is required for analysis [4]. The drop of blood is placed on a glass slide and inserted in a fluorometer. The sample is excited at 424 nm and fluoresces at 625 nm in the red region. The fluorescence originates from a zinc protoporphyrin compound produced in the blood by the action of the lead(II) ion. Toxic levels of 0.07 mg of lead per 100 ml are readily detected by the method (Sec. 14-1).

22-2 | CLINICAL ANALYSIS OF BLOOD

Blood is analyzed for substances other than electrolytes in the clinical laboratory; it is analyzed for the levels of certain compounds, for total protein, for certain enzymes, and for elements such as protein-bound iodine. Before discussing selected analyses, we will discuss sampling.

Sampling of Blood

Samples of blood are frequently taken after a patient has fasted for a certain time—for example, before breakfast. This is particularly important for glucose analysis, but not as important for analysis of other species. Since only part of the blood is sometimes analyzed, it is important for you to understand the composition of whole blood. This can be grasped easily from the following diagram:

Whole Blood

Plasma	Cells
Serum + Fibrinogen	Platelets + Erythrocytes + Leukocytes

The liquid portion of blood is the plasma; the majority of analyses are performed on the serum, which is plasma from which the fibrinogen has been removed.

When the blood is collected, the sample is drawn as quickly as possible using syringes that are chemically clean and dry to avoid contaminating or coagulating the blood. The container for the blood sample should also be clean and dry for the same reasons. The sample is taken to the laboratory immediately for analysis on the same day the sample is collected.

The above procedure is followed where serum is to be analyzed. If plasma or whole blood is required for analysis, then the blood is put into a container having an anti-coagulant in it. Potassium oxalate is one such anticoagulant; the oxalate ion precipitates blood calcium:

$$C_2O_4^{-2} + Ca^{+2} \longrightarrow CaC_2O_4(s)$$

preventing the calcium ion from initiating the clotting mechanism. Another anti-coagulant is EDTA (Sec. 8-2), which chelates the calcium ion and has the same effect as the oxalate ion.

In the laboratory, serum is obtained from whole blood by allowing the blood to clot. Then the serum is separated from the cells and the fibrinogen by centrifuging the solid matter down, leaving the clear serum in liquid form. Centrifuging is done as quickly as possible to avoid *hemolysis*—the destruction of the red blood cells, which releases hemoglobin and other substances into the serum or plasma. (Since the red blood cells contain larger amounts of potassium, iron, magnesium, urea, etc., if hemolysis has taken place, the analysis of the serum gives incorrect high results.) The sample of serum is then ready for analysis.

Special techniques are needed if the sample cannot be analyzed on the same day that it is taken [3], but these are beyond the scope of this text.

Substances Determined in the Blood

Of the possible substances in the blood that can be determined, only a dozen or so are routinely determined. These include electrolytes, such as sodium, potassium, calcium, magnesium, chloride, and bicarbonate ions, as previously discussed. Other ions that are determined are phosphate (as inorganic phosphate), iron, and iodine (as protein-bound iodine or PBI). Compounds that are often determined include blood urea nitrogen (BUN), glucose, carbon dioxide, and uric acid. Other substances commonly determined include enzymes such as alkaline phosphatase and serum glutamic-oxalacetic transaminase (SGOT), total protein, albumin, and total bilirubin. Table 22-3 lists the normal ranges for some of these substances. The normal ranges listed in this table cover only the great majority of people, not all people. For example, Figure 22-1 shows that the serum chloride of some people falls below 98 mmole/liter or above 108 mmole/liter [5].

TABLE 22-3. Normal Ranges of Substances in the Blood

Substance determined	Normal range in 100 ml of serum
Albumin	3.8–5.0 g
Bilirubin	Up to 0.4 mg (direct)
	Up to 1.0 mg (total)
Calcium	2.25–2.75 mmole/liter
Carbon dioxide	25–32 meq/liter
Chloride, serum	98–108 mmole/liter
Cholesterol, total	100–350 mg
Iron, serum	0.050–0.180 mg
Magnesium	0.75–1.25 mmole/liter
Phosphatase, alkaline	Up to 4 Bodansky units
Phosphorus, inorganic	3–4.5 mg
Potassium	3.8–5.0 mmole/liter
Protein, total serum	6.5–8 g
Portein-bound iodine (PBI)	0.0035–0.008 mg
Sodium	138–146 mmole/liter
Sugar (glucose)	65–90 mg
Transaminase (SGOT)	Up to 40 units
Urea nitrogen (BUN)	Up to 20 mg
Uric acid	3–6 mg

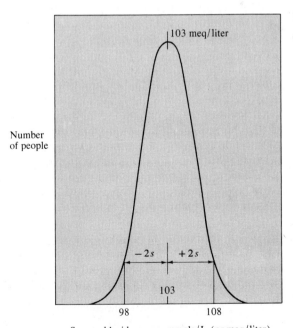

FIGURE 22-1. Frequency distribution of serum chloride levels of a particular population of people [5]. The normal range of chloride is 98–108 mmole/liter (meq/liter).

The determination of serum electrolytes has already been discussed in Section 22-1. Of the remaining substances in Table 22-3, most are determined spectrophotometrically (colorimetrically). In some cases, a digestion is necessary to convert organic material to ionic material before the determination can be carried out. For example, in the determination of protein-bound iodine (PBI), the organic iodine compounds are decomposed to iodide ion and carbon dioxide first. This is followed by addition of an aqueous solution containing the iodide ion to a cerium(IV)-arsenic(III) mixture (Sec. 13-4). The iodide ion catalyzes the oxidation-reduction reaction of these two ions.

$$2\,Ce(IV) + As(III) \xrightarrow{\;I^-\;} 2\,Ce(III) + As(V)$$
(yellow)

The disappearance of the yellow color of cerium(IV) is followed spectrophotometrically and related to the concentration of the iodide.

Determination of Serum Glucose

Since the determination of serum glucose is the analysis most frequently done in the clinical laboratory, we will discuss the various methods for glucose (Table 22-4) in some detail. Each of the methods in Table 22-4 has advantages and disadvantages, so there is not always an overwhelming preference for one method in any area. For example, the method used in the largest number of clinical labs in Connecticut in the 1970s was used by less than a majority (43%) of these labs [5]. We will discuss this method first.

o-Toluidine Method for Glucose

The o-toluidine method depends on the fact that glucose possesses a very reactive aldehyde carbonyl group compared to other sugars in body fluids. Addition of o-toluidine, an aromatic primary amine, to glucose is followed by a condensation reaction of the primary amino group of o-toluidine ($R-NH_2$) with the carbonyl group of glucose.

$$R-NH_2 + C_5H_{11}O_5-\underset{\underset{H}{|}}{C}=O \xrightarrow{\text{heat}} C_5H_{11}O_5-\underset{\underset{H}{|}}{C}=NR + H_2O \quad (22\text{-}3)$$

(glucose)   (Schiff base)

Fructose and sucrose do not interfere significantly because they do not react at the same rate as glucose. Fructose is a ketone sugar; carbonyl groups of ketone sugars (ketoses) are less reactive than carbonyl groups of aldehyde sugars (aldoses).

The reaction produces a Schiff base which has an absorption band at 630–40 nm in the red region. The absorption band covers the yellow, orange, and red regions, so the eye perceives a green color (Table 12-2).

TABLE 22-4. Some Methods for Serum Glucose Determination

Reaction	Reagent, wavelength
Condensation with o-toluidine	$CH_3C_6H_4NH_2$, 630–40 nm
Oxidation by ferricyanide	$Fe(CN)_6^{-3}$, 420 nm
Oxidation with copper(II)-neocuproine	Cu(I) neocuproine, 460 nm
Reaction with oxygen using glucose oxidase enzyme	O_2, glucose oxidase (polarographic)

Although glucose is more reactive than other sugars toward o-toluidine, the reaction in **22-3** does require a short heating period. Directions for the determination are given in Part Four, Experiment 13.

Ferricyanide Method for Glucose

The ferricyanide method is based on the oxidation of glucose to gluconic acid by ferricyanide.

$$C_5H_{11}O_5-CHO + 2Fe(CN)_6^{-3} + H_2O$$

$$\xrightarrow{95°} C_5H_{11}O_5-CO_2H + 2Fe(CN)_6^{-5} + 2H^+$$

Note that the aldehyde group of glucose is oxidized by a carboxylic acid group in a two-electron oxidation step (Sec. 11-4). This method has been criticized because it is empirical and lacks specificity, but it has been claimed to be the most reliable method for evaluating serum glucose [6].

The ferricyanide ion is an intense yellow color, so the reaction can be followed by measurement of the loss of absorption at 420 nm. Although the ferrocyanide ion also absorbs in that region, its molar absorptivity is only about 1.0 as compared to a molar absorptivity of the order of 10^3 for ferricyanide. Thus the absorption of the ferrocyanide ion is negligible. This reaction has been used in the automated determination of glucose (Fig. 13-10).

Copper(II)-Neocuproine Method for Glucose

The copper(II)-neocuproine method is also based on the oxidation of glucose to gluconic acid, as in the previous method. The reagent used is a mixture of copper(II) ion and neocuproine, which is the trivial name for 2,9-dimethyl-1,10-phenanthroline.

Once copper(II) has been reduced to copper(I) by glucose, it reacts with neocuproine.

$$Cu(I) + 2C_{14}H_{12}N_2 \longrightarrow Cu(C_{14}H_{12}N_2)_2^+$$

In this complex ion, the copper(I) is chelated to each neocuproine through the two pairs of electrons on the two nitrogen atoms of each neocuproine, giving it a coordination number of four.

The complex ion, or chelate, has a strong absorption band at about 460 nm with a molar absorptivity of over 7.9×10^3 [7]. Because the complex is so stable and so intensely colored, this reaction has also been used in the automated determination of glucose (Fig. 13-10).

Enzyme Kinetics and Determination of Glucose

Since glucose can also be determined with an enzyme (glucose oxidase), we will first review briefly the kinetics of enzyme-catalyzed reactions as background for the determination.

The simplest enzyme reaction involves a reaction between an enzyme, E, and a substrate, S, to produce a temporary complex, ES, which then reacts further to give a product, P, and regenerate the original enzyme.

$$E + S \underset{k_2}{\overset{k_1}{\rightleftharpoons}} ES \xrightarrow{k_3} P + E \qquad (22\text{-}4)$$

In the above reaction, k_1 is the rate constant for the reaction of E and S to produce ES, k_2 is the rate constant for the reverse reaction in which ES decomposes back to E and S, and k_3 is the rate constant for the decomposition of ES to product.

The rate of production of P from E and S is symbolized as $rate_p$; by a specialized kinetic derivation, it can be shown that when equilibrium is reached (the concentration of ES has reached a constant value at this point also), the $rate_p$ is given by the following:

$$rate_p = k_3[E_0]\frac{[S]}{K_m + [S]} \tag{22-5}$$

where $[E_0]$ is the enzyme concentration at the start of the reaction, $[S]$ is the concentration of the substrate at any time, and K_m is related to the rate constants.

$$K_m = \frac{k_2 + k_3}{k_1}$$

The K_m term is also known as the Michaelis constant. The $k_3[E_0]$ product is usually designated as V_{max}, the maximum obtainable rate under specified conditions for a reaction catalyzed by a particular enzyme. Substituting V_{max} for this product in **22-5** gives the following:

$$rate_p = V_{max}\frac{[S]}{K_m + [S]} \tag{22-6}$$

Since an enzymatic determination of S is carried out by measuring $rate_p$, the measurement of this rate must be performed at a concentration of S at which the rate is directly proportional to $[S]$. Since such concentrations of S are quite low, K_m will be much larger than $[S]$, and **22-6** is simplified.

$$rate_p = V_{max}\frac{[S]}{K_m} \tag{22-7}$$

Equation **22-7** indicates that the rate of the enzyme-catalyzed reaction under these conditions depends directly on only the concentration of the substrate.

The *enzymatic determination of glucose* is in fact performed under conditions where **22-7** is valid. In this method, glucose oxidase enzyme is added to a solution of glucose containing an equilibrium concentration of dissolved oxygen. The glucose oxidase catalyzes the oxidation of the glucose by the dissolved oxygen gas.

$$C_5H_{11}O_5\text{—CHO} + H_2O + O_2(g) \xrightarrow{\text{glucose oxidase}} C_5H_{11}O_5\text{—CO}_2H + H_2O_2 \tag{22-8}$$

In this reaction, the rate of formation of product ($rate_p$) is equivalent to the rate of decrease in concentration of dissolved oxygen ($-rate_{O_2}$) so from **22-7**,

$$-rate_{O_2} = rate_p = V_{max}\frac{[\text{glucose}]}{K_m} \tag{22-9}$$

If the rate of decrease in dissolved oxygen concentration is measured during the early part of the reaction, it will be a measure of [glucose], the initial concentration of glucose.

A commercial *glucose analyzer* instrument has been designed to measure glucose by using the relationship in **22-9**. It utilizes a polarographic measurement (Sec. 17-4) of the dissolved oxygen by reduction at the cathode.

$$\tfrac{1}{2}O_2 + 2H^+ + 2e^- = H_2O$$

The instrument is calibrated using standard glucose solutions containing, for example, 150 mg and 300 mg of glucose per 100 ml. The instrumental readout after 10 seconds of reduction is set to read exactly 150 mg, and then 300 mg, of glucose. Then the electrodes are rinsed and inserted in the glucose samples, one at a time. After 10 seconds of reduction of the oxygen in each, the readout from the instrumental scale is recorded directly as milligrams of glucose per 100 ml.

Because the enzymatic determination of glucose is specific for it, the enzymatic method is said to be the only method to give a true measure of glucose. Other methods give slightly high results for glucose because of interferences.

Medical Significance of Glucose Analyses

Some of the dysfunctions associated with low and high serum glucose values are listed in Table 22-5 [5]. The physician may diagnose a high glucose level as *hyperglycemic* and a low glucose level as *hypoglycemic*. Diabetes is the most common condition associated with the hyperglycemic glucose levels. Lately, it has been recognized that hypoglycemia is also fairly common; people who are afflicted with it attempt to raise their blood sugar by eating sweets, which unfortunately enhances body consumption of glucose to the point that blood sugar is lowered rather than raised.

The Glucose Tolerance Test

The glucose tolerance test has already been described in Section 4-1. This test is used to detect both hyperglycemia and hypoglycemia by measurement of the glucose level over a period of time after ingestion of a standard amount of glucose. Figure 22-2 shows the tolerance curves for a normal person and a diabetic person.

Determination of Blood Urea Nitrogen (BUN)

Another substance commonly determined in the blood is urea. One method for the determination of urea is based on an enzyme-catalyzed reaction.

$$NH_2CONH_2 + H_2O \xrightarrow{\text{urease}} 2NH_3 + CO_2$$

The ammonia produced in the reaction can be determined with Nessler's reagent; the product is measured spectrophotometrically at 480 nm.

A high BUN value is found in cases of kidney malfunction, carbon tetrachloride or mercuric chloride poisoning, and heart attack. A low BUN value is often found with acute hepatic insufficiency and in normal pregnancy.

TABLE 22-5. Dysfunctions Associated with Low and High Glucose Levels

Hypoglycemic dysfunctions (low glucose levels)	Hyperglycemic dysfunctions (high glucose levels)
Postprandial causes	Endocrine
Tumor of the pancreas	Metabolic
Glucagon deficiency	Pancreatic
Liver disease	Ineffective insulin
	Stress

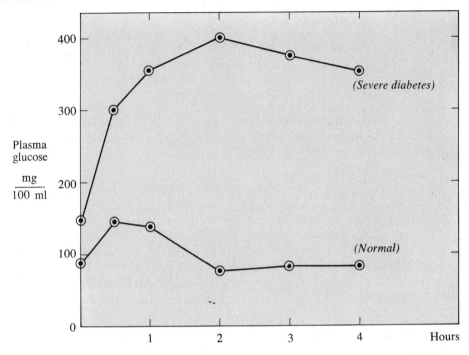

FIGURE 22-2. Glucose tolerance curves for a normal individual (lower curve) and a diabetic individual (upper curve).

| QUESTIONS AND PROBLEMS

(Answers to even-numbered problems are in Appendix 5.)

Definitions and Concepts

1. Define clinical analysis.
2. What is the difference between a major electrolyte ion and a trace metal ion?
3. Indicate whether each of the following groups of analyses gives information on the serum electrolyte pattern or on possible toxicity levels:
   a. Determination of serum iron, copper, and lead.
   b. Determination of serum chloride, potassium, and sodium.
   c. Determination of serum pH, bicarbonate, calcium, and magnesium.
4. What is the difference in electrolyte levels between:

   a. Respiratory alkalosis and respiratory acidosis?
   b. Respiratory alkalosis and metabolic acidosis?
   c. Dehydration and respiratory alkalosis?
5. List several of the important compounds and elements determined in a clinical blood serum sample.
6. Compare the EDTA titration method with the atomic absorption spectrometric method for the determination of calcium and magnesium in blood serum.

Elements in the Body

7. Analysis gives ppm of the following elements in

a blood serum sample. Convert each to molarity.

   a. 3050 ppm sodium ion

   b. 200 ppm potassium ion

   c. 100 ppm calcium ion

8. Decide whether any of the results in each set below should be checked by further analysis by comparing the sum of the chloride and bicarbonate with the molarity of the sodium ion, and by comparing values with those in Table 22-1.

   a. Na^+: 144 mmole/liter;
   HCO_3^-: 29 mmole/liter;
   Cl^-: 112 mmole/liter

   b. Na^+: 140 mmole/liter;
   HCO_3^-: 30 mmole/liter;
   Cl^-: 101 mmole/liter

   c. Na^+: 136 mmole/liter;
   HCO_3^-: 29 mmole/liter;
   Cl^-: 101 mmole/liter

9. Indicate whether or not the following blood samples contain toxic amounts of lead(II) ion:

   a. 0.05 mg of lead per 100 ml.

   b. 0.08 mg of lead per liter.

   c. 0.8 ppm of lead.

   d. $1.0 \times 10^{-5} M$ lead.

Clinical Analysis of Blood

10. Describe what relationship each of the following substances has to whole blood:

   a. Plasma.

   b. Platelets.

   c. Serum.

   d. Fibrinogen.

11. Indicate whether the serum chloride values given are high or low, and whether they could possibly be diagnosed by a physician as abnormal.

   a. 94 mg/100 ml

   b. 99 mg/100 ml

   c. 109 mg/100 ml

   d. 120 mg/100 ml

12. State whether the serum glucose values given could be diagnosed as hypoglycemia or hyperglycemia and whether the dysfunction is serious.

   a. 60 mg/100 ml

   b. 50 mg/100 ml

   c. 400 mg/100 ml

   d. 95 mg/100 ml

13. What effect would a high concentration of glucose have on **22-6** in terms of possible enzymatic determination of glucose?

| NOTES

[1] R. J. Henry, D. C. Cannon, and J. W. Winkelman, *Clinical Chemistry: Principles and Techniques*, Harper and Row, New York, 1974.

[2] G. D. Christian, *Anal. Chem.* **41** 24A (Jan., 1969).

[3] J. S. Annino, *Clinical Chemistry*, Little, Brown, Boston, 1964, pp. 122–31.

[4] *Chem. & Eng. News* **53**, 18 (Feb. 3, 1975).

[5] J. C. MacDonald, *Amer. Lab.* **7**, 61 (July, 1975).

[6] B. R. Sant, *Talanta* **3**, 261 (1960).

[7] H. Diehl and G. F. Smith, *The Copper Reagents: Cuproine, Neocuproine, Bathocuproine*, G. F. Smith Chemical Co., Columbus, Ohio, 1958.

PART FOUR | Selected Experiments

We include here a number of selected experiments to be used in connection with this text. The theory for essentially all of these experiments has been covered in Parts One, Two, and Three, so there will be little discussion preceding each experiment. You should refer to the relevant material in the first three parts of the book for background.

To begin with, you should acquire a notebook of sufficient size to record all your laboratory data. General directions for the proper recording of data have been given in Section 3-5. Your instructor will probably wish to supplement this with additional directions.

The first experiment in the laboratory might very well be an experiment involving a simple weighing. We have included such an experiment at the beginning of this part.

There are three experiments in this part concerned with the determination of the chloride ion by three different methods—gravimetric, titrimetric, and density determination. These three experiments provide a useful comparison of the accuracy and precision obtainable by three different experimental approaches. The experiments are written to accommodate either a liquid unknown or a solid unknown.

Experiment 1 | TREATMENT OF DATA FROM WEIGHING

This experiment illustrates the statistical treatment of data from the weighing of ten nearly identical objects, such as ten coins or ten pencils. Assume that all ten objects have been made by the same machine at the same time and that they constitute a representative sample. Review Chapter 3 before making the calculations.

Procedure
1. Obtain ten identical objects to weigh from your instructor and record the sample number.
2. Check and, if necessary, adjust the zero point of the balance before making any weighings. This is necessary because the weighings are absolute.
3. Weigh each object to the nearest 0.1 mg and record each weight in your notebook.
4. Calculate the range of your weighings and use the Q test (Ch. 3) to see whether any result should be discarded because of a gross error.
5. Calculate the mean, the absolute and relative values of the average deviation, and the absolute and relative values of the standard deviation.
6. Report all data in tabular form and indicate the result of the application of the Q test to the most suspicious weighing of the ten weighings.

Experiment 2 | USE OF THE VOLUMETRIC PIPET

The objectives of this experiment are to acquaint you with the proper procedure in using a 25-ml volumetric pipet and to check the precision of your measurement of a 25-ml volume three times with the same pipet. If your pipetting technique is reproducible, the weight of 25 ml of distilled water that you measure with the pipet should agree each time within the limits of the tolerance for the 25-ml pipet.

Procedure
1. Weigh three weighing bottles to the nearest 0.01 g. (It is not necessary to weigh them more accurately because you will be checking the weight of water delivered from a pipet to within ± 0.01 g.)
2. Rinse a 25-ml pipet with distilled water and allow it to drain to check for cleanliness. If drops cling to the inside of the pipet, clean as directed in Section 6-3.
3. Fill a 100-ml beaker with distilled water to use in pipetting.
4. Obtain a rubber suction bulb to fill the pipet. Review steps 1 to 5 in Section 6-3 for using the pipet. Then using the rubber bulb, fill the pipet to about an inch above the etched line. Dry the outside of the pipet with a tissue. Then place the tip of the pipet against the wall of the beaker and drain it carefully until the meniscus just touches the etched line. Then place the tip of the pipet against the inside of one of the weighing bottles and allow the water to drain. Keep the tip against the side of the bottle for 20 seconds after the pipet has emptied for complete drainage. Repeat the process two more times using the other two weighing bottles.
5. Weigh each weighing bottle again to obtain the weight of water by difference to the nearest 0.01 g.
6. Report the weight of water delivered for all three times. Compare the difference in weights with the tolerance of the 25-ml pipet (Table 6-1) to see whether or not your precision is acceptable.

Experiment 3 | THREE METHODS FOR THE DETERMINATION OF CHLORIDE

This experiment is actually three different experiments; that is, it consists of three different methods for the determination of chloride in the form of sodium chloride. These methods are 3A, the determination of chloride by density measurement; 3B, the gravimetric determination of chloride; and 3C, the volumetric determination of chloride.

Each experiment can be performed on a liquid sodium chloride unknown containing 0.05–0.17M sodium chloride. If will be instructive for you to analyze a liquid unknown by all three methods and compare their accuracy and precision. Of course, the gravimetric determination should be the most accurate and precise, but you may not be able to guess which of the other methods is second to the gravimetric method in accuracy and precision.

Each experiment is written so that you can also analyze a solid chloride sample using just that one method of analysis.

Experiment 3A | DETERMINATION OF CHLORIDE BY DENSITY MEASUREMENT

This experiment consists of two parts: an optional part involving repetitive filling of a volumetric flask (and treating the resulting data), and the main part consisting of the determination of the density of a sodium chloride solution using a volumetric flask. The optional part of the experiment is designed to provide experience in filling a volumetric flask carefully, if you have not used such a flask before.

As a preparation for the determination of density, recall that the density of a substance is defined as its mass (weight) per unit volume, or stated mathematically $D = $ weight/volume. In metric measurements $D = $ g/ml; hence density has the dimensions of grams per milliliter (or cubic centimeter). To determine the density of a substance experimentally, take a known volume of it, weigh it, then compute its density from the above formula. In the following experiment, the density of a sodium chloride solution is determined using a volumetric flask to measure the exact volume (Fig. 2-10). First weigh the empty flask and then the filled flask to determine the weight of the solution. In this manner, the density can be determined very accurately.

A pycnometer is also used for density measurement. The density of an irregularly shaped solid can be determined by weighing it in air, then weighing it while it is suspended in water. The loss in weight in water is a measure of its volume (Archimedes' principle); from that its density is computed.

If you have a pure solution of a single substance, the measurement of its density is a fairly accurate method to determine its concentration. After the density is determined, consult a table or a graph (see Fig. A) and read off the concentration corresponding to the density. This technique cannot be used with solutions containing more than one solute. In certain instances, as in the case of urine, one is interested only in a measure of total dissolved solids to determine the concentrating ability of the kidney. In this case, the density gives important information, but does not give the individual concentrations of urea, sodium chloride, or other dissolved solids.

Procedure   *Optional Repetitive Filling of a Volumetric Flask*

1. Obtain a suction bulb and a Pasteur pipet with a constricted tip at least 4 cm long. Also obtain a 25-ml volumetric flask with stopper. Clean if necessary. If the

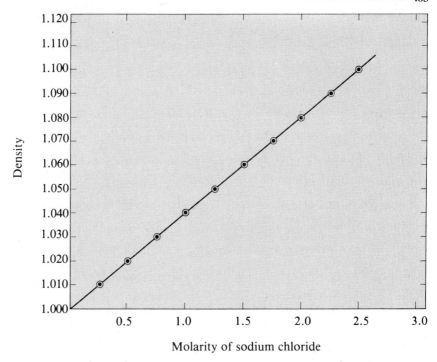

FIGURE A. Density of sodium chloride solutions at 20°C.

instructor directs, place a 2-cm length of black tape just below the etched line to sharpen the meniscus when filling (Fig. B).

2. To fill the flask for the first time, add distilled water to the flask until the level is about 2 cm below the etched line.   Then add distilled water dropwise using a Pasteur pipet (Fig. B) until the bottom of the meniscus just touches the line at eye level. Do this against a white background using a black card or black tape (Step 1).   Stopper the flask.

3. Wipe the outside of the flask with a tissue to remove any water before weighing.

4. Center the flask on the balance pan; find and record the weight to four digits to the right of the decimal point.

5. After removing the flask from the balance, withdraw about 2 cm of water from the flask with the Pasteur pipet.   Refill the flask by adding water dropwise until the bottom of the meniscus just touches the etched line.   Take care that no drops of water cling to the neck of the flask.   Stopper and reweigh.

6. The difference in weight between the first and second fillings typically should be from 0.01 to 0.03 g, but no more than 0.040 g.   Repeat the procedure until six good weighings with a range no greater than 0.040 g are obtained.

7. Turn the volumetric flask upside down and allow it to air dry in the drawer for the next part of the experiment (or dry it in the oven if you are continuing with the next part of the experiment right away).

8. Calculate the mean and the absolute and relative values of: the average deviation, the standard deviation, and the range.

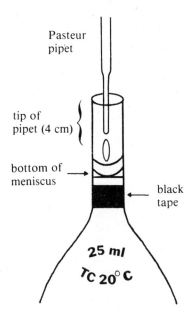

FIGURE B. Careful filling of a volumetric flask with a Pasteur pipet.

Determination of Density

1. Obtain a 25-ml volumetric flask and clean it if necessary. Dry for 30 minutes unless the instructor directs otherwise. After allowing it to cool, weight it on the analytical balance with its stopper.
2. Fill the flask with distilled water to about 2 cm below the etched line. Then add distilled water dropwise using a Pasteur pipet (Fig. B) until the bottom of the meniscus just touches the line at eye level. Do this against a white background using a black card or black tape just below the line. Stopper the flask and weigh it again.
3. The difference between the weight of the filled flask and the weight of the empty flask is the weight of water corresponding to the true volume of the flask. Record the temperature of the laboratory and look up the density of water to at least five significant figures. Divide the weight of the water by its density in g/ml to obtain the true volume.
4. Pour out the distilled water from the flask, then rinse it three times using 5-ml portions of your unknown solution, discarding each 5-ml portion before adding the next. Using this technique it is unnecessary to redry the flask. Carefully fill the flask to the mark with your unknown solution, stopper it, then weigh.
5. From the weight of the unknown solution and the volume of the flask, calculate the density of your solution.
6. Plot a graph like Figure A on good graph paper, using the data from Table A-1. Unless the instructor directs otherwise, enter only the first four points on the paper: $0.000M$, $0.0855M$, $0.170M$, and $0.255M$ sodium chloride. If only these four points are used, then molarities should be plotted on the horizontal axis (parallel to shorter side) of the paper to obtain a reasonable spread between points. On most graph paper, the $0.100M$ point should be located about 5 cm to the right of the origin. Report molarity of sodium chloride and weight percentage of sodium chloride in the solution.

TABLE A-1. **Density of Sodium Chloride Solutions, 20°C**

Molarity (wt %)	Density, 20° (g/ml)	Refractive index (n_D^{20})
0.000 (0.0%)	0.9982	1.3330
0.0855 (0.5%)	1.0017	1.3339
0.170 (1.0%)	1.0053	1.3347
0.255 (1.5%)	1.0089	1.3356
0.340 (2.0%)	1.0125	1.3365
0.425 (2.5%)	1.0160	1.3374
0.513 (3.0%)	1.0196	1.3382

Experiment 3B | GRAVIMETRIC DETERMINATION OF CHLORIDE

The basis for the gravimetric method is the precipitation of chloride ion by silver(I) nitrate.

$$Cl^- + AgNO_3(\text{excess}) \longrightarrow AgCl(s) + NO_3^-$$

Either a liquid or a solid sample can be used. Complete directions for the experiment follow. You should read Chapter 5 for a complete background on the experimental details.

Crucibles

1. Sintered-glass crucibles of medium (M) or fine (F) porosity are recommended for filtering silver chloride (do not use crucibles marked C for coarse). Porcelain (Gooch) crucibles with an asbestos mat or glass fiber filter paper may be used, but as their chief advantage is withstanding temperatures above 500°C, they are not necessary. If your sample is a solid, dry it (Step 2, *Procedure*) while cleaning the crucibles.

2. If the crucibles are not new, remove any visible dirt with detergent solution and brush and rinse. Whether crucibles are new or dirty, assemble filter flask, rubber crucible holder, and crucible. Connect filter flask with rubber suction tubing to aspirator.

   Fill the crucible about halfway with concentrated nitric acid and, using gentle suction, draw the acid slowly through the crucible. Fill halfway again, and interrupt the suction briefly to allow a few minutes for the acid to remain in contract with the crucible. Wash the crucible several times with distilled water. Number each crucible. If the crucibles are clean, proceed to Step 3.

   If silver chloride is apparently not removed by this procedure, draw 6M ammonium hydroxide through the crucible after the acid has been rinsed out with distilled water. Again, rinse the ammonium hydroxide out with distilled water.

3. Dry crucibles for 1 hr in an oven at 120°C using a beaker, glass hooks and a cover glass. Remove crucibles to the desiccator. Cool for 0.5 hr or longer. Weigh. Dry at 120°C for 0.5 hr and cool for 0.5 hr. Reweigh. Constant weight is attained if the second weighings agree with the first weighings within ±0.4 mg. It is advisable to store crucibles in a desiccator until time to use.

Procedure

1. *Liquid chloride sample.* If the chloride sample is a solution, pipet out exactly 25 ml into each of three 400-ml beakers numbered with numbers corresponding to those on your crucibles. Proceed to Step 3.

2. *Solid chloride sample.* If the chloride sample is a solid, put it in a weighing bottle, place the open weighing bottle and stopper in a beaker marked with your name, cover the beaker with a cover or watch glass on glass hooks, and dry in the oven at 120°C for 1–2 hr. Let cool in desiccator. While the sample is drying, the crucibles may be cleaned. Next, weigh three samples of from 0.4 to 0.7 g (see the instructor's directions) into clean 400-ml beakers numbered with numbers corresponding to those on your crucibles.

3. Add 150 ml of distilled chloride-free water to each beaker and acidify with 0.5 ml (10 drops) of concentrated nitric acid. Stir. Use a separate stirring rod for each beaker, and leave it in the beaker throughout the procedure.

4. Consult the instructor's directions for the amount of silver nitrate to add to the liquid sample; it should be a 10 mole % excess. For the solid sample, assume the sample is pure sodium chloride and calculate the mmoles of silver nitrate needed to precipitate it. For instance, 410 mg of sample would be $410/58.5 = 7$ mmole $NaCl = 7$ mmole $AgNO_3$. If a $0.5M$ silver nitrate solution is available, use $7/0.5 = 14$ ml, etc.

5. Add slowly, with stirring, the calculated amount of silver nitrate. Then add a 10% excess over the calculated amount. Heat the beakers almost to boiling, keeping the beakers out of the direct sunlight (a hood is preferred).

6. Stir to aid coagulation. Once the precipitate has coagulated on the bottom of the beaker, test for complete precipitation by adding a few drops of silver nitrate. If a cloud of precipitate appears, add about 5% more silver nitrate and stir. After complete coagulation, test again for complete precipitation, etc.

7. Continue heating, with occasional stirring, until the solution is crystal-clear, indicating that all fine particles of silver chloride have deposited on the coagulated silver chloride on the bottom of the beaker.

8. Remove from the heat, cover with a watch glass, and allow to cool for at least 1 hr in a hood or darkened area. Silver chloride is considerably more soluble in hot water than in cold. (The beakers may be left to cool overnight if you protect them from hydrochloric acid fumes by storing in a desk and covering with watch glasses.)

Filtration and Weighing

1. Place weighed filter crucible in the suction filtration apparatus and apply gentle suction. Decant the clear supernate from the corresponding beaker through the crucible, retaining the precipitate in the beaker.

2. Add about 10 ml of wash solution (0.6 ml of concentrated nitric acid per 200 ml of distilled water) to the beaker, and agitate the precipitate. Let it settle, and pour the wash solution through the crucible. Repeat twice.

3. Hold the stirring rod across the lip and top of the beaker with one hand, and direct the stream of wash solution from a wash bottle at the precipitate, washing it into the crucible. If some precipitate sticks to the beaker, loosen it by scrubbing with a rubber policeman. Wash the precipitate completely into the crucible and wash it in the crucible with several 5-ml portions of wash solution. Catch the last washing by itself, and test it for excess silver ion by adding a drop of $12M$ hydrochloric acid. Wash again, if a precipitate is noted.

4. Completely drain the crucible with strong suction, place the precipitate in a marked beaker, cover with a watch glass on hooks, and dry in an oven at 120°C for 2 hr. Cool in a desiccator for 0.5 hr or longer. Weigh. Dry again for 45 min and cool 0.5 hr or longer. Reweigh. Constant weight is attained if the second weighings

agree with the first weighings within ± 0.4 mg. Repeat the process, if constant weight is not attained.

5. For the liquid chloride sample, use the following formula to calculate the molarity of the chloride:

$$M = \frac{\text{(mg silver chloride)}/(143.32 \text{ mg/mmole})}{\text{ml sample}}$$

To calculate the percentage chloride in a solid sample, use 35.45 as the atomic weight of chloride and 143.32 as the formula weight of silver chloride. Report the percentage for each sample and the mean of the three percentages. If one of the results appears questionable, test it, using the Q test (Ch. 3). If the questionable result cannot be rejected, analyze a fourth sample and apply the Q test again. If this does not reject the questionable result, report the median (average of the two middle results) after consultation with the instructor.

Experiment 3C | VOLUMETRIC DETERMINATION OF CHLORIDE BY THE MOHR AND FAJANS'S METHODS

The reaction involved in the determinations of the chloride ion by the Mohr and Fajans's methods is identical to that in the gravimetric determination of chloride. Silver nitrate is added to the chloride solution, forming a precipitate of silver chloride. The amount of chloride is determined by measuring the amount of a standard solution of silver nitrate required to exactly precipitate all of the chloride ion. In the gravimetric determination, the exact concentration of the silver nitrate and the exact amount used were not important; an excess of silver nitrate was used to completely precipitate the chloride ion.

In the volumetric determination, however, it is necessary to know the exact point at which the silver nitrate has completely precipitated all of the chloride. Two methods are used to detect this point (i.e., the end point of the titration). The first method was developed by C. F. Mohr, a German chemist, over a century ago; hence this very ingenious method is called the *Mohr method* for chloride. A solution of potassium chromate is used as indicator. A known, exact amount of this is added to the unknown chloride solution. When this mixture is titrated with standard silver nitrate, the silver nitrate preferentially reacts with the chloride ion to produce a white insoluble precipitate of silver chloride. This reaction continues until all of the chloride ion is precipitated as silver chloride. At this point, the silver nitrate beings to react with the potassium chromate to form a pink-brown precipitate of silver chromate. The formation of this pink-brown color in the solution indicates the complete precipitation of all the chloride, so the buret is then read and the amount of chloride computed from the exact concentration and volume of the standard silver nitrate used.

The second method was developed by Kasimir Fajans, professor at the University of Michigan. It employs a so-called *adsorption indicator*. Before the end point, the silver chloride particles are negatively charged, owing to the presence of an excess of chloride ions. As soon as precipitation is complete and with the first excess of silver nitrate, the silver chloride particles become positively charged, owing to the excess silver ions. The positively charged particles adsorb a dye, dichlorofluorescein, on the surface and cause a change in color from yellow to pink. This represents the end point of the titration, and the buret is then read. Dextrin solution is added in this titration to keep the silver chloride particles from coagulating, thus giving a better end point.

Both methods are very sensitive to pH variations and work best in neutral or very slightly acid solution. If the pH of the solution is unknown, it should be measured before the titration is performed and the pH brought in the range of 6–8 by adding dilute acetic acid or ammonium hydroxide as required (pH paper).

Both methods require the use of a standard silver nitrate solution. Solid silver nitrate, $AgNO_3$, is a chemical that can be purchased in a very pure form; hence it can be used as a primary standard. A solution of very exact concentration can be prepared by weighing out an exact known weight of solid silver nitrate and diluting it to an exact known volume in a volumetric flask.

Preparation

Standard Silver Nitrate Solution
1. Clean your weighing bottle. Then dry it in the oven for about 15 min with the top off.
2. Weigh the empty, dry weighing bottle and top and record the weight.
3. Obtain about 8.5 g of pure, primary standard silver nitrate crystals from the instructor. Place them in your weighing bottle, then dry in the oven for 30–60 min. (Be sure to leave the top off the weighing bottle.)
4. Remove bottle with silver nitrate from oven, replace the top, and allow to cool; then weigh. Note exact weight of silver nitrate.
5. Quantitatively transfer the solid silver nitrate to a clean 500-ml volumetric flask, using a transfer funnel. Wash out weighing bottle and funnel with distilled water and pour washings into the volumetric flask. Remove funnel and dilute to the mark with distilled water. Shake and mix until all solid $AgNO_3$ dissolves.
6. Calculate the exact normality of your silver nitrate solution, using 169.88 as the formula weight.

Samples
1. *Liquid chloride sample.* If the chloride sample is a solution, pipet exactly 25 ml into each of three 250-ml Erlenmeyer flasks, then proceed to the first step of either Fajans's method or the Mohr method.
2. *Solid chloride sample.* If the chloride sample is a solid, put it in a weighing bottle, place the open weighing bottle and stopper in a beaker marked with your name, cover the beaker with a cover or watch glass on glass hooks, and dry in the oven at 120°C for 1–2 hr. Let cool in desiccator. Weigh three samples of from 0.4 to 0.7 g (see instructor's directions) into clean 250-ml Erlenmeyer flasks and add 25 ml of distilled water. Swirl until the sample is completely dissolved, then proceed to the first step of either Fajans's method or the Mohr method.

Determination of Chloride

Fajans's Method
1. Choose a place away from sunlight for the titration. Add 0.1 g of dextrin (or 5 ml of a 2% solution) and 10 drops of a 0.1% solution of dichlorofluorescein to the first flask.
2. Titrate with 0.1M silver nitrate at a fairly rapid rate, using continuous shaking or magnetic stirring. The suspended precipitate will have a pink tinge because of some premature displacement of chloride ion by the dichlorofluorescein ion. Detect the end point by setting the buret so that it delivers titrant at a slow dropwise rate, shaking constantly, and continuously observing the color of the suspension until it changes from a light pink to a dark pink.

3. If the color change to dark pink is observed, repeat Steps 1 and 2 for the remaining two samples. Note the buret reading and calculate the chloride molarity of the sample using the known normality and observed volume of the standard silver nitrate.

The Mohr Method
1. Add exactly 1.0 ml of 0.25M potassium chromate to each sample.
2. Titrate with your standard silver nitrate solution from a 50-ml buret, adding the solution rapidly at first while rotating the flask continuously to ensure mixing. As the end point approaches, the red-brown color of silver chromate becomes more widespread. The silver nitrate should then be added dropwise until a faint red-brownish color of silver chromate persists throughout the yellow solution and suspension. It should not be a dark brown, which indicates the end point is over-stepped; the color change should be faint but noticeable. This end point is not as distinct as in Fajans's method, but the observation improves with practice. Note the buret reading and calculate the chloride molarity of the sample using the known normality and observed volume of the standard silver nitrate.

Experiment 4 | PREPARATION AND STANDARDIZATION OF NaOH

Sodium hydroxide titrant is best standardized against primary standard potassium acid phthalate (Sec. 6-1). The reaction is

$$NaOH + KHC_8H_4O_4 \longrightarrow C_8H_4O_4^{-2} + H_2O + K^+ + Na^+$$
$$\text{(phthalate ion)}$$

The phthalate anion produced makes the solution slightly basic at the end point, the pH being about 9. This pH is roughly in the middle of the pH transition range of phenolphthalein and thymol blue indicators (Table 10-1). Phenolphthalein is generally used as the indicator; the first perceptible pink is taken as the end point.

Directions are given below for the preparation and standardization of both the usual 0.1M sodium hydroxide and the 0.25M sodium hydroxide needed for the analysis of milk of magnesia (Exp. 6).

Preparation
1. If it is not available in the laboratory, prepare a saturated (1 : 1) solution of sodium hydroxide by mixing carefully 50+ g of sodium hydroxide pellets with 50 ml of water in a beaker. When the solution has partly cooled, transfer to a polyethylene bottle, stopper, and let stand for about a week to allow the insoluble sodium carbonate to settle.
2. Obtain about a liter of distilled water and a polyethyelene bottle. If the instructor directs, boil the water for 5 min to remove carbon dioxide, and allow to cool to about 40°C before transferring to the polyethylene bottle. If the instructor does not wish the water to be boiled, transfer it directly to the polyethylene bottle.
3. *0.25M sodium hydroxide for milk of magnesia analysis.* Pipet about 17 ml of a *clear* saturated solution of sodium hydroxide into the polyethylene bottle containing about a liter of distilled water, using a *transfer* pipet and a rubber bulb. Mix thoroughly and keep stoppered.
4. *0.1M sodium hydroxide.* Pipet about 7 ml of a *clear* saturated solution of sodium hydroxide into the polyethylene bottle containing the liter of distilled water, using a *transfer* pipet and a rubber bulb. Mix thoroughly and keep stoppered.

Standardiza-
tion

1. Dry 6–9 g of primary standard potassium acid phthalate in a weighing bottle at 110°C for 2 hr. Allow to cool 30 min in a desiccator before weighing.

2. If your buret is being used for the first time in the term, fill it with distilled water and allow it to drain to check for dirt. If it does not drain uniformly, it must be cleaned (Sec. 6-3).

3. *0.25M sodium hydroxide.* Weigh out 1.2–1.4 g of potassium acid phthalate into each of three 250-ml flasks. (This is about 6–7 mmoles and should require about 24–28 ml of the sodium hydroxide.) Dissolve in about 50–75 ml of distilled water, warming if necessary. Proceed to step 5.

4. *0.1M sodium hydroxide.* Weigh out 0.8–0.9 g of potassium acid phthalate into each of three 250-ml flasks. Dissolve in about 50 ml of distilled water, warming if necessary.

5. Rinse the buret thoroughly with the titrant (Sec. 6-3), fill, and take an initial reading using the meniscus illuminator (Fig. 6-2).

6. Add three to four drops of phenolphthalein indicator to the first flask and titrate to the first faint pink that persists for 20 seconds. Titrate dropwise near the end point, splitting drops at the end point (Sec. 6-3). Add the same volume of phenolphthalein indicator to the remaining two flasks, and titrate those solutions also. Try to achieve the same shade of pink color for all end points.

7. Calculate the molarity of the sodium hydroxide by using the formula weight of 204.22 for potassium acid phthalate (Sec. 6-2). The molarity is the same as the normality in this case. If one of the values deviates significantly from the others, apply the Q test (Sec. 3-4).

Experiment 5 | PREPARATION AND STANDARDIZATION OF HCl

Hydrochloric acid titrant can be standardized against a standard solution of sodium hydroxide or against primary standard sodium carbonate. The equations for these reactions are

$$HCl + NaOH \longrightarrow H_2O + Na^+ + Cl^-$$

$$2HCl + Na_2CO_3 \longrightarrow H_2CO_3 + 2Na^+ + 2Cl^-$$

When both sodium hydroxide and hydrochloric acid must be used for a method, such as the analysis of milk of magnesia (Exp. 6), it is recommended that the sodium hydroxide be standardized first (Exp. 4), and then used to standardize the hydrochloric acid. If sodium hydroxide is not needed, then it is recommended that the hydrochloric acid be standardized against primary standard sodium carbonate. Since the end point is not sharp, two indicator procedures are given. In the methyl orange procedure, the end point is not sharp, but the solution need not be boiled to remove the carbonic acid product as carbon dioxide. In the methyl red procedure, the end point is sharp, but the solution must be boiled.

Preparation

0.50M Hydrochloric Acid for Milk of Magnesia Analysis. Fill a 1-liter ground-glass stoppered bottle with almost a liter of distilled water. Add about 45 ml of concentrated hydrochloric acid from a graduated cylinder and mix well. Allow to cool before using.

0.1M Hydrochloric Acid. Fill a 1-liter ground-glass stoppered bottle with about a liter of distilled water. Add about 9 ml of concentrated hydrochloric acid from a measuring pipet and mix well.

Standardiza-
tion

0.50M Hydrochloric Acid vs. Sodium Hydroxide

1. Standardize 0.25M sodium hydroxide as directed in Experiment 4. The sodium hydroxide titrant should be standardized and used within a period of two weeks. If a longer time elapses, it is recommended that you recheck the molarity of the sodium hydroxide by titration.
2. Using a volumetric pipet, transfer exactly 20 ml of 0.50M hydrochloric acid to each of three 250-ml flasks.
3. To the first flask add 2–3 drops of phenolphthalein indicator, and titrate to the first faint pink color that persists for at least 15 seconds, using 0.25M sodium hydroxide titrant. Use a white paper background to detect the pink color. If necessary, adjust the volume of phenolphthalein indicator for the titration of the remaining two flasks, and titrate these also.
4. Calculate the molarity of the hydrochloric acid as follows:

$$M_{HCl} = \frac{(M_{NaOH})(ml_{NaOH})}{ml_{HCl}}$$

0.1M Hydrochloric Acid vs. Sodium Carbonate

1. Dry 1–1.5 g of primary standard sodium carbonate at 110°C for 2 hr in a weighing bottle.
2. Weigh three 0.2-g samples by difference (to avoid absorption of water from the air) into 250-ml flasks and dissolve in 50–100 ml of water.
3. *Methyl orange procedure.* Add 2–3 drops of modified methyl orange indicator to each flask. Titrate with 0.1M hydrochloric acid, taking the end point as the point of color change from green to grey. If the color change is difficult to observe, record the color for each increment of titrant added and proceed dropwise past the grey to a purple. By comparison, the change to grey should now be evident. If the color change to purple appears sharper, take this to be the end point, but use the same color change for the unknown analysis so that the errors cancel out.
4. *Methyl red procedure.* Add 2–3 drops of methyl red indicator to each flask. Titrate with 0.1M hydrochloric acid until the indicator has changed color gradually from a yellow to a definite red. Then boil the solution gently for 2 min (the color should revert to yellow). Cover the flask with a watch glass, cool to room temperature, and continue the titration to a sharp change to red.
5. Calculate the molarity or normality of the acid as follows:

$$N_{HCl} = M_{HCl} = \frac{(mg\ of\ Na_2CO_3/53.0\ mg\ per\ meq\ Na_2CO_3)}{ml\ of\ HCl}$$

Experiment 6 | ANALYSIS OF MILK OF MAGNESIA

According to the *United States Pharmacopeia* (*U.S.P.*), medicinal preparations of milk of magnesia should contain a minimum of 7% of magnesium hydroxide by weight. Since such solutions contain a suspension of magnesium hydroxide, as well as being saturated with it, an accurate analysis must measure both the suspended and dissolved magnesium hydroxide. To take a representative sample for analysis, it is necessary that the bottle of milk of magnesia be well shaken just before the sample is taken.

Since the sample is a white opaque suspension, it poses an analytical problem if a *direct* titration is to be used. The suspended magnesium hydroxide particles may cause errors by adhering to the walls of the flask out of contact with the titrant. The opaque sample may also cause difficulties in perceiving an accurate end point color change.

A much better analytical approach is therefore to add a measured excess of a standard hydrochloric acid to neutralize and dissolve all of the suspended particles, giving a clear solution. The unreacted acid is then back titrated with standard base. The equations are

$$Mg(OH)_2(s) + 2HCl(excess) \longrightarrow 2H_2O + Mg^{+2} + 2Cl^-$$

$$\text{unreacted HCl} + \underset{\text{(titrant)}}{NaOH} \longrightarrow H_2O + Na^+ + Cl^-$$

The *U.S.P.* procedure calls for the use of $1M$ acid and $1M$ sodium hydroxide solutions, which can involve large errors for beginning students. The procedure below utilizes $0.5M$ hydrochloric acid and $0.25M$ sodium hydroxide to reduce those errors.

Procedure
1. Prepare and standardize $0.25M$ sodium hydroxide (Exp. 4) and $0.5M$ hydrochloric acid (Exp. 5).
2. Obtain a sample of about 20 g of commercial milk of magnesia [1].
3. Obtain three weighing bottles, each of which should be large enough to contain over 5 ml of solution. Weigh each dry to the nearest mg on the analytical balance.
4. Shake the sample of milk of magnesia thoroughly, and immediately transfer about 5–6 g of the sample to each weighing bottle. Again weigh each bottle to the nearest mg on the analytical balance.
5. Using a wash bottle, rinse each sample quantitatively out of the weighing bottle into a 250-ml flask. Using a 50-ml volumetric pipet or 50-ml buret, add exactly 50 ml of $0.5M$ standard hydrochloric acid to each of the three flasks. Shake to ensure complete reaction. The samples should dissolve completely. If any cloudiness or precipitate remains, this indicates that not enough acid has been added, and a measured additional amount should be added.
6. Add 3–4 drops of methyl red indicator to each flask and titrate the unreacted hydrochloric acid with $0.25M$ standard sodium hydroxide to the yellow (basic) color. The back titration should require at least 20 ml of sodium hydroxide for accurate analysis; if too large a milk of magnesia sample has been used, less than 20 ml will be required.
7. Calculate the weight percent of magnesium hydroxide in the sample as follows:

$$\text{mmole Mg(OH)}_2 = \tfrac{1}{2}[(M \text{ HCl})(\text{ml HCl}) - (M \text{ NaOH})(\text{ml NaOH})]$$

$$\% \text{ Mg(OH)}_2 = \frac{\text{mmole Mg(OH)}_2 \times 58.34 \text{ mg/mmole} \times 100}{\text{mg sample}}$$

Note   [1] Milk of magnesia preparations from any drug store make suitable samples. The authors have found that the Phillips's brand assays at significantly above 7% of $Mg(OH)_2$.

Experiment 7 | ACID-NEUTRALIZING CAPACITY OF AN ANTACID TABLET

Antacid tablets are made from a variety of substances that react with the hydrochloric acid of the stomach and neutralize it. Most commonly used are solid sodium bicarbonate, magnesium hydroxide, calcium carbonate, aluminum oxide, and magnesium trisilicate. Other substances, such as flavors, are added as well as salicylates and aspirin in small amounts.

The acid-neutralizing capacity is measured by adding an excess of standard hydrochloric acid to a previously weighed tablet, allowing it to react, then back titrating the excess acid with standard sodium hydroxide solution. The preparation and standardization of these solutions have been discussed previously in Experiments 4 and 5.

The measurement of the pH of a mixture of a tablet with water may give some clue as to its composition. Sodium bicarbonate and magnesium hydroxide form basic solutions of pH 8–9, whereas the other compounds are less basic, forming solutions of pH 6–8.

Measurement of pH

1. Grind up a tablet using a mortar and pestle, then mix it with 25 ml of distilled water in a small beaker.
2. Stir for about 5 min, then measure the pH of the solution, first using indicator paper, then using a glass pH electrode.

Acid-Neutralizing Capacity

1. Weigh a single tablet on the analytical balance, then transfer the tablet to a 250-ml Erlenmeyer flask.
2. From a buret, add 50.00 ml of standard (about 0.5 N) hydrochloric acid to the tablet in the flask.
3. Let the mixture stand for about one-half hour with occasional swirling, then add 4–5 drops of phenolphthalein indicator. If the solution turns pink, add another 50.00 ml of standard hydrochloric acid. Repeat this operation until a colorless solution is obtained.
4. Titrate the mixture with standard 0.25 N sodium hydroxide solution until a faint pink color is obtained. Note the volume of standard sodium hydroxide required.

Calculations

1. Calculate the number of milliequivalents of hydrochloric acid neutralized per tablet.
2. Calculate the number of milliequivalents of hydrochloric acid neutralized per gram of tablet.
3. If normal gastric juices contain 0.4% hydrochloric acid, calculate the number of milliliters of gastric juice neutralized by one tablet.
4. Write balanced equations representing the reaction of hydrochloric acid with $NaHCO_3$, $CaCO_3$, MgO, Al_2O_3, and $MgSi_3O_7$.

Experiment 8 | DETERMINATION OF WATER HARDNESS

The determination of water hardness using EDTA is discussed in Section 8-2. This analysis involves the determination of the sum of calcium and magnesium carbonate in any type of hard water. If EDTA is not available in primary standard form, it must be standardized against reagent-grade calcium carbonate. Either Eriochrome Black T or Calmagite indicator can be used for the titration; it is recommended that Calmagite be used because it is more stable.

Reagents

1. Prepare 0.01 M EDTA titrant by dissolving about 1.9 g of the reagent-grade disodium salt ($Na_2H_2Y \cdot 2H_2O$, form wt = 372) in 500 ml of distilled water. Add about 0.5 g (6 pellets) of sodium hydroxide and about 0.1 g of magnesium chloride ($MgCl_2 \cdot 6H_2O$). Mix well. Store in a Pyrex bottle. If A.C.S.-grade EDTA (99.0% min $Na_2H_2Y \cdot 2H_2O$) is available, the instructor may direct you to use it as a

primary standard. In that case, omit the addition of magnesium chloride and follow [1] below.

2. Prepare standard 0.0100M calcium(II) solution by weighing 0.500 g $\pm$ 0.2 mg of 99+% calcium carbonate into 20 ml of distilled water in a 250-ml beaker. Pipet about 1 ml of concentrated hydrochloric acid down the side of the beaker, and cover with a watch glass placed directly on top of the beaker. When the calcium carbonate has dissolved, rinse the watch glass into the beaker, raise it on glass hooks, and evaporate the solution to a volume of about 2 ml to expel most of the carbon dioxide. Add 50 ml of distilled water, transfer to a 500-ml volumetric flask, and dilute to the mark.

3. Prepare the pH 10 ammonia buffer by diluting 32 g of ammonium chloride and 285 ml of concentrated ammonium hydroxide to 500 ml with distilled water. Store in a polyethylene bottle. The buffer is made as concentrated as possible, to avoid contaminating the solutions with interfering metal ions. It is not stored in glass, because metal ions would be leached from the glass.

4. Prepare a dilute, aqueous solution of Calmagite. If Eriochrome Black T indicator is to be used instead, prepare a solid mixture by grinding 50 mg of Eriochrome Black T into 5 g of sodium chloride and 5 g of hydroxylamine hydrochloride.

Procedure

1. Pipet exactly 20 ml of 0.0100M calcium(II) solution into each of three clean 250-ml flasks [2], and add about 1 ml of the pH 10 ammonia buffer to each flask. Add enough Calmagite indicator to the first flask to give a wine-red color [3]. Alternatively, add a small scoop of the solid Eriochrome Black T indicator mixture.

2. Titrate with 0.01M EDTA from a 25-ml buret until a color change from wine-red through purple to clear blue is observed. The color change is somewhat slow, so titrant must be added slowly near the end point [4]. Repeat the titration for the remaining two samples. Calculate the molarity of the EDTA, using the 0.0100M of the calcium(II) solution.

3. Determine the water hardness of the unknown hard water by pipetting exactly 50 ml of the sample into each of three clean 250-ml flasks. Add 1 ml of ammonia buffer and Calmagite indicator to the first sample and titrate with 0.01M EDTA from a 25-ml buret to a clear blue end point [4, 5].

4. Calculate the parts per million of calcium carbonate in the unknown, using the molarity of EDTA and 100.1 mg/mmole as the formula weight of calcium carbonate.

Notes

[1] Magnesium chloride is added to the titrant to ensure sufficient magnesium(II) for a sharp end point, since calcium(II) does not form a strong enough chelate with the indicator. Instead of this, about 0.5 ml of 0.0005M magnesium-EDTA chelate may be added to each flask. Prepare this by mixing equal volumes of 0.010M EDTA and 0.010M magnesium(II) solutions.

[2] Remove traces of interfering metal ions on the sides of the titration flasks by rinsing each flask with about 10 ml of 1 : 1 nitric acid. Tilt and rotate the flask to contact the entire inside. Rinse well with distilled water, and allow to drain upside down.

[3] Possible air oxidation makes it advisable to add the indicator just before the titration. Traces of metal ions such as manganese(II) can catalyze this oxidation; this is prevented by addition of a bit of ascorbic acid. Avoid adding too much indicator, because the end point color change will be too gradual.

[4] To avoid the necessity of titrating slowly near the end point, the solutions may be warmed to about 60°C.

[5] If the end point color change is to a violet and not to a clear blue, a high level of iron in the water may be responsible. Avoid this in succeeding samples by adding a few crystals of

potassium cyanide (as much as will fit on the tip of a spatula) *after* the buffer has been added. *Caution!* Potassium cyanide is a poison; on contact with acid it forms hydrogen cyanide gas which is especially dangerous.

Experiment 9 | PREPARATION AND STANDARDIZATION OF IODINE

A standard solution of iodine can be prepared by weighing the iodine exactly on an analytical balance or by standardizing it against primary standard arsenic(III) oxide. Since it is difficult to handle iodine without losing some of it, it is usually weighed out approximately and standardized. The arsenic(III) oxide is first dissolved in base and then neutralized to arsenious acid.

$$As_2O_3(s) + 2OH^- + H_2O \longrightarrow 2H_2AsO_3^- + 2H^+ \longrightarrow 2H_3AsO_3$$

The arsenious acid is then oxidized by the iodine titrant to arsenic acid in a solution buffered at about pH 8 with sodium bicarbonate.

$$I_2 + H_3AsO_3 + H_2O \longrightarrow 2I^- + H_3AsO_4 + 2H^+$$

The end point is detected by the formation of the deep blue starch-triiodide color. Since iodine exists as the triiodide (I_3^-) ion in aqueous solution, the first excess of iodine titrant added will form the starch-triiodide complex.

Preparation

Starch Solution. Make a paste of 2 g of soluble starch and 25 ml of water, and pour with stirring into 250 ml of boiling water. Boil the mixture for 2 min, add 1 g of boric acid as preservative, and allow to cool. Store in a glass-stoppered bottle.

0.03N Iodine Solution for Vitamin C Analysis. Weigh about 3.8 g of reagent-grade iodine on a trip scale and transfer it to a 100-ml beaker containing 20 g of potassium iodide dissolved in 25 ml of water. Stir carefully to dissolve all the iodine (it sometimes dissolves slowly). Pour the entire contents of the beaker into a glass-stoppered amber liter bottle. Rinse the beaker with 50 ml of distilled water and pour this into the bottle also. Dilute to a liter using distilled water and shake several times. (If there is any doubt whether all of the iodine dissolved, allow to stand until the next lab period.)

0.1N Iodine Solution. Weigh about 12.7 g of reagent-grade iodine on a trip scale and transfer it to a 250-ml beaker containing 40 g of potassium iodide dissolved in 25 ml of water. Stir carefully to dissolve all the iodine. Pour the solution into a glass-stoppered amber liter bottle. Rinse the beaker with 50 ml of distilled water and pour this into the bottle also. Dilute to a liter using distilled water and shake several times.

Standardization

0.03N Iodine

1. If a 0.0300N arsenious acid solution is not available in the lab, prepare it by weighing exactly 0.370 g of primary standard arsenic(III) oxide into a 250-ml beaker. Add a solution of 1 g of sodium hydroxide pellets freshly dissolved in 20 ml of distilled water. Swirl until the oxide dissolves completely, warming if necessary. Add 50 ml of water and 2 ml of 12M hydrochloric acid (concentrated). Transfer quantitatively to a 250-ml volumetric flask and dilute to volume.

2. Pipet 25 ml of 0.0300N arsenious acid into each of three flasks. Add 25 ml of water to each. Then add about 3.5 g (trip scale) of sodium bicarbonate to each, and

check that the pH is 7–8 with pH paper. If the pH is not in this range, add more sodium bicarbonate.

3. To each flask add about 5 ml of starch solution and titrate with $0.03N$ iodine to the *first* appearance of the blue starch-triiodide color. (For the first flask it is advisable to check that the pH is in the 7–8 range. If not, add more sodium bicarbonate.) Since the starch-triiodide color is so intense, the depth of the color does not necessarily mean you have overshot the end point. Simply approach the end point more cautiously with the other flasks (all of them should require the same volume of iodine).

4. Calculate the normality of the iodine using the *average* volume consumed, as follows:

$$N_{I_2} = \frac{(N_{H_3AsO_3})(ml_{H_3AsO_3})}{ml_{I_2}}$$

0.1N Iodine

1. If directed, dry 1.5 g primary standard arsenic(III) oxide for 1 hr. Before weighing out the oxide, put at least 1 g of sodium hydroxide pellets into each of three 250-ml flasks, and add no more than 20 ml of water to one flask at a time. Swirl to dissolve the sodium hydroxide, and then immediately weigh 0.2–0.25 g of arsenic(III) oxide, exactly, into this flask. This utilizes the heat released from dissolution of sodium hydroxide to help the arsenic(III) oxide to dissolve.

2. Swirl to dissolve the arsenic(III) oxide and warm the flasks if necessary to complete dissolution. Inspect each solution for undissolved particles, washing down the sides of the flasks if necessary.

3. To each flask add 50 ml of water and exactly 2.5 ml of concentrated hydrochloric acid from a measuring pipet. The solution should almost be neutral at this point. Add about 3.5 g of sodium bicarbonate to each flask and check that the pH is 7–8.

4. To each flask add 5 ml of starch solution and titrate with $0.1N$ iodine to the first appearance of the deep blue starch-triiodide color. (For the first flask, it is advisable to check that the pH is in the 7–8 range. If not, add more sodium bicarbonate.) Since the indicator color is so intense, the depth of the color does not necessarily mean you have overshot the end point. Simply approach it more cautiously with the other samples.

5. Calculate the normality of the iodine of each sample as follows:

$$N = \frac{mg\ As_2O_3/(197.8/4)}{ml_{I_2}}$$

where 197.8/4 is the equivalent weight of arsenic(III) oxide in mg/meq.

Experiment 10 | ANALYSIS OF VITAMIN C TABLETS USING IODINE

Vitamin C, or ascorbic acid, is oxidized to dehydroascorbic acid.

As shown in the self-test on p. 236, the oxidation is a two-electron oxidation process. Iodine is a mild enough oxidizing agent to oxidize the ascorbic acid to dehydroascorbic acid and no further. The titration reaction is

$$I_2 + \underline{C_4H_6O_4}(OH)C=COH$$

$$\longrightarrow 2I^- + 2H^+ + \underline{C_4H_6O_4}C(=O)-C=O$$

Note that we have simplified the above reaction by writing the atoms of both molecules which undergo no electron change as the $C_4H_6O_4$ group and underlining it, as recommended in Sec. 10-1.

Vitamin C in solution is readily oxidized by dissolved oxygen, so samples should be analyzed as soon as they are dissolved. The titration flask should be covered during the titration to prevent absorption of additional oxygen from the air. A small amount of oxidation from oxygen already dissolved in the solution is usually not significant, but continuous shaking of an open flask will bring about enough absorption of oxygen to cause error [1].

Procedure

1. Prepare and standardize $0.03N$ iodine titrant (Exp. 9).
2. Obtain a sample of 1–4 vitamin C tablets, equivalent to 400–500 mg of ascorbic acid. (Anything less will require too small a volume of iodine titrant.)
3. Because a whole tablet may dissolve slowly, it is recommended that you carefully cut the tablet(s) into 5–6 pieces before weighing. (If the instructor wishes you to report the weight of vitamin C per tablet, be careful not to lose any of the tablet in cutting it. If you are required to report only the percentage of vitamin C, then any loss is not a problem.)
4. Weigh a sample equivalent to 400–500 mg of ascorbic acid by difference into a clean, dry 100-ml volumetric flask. If you have prepared the iodine titrant and are ready to titrate, dissolve the tablets by adding about 50 ml of water, *stoppering the flask*, and shaking vigorously until the tablets have dissolved. (A small amount of binder in the tablets may not dissolve and may be visible as fine particles but should not cause any error.) Then fill to the mark with distilled water.
5. Fill a buret with $0.03N$ iodine. Pipet exactly 25 ml of the vitamin C solution into a 250-ml flask and add 5 ml of starch indicator (Exp. 9). Cover the opening of the flask with a piece of cardboard having a small hole for the buret tip.
6. Insert the buret tip through the cardboard covering the flask and titrate with deliberate speed to the first appearance of the blue starch-triiodide color.
7. Repeat Steps 5 and 6 for each of the two remaining 25-ml aliquots of the vitamin C sample. Do not pipet a sample into a flask until you are ready to titrate it.
8. Calculate the results of your titrations. If you are to report the percentage of ascorbic acid in a certain weight, then consult the example on p. 236 for the calculation setup. If you are to report the weight of vitamin C in one tablet, use a setup similar to the following.

$$\text{mg/tablet} = \frac{(\text{ml of } I_2)(N \text{ of } I_2)(88.06 \text{ mg/meq})(100/25)}{\text{no. tablets}}$$

Notes

[1] C. E. Moore, *J. Chem. Educ.* **25**, 671 (1948).

Experiment 11 | IODOMETRIC DETERMINATION OF ARSENIC

Arsenic in the form of arsenic(III) oxide can be determined by oxidation with iodine after the oxide has been dissolved to form arsenious acid.

$$As_2O_3(s) + 2OH^- + H_2O \longrightarrow 2H_2AsO_3^- + 2H^+ \longrightarrow 2H_3AsO_3$$

Before the titration, the pH is adjusted to between 5 and 9 because the oxidation is quantitative and rapid only in this range. Sodium bicarbonate is added to buffer the solution at a pH between 7 and 8. Then iodine is used to titrate the arsenious acid to arsenic acid.

$$I_2 + H_3AsO_3 + H_2O \longrightarrow 2I^- + H_3AsO_4 + 2H^+$$

Note that acid is released during the titration, which is another reason for adding sodium bicarbonate buffer. The end point is detected by the formation of the deep blue starch-triiodide color.

Procedure
1. Prepare $0.1N$ iodine and starch solutions (Exp. 9).
2. Only if directed, dry the arsenic(III) unknown for 1 hr. Longer drying may sublime arsenic(III) from some samples.
3. Before weighing out the unknown, put at least 1 g of sodium hydroxide pellets into each of three 250-ml flasks. When you are ready to weigh the first portion of unknown into the flask, add no more than 20 ml of water to the first flask. Then weigh out a 0.3–0.35 g sample (see instructor's directions) immediately into the first flask and swirl to use the heat released from the dissolution of sodium hydroxide. Swirl to dissolve the arsenic(III) oxide and warm the flask if necessary to complete the dissolution.
4. When the first sample has dissolved, repeat the process, adding water to dissolve the sodium hydroxide, and then weighing the arsenic(III) sample into the flask.
5. To each flask add 50 ml of water and not less than 2.5 ml of concentrated hydrochloric acid from a measuring pipet. The solution should be almost neutral at this point. Add about 3.5 g of sodium bicarbonate to each flask and check that the pH is 7–8.
6. To each flask add 5 ml of starch solution and titrate with $0.1N$ iodine to the first appearance of the deep blue starch-triiodide color. (For the first flask it is advisable to check that the pH is in the 7–8 range. If it is not, add more sodium bicarbonate.)
7. Calculate the percentage of arsenic using an equivalent weight of 74.92/2 mg per meq of arsenic(III). Follow the method of the example on p. 105.

Experiment 12 | USING A SPECTROPHOTOMETER TO OBTAIN AN ABSORPTION SPECTRUM

The objectives of this experiment are to become familiar with the operation of a simple spectrophotometer such as the Bausch and Lomb Spectronic 20 and to obtain an absorption spectrum manually. The type of spectrophotometer used will depend on the equipment available in the laboratory. Regardless of the type, you should become familiar with the following controls on the spectrophotometer—the wavelength control, the on-off control, the control to adjust $0\%T$, the control to adjust $100\%T$, and the place where cells are inserted.

If you have not already done so, you should read all of the sections in Chapter 12 pertaining to the Beer–Lambert law and the various spectrophotometers.

The absorption spectrum of any species may be measured; we have chosen to have you obtain the spectrum of the $Cu(H_2O)_6^{+2}$ ion as an example of a species that absorbs primarily in the red and near-infrared regions. This ion is formed when copper(II) sulfate is added to water. The copper(II) ion is complexed by six water molecules and forms the light blue-green hexaquocopper(II) complex ion.

Preparation

1. Each student will be assigned one of the following weights of $CuSO_4 \cdot 5H_2O$ to weigh to ± 0.001 g: 0.499 g, 0.624 g, 0.749 g, 0.811 g, 0.874 g, 0.936 g, 0.999 g, 1.06 g, 1.124 g, 1.186 g, 1.249 g, 1.311 g, 1.374 g, 1.436 g, 1.498 g, or 1.622 g.
2. Transfer your assigned amount of $CuSO_4 \cdot 5H_2O$ to a 100-ml volumetric flask. Add distilled water carefully to the mark on the flask and shake until the copper sulfate is completely dissolved.
3. Make sure the Spectronic 20 or other spectrophotometer is equipped with a red measuring phototube and a red filter. (See Fig. C for the optical diagram of the Spectronic 20.) The following procedure incorporates directions for operating the Spectronic 20, but other spectrophotometers may be used in its place.

Procedure

1. Turn on the Spectronic 20 or other spectrophotometer, and allow it to warm up 10 min (Spectronic 20) or longer (instructor's directions) to stabilize. As shown in Figure C, the Spectronic 20 should be equipped with a red measuring phototube and a red filter for measurements above 620 nm.
2. Obtain two test tube cells (Spectronic 20) and clean them before using them—rinse each twice with distilled water. Wipe off the outside of each with a tissue rather than cloth or anything that might scratch the cells.
3. Fill one cell with distilled water for the *blank*. Fill the other with the solution of copper(II) sulfate, after rinsing the cell at least once with the copper(II) sulfate solution. Wipe off any liquid on the outside of each cell with a tissue.

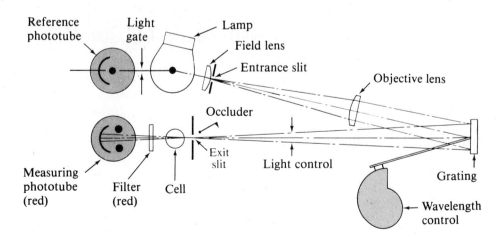

FIGURE C. Optical diagram of the Spectronic 20, top view. (Courtesy of Bausch and Lomb, Inc., Rochester, N.Y.)

4. *Setting 0% transmittance (absorbance = ∞).* The 0% transmittance setting corresponds to no light passing through the solution; it is made on the Spectronic 20 without inserting the cell into the instrument. Simply adjust the left-hand knob so that the meter needle on the scale reads 0 on the percent transmittance scale. Be sure the reading is stable. Check it periodically throughout the experiment and readjust it if necessary.

5. *Setting the wavelength.* Adjust the wavelength to the desired setting, starting at 620 nm. Because the $Cu(H_2O)_6^{+2}$ ion does not absorb appreciably until the start of the red region, it is not necessary to measure below 620 nm. For the Spectronic 20, adjust the wavelength using the knob on the top of the instrument.

6. *Setting 0 absorbance (100% transmittance).* Turn the right-hand knob of the Spectronic 20 counterclockwise almost to its limit. Insert the test tube cell containing distilled water into the cell holder. Match the line on the test tube cell with the index line on the holder. Close the top of the holder. Adjust the right-hand knob clockwise until the meter needle on the scale reads 0 absorbance. Remove the test tube to avoid instrument fatigue. Check the 0% transmittance setting to be sure it has not changed. If it has, both the 0% transmittance and 0 absorbance settings must be readjusted.

7. *Measurement of absorbance of the $Cu(H_2O)_6^{+2}$ ion.* Insert the test tube cell containing the copper(II) sulfate solution into the holder, again matching the marks. Allow the needle to stabilize, and read the absorbance of the solution to two significant figures; for example, 0.05, 0.10, 0.21. (Recall that for log terms such as absorbance all digits, including zeroes to the right of the decimal point, are significant.) Remove the test tube to avoid instrument fatigue.

8. Repeat Steps 5 through 7 for each wavelength used. It is suggested that the absorbance values be measured every 40 nm, starting at 620 nm. Near the location of the maximum absorbance, absorbance values must be measured every 10 nm, since you are required to locate the absorbance maximum within 10 nm.

9. After you have finished your measurements to the limits of the instrument, empty you test tube cells and rinse them thoroughly with distilled water. Turn them upside down to drain. Do not turn off the spectrophotometer unless directed to do so.

10. Using graph paper, make a plot of absorbance of $Cu(H_2O)_6^{+2}$ against wavelength, similar to Figure 12-1. Mark the wavelength of maximum absorbance.

11. Calculate the molarity of the copper(II) sulfate solution. Using $b = 1.1$ cm for the diameter of the cells, calculate the molar absorptivity of $Cu(H_2O)_6^{+2}$ and report it on the graph. Turn in the graph as your report.

Experiment 13 | COLORIMETRIC DETERMINATION OF GLUCOSE

The colorimetric determination of glucose has been discussed in Section 22-2. The method to be used in this experiment is the *o*-toluidine method, which is based on the condensation of glucose, an aldehyde sugar (aldose), and *o*-toluidine, a primary aromatic amine.

$$C_5H_{11}O_5\text{---}\overset{\overset{\displaystyle H}{\displaystyle |}}{C}\text{==}O + C_7H_8\text{---}NH_2 \xrightarrow{\text{heat}} C_5H_{11}O_5\text{---}\overset{\overset{\displaystyle H}{\displaystyle |}}{C}\text{==}N\text{---}C_7H_8 + H_2O$$

The product is a green-colored Schiff base, which has an absorption band at 630–640 nm in the red region.

Because the absorption maximum is close to the borderline for the blue-sensitive and red-sensitive phototubes of most simple spectrophotometers, measurements can be made at 620 nm using the blue-sensitive phototube or at 630 (or 640) nm using the red-sensitive phototube. Directions given below are for 620 nm using the blue-sensitive phototube, but the method may be used equally as well with the red-sensitive photo-tube.

Procedure

1. Prepare fresh each day a solution of *o*-toluidine reagent consisting of 0.28 g of thiourea and 10 ml of pure *o*-toluidine, diluted to 100 ml with glacial acetic acid.
2. Prepare four standard solutions of glucose containing 50, 100, 150, and 200 mg of glucose, each in 100 ml of saturated aqueous benzoic acid.
3. Into five small flasks or test tubes, pipet exactly 5 ml of the freshly prepared *o*-toluidine reagent. Using a 100-microliter (lambda) pipet, add 100 microliters of each one of the four standards to four of the five containers. (Be sure to blow out the pipet to transfer all of the solution.) Then add 100 microliters of the glucose unknown to the remaining flask or test tube. Mark each container as to what solution it contains.
4. Heat each solution for exactly 8 min in boiling water. Then remove from the heat and cool quickly under tap water for about 30 seconds. Allow to cool under tap water and in the air a total time of 10 min. During this time, turn on the Spectronic 20 and set the wavelength at 620 nm with the blue-sensitive phototube in the instrument. After it has stabilized, set the 0% transmittance and 0 absorbance settings.
5. After 10 min has elapsed for the first glucose reaction mixture, insert it into the spectrophotometer and read the absorbance quickly. Remove and repeat the measurement for the other reaction mixtures.
6. Plot the absorbance vs. concentration for each of the four standard glucose solutions. Enter the absorbance reading of the glucose unknown on the graph and read the concentration of the glucose from it in mg glucose per 100 ml of solution.

Experiment 14 | SPECTROPHOTOMETRIC DETERMINATION OF MANGANESE

Manganese is determined spectrophotometrically by dissolving it in nitric acid as manganese(II), and then oxidizing it with potassium periodate to the permanganate ion. This ion has a visible absorption band at about 520 nm, so it can be measured with a spectrophotometer or a colorimeter.

A more challenging analytical problem is the spectrophotometric determination of manganese in steel. The manganese steel is first dissolved in nitric acid, yielding manganese(II) and iron(III).

$$Mn^\circ + Fe^\circ + excess\ HNO_3 \longrightarrow Mn^{+2} + \underset{\text{(yellow)}}{Fe^{+3}} + NO_2 + NO$$

Since the NO_2 and NO produced may interfere later by reacting with the potassium periodate, they are removed partially by boiling and partially by oxidation with ammonium persulfate.

$$NO + NO_2 + (NH_4)_2S_2O_8 \longrightarrow 2NO_3^- + 2SO_4^{-2} + 2NH_4^+$$

Although the persulfate anion has the potential to oxidize manganese(II) to permanganate, the reaction is too slow. Hence potassium periodate is added to oxidize the manganese(II).

$$2\,Mn^{+2} + 5\,IO_4^- + 3\,H_2O \xrightarrow{\text{heat}} 2\,MnO_4^- + 6\,H^+ + 5\,IO_3^-$$

The reaction is carried out at the boiling point to ensure a rapid rate of reaction.

Note that at this point, the yellow iron(III) is still in the solution and can interfere somewhat by absorbing at 520 nm. Therefore, phosphoric acid is added to form a colorless phosphate complex ion with the iron(III).

Procedure This procedure will be performed with both a *standard steel* containing a known percentage of manganese and an unknown steel. The final calculation of percentage is very simple if the two sample weights are kept the same within ±0.01 g, as directed below.

1. Weigh out *equal* weights of the standard steel (known percentage of manganese) and the unknown steel, using 1.00–1.50 g (instructor's directions), to within ±0.01 g. Because of the accuracy limitation of the spectrophotometer, it is necessary to weigh only to the nearest 0.01 g.

2. Place the steels in separate 400-ml beakers, add 50–60 ml of dilute (1 : 3) nitric acid, and heat both in a hood to dissolve. Finally, cover and boil gently for 2 min to remove most of the nitric oxide. A black residue of carbon may remain at this point.

3. Remove the beakers from the heat and carefully sprinkle 1 g of ammonium persulfate into each beaker. Boil gently for 10–15 min to oxidize the carbon and destroy the excess ammonium persulfate. (Any precipitated manganese dioxide at this point can be removed by adding a few drops of dilute sodium sulfite solution, followed by boiling to expel the excess sulfur dioxide.)

4. Dilute each solution to about 100 ml, add 15 ml of 85% phosphoric acid or 40 ml of 6M phosphoric acid, and 0.5 g of potassium periodate. Boil gently for about 3 min to effect oxidation to the permanganate ion. Remove the heat and allow to cool below boiling, and then add an additional 0.2-g portion of potassium periodate. Boil for another minute or two.

5. Transfer both solutions to 500-ml volumetric flasks, and dilute to volume when cool to avoid oxidation of any organic matter in the distilled water used for dilution. It is recommended that the solutions be measured spectrophotometrically on the same day they are prepared. (Solutions may last for a few days if carefully prepared.)

6. Turn on the Spectronic 20 spectrophotometer and allow it to stabilize. Obtain two test tube cells; fill one with distilled water and the other with a solution of permanganate from the standard steel, after rinsing it once with the solution.

7. Adjust the wavelength to 520 nm and set the 0% transmittance setting. Insert the distilled water cell and set the 0 absorbance setting. Remove the cell containing the distilled water and insert the cell containing the solution from the standard steel. Read its absorbance.

8. Pour the standard steel solution out of the cell and rinse it with the solution containing the unknown steel. Fill the cell with that solution and insert it into the Spectronic 20. Read the absorbance of the solution containing the unknown steel.

9. Rinse both test tube cells with distilled water and allow them to drain upside down. Leave the Spectronic 20 on unless directed otherwise.

10. Calculate the percentage of manganese in the unknown steel by comparing its absorbance with the absorbance of the standard steel. If the weights are the same within ± 0.01 g, equation **13-2** can be used after the weights of the steels have been omitted.

Equilibrium Constants

∣ SOLUBILITY PRODUCT CONSTANTS

The values below are given for *pure* water where the concentration of other ions is zero; hence the ionic strength $\simeq 0$. Values in parentheses are for an ionic strength of 0.1.

Compound	K_{sp} *(Values in parentheses are for $\mu = 0.1$)*
Silver salts	
$AgBr$	$4.9 \times 10^{-13} M^2$ $(8.7 \times 10^{-12} M^2)$
Ag_2CO_3	$8.1 \times 10^{-12} M^3$
Ag_2CrO_4	$1.1 \times 10^{-12} M^3$ $(5.0 \times 10^{-12} M^3)$
$AgCl$	$1.8 \times 10^{-10} M^2$ $(3.2 \times 10^{-10} M^2)$
AgI	$8.3 \times 10^{-17} M^2$ $(1.5 \times 10^{-16} M^2)$
$AgIO_3$	$3.0 \times 10^{-8} M^2$
AgN_3	$2.9 \times 10^{-9} M^2$
Ag_3PO_4	$1.3 \times 10^{-20} M^4$
$AgSCN$	$1.1 \times 10^{-12} M^2$ $(2.0 \times 10^{-12} M^2)$
Ag_2HVO_4	$2 \times 10^{-14} M^3$
Aluminum salts	
$AlAsO_4$	$1.6 \times 10^{-16} M^2$
$Al(OH)_3$	$4.6 \times 10^{-33} M^4$
$AlPO_4$	$5.8 \times 10^{-19} M^2$
Gold salts	
$Au(OH)_3$	$5.5 \times 10^{-46} M^4$
$AuK(SCN)_4$	$6 \times 10^{-5} M^6$
Barium salts	
$BaCO_3$	$4.9 \times 10^{-9} M^2$ $(3.2 \times 10^{-8} M^2)$
$Ba(oxinate)_2$	$5.0 \times 10^{-9} M^3$
$BaSO_4$	$1.1 \times 10^{-10} M^2$ $(6.3 \times 10^{-10} M^2)$

Compound	K_{sp} (*Values in parentheses are for* $\mu = 0.1$)
Calcium salts	
$CaCO_3$	$4.8 \times 10^{-9} M^2$
$CaMg(CO_3)_2$	$1 \times 10^{-11} M^4$
CaC_2O_4	$2.3 \times 10^{-9} M^2$ ($1.6 \times 10^{-8} M^2$)
$Ca_3(PO_4)_2$	$1 \times 10^{-26} M^5$ ($1 \times 10^{-23} M^5$)
$CaSO_4$	$2.4 \times 10^{-5} M^2$ ($1.6 \times 10^{-4} M^2$)
Iron salts	
$Fe(OH)_3$	$2.5 \times 10^{-39} M^4$ ($1.3 \times 10^{-38} M^4$)
$Fe_4[Fe(CN)_6]_3$	$3.0 \times 10^{-41} M^7$
Magnesium salts	
$MgNH_4PO_4$	$2.5 \times 10^{-13} M^3$
$Mg(OH)_2$	$1.8 \times 10^{-11} M^3$
Mercury salts	
Hg_2Cl_2	$1.3 \times 10^{-18} M^3$
Lead salts	
$PbCl_2$	$1.6 \times 10^{-5} M^3$ ($8.0 \times 10^{-5} M^3$)
$PbSO_4$	$1.66 \times 10^{-8} M^2$ ($1.0 \times 10^{-7} M^2$)
Potassium salts	
$K_2NaCo(NO_2)_6$	$2.2 \times 10^{-11} M^4$
KUO_2AsO_4	$2.5 \times 10^{-33} M^3$
Thorium salts	
$Th(OH)_4$	$4 \times 10^{-45} M^5$
$Th(IO_3)_4$	$2.5 \times 10^{-15} M^5$
Uranium salts	
UO_2NaAsO_4	$1.3 \times 10^{-22} M^3$

IONIZATION CONSTANTS OF ACIDS

Acid	K_a at $\mu = 0$	K_a at $\mu = 0.1$
Acetic acid, CH_3COOH	1.74×10^{-5}	2.24×10^{-5}
Arsenic acid, H_3AsO_4		
K_1	6.5×10^{-3}	8.0×10^{-3}
K_2	1.3×10^{-7}	2.0×10^{-7}
K_3	3.2×10^{-12}	6.3×10^{-12}
Arsenious acid, H_3AsO_3		
K_1	6.0×10^{-10}	8.0×10^{-10}
K_2	—	8.0×10^{-13}
K_3	—	4.0×10^{-14}
Benzoic acid, C_6H_5COOH	6.3×10^{-5}	8.0×10^{-5}
Boric acid, H_3BO_3	5.9×10^{-10}	8.0×10^{-10}
Carbonic acid, H_2CO_3		
K_1	4.3×10^{-7}	5.0×10^{-7}
K_2	4.8×10^{-11}	8.0×10^{-11}

(*Continued*)

Ionization Constants of Acids (cont.)	*Acid*	K_a *at* $\mu = 0$	K_a *at* $\mu = 0.1$
	Chloroacetic acid, $CH_2ClCOOH$	1.38×10^{-3}	2.0×10^{-3}
	o-Chlorobenzoic acid, $C_6H_4ClCOOH$	1.14×10^{-3}	—
	m-Chlorobenzoic acid, $C_6H_4ClCOOH$	1.50×10^{-4}	—
	p-Chlorobenzoic acid, $C_6H_4ClCOOH$	1.03×10^{-4}	—
	o-Chlorophenol, C_6H_4ClOH	3.33×10^{-9}	—
	m-Chlorophenol, C_6H_4ClOH	9.48×10^{-10}	—
	p-Chlorophenol, C_6H_4ClOH	4.19×10^{-10}	—
	Dichloroacetic acid, $CHCl_2COOH$	5.5×10^{-2}	8.0×10^{-2}
	EDTA, H_4Y		
	K_1	8.5×10^{-3}	—
	K_2	1.8×10^{-3}	—
	K_3	5.8×10^{-7}	—
	K_4	4.6×10^{-11}	—
	Formic acid, $HCOOH$	1.70×10^{-4}	2.24×10^{-4}
	Fumaric acid, $C_2H_2(COOH)_2$		
	K_1	9.6×10^{-4}	1.3×10^{-3}
	K_2	4.1×10^{-5}	7.9×10^{-5}
	Hydrazoic acid, HN_3	1.9×10^{-5}	2.5×10^{-5}
	Hydrogen cyanide, HCN	4.9×10^{-10}	6.3×10^{-10}
	Hydrogen fluoride, HF	6.75×10^{-4}	8.9×10^{-4}
	Hydrogen peroxide, H_2O_2	1.8×10^{-12}	2.5×10^{-12}
	Iodic acid, HIO_3	1.67×10^{-1}	—
	Lactic acid, $CH_3CH(OH)COOH$	1.32×10^{-4}	1.74×10^{-4}
	Maleic acid, $C_2H_2(COOH)_2$		
	K_1	1.2×10^{-2}	1.6×10^{-2}
	K_2	6.0×10^{-7}	1.3×10^{-6}
	Glycine, H_2NCH_2COOH	1.65×10^{-10}	2.0×10^{-10}
	Malonic acid, $CH_2(COOH)_2$		
	K_1	1.4×10^{-3}	2.0×10^{-3}
	K_2	2.2×10^{-6}	4.0×10^{-6}
	o-Nitrophenol, $C_6H_4(NO_2)OH$	6.2×10^{-8}	—
	m-Nitrophenol, $C_6H_4(NO_2)OH$	4.0×10^{-9}	—
	p-Nitrophenol, $C_6H_4(NO_2)OH$	5.2×10^{-8}	—
	Nitrous acid, HNO_2	5.1×10^{-4}	6.3×10^{-4}
	Oxalic acid, $(COOH)_2$		
	K_1	8.8×10^{-2}	8.0×10^{-2}
	K_2	5.1×10^{-5}	1.0×10^{-4}
	Periodic acid, HIO_4	2.3×10^{-2}	—
	Phenol, C_6H_5OH	1.4×10^{-10}	1.6×10^{-10}
	Phosphoric acid, H_3PO_4		
	K_1	7.5×10^{-3}	1.0×10^{-2}
	K_2	6.2×10^{-8}	1.26×10^{-7}
	K_3	4.8×10^{-13}	2.0×10^{-12}
	Phosphorous acid, H_2PHO_3		
	K_1	7.1×10^{-3}	1.0×10^{-2}
	K_2	2.0×10^{-7}	4.0×10^{-7}

Acid	K_a at $\mu = 0$	K_a at $\mu = 0.1$
o-Phthalic acid, $C_6H_4(COOH)_2$		
K_1	1.20×10^{-3}	1.6×10^{-3}
K_2	3.9×10^{-6}	8×10^{-6}
Propionic acid, CH_3CH_2COOH	1.34×10^{-5}	—
Pyridinecarboxylic acid, C_5H_4NCOOH	5.0×10^{-6}	—
Salicylic acid, $C_6H_4(OH)COOH$		
K_1	1.05×10^{-3}	1.3×10^{-3}
K_2 (OH)	—	8×10^{-14}
Sulfurous acid, H_2SO_3		
K_1	1.3×10^{-2}	1.6×10^{-2}
K_2	6.3×10^{-8}	1.6×10^{-7}
Tartaric acid, $[CH(OH)COOH]_2$		
K_1	9.1×10^{-4}	1.3×10^{-3}
K_2	4.3×10^{-5}	8.0×10^{-5}
Trichloroacetic acid, CCl_3COOH	2.2×10^{-1}	3.2×10^{-1}

| IONIZATION CONSTANTS OF BASES

Base	K_b at $\mu = 0$	K_b at $\mu = 0.1$
Ammonia, NH_3	1.78×10^{-5}	2.34×10^{-5}
Aniline, $C_6H_5NH_2$	4.2×10^{-10}	5.0×10^{-10}
Butylamine, $C_4H_9NH_2$	4.09×10^{-4}	5.1×10^{-4}
Ethanolamine, $HOCH_2CH_2NH_2$	2.8×10^{-5}	—
Ethylamine, $C_2H_5NH_2$	4.7×10^{-4}	—
Ethylenediamine, $H_2NCH_2CH_2NH_2$		
K_1	1.28×10^{-4}	—
K_2	2.0×10^{-7}	—
Glycine, H_2NCH_2COOH	2.2×10^{-12}	3.2×10^{-12}
Hydrazine, NH_2NH_2	1.0×10^{-6}	1.3×10^{-6}
Hydroxylamine, NH_2OH	1.23×10^{-8}	1.6×10^{-8}
Methylamine, CH_3NH_2	4.4×10^{-4}	—
Pyridine, C_5H_5N	1.5×10^{-9}	1.9×10^{-9}
THAM, tris(hydroxymethyl)aminomethane,		
$(HOCH_2)_3CNH_2$	1.26×10^{-6}	1.6×10^{-6}
Triethanolamine, $(HOCH_2CH_2)_3N$	6.6×10^{-7}	8×10^{-7}
Trien, $(CH_2NHCH_2CH_2NH_2)_2$		
K_1	1.0×10^{-4}	—
K_2	1.9×10^{-5}	—
K_3	5.6×10^{-8}	—
K_4	2.5×10^{-11}	—
Urea, NH_2CONH_2	1.5×10^{-14}	—

Standard Electrode Potentials

Oxidizing agent + ne⁻ → reducing agent	Standard potential, $E°$, in volts (1M H^+)
$Ce(IV) + e^- = Ce^{+3}$ (in 1M $HClO_4$)	1.70[a]
$KMnO_4 + 4H^+ + 3e^- = MnO_2 + 2H_2O + K^+$	1.7 (neutral solution)
$H_5IO_6 + H^+ + 2e^- = IO_3^- + 3H_2O$ (periodic acid)	1.60
$KMnO_4 + 8H^+ + 5e^- = Mn^{+2} + 4H_2O + K^+$	1.51
$Ce(IV) + e^- = Ce^{+3}$ (in 1N H_2SO_4)	1.44[a]
$Cl_2 + 2e^- = 2Cl^-$	1.36
$Cr_2O_7^{-2} + 14H^+ + 6e^- = 2Cr^{+3} + 7H_2O$	1.33
$O_2 + 4H^+ + 4e^- = 2H_2O$	1.229 (= +0.816 at pH 7)
$IO_3^- + 6H^+ + 5e^- = \frac{1}{2}I_2 + 3H_2O$	1.195
$Br_2 + 2e^- = 2Br^-$	1.065
$Cu^{+2} + I^- + e^- = CuI(s)$	0.86
$Ag^+ + e^- = Ag(s)$	0.800
$Fe^{+3} + e^- = Fe^{+2}$	0.771
$O_2 + 2H^+ + 2e^- = H_2O_2$	0.682
$H_3AsO_4 + 2H^+ + 2e^- = H_3AsO_3 + H_2O$	0.559 (=0.004 at pH 7)
$I_2 + 2e^- = 2I^-$	0.535
Dehydroascorbic + $2H^+ + 2e^- =$ Ascorbic acid	0.39 (= −0.023 at pH 7)
$Fe(CN)_6^{-3} + e^- = Fe(CN)_6^{-4}$	0.36
Cytochrome-a(Fe^{III}) + $e^- =$ Cytochrome-a(Fe^{II})	0.290
$Hg_2Cl_2 + 2e^- = 2Hg + 2Cl^-$	0.268
$AgCl(s) + e^- = Ag(s) + Cl^-$	0.222
$Sn^{+4} + 2e^- = Sn^{+2}$	0.15
$S_4O_6^{-2} + 2e^- = 2S_2O_3^{-2}$ (thiosulfate)	0.15
$TiO^{+2} + 2H^+ + e^- = Ti^{+3} + H_2O$	0.1
$2H^+ + 2e^- = H_2(g)$	0.000 (exact no.)
$Pb^{+2} + 2e^- = Pb(s)$	−0.126
$Cr^{+3} + e^- = Cr^{+2}$	−0.41

Oxidizing agent + ne^- → reducing agent	Standard potential, $E°$ in volts ($1M$ H^+)
$2CO_2(g) + 2H^+ + 2e^- = H_2C_2O_4$	-0.49
$Zn^{+2} + 2e^- = Zn(s)$	-0.763
$Mg^{+2} + 2e^- = Mg(s)$	-2.37

[a] Formal Potentials

APPENDIX
THREE

Using the Electronic Calculator to Find the Standard Deviation

If your electronic calculator does not have a key for calculating the standard deviation, you can still use it for rapid calculation of this parameter. Directions are given for using both a calculator with a memory and a calculator without a memory.

For each of the methods below, the same raw data will be used—the weights of five pennies: 3.0220 g, 3.1200 g, 3.0440 g, 3.1420 g, and 3.0420 g.

I USING A CALCULATOR WITH A MEMORY

1. Calculate $\bar{X}$ by summing and dividing by 5: $\bar{X} = 3.0740$.
2. Use 3-3 (p. 46) to calculate s, the standard deviation. Activate the memory of the calculator to sum up $(X_i - \bar{X})^2$ for each value of X_i. These operations for the five pennies above are as follows:

X_i	$\bar{X}$	$(X_i - \bar{X})$	$(X_i - \bar{X})^2$	$\sum(X_i - \bar{X})^2$—summed in memory
3.0220	3.0740	-0.0520	0.002704	0.002704
3.1200	3.0740	$+0.0460$	0.002116	0.004820
3.0440	3.0740	-0.0300	0.000900	0.005720
3.1420	3.0740	$+0.0680$	0.004624	0.0103440
3.0420	3.0740	-0.0320	0.001024	0.0113680

3. Recall $\sum(X_i - \bar{X})^2$ from memory. $\sum(X_i - \bar{X})^2 = 0.0113680$.
4. Divide by $(n - 1)$:

$$\frac{0.0113680}{5 - 1} = 0.002842$$

5. Take the square root:

$$\sqrt{0.002842} = 0.0533_{104} = s$$

510

USING A CALCULATOR WITH NO MEMORY BUT WITH SQUARE AND SQUARE ROOT KEYS

1. Calculate $\overline{X}$ for five pennies given as raw data above: $\overline{X} = 3.0740$.
2. Subtract $\overline{X}$ from each weight and square; write down each square before clearing:

X_i	$\overline{X}$	$(X_i - \overline{X})^2$—write this result down before clearing
3.0220	3.0740	0.002704
3.1200	3.0740	0.002116
3.0440	3.0740	0.00900
3.1420	3.0740	0.004624
3.0420	3.0740	0.001024

3. Enter each $(X_i - \overline{X})^2$ value above and sum up. $\sum(X_i - \overline{X})^2 = 0.0113680$.
4. Divide $\sum(X_i - \overline{X})^2$ by $(n - 1)$:

$$\frac{0.0113680}{5 - 1} = 0.002842$$

5. Take the square root:

$$\sqrt{0.002842} = 0.0533_{104} = s$$

FOUR

Area under One Half of the Standard Normal Curve

To find the percentage of the area under the standard normal curve, double the entry in the table. To find the percentage of the area outside of one half of the curve, subtract the table entry from 50.00%.

Distance from the mean	Percentage of the area under one half of the standard normal curve
0.50σ	19.15%
1.00σ	34.13%
1.10σ	36.43%
1.20σ	38.49%
1.30σ	40.32%
1.40σ	41.92%
1.50σ	43.32%
1.60σ	44.52%
1.70σ	45.54%
1.80σ	46.41%
1.90σ	47.13%
2.00σ	47.73%
2.10σ	48.21%
2.20σ	48.61%
2.30σ	48.93%
2.40σ	49.18%
2.50σ	49.38%
2.60σ	49.53%
2.70σ	49.65%
2.80σ	49.74%
2.90σ	49.81%
3.00σ	49.865%
4.00σ	49.9968%

Answers to
Even-Numbered
Problems

Where possible, the correct number of significant figures has been used to express the answers. Answers to a few odd-numbered problems are also given.

Chapter 1
3. a. Mercury(I) chloride   d. Potassium hexacyanoferrate(II)
4. 0.85_5 mmoles; 85_5 μmoles. $8.5_5 \times 10^{-4}M$
6. 1 mg; 1×10^3 μg. $1.00 \times 10^{-3}M$
8. Using **1-3** for activity coefficient, both = 0.004. Using Table 1-1 for activity coefficient estimation, $a_{Mg} = 0.005_6$; $a_{SO_4} = 0.005$.

Chapter 2
2. Container weight is subtracted from weight of sample plus container.
4. Weighing is at constant load, so that sensitivity is constant at all weights.
6. Use a paper loop, spatula, or scoop.
9. b. 0.06_7 pph, 0.1_3 pph (by dif.)
10. a. 10 mg   b. 20 mg
12. Use a balance.
14. 100 mg
16. 1.594 g/ml
18. 8.0% NaCl
20. Use weights of 1, 2, 4, 8, 16, and 32 g.

Chapter 3
4. a. Two-fourths results used for $n = 4$, only one-third for $n = 3$.
    b. Two-fourths results used for $n = 4$, only two-sixths for $n = 6$.
6. More convenient and faster
8. a. 1.0×10^{-7} b. 1.00×10^{-7} c. 1.004×10^{-7}
10. a. 2.5×10^2 b. 2.0×10^2 c. $_2 \times 10^{10}$ (no sig. fig.)
12. a. Obtain more results if possible; if not, report $\bar{X}$ of 9.8%.
    b. Obtain more results if possible; if not, report M of 9.6_5%.
    c. No gross error. Use $\bar{X} = M = 9.70$%.
14. a. $\bar{X} = 5._{08}$, $M = 4._5$, mode = 4.
    b. Data are skewed to the left; M and mode are to the left of the mean, as usual.
16. Report $\bar{X}$ of 20.36_{67}%.
18. a. Relative range = $41._7$ pph; relative $s = 18._0$ pph.
    b. Relative range = $41._7$ pph; relative $s = 18._0$ pph.

20. $\pm 0.068_{67}\%$
22. a. Reject 1.802.   b. $\bar{X} = 1.705$.   c. No.
24. a. Reject 1.805 on basis of past experience.   b. Report $\bar{X} = 1.705$.
26. a. No   b. $80.53_4\%$ ($\bar{X}$ of all 7 results. Conservative viewpoint)
28. Don't apply Q test; 70.00 and 70.01% are fortuitously close. Report $\bar{X}$ of 70.34%.
30. The agreement indicates the method is good to > 2 significant figures; rounding off is not justified. Fish are safe to eat by FDA standards.
32. $\bar{X} = 5.5$, $M = 5.5$, mode $= 8$

Chapter 4   2. Aspirin capsules near the top of the bottle probably do not react to a greater degree with water than those elsewhere in the bottle; aspirin tablets will.
4. a. Shake, and sample before homogeneity is lost.
   b. Sample at different depths.
6. Where to sample is always one difficulty.
8. Variation in the amount of water causes a variation in sample weight, which causes the weight percentage of a constituent to vary.
10. Hydrochloric acid forms insoluble chlorides with these metal ions.
11. a. Most alcohols   b. Water

Chapter 5   1. a. Simplicity, accuracy, precision
   b. Exacting, lengthy, subject to possible serious error
4. Crystalline, granular (curdy), and gelatinous
6. a. More convenient   b. Can be heated to higher temps   c. Can be used for silver halides   d. Same as b
8. The formula weight of $AlCl_3$ contains 3 formula weights of Cl.
12. a. 0.6994_{36} b. None needed $(= 1)$ c. 0.9665_{62} d. 1.056_3
14. 4.0_2 ppm
16. a. $59.80_{85}\%$ Cl^- (impure NaCl)   b. $63.54_{47}\%$ Cl^- (cannot be NaCl)
18. a. $26.71_{67}\%$ S (wet)   b. $28.12_{28}\%$ S (dry)
20. a. 2.609_{91} mmole AgCl or NaCl   b. $0.1043_{96}M$ NaCl
22. $0.1048_{15}M$ $MgCl_2$
24. Actual %Cl $= 66.62\%$; $x = 5$
26. a. NaBr(xs) $+$ AgNO$_3$ $\rightarrow$ $\begin{array}{l} \text{AgBrAg:Br}^- \cdot \cdot \text{Na}^+ \\ \text{BrAgBr} \\ \text{AgBrAg:Br}^- \cdot \cdot \text{Na}^+ \end{array}$ $+$ Na$^+$ $+$ NO$_3^-$

   b. AgClO$_4$(xs) $+$ KBr $\rightarrow$ $\begin{array}{l} \text{BrAgBr:Ag}^+ \cdot \cdot \text{ClO}_4^- \\ \text{AgBrAg} \\ \text{BrAgBr:Ag}^+ \cdot \cdot \text{ClO}_4^- \end{array}$ $+$ K$^+$ $+$ ClO$_4^-$

28. a. Good   b. Not good, weakly ionized   e. Not good, weakly ionized
30. a. $6.5_0 \times 10^{-5} M$ b. $6.8_8 \times 10^{-7} M$ c. $1._{26} \times 10^{-3} M$ d. $6.2_{99} \times 10^{-5} M$
32. a. $[Cl^-] = S = 1.8 \times 10^{-8} M$ b. Yes
34. $10.3_{46}\%$ Fe_2O_3
36. a. Empirical formula $= C_4O_3$ b. $C_{12}O_9$
38. 61.05% NaCl; 38.95% KCl

Chapter 6   2. Requirements: known composition, known purity, large molecular weight, stable at room temperature, etc.; at least 99.9% pure.
4. TC $=$ to contain, TD $=$ to deliver
6. See text—Section 6-3.
8. a. $0.1004_{49}M$ b. $0.1004_{49}M$
10. a. $0.0215_{77}M$ b. $239._{29}$ mg/dL
12. 0.1268_7M or N
14. 0.0933_2M or N

16. $41.97_6\%$
18. a. $0.1599_{68}M$ b. $0.3199_{35}N$
20. a. $22.66_2\%$ b. $13.48_6\%$
22. 1.6_{03} mg/dL
24. 14.08_0 ml
26. 108 meq/L Cl^-; 29 meq HCO_3^-/L; 143 meq Na^+/L

Chapter 7   2. 0%: pH = 1.00; 90.00%: pH = 2.00; 99.00%: pH = 3.00; 100%: pH = 7.0
   4. 0%: pH = 3.00; 90.00%: pH = 4.00; 99.00%: pH = 5.00; 100%: pH = 7.0
   6. 50%: $E = 0.771$ V; 100%: $E = 1.23_{55}$ V; 200%: $E = 1.70$ V
   8. a. 1.000   b. 2.000   c. 3.000   d. 4.000   e. 4.87_{29} f. 5.74_{47}
  10. a. $[Cl^-]$ before add'n. of KSCN $= 4.7_6 \times 10^{-9}M$; pCl $= 8.32_2$
      b. $[Ag^+]$ before add'n. of KSCN $= 3.7_8 \times 10^{-2}M$; pAg $= 1.42_{28}$
      c. $[Ag^+] = 1.0_5 \times 10^{-6}M$; pAg $= 5.97_9$
      d. $[Ag^+]$ 1 ml after eq pt $= 9.9 \times 10^{-10}M$; pAg $= 9.00$
  12. a. Quant.—K_{rxn} greater than min theo
      b. Not quant.—K_{rxn} less than min theo
      c. Quant.—K_{rxn} greater than min theo
      d. Quant.—K_{rxn} greater than min theo
  14. a. Quant.—$\%$ diss'd $= 0.006_7\%$
      b. Quant.—$\%$ diss'd $= 0.06_7\%$
      c. Quant.—$\%$ diss'd $= 0.09_6\%$
      d. Not quant.—$\%$ diss'd $= 0.10_3\%$
      e. Not quant.—$\%$ diss'd $= 0.11_{17}\%$
  16. S of $Ag_2CO_3 = 1.2_{65} \times 10^{-4}M$
      a. No error; S of AgCl of $1.3_4 \times 10^{-5}M$ less than above S
      b. No error; S of AgBr of $7.0 \times 10^{-7}M$ less than above S
      c. No error; K_{sp} of Ag_2MoO_4 less than K_{sp} of Ag_2CO_3
      d. No error; K_{sp} of Ag_2S less than K_{sp} of Ag_2CO_3
      e. No error; S of Ag_3AsO_4 of $4.4_9 \times 10^{-6}M$ less than above S
      f. No error; S of $Ag_4Fe(CN)_6$ of $2._{29} \times 10^{-9}M$ less than above S
  18. For AgN_3, $[Ag^+]_{ep} = 5.3_8 \times 10^{-5}M$.
      a. $[Ag^+]_{ep}$ is greater than $[Ag^+]_{min}$ of $2.1 \times 10^{-5}M$; can't use.
      b. $[Ag^+]_{ep}$ is greater than $[Ag^+]_{min}$ of $4.6_9 \times 10^{-5}M$; can't use.
      c. $[Ag^+]_{ep}$ is greater than $[Ag^+]_{min}$ of $1.6 \times 10^{-6}M$; can't use.
      d. $[Ag^+]_{ep}$ is less than $[Ag^+]_{min}$ of 2×10^{-4}; VO_3^- can be used if salt is colored (check).
  20. Volhard method is more applicable because excess Ag^+ is added, which precipitates even the more soluble silver salts quantitatively.
  22. $3._{16} \times 10^{27}$
  24. a. Borderline case: K_{rxn} of $5.0_{07} \times 10^1$ predicts slight error (correct); molar solubility of Ag_3PO_4 of $4.6_{84} \times 10^{-6}$ is somewhat smaller than that of Ag_2SO_3 but is incorrect for a borderline case like this.
      b. No error: K_{rxn} of $7.6_{99} \times 10^{-6}$ is less than one; molar solubility of Ag_3PO_4 (above) is significantly less than that of Ag_2MoO_4.

Chapter 8   2. Mohr: no. Adsorption indicator: yes, if positively charged indicator is used. Volhard: yes.
   4. a. A one-atom donor ligand ($:NH_3$)
      b. A six-atom donor ligand (EDTA)
      c. A two-atom donor ligand (ethylenediamine)
   6. Stability constants for hexadentate ligands are much larger.
   8. a. Not quant.—$\%$ diss'd $= 0.17\%$; $K_{rxn} = 3.3_3 \times 10^7$
      b. Not quant.—$\%$ diss'd $= 0.13_4\%$; $K_{rxn} = 5.5_6 \times 10^9$

c. Not quant.—% diss'd $= 0.10_{49}\%$; $K_{rxn} = 9.0_9 \times 10^{11}$

d. Not quant.—% diss'd $= 0.7\%$; $K_{rxn} = 2.0_4 \times 10^{12}$

10. $[Ag^+]_{ep}$ for AgCl $= 1.3_4 \times 10^{-5}M$

    a. No, $[Ag^+]_{ep}$ is greater than $[Ag^+]_{min}$ of $1.9_6 \times 10^{-10}M$

    b. No, $[Ag^+]_{ep}$ is greater than $[Ag^+]_{min}$ of $4._{899} \times 10^{-24}$

    c. No, $[Ag^+]_{ep}$ is greater than $[Ag^+]_{min}$ of $1.6_{39} \times 10^{-6}$

    d. No   e. No   f. No

12. $[Ag^+]_{min} = 1.2_{54} \times 10^{-5}M$

    a. No, $[Ag^+]_{ep}$ of $1.3_4 \times 10^{-5}M$ is greater than $[Ag^+]_{min}$

    b. No, $[Ag^+]_{ep}$ of $3.1_{07} \times 10^{-5}M$ is greater than $[Ag^+]_{min}$

    c. Feasible, $[Ag^+]_{ep}$ of $1.2_{47} \times 10^{-5}M$ is less than $[Ag^+]_{min}$

    d. Feasible, $[Ag^+]_{ep}$ of $1.4_{05} \times 10^{-5}M$ is more than $[Ag^+]_{min}$

14. $47.20_7\%$

16. $45.72_{55}\%$

18. a. $10^{-2.79}$ b. $10^{-1.77}$ c. $10^{-0.83}$

20. The pH range is from 10 to the pH of saturated $Ca(OH)_2$.

22. $206._{44}$ ppm Ca; $515._{59}$ ppm $CaCO_3$

24. $K_{Cu-EDTA} = 10^{8.88}$ (quantitative)

26. a. Detn: Mohr and Volhard without nitrobenzene

    b. Detn: Volhard with nitrobenzene

    c. Dent: Volhard with nitrobenzene

    d. Detn: Mohr and Volhard with nitrobenzene

28. 5.01 mg/dL

Chapter 9   2. $[H^+] = [OH^-]$

4. a. 7.80 b. 7.47   c. 6.51

6. a. 1.70   b. 3.82_{39} c. Digestion consumes H^+

8. a. 0.7   b. 0.679_8 (3 sig. figs.)

10. $3.9_{84} \times 10^{-6}M$

12. 10.25_{53}

14. a. Acetate/acetic acid $= 0.72_{44}/1.0$

    b. Ammonium/ammonia $= 1.0/1.0$

    c. Ammonium/ammonia $= 1.4_{45}/1.0$

16. 4.33_{38}

18. a. 8.34_{26} b. 4.78_{78}

20. The pH is approximately that of pure water, about 7.0.

22. 113.5 mmole H_2CO_3, 2270 mmole HCO_3^-

Chapter 10   2. Easier to locate end point.

4. Neither is available with a known composition.   Yes, as a constant boiling substance.

6. Exhibits most perceptible indicator color change

10. a. $1.4_{82} \times 10^{-11}$; quantitative

    b. $1.6_{949} \times 10^{-5}$; not quantitative

    c. $1.9_{23} \times 10^{-7}$; not quantitative

    d. $2.5_{64} \times 10^{-9}$; quantitative

12. a. Actual $K_a = 5.6 \times 10^{-10}$; max theo $K_a = 1 \times 10^{-7}$; quantitative

    b. Actual $K_a = 5.6 \times 10^{-10}$; max theo $K_a = 1 \times 10^{-10}$; not quantitative

    c. Actual $K_a = 8.1_3 \times 10^{-7}$; max theo $K_a = 1 \times 10^{-6}$; quantitative

14. a. 2.79_9 b. 3.72_{35} c. 4.20_{07} d. 8.27_3 e. 11.69_9

16. a. pH $= 4.07_{13}$ b. 5.74_{47} c. 6.22_{18} d. 9.28_{43} e. 11.69_9

18. $50.0_8\%$

20. a. NaOH and Na_2CO_3 b. 44.0 ml

22. pH $= 8.96$, phenolphthalein

24. a. Diprotic   b. 8.0×10^{-6} c. 166

26. a. Visual, sulfathiazole
    b. Visual, sulfathiazole
    c. Potentiometric, acetic acid
    d. Potentiometric, n-butylamine
    e. Potentiometric, 2,4-dinitrophenol

Chapter 11

2. a. 6-electron change   b. 2-electron change   c. 2-electron change
   d. 1-electron change   e. 2-electron change
3. a. 4-electron change   b. 4-electron change   c. 2-electron change
   d. 2-electron change   e. 2-electron change
6. a. 0.66_9 V (quantitative)   b. 0.029 V (not quantitative)
   c. -0.024 V (not quantitative)   d. 0.531 V (quantitative)
   e. 0.558 V (quantitative)   f. 0.32_5 V (quantitative)
   g. 0.38_5 V (quantitative)   h. 0.62_1 V (quantitative)
8. $\log K = 3.8_8$, $K = 7._6 \times 10^3$
10. $79.43_0\%$
12. $7.688_3\%$
14. $M = 0.0155_{60}$, $N = 0.0311_{20}$
16. $M = 0.0660_{69}$, $N = 0.1321_{29}$
18. 502.6_{46} mg/tab; label should read 500 ± 3 mg (500 implies 500 ± 1 mg)
20. $250._{44}$ mg/tab
22. $0.1003_{40}N$
24. $4.779_2\%$
26. a. Reduce with I^- to $CuI(s) + I_2$ and titrate I_2 with $Na_2S_2O_3$.
    b. Dissolve, reduce $Ti(IV)$ to $Ti(III)$; titrate with standard $Fe(III)$.
28. a. Error   b. No error   c. No error   d. Error   e. Error
30. a. $[Ag^+] = 6.3_1 \times 10^{-7}M$
32. $+0.423$ V
34. $37.33_8\%$

Chapter 12

2. a. A is an instrumental measure of absorption.
   b. Positive: solution absorbs more than predicted; negative: less.
5. a. Purple (red + blue)   b. Purple   c. Orange   d. Yellow   e. Green
6. a. Red (weak response)   b. Violet-blue (violet stronger at night than day)
   c. Violet   d. Violet-blue (violet stronger at night than day)
7. a. 0.69_9 b. 1.00
8. a. 0.69_9 b. 1.00
9. a. 0.08_{62} b. 0.01_{32}
10. a. 0.40_0 b. 0.22_{11}
11. a. 0.60_0 b. 1.0_{13}
12. a. 0.35_{22} b. 0.35_{22}
13. a. 4.30_1 b. -1.7_0
14. a. $1._0 \times 10^4$ b. $_1 \times 10^4$ (no sig. figs.)
15. a. $2.0_0 \times 10^3$ b. 0.03_{50}
16. a. 0.030 (only two sig. figs. because A is a product rather than a log term here)
    b. 0.059_{86} (as above)
17. a. $5._0 \times 10^{-4}M$; $10._0$ mg/dL
    b. $5.0_1 \times 10^{-5}M$; 1.0_{02} mg/dL
18. a. 12.0_0 cm^{-1} M^{-1} b. $10._{03}$ cm^{-1} M^{-1}
19. a. $1.9_{70} \times 10^2$ b. $2.0_{34} \times 10^2$
20. a. $3.5_0 \times 10^{-4}M$ b. $6.5_{34} \times 10^{-4}M$
22. $c_{50\ ml} = 8.8_{84} \times 10^{-6}M$
    a. $4.4_2 \times 10^{-5}M$ b. 2.4_{81} ppm Fe

24. Negative deviation
26. a. UV spectrophotometer
   b. $1 \times 10^{-6}M$ (203 nm); $1._3 \times 10^{-3}$ (300 nm)
28. Tungsten, hydrogen, deuterium, and quartz-halogen lamps.
30. 4th—hydrogen; 3rd—globar; 2nd—Nernst glower; 1st—tungsten
32. a. Blue-sensitive phototube   b. Red-sensitive phototube   c. Silicon photodiode
34. Red + some violet
36. a. $A_1 - A_2 = \log(0.79_{43}/0.10) = 0.90$
38. a. 0.12_{49} b. 0.04_{58} c. 00_{87} d. 0.00_{48} e. 00_{46}
40. $0.37_4 \text{ cm}^{-1}(M \text{ Tb}^{+3})^{-1}$

Chapter 13

2. a. The e^- absorbs a photon and jumps to a higher orbital.
   b. A bond absorbs an infrared photon and undergoes higher energy vibrations.
4. a. C—H   b. O—H   c. C=O   d. C=O   e. Ester carbonyl
6. a. To an empty d orbital of Mn(VI)
   b. The Mn(VI) has no d electrons.
8. a. Yes; it has π electrons.
   b. NO and atomic O
   c. Alkenes + atomic O → C—C=O
10. a. Will; O—H   b. Will; aldehyde C—H   c. Will; O—H   d. Will not
12. It also exists in the enol form having O—H absorption at 3.3 μ.
14. a. 404 mg/tab   b. 324 mg/tab   c. $0.037_{04}\%$
16. UV: benzene ring, C=O (CO_2H); IR: $(H-)_3CN$, C=O (CO_2H)
18. Barbiturate
19. a. $0.51_{40}\%$ Mn without weights (0.68% error by omitting weights)
   b. $0.96_{14}\%$ Mn with weights ($1.1_{10}\%$ Mn without weights)
20. a. 0.185% Mn   b. 0.720% Mn
21. a. $80._{49}$ mg/dL   b. $1.5_{75} \times 10^2$ mg/dL
22. a. 0.57_{38} mg Fe/tab   b. 1.7_{21} mg Fe/tab
23. a. $7.7_{25} \times 10^{-5}M$
24. Use 460 and 560 nm; in the mixture with I, $c_b = 2.5_{48} \times 10^{-4}M$; in the mixture with II, $c_b = 2.5_{48} \times 10^{-4}M$; in the mixture with III, $c_b = 2.7_{76} \times 10^{-4}M$.
25. a. $c_o = 1.77_8 \times 10^{-4}M$; $c_p = 7.48_{36} \times 10^{-5}M$
26. a. $5.0_0 \times 10^{-2}M \text{ Co}^{+2}$; $3._0 \times 10^{-2}M \text{ Cr}^{+3}$
   b. $2.0_0 \times 10^{-2}M \text{ Co}^{+2}$; $4._0 \times 10^{-2}M \text{ Cr}^{+3}$
   c. $3.0_0 \times 10^{-2}M \text{ Co}^{+2}$; $5._0 \times 10^{-2}M \text{ Cr}^{+3}$
28. 0.39_{05} mg/ml
30. a. Salicylic acid   b. Hexatriene   c. Octatetrene
   d. Naphthalene; median of $375_{anth} + 254_{bz} = 30_5$ nm (actual = 314 nm)
   e. Tetracene; $375_{anth} + 60$ (half of above difference) $= 44_0$ nm (actual = 480 nm)
32. Twice as many SCNs in $Fe(SCN)_2{}^+$ to absorb photons
34. Two of the three are $5.0 \times 10^{-2}M \text{ Co}^{+2}$ and $4.0 \times 10^{-2}M \text{ Cr}^{+3}$.
36. a. A nonbonding electron from I, Br, or Cl jumps to an empty σ^* orbital of the carbon-halogen bond.
   b. As the halogen is changed from Cl to Br to I, the energy needed to remove the excited electron from the halogen decreases.
38. a. $1.361_{89}/1$ b. $2188._{14} \text{ cm}^{-1}$

Chapter 14

2. Key differences are that the spectrofluorometer can select excitation and emission wavelengths because it uses gratings and a xenon arc. The filter fluorometer has a line source and filters that cannot select a given wavelength.
4. The 254-nm radiation travels through the open top of the cell.
6. Turbidimetric measures decrease in P_0; nephelometric measures scattered P_0.

8. 2.5_2 mg/100 ml
10. 9/1
11. a. Cerium(III) has outer electronic structure of $4f^1$.

Chapter 15   2. a. Flame excitation should be maximized for flame emission but minimized for atomic absorption.
    b. Loss of excitation energy by excited state is immaterial in atomic absorption, but in flame emission radiative loss should be optimized as opposed to nonradiative loss.
    c. Measurement step in flame emission involves photons emitted; in atomic absorption it involves photons absorbed.
4. a. I_u is compared directly to I_s of the standard(s).
    b. The c_u is multiplied by a correction factor to obtain the true c_u.
    c. Ratios of I_u/I for unknown and internal standard are plotted against the c_u.
6. Flame emission is simpler and can measure lower concentrations.
8. a. $2.1_0 \times 10^{-1} M$ b. $1.3_8 \times 10^{-1} M$ c. $1.2_6 \times 10^{-1} M$
10. 0.14% Na_2O
11. 7.45 ppm Cd^{+2}

Chapter 16   2. The indicating electrode responds to the concentration (activity) of one or more ions in the sample; the reference electrode does not respond to the sample solution.
4. It consists of a platinum wire dipping into an inner tube holding mercury, Hg_2Cl_2, and KCl. The inner tube is positioned in an outer tube containing a solution of potassium chloride.
6. It is the error caused by the response of the electrode to sodium ions, as well as to hydrogen ion, in alkaline solution.
8. Because it does not draw an appreciable current, the potentiometer can be used to measure concentrations; the voltmeter draws an appreciable current and cannot be so used.
10. Sweating is induced electrically after chemical treatment. A flat-headed chloride indicating electrode is applied directly to the skin for measurement.
11. a. $1.0_{96} \times 10^{-2}$ b. $1.15_{08} \times 10^{-1}$ c. 1.0_{96}
12. a. Larger   b. Larger   c. Larger   d. Larger   e. Smaller   f. Smaller
13. a. 12.04_0 b. 13.061
14. a. 1.0 pH unit   b. 0.2 pH unit
16. a. $5.8_{88} \times 10^{-2}$ b. $1.0_{97} \times 10^{-11}$ c. $6.3_{09} \times 10^{-1}$ d. 1.99_{99}

Chapter 17   2. a. Measurement step in electrogravimetry is weighing; in coulometry it is electrical measurement.
    b. In polarography, the measurement is done with a polarogram.
    c. In an amperometric titration, the measurement is done by plotting a titration curve.
    d. In direct, the species determined is measured directly by electrical measurement; in indirect, the generated species is measured electrically.
4. Oxygen is reduced to H_2O_2 or H_2O at the cathode. In blood it can be determined amperometrically.
6. a. Yes   b. No   c. Yes   d. No   e. No   f. Yes
8. a. A reagent that reacts quantitatively with the sample species
    b. Direct: electrons; indirect: species generated at the electrode
10. a. $0.1888_{57} M$ b. $0.0629_{53} M$
12. $44.1_{17}\%$

Chapter 18   2. $(t_{3/4})/(t_{1/2}) = (\log 100/25)/(\log 100/50) = 2.0000$
4. a. 0.81_{23} day ($19._{49}$ hr)   b. 0.117_{20} day (2.81_{29} hr)
6. a. 50% Ag bound   b. 43.75% Ag bound   c. 37.5% Ag bound
    d. 25% Ag bound   e. 20% Ag bound   f. 17.5% Ag bound

Chapter 19
2. The separation need not be quantitative.
4. The method measures the sum of cholesterol and cholesterol esters.
6. They are (1) distribution between a stationary phase and a carrier phase, (2) a bed of large surface area for a stationary phase, (3) a fluid of liquid or gas as the carrier, and (4) different rates of elution for the components of the sample.
8. The stationary phase in one is paper; in the other, it is a finely divided adsorbent.
10. The carrier is a gas; the stationary phase is a liquid coating on a solid support.
12. They are immiscible, and one is significantly heavier than the other.
14. The partition coefficient pertains to only one form of the extracted component; the ratio refers to all forms of this component.
16. Two different equilibria are established.
20. a. $16.6_6\%$ b. $90.9_1\%$ c. $98.3_6\%$ d. $99.75_{06}\%$
22. One extraction ($\% \text{ Ex} = 98.0_4\%$)

Chapter 20
2. R is a measure of the overlap of any two chromatographic zones in general; R_r refers to the fraction of time spent in the mobile phase of a chromatographic column; and R_f refers to the fraction of the distance traveled by the sample as compared to the solvent on a thin-layer chromatogram.
4. The R_f should reflect the distance traveled by the average sample constituent molecule, not the fastest-moving molecules.
6. The time t_r is larger because it includes t_m, as well as the time the average constituent molecule spends on the stationary phase.
8. The larger R is, the better the separation between the constituent zones; the longer the column, the better the resolution is, in proportion to the square root of the length.
10. a. 97.73% A and 2.27% B
    b. 95.46% B, 2.27% A, and 2.27% C
    c. 97.73% C and 2.27% B
12. a. R_r of $A = 0.50$ b. R_r of $B = 0.33$
14. a. 3.2 min   b. 8.0 min

Chapter 21
2. a. Silica gel is SiO_2 and alumina is Al_2O_3.
    b. Cellulose is a natural glucose polymer $(C_6H_{10}O_5)_x$; silica gel is SiO_2.
    c. Kieselguhr is a purified form of diatomaceous earth; alumina is Al_2O_3.
4. a. Faster   b. More accurate, faster
5. Partitioning occurs between the stationary liquid phase held on the solid support and the mobile liquid phase which passes through the support and the column.
8. Blue spot is the more polar salicylic acid.
10. a. n-Propanol or ethanol   b. Ethyl acetate or acetone   c. Carbon tetrachloride
12. a. NaCl   b. $MgCl_2$
14. 1-Butanol, 2-butanol, $tert$-butyl alcohol, and 2-methyl-1-propanol

Chapter 22
2. A major electrolyte ion is present in major concentrations (0.01–$0.1M$) in body fluids whereas a trace metal is present at concentrations lower than $0.01M$.
4. a. Alkalosis vs. acidosis: high vs. low pH; normal vs. low chloride; low total carbonate vs. high total carbonate.
    b. Respiratory alkalosis vs. metabolic acidosis: high vs. low pH; normal vs. low chloride.
    c. Dehydration vs. alkalosis: normal vs. high pH; high vs. normal sodium; high vs. normal chloride; normal vs. low total carbonate.
6. EDTA titration requires little instrumentation but is slower for many analyses.
8. a. $0.141M$ sum of chloride plus bicarbonate is not $0.01M$ less than sodium; check
    b. No checking
    c. $0.130M$ sum of chloride plus bicarbonate is not $0.01M$ less than sodium; check

10. a. Whole blood less cells
    b. Whole blood less plasma; cells less erythrocytes and leukocytes
    c. Whole blood less cells; plasma less fibrinogen
    d. Whole blood less cells; plasma less serum
12. a. Marginal hypoglycemia   b. Hypoglycemia   c. Serious hyperglycemia
    d. Hyperglycemia

Index

(The letter t, followed by a number, indicates the entry is found in a table on that number.)

FOUR-PLACE LOGARITHMS

No.	0	1	2	3	4	5	6	7	8	9
10	0000	0043	0086	0128	0170	0212	0253	0294	0334	0374
11	0414	0453	0492	0531	0569	0607	0645	0682	0719	0755
12	0792	0828	0864	0899	0934	0969	1004	1038	1072	1106
13	1139	1173	1206	1239	1271	1303	1335	1367	1399	1430
14	1461	1492	1523	1553	1584	1614	1644	1673	1703	1732
15	1761	1790	1818	1847	1875	1903	1931	1959	1987	2014
16	2041	2068	2095	2122	2148	2175	2201	2227	2253	2279
17	2304	2330	2355	2380	2405	2430	2455	2480	2504	2529
18	2553	2577	2601	2625	2648	2672	2695	2718	2742	2765
19	2788	2810	2833	2856	2878	2900	2923	2945	2967	2989
20	3010	3032	3054	3075	3096	3118	3139	3160	3181	3201
21	3222	3243	3263	3284	3304	3324	3345	3365	3385	3404
22	3424	3444	3464	3483	3502	3522	3541	3560	3579	3598
23	3617	3636	3655	3674	3692	3711	3729	3747	3766	3784
24	3802	3820	3838	3856	3874	3892	3909	3927	3945	3962
25	3979	3997	4014	4031	4048	4065	4082	4099	4116	4133
26	4150	4166	4183	4200	4216	4232	4249	4265	4281	4298
27	4314	4330	4346	4362	4378	4393	4409	4425	4440	4456
28	4472	4487	4502	4518	4533	4548	4564	4579	4594	4609
29	4624	4639	4654	4669	4683	4698	4713	4728	4742	4757
30	4771	4786	4800	4814	4829	4843	4857	4871	4886	4900
31	4914	4928	4942	4955	4969	4983	4997	5011	5024	5038
32	5051	5065	5079	5092	5105	5119	5132	5145	5159	5172
33	5185	5198	5211	5224	5237	5250	5263	5276	5289	5302
34	5315	5328	5340	5353	5366	5378	5391	5403	5416	5428
35	5441	5453	5465	5478	5490	5502	5514	5527	5539	5551
36	5563	5575	5587	5599	5611	5623	5635	5647	5658	5670
37	5682	5694	5705	5717	5729	5740	5752	5763	5775	5786
38	5798	5809	5821	5832	5843	5855	5866	5877	5888	5899
39	5911	5922	5933	5944	5955	5966	5977	5988	5999	6010
40	6021	6031	6042	6053	6064	6075	6085	6096	6107	6117
41	6128	6138	6149	6160	6170	6180	6191	6201	6212	6222
42	6232	6243	6253	6263	6274	6284	6294	6304	6314	6325
43	6335	6345	6355	6365	6375	6386	6395	6405	6415	6425
44	6435	6444	6454	6464	6474	6484	6493	6503	6513	6522
45	6532	6542	6551	6561	6571	6580	6590	6599	6609	6618
46	6628	6637	6646	6656	6665	6675	6684	6693	6702	6712
47	6721	6730	6739	6749	6758	6767	6776	6785	6794	6803
48	6812	6821	6830	6839	6848	6857	6866	6875	6884	6893
49	6902	6911	6920	6928	6937	6946	6955	6964	6972	6981
	0	1	2	3	4	5	6	7	8	9